ROUTLEDGE HANDBOOK OF URBAN WATER GOVERNANCE

This handbook provides a comprehensive, state-of-the-art overview of urban water governance.

Of the many growing challenges presented by rapid urbanisation, water governance is a critical one and while urban water governance is now regarded as a critical field of research, the literature is fragmented. For the first time, this handbook brings together urban water governance research, containing interdisciplinary contributions from established and emerging scholars, practitioners, and policymakers. It addresses the key questions of how urban water governance works, how is it shaped, and what the impacts are. The handbook's structure offers a progressive entry into the complexity of urban water governance. Starting with technical dimensions, the handbook addresses supply and demand, wastewater, and sanitation. It then considers regulation and economic factors, examining water utilities and services. Political processes, and the actors involved, are addressed and the handbook finishes with a part focusing on governance and sustainability, where chapters address critically important topics such as access to water, water safety, and water security.

This handbook is essential reading for students, scholars, and professionals interested in urban water governance, urban studies, and water resource management and sustainability more broadly.

Thomas Bolognesi is a researcher at the Geneva School of Business Administration, HES-SO. His research investigates the processes of social-ecological systems evolution, emphasizing non-linearities and patterns diversity. He combines economics and public policy analysis to study the organization and effects of urban water services regulation, the development of water policy regimes, and water security.

Francisco Silva Pinto is an Assistant Professor at Lusofona University (LU) and researcher at EIGeS, FE-LU, and CERIS, IST-UL. His research interests cover the application of numerical modelling and analytics to support decision making in governance, pricing, and finance of utilities (mainly water supply, wastewater, and waste) under critical socio-economic and environmental situations.

Megan Farrelly is an Associate Professor in the School of Social Sciences at Monash University. Her research explores the intersection of urban water governance and sustainability transitions, focusing on processes and pathways for delivering practical and socio-institutional change towards sustainable urban transformations.

ROUTLEDGE HANDBOOK OF URBAN WATER GOVERNANCE

Edited by Thomas Bolognesi, Francisco Silva Pinto and Megan Farrelly

Routledge
Taylor & Francis Group
LONDON AND NEW YORK
earthscan
from Routledge

Cover image: Getty Images

First published 2023
by Routledge
4 Park Square, Milton Park, Abingdon, Oxon OX14 4RN

and by Routledge
605 Third Avenue, New York, NY 10158

Routledge is an imprint of the Taylor & Francis Group, an informa business

British Library Cataloguing-in-Publication Data
A catalogue record for this book is available from the British Library

Library of Congress Cataloging-in-Publication Data
A catalog record has been requested for this book

ISBN: 978-0-367-52353-4 (hbk)
ISBN: 978-0-367-52354-1 (pbk)
ISBN: 978-1-003-05757-4 (ebk)

DOI: 10.4324/9781003057574

Typeset in Bembo
by Deanta Global Publishing Services, Chennai, India

CONTENTS

CONTRIBUTORS

Christopher Gasson is an authority on water finance and markets. After obtaining a degree in politics and economics at Oxford University, he worked as a financial journalist and investment banker before acquiring Global Water Intelligence in 2002.

Raziyeh Farmani (ORCID: 0000-0001-8148-0488) is an Associate Professor of Water Engineering and Industrial Fellow of the Royal Academy of Engineering at the Centre for Water Systems, University of Exeter (UK). She is the Chair of IWA's Intermittent Water Supply Specialist Group. She specialises in urban water systems modelling, asset management, water resources management, many-objective optimisation, uncertainty and risk assessment, and decision aid. Her research interests cover the interdisciplinary field of hydroinformatics, including artificial intelligence (AI), data mining, and optimisation techniques and their application for real-time control for smart water systems, asset deterioration modelling and management of water supply and distribution systems, leakage management, energy management, and sustainability and resilience issues.

Chris Sweetapple (ORCID: 0000-0002-9329-5367) is a Research Fellow at the Centre for Water Systems, University of Exeter (UK). His interests lie broadly in improving the resilience and sustainability of urban water systems and, in particular, the development of tools that can inform the design and implementation of interventions.

Casey Furlong (ORCID: 0000-0003-0594-1179) has a decade of experience in water security and urban livability. His employment has spanned water utilities, government, consulting, and academia. He has published extensively on various topics related to Integrated Water Management, including decision making for alternative water source projects and water sector support for urban greening and cooling.

Ryan Brotchie (ORCID: 0000-0002-5903-2360) is a water strategy consultant with GHD in Melbourne. He is passionate about developing sustainable water solutions and helping the water industry plan for future uncertainty and adapt to climate change. He has particular experience in multidisciplinary and multi-stakeholder strategic planning and advisory projects across the whole water cycle as well as across Australia and overseas.

Greg Finlayson (ORCID: 0000-0002-5468-0545) has more than 30 years of experience in the water sector. As a water treatment and desalination specialist, he has held senior technical and commercial roles on a number of large PPPs and DBOs, including a central role in the Victorian Desalination Project. He is now focused on leading strategic planning and integrated water management in the water industry.

Peter Morison (ORCID: 0000-0003-2715-9911) is an environmental and social scientist with over 25 years of experience working in water services, natural resource management, local governance, and as a consultant to various government organisations. He maintains collaborative research as a Senior Fellow at the University of Melbourne's School of Ecosystem and Forest Services.

Lindsey Brown (ORCID: 0000-0003-0420-2542) is a water industry strategist specialising in urban water management and industry modernisation. With experience spanning three continents, her passions for collaborative leadership, governance, and stakeholder engagement see her involved in important conversations across the Australian water industry on how it can continue to evolve to meet the needs of current and future generations while delivering best value. She is a renowned commentator and speaker on industry issues and trends, particularly relating to ESG.

Raphael Ricardo Zepon Tarpani (ORCID: 0000-0001-6774-458X) has a PhD in chemical engineering from the University of Manchester (UK). Research interests are related to water supply, circular economy, food–energy–water nexus, and life cycle sustainability assessment.

Alejandro Gallego Schmid (ORCID: 0000-0002-0583-2143) works as a Senior Lecturer in Circular Economy and Life Cycle Sustainability Assessment at the University of Manchester (UK). Research interests are related to the circular economy, renewable energy, climate change, informal waste sector, digitalisation, and life cycle sustainability assessment.

Yves Le Gat is a civil engineer and PhD researcher at the French National Research Institute for Agriculture, Food and Environment. He has been involved since 1995 in applied research works, in the domain of water infrastructure asset management, and in mobilizing approaches that combine reliability engineering with probability theory and statistics.

Bernard O. Barraqué (ORCID: 0000-0003-4638-3708) is a civil engineer with a PhD in urban socioeconomic issues from the University of Paris. After working as a consultant in urban environmental policies, he turned to teaching and research, increasingly focusing on water policies in Europe (resources allocation, urban water). He is CNRS research director (emeritus) at the Centre International de Recherches sur l'Environnement et le Développement (CIRED).

Jennifer R. McConville (ORCID: 0000-0003-0373-685X) is an Associate Professor at the Swedish University of Agricultural Sciences. She teaches and undertakes research on design, planning, and decision-making processes for sanitation systems, with a special focus on sustainability assessments and resource recovery.

Max Maurer (ORCID: 0000-0002-5326-6035) is Professor of Urban Water Systems at ETH Zürich in Switzerland, and heads the department of urban water management at the Swiss Federal Institute of Aquatic Science and Technology (Eawag) in Dübendorf, Switzerland.

Nelson Carriço (ORCID: 0000-0002-2474-7665) is a water resources engineer and holds a PhD in civil engineering from the University of Lisbon, Portugal. He is an assistant professor at the Portuguese Polytechnic Institute of Setúbal where he lectures on urban hydraulics. His research interests focus on urban water infrastructure asset management methods, multicriteria decision analysis, and hydroinformatics.

Maria do Céu Almeida (ORCID: 0000-0001-8488-2474) is a civil engineer and holds a PhD in civil and environmental engineering, Imperial College London (UK). She is a researcher at the Portuguese National Laboratory for Civil Engineering (LNEC). Her research is on urban water systems, focusing on infrastructure asset management, risk management and resilience, systems modelling and analysis (quantity and quality), undue inflows and water losses control, inspection and monitoring of systems, and performance assessment of urban water services.

João Paulo Leitão (ORCID: 0000-0002-7371-0543) is an environmental engineer and holds a PhD in civil and environmental engineering from the Imperial College, London (UK). He is a senior researcher (tenured) at the Swiss Federal Institute of Aquatic Science and Technology (Eawag) and has a research interest in infrastructure asset management methods aiming at improving the industry's perennial need for more efficient infrastructure, geared to reducing costs and risks while increasing its performance and flexibility.

Jonas Hallström (ORCID: 0000-0003-0829-3349) is Professor of Technology Education at TESER, Technology and Science Education Research, Dept. of Behavioural Sciences and Learning, Linköping University, Sweden. His research interests primarily focus on the history of technology, philosophy of technology, technology education, STEM education, and the history of education. The history of water and sewerage is a salient theme in Hallström's research and teaching.

Martin V. Melosi (ORCID: 0000-0001-7726-8062) is Cullen Professor Emeritus of History and founding director of the Center for Public History at the University of Houston. His major areas of interest are the urban environment and technology, energy history, and the environmental justice movement. He is the author or editor of 20 books and more than 100 articles and book chapters. His most recent book is *Fresh Kills: A History of Consuming and Discarding in New York City* (New York: Columbia University Press, 2020).

Sylvain Barone is a political scientist at the French National Institute for Agriculture, Food, and Environment (INRAE – G-EAU laboratory, University of Montpellier). His research mainly focuses on water and coastal risk policies. He is particularly interested in the political authority of the state and its capacity to govern (in) the Anthropocene.

Pierre-Louis Mayaux is a political scientist at the International Center for Agricultural Research and Development (CIRAD – G-EAU laboratory, University of Montpellier). His research focuses on comparative water policies in North Africa and Latin America. His main theoretical interests lie in the political effects of resource scarcity, including on state legitimacy and state capacities.

Lee Godden (ORCID: 0000 0001 95345148) is Director, Centre for Resources, Energy and Environmental Law, Melbourne Law School, the University of Melbourne, Australia. Her research interests include water law and governance, environmental law and policy, Indigenous peoples' water rights, and energy transition.

Rui Cunha Marques (ORCID: 0000-0003-0344-5200) is Professor of the Scientific Area of Systems and Management of Infrastructure in the Department of Civil Engineering, Architecture and Geo-resources at the Instituto Superior Técnico of the University of Lisbon (IST), Portugal. He is a researcher at the Civil Engineering Research and Innovation for Sustainability of IST

(CERIS / CESUR), of the Public Utility Research Center (PURC) at the University of Florida and at the Center of Local Government (CLG) at the University of New England in Australia. His areas of expertise include regulation of public services, performance assessment, project management, procurement (particularly public–private partnerships), and infrastructure services.

Germà Bel Queralt (ORCID ID: 0000-0002-1330-8085) is Professor of Economics and Public Policy at the University of Barcelona (UB), Catalonia, Spain, and Director of the Observatory of Public Policies Analysis and Evaluation at UB. His research focuses on public sector reform. Within this field, his main interests are local governments; local public services delivery; and transport infrastructures and mobility. In the past, he has been a visiting professor in graduate teaching at Cornell University and Princeton University. He has been visiting researcher at Cornell, Harvard, Paris I-Sorbonne, EUI-Firenze, KU Leuven, State University of Saint Petersburg, and University College London. Outside academia, between 1990 and 1993, he served as advisor to the Spanish Ministries of Public Affairs and of Public Works and Transportation. He was a member of the Spanish Parliament (2000–2004) and a member of the Catalan Parliament (2015–2017).

Ellis A. Adams (ORCID: 0000-0003-3783-9005) is Assistant Professor of Geography and Environmental Policy at the Keough School of Global Affairs, University of Notre Dame. Adams's work examines the social, political, and institutional dimensions of water resources. He has conducted field research in Ghana, Malawi, Uganda, Kenya, and the United States.

William F. Vásquez (ORCID: 0000-0001-6366-9535) is Professor at Fairfield University in the United States, with expertise in development and environmental economics with a regional emphasis on Latin America. He has field research experience in Brazil, Ecuador, Guatemala, Mexico, Nicaragua, and the United States.

Manuel Fischer is a research group leader in Policy Analysis and Environmental Governance at the Department of Environmental Social Sciences at the Swiss Federal Institute of Aquatic Science and Technology (Eawag) in Dübendorf, Switzerland, and an adjunct professor (titular professor) at the Institute of Political Science at the University of Bern. He studies environmental governance processes and networks across multiple levels of decision making, across multiple policy sectors, and between different types of actors.

Karin Ingold is Professor at the Institute of Political Science and the Oeschger Center of Climate Change research at the University of Bern, Switzerland. She also co-leads the research group "Policy Analysis and Environmental Governance" at the Swiss Federal Institute of Aquatic Science and Technology (Eawag) in Dübendorf, Switzerland. In her research, she often applies techniques and models of Social Network Analysis to study complex governance arrangements in energy, water, climate, and environmental politics.

Mert Duygan holds a PhD in Environmental Systems Science from ETH Zurich. He is a postdoc in Environmental Social Sciences at the Swiss Federal Institute of Aquatic Science and Technology (Eawag) and in the Laboratory for Human-Environment Relations in Urban Systems at the Swiss Federal Institute of Technology in Lausanne (EPFL). He conducts conceptual and empirical research that incorporates cross-disciplinary insights from transition and innovation studies, political science, urban studies, and institutional sociology.

Liliane Manny is a doctoral candidate at the Swiss Federal Institute of Aquatic Science and Technology (Eawag) in Dübendorf, Switzerland, and at ETH Zurich in Switzerland. Her research addresses social and political aspects of the digital transformation of urban wastewater systems. Using network analysis concepts, she aims at identifying potential social and sociotechnical barriers towards smart urban wastewater systems.

Katrin Pakizer is a PhD candidate in the research group Natural Resource Policy (NARP) at the Institute for Environmental Decisions, Department of Environmental Systems Science at ETH Zurich in Switzerland. She works on the research project COMIX: Challenges and Opportunities of Modular Water Infrastructures for Greening the Swiss Economy, focusing on alternative governance arrangements for integrating decentralised water technologies in urban water systems.

Jannes J. Willems is an Assistant Professor of Urban Planning at the Department of Geography, Planning, and International Development, Faculty of Social and Behavioural Sciences, University of Amsterdam, the Netherlands. He researches how urban water authorities are creating climate-sensitive cities.

Ellen Minkman is a lecturer and researcher at the Multi Actor Systems Department of the Faculty of Technology, Policy, and Management of Delft University of Technology in the Netherlands. Her research concentrates on the governance of transitions in water management.

William Veerbeek is senior lecturer and co-founder of the flood resilience chair group at IHE-Delft, the Netherlands. Strengthening IHE's mission in capacity development he has been leading the city-to-city learning network in the BEGIN project.

Richard M. Ashley is Emeritus Professor of Urban Water at the University of Sheffield, Department of Civil and Structural Engineering (UK) and Director of EcoFutures Ltd. He has worked in the field of urban water for more than 50 years.

Arwin van Buuren is Professor in Public Administration at the Department of Public Administration and Sociology, Erasmus University Rotterdam, the Netherlands. His recent research interests include creating societal impact through the use of design.

Eva Lieberherr (ORCID: 0000-0001-5985-0809) leads the research group Natural Resource Policy (NARP) at the Institute for Environmental Decisions, Department of Environmental Systems Science at ETH Zurich in Switzerland. Her research addresses socio-ecological challenges, such as the use and protection of the natural resources of water, forest, and landscapes.

Frank Hüesker (ORCID: 0000-0002-1412-8588) is a political scientist and post-doc researcher on water governance in household and third-party-funded projects. His qualitative work focuses on changing actor constellations, and power and problem perceptions in sustainability transitions. Currently he is teaching at the University of Jena in Germany.

Katrin Pakizer (ORCID: 0000-0003-4346-0202) is a PhD candidate at NARP, working on the research project COMIX: Challenges and Opportunities of Modular Water Infrastructures for Greening the Swiss Economy. Her research focuses on alternative governance arrangements for integrating decentralised water technologies in urban water systems.

Aaron Deslatte is an Assistant Professor in the O'Neill School of Public and Environmental Affairs at Indiana University in Bloomington. There, he directs the Metropolitan Governance and Management Transitions (MGMT) Laboratory, which focuses on sustainability issues.

Margaret Garcia is an Assistant Professor in the School of Sustainable Engineering and the Built Environment at Arizona State University. Her specialty area is water resources engineering and her research investigates the factors influencing the sustainability and resilience of water systems.

Elizabeth A. Koebele is an Assistant Professor of Political Science at the University of Nevada, Reno. Her work focuses on the role of collaboration in environmental governance processes, with an emphasis on water and natural hazards management.

John M. Anderies is a Professor in the School of Human Evolution and Social Change and the School of Sustainability, as well as the Associate Director of the Center for Behavior, Institutions and the Environment at Arizona State University. He also directs the Behavior, Economics, and Nature Network at the Beijer Institute of Ecological Economics, a research institute under the auspices of the Royal Swedish Academy of Sciences.

Xu Wang (ORCID: 0000-0003-4555-1108) is a Professor of Urban Water Systems Engineering at the School of Civil and Environmental Engineering, Harbin Institute of Technology (Shenzhen), China. Xu's research focuses on water resource recovery and broad areas linked to the food–energy–water nexus.

Julian S. Yates is a Research Fellow at Monash University and a Visiting Scholar at the University of Victoria Centre for Global Studies. His research focuses on the geopolitics of situated knowledges and Indigenous leadership in water governance, conservation management, and local development.

Marc Tadaki is an environmental geographer and social scientist at the Cawthron Institute in Aotearoa New Zealand. His research examines how social values and power relations are enacted through the production and use of science in environmental governance.

Cristy Clark is a Senior Lecturer at the University of Canberra Law School, Australia. Her research focuses on legal geography and the intersection of human rights, neoliberalism, activism, and the environment.

Susana Neto is affiliated with the University of Lisbon (CERIS), and she is also an adjunct professor at the University of Western Australia (Faculty of Science). She has worked in research and policy for integrated water management and territorial planning for the last 30 years. Her fields of expertise cover water and territorial planning, water governance, water policy, regional planning, and sustainable urban development. She has published and presented more than 150 titles on issues related to water and territorial planning. She is deeply involved with the practice of water management, for instance, in the Portuguese public administration and water reforms, at the Portuguese Water Resource Association, and at the Water Governance Initiative Group (OECD).

Joost Buurman (ORCID: 0000-0001-7038-4239) works at the Institute of Water Policy, Lee Kuan Yew School of Public Policy, National University of Singapore. He is an economist with a focus on water, environment, and climate adaptation.

Serge Stoll (ORCID: 0000-0003-3158-4208) received his PhD in 1992 at the University Louis Pasteur, France. He teaches environmental, soft-condensed, and analytical chemistry at the University of Geneva, Switzerland, serves as the head of a group in environmental physical chemistry, and has authored more than 100 publications in refereed international journals and three book chapters. He has also supervised PhD thesis research studies and conducted several international projects in the field of emerging contaminants.

Stéphan Ramseier Gentile received his PhD from the University of Geneva (Switzerland) in environmental analytical chemistry and joined the Industrial Boards of Geneva (SIG), Switzerland, as a chemist. In 1995 he was nominated to head the laboratory for treatment and analysis of drinking waters and was promoted in 2006 to scientific advisor of the direction "Environment" in the fields of drinking water, wastewater, and waste recovery.

Steven J. Kenway is a water leader with senior experience in research, industry, and government. He has undertaken work with the University of Queensland, CSIRO, Brisbane Water, Sydney Water, and in private consulting. He has worked with urban water, wastewater, stormwater, and related energy and greenhouse gas issues since 1990. His work addresses urban water security, water–energy nexus, and circular economy.

Marguerite Renouf is a Senior Research Fellow (Circular Bioeconomy) at the QUT Centre for Agriculture and the Bioeconomy (CAB). She is responsible for promoting research that quantifies the environmental performance and resource efficiency of bio-production and agri-food systems to support the strategic development of sustainable and circular production pathways for the future.

J. Allan is a research scholar at the University of Queensland Centre for Water and Environmental Biotechnology.

KMN Islam is a research scholar at the University of Queensland Centre for Water and Environmental Biotechnology.

N. Tarakemehzadeh is a research scholar at the University of Queensland Centre for Water and Environmental Biotechnology.

M. Moravej is a research scholar at the University of Queensland Centre for Water and Environmental Biotechnology.

B. Sochacka is a research scholar at the University of Queensland Centre for Water and Environmental Biotechnology.

M. Surendran is a research scholar at the University of Queensland Centre for Water and Environmental Biotechnology.

URBAN WATER GOVERNANCE

Approaching a pressing environmental and social challenge

Thomas Bolognesi, Megan Farrelly, and Francisco Silva Pinto

0.1 Introduction

Do you drink water, take showers, water your garden? Do you live in a city? Have you considered whether your water consumption impacts the volume of available water, quality of waterways, the habitat of aquatic animals, or other people's use of water upstream and/or downstream? If you live in a rural area, do you think that the functioning of cities has an impact on your water uses or on the local fauna and flora? One single yes means you are concerned with and connected to urban water governance. Indeed, urban water is a smaller water cycle within the global water cycle, and governance contributes to determining how these two cycles co-evolve. Nevertheless, many cities around the world have experienced a "water crisis," and scientific evidence is growing that substantiates the argument that the current co-evolution is neither sustainable nor satisfactory.

Urban settings have long been important centres of economic, political, social, innovative, and cultural activities. Yet prior to 2005, scholarly attention directed towards understanding the nuances and challenges of urban water governance was limited. This has now shifted, with urban water governance regarded as a critical field of research (see, e.g., Figure 0.1). Access to water and/or wastewater disposal is a complex collective-action dilemma that faces environmental, socio-economic, and technical constraints. The recognition of the human right to water and the global monitoring of water access through the Sustainable Development Goals (SDGs) ([formerly the Millennium Development Goals]) emphasise how vital and inequal this access is worldwide. The decision-making and delivery contexts for urban water are rapidly changing, due, in part, to changes in patterns of precipitation, ageing infrastructures, decreasing water quality (occasionally due to lack of wastewater treatment), increasing water withdrawals/discharges (in volume, quality, and across space), and damage to water ecosystems, among others. The socio-economic and technical delivery of water services is costly, and it rests on network-related industries that are difficult to regulate, which limits the ability to offer universal access to water/wastewater collection. This issue affects both developed and developing countries, the former in renewing infrastructure and the latter in building it.

According to (Pahl-Wostl 2015), "water governance is the social function that regulates development and management of water resources and provisions of water services at different levels of society and guiding the resource towards a desirable state and away from an undesirable

DOI: 10.4324/9781003057574-1

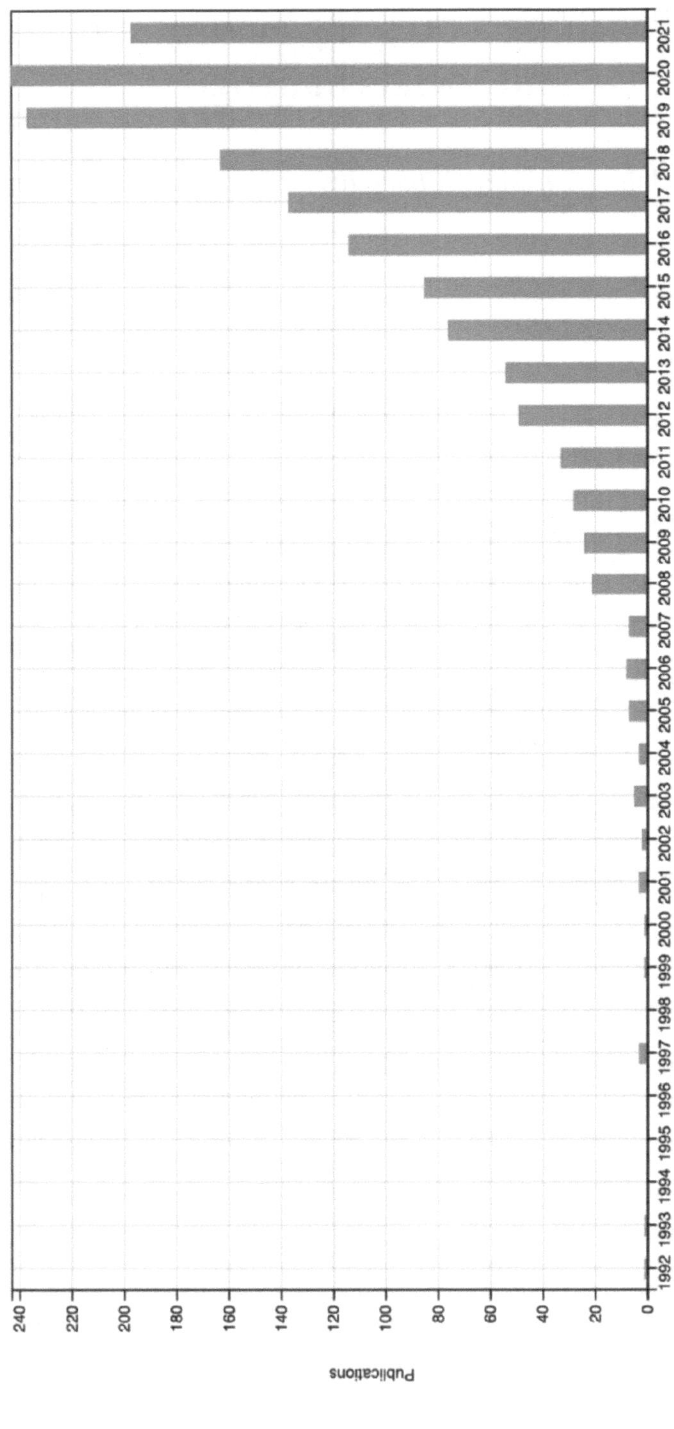

Figure 0.1 Total publications per year about urban water governance – We use data from web of sciences. The request is: TOPIC: (urban water governance); Timespan = 1992–2020; Indexes = SCI-EXPANDED, SSCI, A&HCI, ESCI.

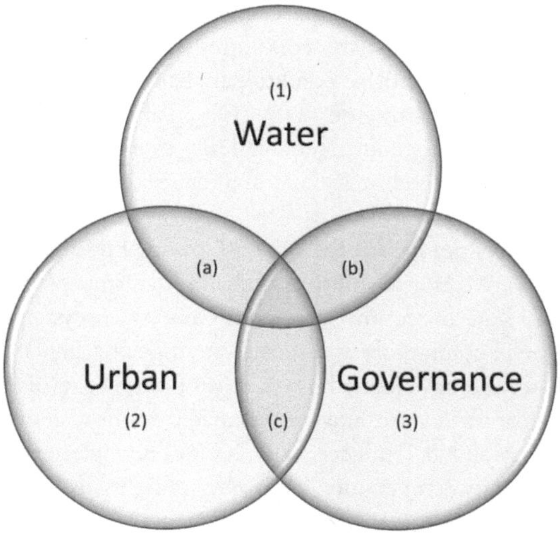

Figure 0.2 Delineation of the dimensions of urban water governance.

state." We define urban water governance as the organisation of the urban water cycle and of its interlinkages with directly interacting uses. We identify an urban water cycle that consists of the technological dimensions of water supply and uses. It is the object of the governance stricto sensu. Governance consists of the political-economic, institutional, and social dimensions that organise the urban water cycle. In this view, governance is the practice of interaction by actors to coordinate the urban water cycle, to engage in political and power relations, and to take into consideration technical and urban planning needs (Bakker 2010; Bolognesi 2018; Varone et al. 2013). This simple identification highlights that the relations are not unilateral; rather, they consist of numerous interlinkages and complex feedback relations.

Figure 0.2 depicts urban water governance as a particular *and* generic object. We use this combination of facets to structure our delineation of urban water governance. In the next section, we define the generic dimensions: (1) water, (2) urban, and (3) governance (Section 0.2). Then, we focus on associations of dimensions a, b, c (Section 0.3). Before concluding, we present the main ongoing challenges in urban water governance (Section 0.4).

0.2 Defining water, urban, and governance

(1) Water: Hydrology, ecology, and society

Water is a natural resource. Water has hydrological characteristics, which the "water cycle" heuristic scheme synthesises. The essential variables of the global water cycle are precipitation, evaporation, infiltration, and runoff. In the *Encyclopedia of Hydrological Sciences*, Beckie (2005) gives an overview of the fundamental principles of hydrological modelling. He writes that the water balance (S) results from precipitation (P) and evaporation and transpiration (ET), the net evolution of groundwater (G), and surface water (Q) for a given time (t) and catchment:

$$\Delta S = \left(P - ET + G_{in} - G_{out} + Q_{in} - Q_{out}\right)\Delta t$$

This equation shows that water quantity varies over time and space, and understanding this variability in the context of changing land, human, and climatic conditions remains a critical challenge for hydrology (Blöschl et al. 2019). Because of these conditional changes, the water cycle intensifies, i.e., more variation and extremes (Huntington 2006). Observations indicate that the subsequent changes in trends of water availability contrast dramatically over geographical areas (Rodell et al. 2018). The distribution of water scarcity and abundance, or events as floods and droughts, evolves significantly.

Water has ecological characteristics. Water serves as a habitat for 140,000 specialist freshwater species and is a condition for life (WWF 2018). Water availability and quality determine how liveable the habitat is, and, thus, they affect the state of and the interactions among ecosystems and finally biodiversity properties, i.e., ecosystem, community and species, intraspecific, genetic and functional diversity (Geist 2011). Freshwater ecology studies how these properties co-evolve within aquatic ecosystems. Aquatic ecosystems are diverse and are estimated to host about 6–10% of the existing animal species (Balian et al. 2008; Dudgeon et al. 2006). The diversity of water habitat and life depends primarily on "salinity, temperature, availability of light, dissolved gases and nutrients, along with biogeographic processes" (Geist 2011, p. 1508).

Water has social characteristics. Like other animals, humans use water and contribute to both the stability and the alteration of freshwater habitats. Even more, since the Anthropocene, humans are a primary driver of water characteristics, and, as such, water characteristics are both a social construct and a social context. For instance, dams are a significant source of disturbance for freshwater ecology and hydrology (Dudgeon 2019; Piégay et al. 2020); and they exist because the local water context was considered to limit local development (Flaminio, Piégay, and Le Lay 2021). Economic activity (production and consumption), demographic trends, technology, and models of organisation and coordination, among other social phenomena, affect water and vice versa. Concepts like social-ecological systems, socio-hydrology, and co-evolution place these nuanced relationships at the forefront of analysis (Baldassarre et al. 2019; Kallis and Norgaard 2010; Ostrom 2009). We place ourselves, and this handbook, in this tradition.

(2) Urban: People, infrastructures, and land

An urban area consists of a concentration of people in a built-up area. It consists of a centre and a periphery, the former being denser. Most of the world's population lives in urban areas. In 2020, urban population is estimated to represent 56.2% of the world's total population, and it is forecast to reach 62.5% by 2035 (UN Habitat 2020). Increasing urban populations typically result in the growth of urban areas. In 1950, there were 306 metropolises with more than 300, 000 inhabitants, including 1 with more than 10 million inhabitants. In 2020, 1,934 metropolises were counted with 34 hosting 10 million people. Thus, the share of 10-million-inhabitants metropolises within metropolises has been multiplied by more than five. Worldwide, urbanisation means a change in economic activity, way of life, and concentration of people. Urbanisation is not a homogeneous phenomenon (Figure 0.3). The location and size of cities are not uniformly distributed around the globe. This spatial heterogeneity comes with significant environmental, economic, and political consequences.

The high human density in urban areas stresses carrying capacity, i.e., the ability of an environment to sustain its population size. Urban ecological footprint indicators aim at measuring the extent to which human pressures affect carrying capacity. Such indicators compare the city surface to the land surface theoretically required to support its population. The discrepancy is sizeable. For instance, Rees and Wackernagel (1996) estimated that the ecological footprint of Vancouver in 1991 was 180 times larger than its administrative size. As per capita urban

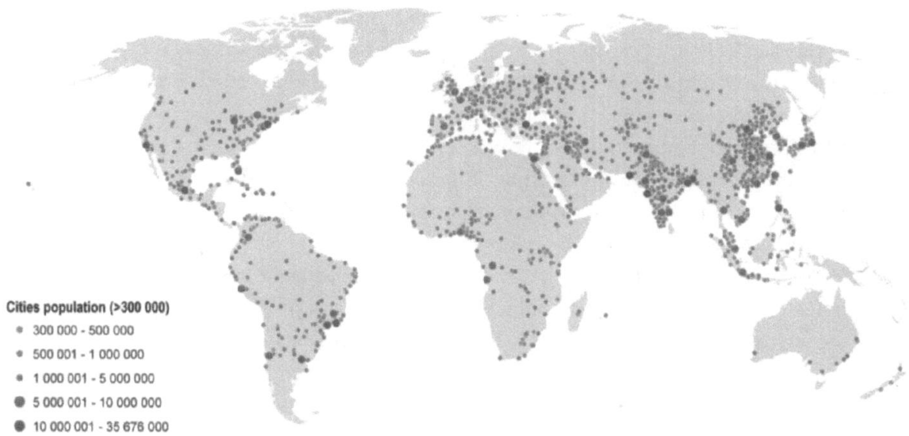

Figure 0.3 Urbanisation of the world: location and size of cities with more than 300 000 inhabitants (Authors' elaboration with the support of Stéphane Kluser).

ecological footprint tends to be greater than per capita national ecological footprint, urbanisation poses severe sustainability challenges (Ortega-Montoya and Johari 2020). Consequently, more and more research focuses on the factors favouring the introduction of sustainability measures in the policy agendas of cities (Reckien et al. 2018; Swann and Deslatte 2018).

Heterogeneity in urbanisation also reveals asymmetric development dynamics and power distribution, reflecting the multiplicity of embedded urban systems (e.g., energy, water, mobility, etc.) (Nielsen and Farrelly 2019). Not all cities are equal, nor are the spaces and people within an urban area. Some cities concentrate economic or political power, which affects their own capacities to organise their development. Consequently, cities are not isolated; rather, they are part of networks and systems that structure cooperation and competition. Zipf's law ranks cities according to their size, facilitating a determination of the hierarchical distribution of cities within national urban systems (Cura et al. 2017; Gabaix 1999).

(3) Governance: Problems, coordination, and strategies

Governance involves multiple actors and typically involves a coordinative approach, thus implying the need to examine coordination mechanisms through their effectiveness, legitimacy, and power struggles (Ostrom 2005; Pierre and Peters 2020; Rhodes 2007; Williamson 1996). The notion is polysemic, but we can distinguish conventional (e.g., top-down and market-based modes of governance) from non-conventional (e.g., network and/or hybrid governance) approaches. Consequently, numerous approaches alternatively emphasise economic, political, or sociological processes. To navigate this diversity and make sense of it, instead of rejecting some approaches, we refer back to Knight's (1992) recommendation for considering a "building block approach" to governance, starting from a simple perspective to an increasingly more sophisticated position that includes significant, but hard-to-measure, phenomena, such as power and/or cultural mechanisms.

A governance system consists of actors that create coordination using formal and informal institutions with a particular allocation of decision and control power. The actors in the system are people or public/private entities that make choices to achieve a goal (e.g., drinking water) and that interact through transactions (e.g., water delivery) (Commons 1924). Actors are boundedly rational (Simon 2000). They do not intend to maximise any utility optimally but rather choose a satisfying option using imperfect information and values (i.e., they primarily focus on drinking water, not on drinking the cheapest water with equal quality). To enable (water) transaction and (water) use satisfaction, actors coordinate by creating and crafting formal institutions (like laws, contracts, and any written rules) as well as informal institutions (like habits, values, and unwritten codes of conduct). These institutions are multilevel, which means they concern more or fewer people, overlap, are complex, and rely on different enforcement mechanisms, from trust to policing.

There are a plethora of governance system structures. A governance system structure is qualified as a complex and hybrid form that mixes the characteristics of three ideal types: hierarchy, market, and self-organisation (Ostrom 2010). These institutions are nested across multiple levels from the local to the global, shaping a polycentric system (Carlisle and Gruby 2019). A polycentric system combines multiple decision centres that are interdependent, more or less autonomous, and hierarchical. Interdependency and nestedness are critical characteristics because they imply complementarities, conflicts, and externalities between institutions. Institutional complementarity, conflicts, and externalities cause consequential non-linearities in coordination processes, and thus, they are pivotal to understanding governance systems' functioning and effects (Aoki 2007; Bolognesi and Nahrath 2020). From a dynamic perspective, power struggles and contextual changes contribute to the evolution of governance systems, with the components evolving at varying frequencies and speeds (Mahoney and Thelen 2009; North 2005; Roland 2004).

0.3 Urban water governance: Assembling water governance, urban water, and urban governance

(a) Urban water

Studying urban water relates to three questions: where does water come from? what are the different uses? and how are they delivered? Responding to these questions relates to economic and infrastructure development and spatial interlinkages between cities and their hinterland. Urban water withdrawals are largely dedicated to municipal water, i.e., mainly households' consumption of water for drinking, showering, and washing as well as other uses depending on the public distribution network. On average, in 2018, countries used 28.25% of their total water withdrawal for municipal water use (Figure 0.4). There is a large diversity in the share of municipal water use ranging from 0.46% in Afghanistan to 100% in Monaco. Noticeably, 50% of countries use less than 20% of their water withdrawal for municipal use. Two key drivers of national disparities in the share of municipal water are economic development and the importance of agriculture in the national gross domestic product (GDP).

Statistics on municipal water use cover water delivered through pipes, which should not hide the considerable diversity in how people access water for basic needs. SDG 6 monitors the types of access to basic water needs, including sanitation. Despite significant progress since 1990, worldwide "one in three people do not have access to safe drinking water" and "two out of five people do not have a basic hand-washing facility," according to UNICEF and WHO.[1] If, in general, households in developed countries have tap-water access, that is certainly not the case in many developing countries and in slums in many countries. The type of access and required infrastructure strongly affect how people use water, e.g., building new pipes or investing in renewing old infrastructure (Bolognesi 2018; Vasquez et al. 2009). Essential drinking water and sanitation are not the only urban water uses. Fire protection, urban agriculture, hydrothermal

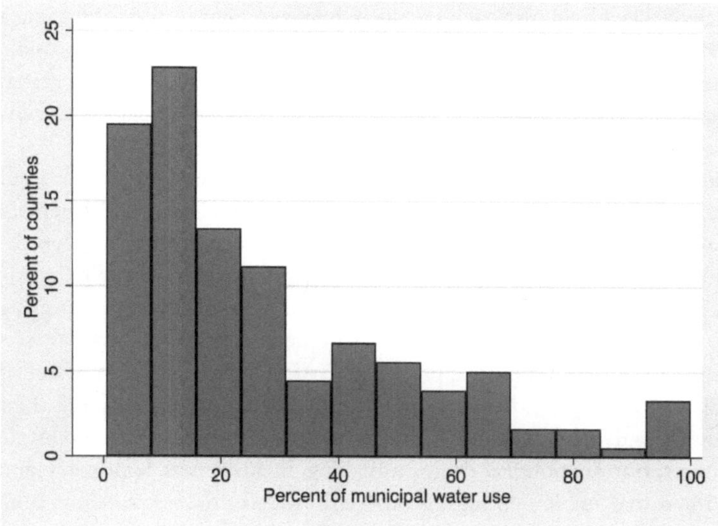

Figure 0.4 Countries distribution regarding the share of municipal water use in total water withdrawal (Authors' elaboration using Aquastat data. The data covers 179 countries in 2018).

power, refreshment, leisure and bathing, and protection against floods are other types of urban water use and are likely to increase in the near future. Due to the increase in anthropogenic use and social-ecological feedback loops, aquatic habitat quality and water flows are likely to be more and more threatened.

As seen with the urban water footprint, cities are not large enough to get their water from themselves. Water used in urban areas has sources, mainly outside the urban areas. Linking sources to the end of the pipes implies the use of significant infrastructure outside cities, generates interdependencies between cities and hinterland, and political and economic factors shape these relationships. Las Vegas is a typical example. Located in a desert, the city pumps water from more than 400 kilometres to the north, limiting water use drastically near the source. Such spatial interlinkages have consequential redistributive effects on water access, economic development, and environmental inequalities. Therefore, determining the volume of water transiting to cities becomes increasingly relevant (Garrick et al. 2019; McDonald et al. 2014). McDonald et al. (2014) estimate that 10–14% of large cities transfer water from another watershed, corresponding to 83,000 million litres per day (±15 000 million litres).

(b) Water governance

Water governance is pivotal in shaping water resource dynamics and uses (Ostrom 2009). While often grouped together, it is of primary importance that governance processes maintain water resources distinct from water uses to improve system sustainability (Bolognesi 2014; Gerber et al. 2009). It allows for defining and monitoring environmental conditions for sustainability and making explicit sociopolitical trade-offs that grant water access and uses. Combining water resources with their uses causes severe difficulties in defining the governance perimeter that are fit for purpose (Ingold et al. 2018; Varone et al. 2013). Should it be based on spatial, sectoral, or actors configurations? Many normative approaches to water governance have emerged depending on the principal perspective, and academia and practitioners' views often meet (van de

Meene, Brown, and Farrelly 2011). A common trend is a search for integration and for finding a fit between environmental and social dimensions because it increases both the range of possible solutions and the soundness of the analysis (Gleick 2018; Hering and Ingold 2012). Integrated water resource management (IWRM), nexus, water security, adaptive management, and economic regulation of infrastructures are among the most influential approaches.

IWRM, a concept that is more than 80 years old, reemerged during the 1990s and was put at the forefront of the international water governance agenda by the Global Water Partnership in 2000 with its technical report defining IWRM as "a process which promotes the coordinated development and management of water, land and related resources, in order to maximize the resultant economic and social welfare in an equitable manner without compromising the sustainability of vital ecosystem" (Biswas 2004). It is a vague definition but two aspects are noteworthy and common to most IWRM approaches. First, IWRM considers water as a complex system and takes into account the different components of the systems and their interrelations jointly. Consequently, it is cross-sectoral and multidimensional (like economic efficiency, environmental preservation, political decision making, inclusiveness, legitimacy, and equity). Second, it is assumed that implementing IWRM will lead to more sustainable outcomes. However, IWRM has been routinely found to be significantly constrained in its implementation (Biswas 2004).

To go beyond the vagueness of IWRM, authors have focused on specific aspects. The nexus approach emphasises the cross-sectoral interlinkages, initially between water and energy with food now increasingly included in the analyses (Bazilian et al. 2011; Hussey and Pittock 2012; Liu et al. 2018). Others wanted to detach from the normative perspective of the IWRM and propose a cohesive and objective measurement of the water multidimensions. Water security aims at simultaneously addressing the water resource and its main uses, e.g., freshwater biodiversity, drinking water consumption, agricultural uses, and floods and droughts. Many place-based studies have been conducted and, nowadays, research is moving towards large N studies and reproducible measurements to identify patterns of water security improvements (Gerlak et al. 2018; Hoekstra, Buurman, and Ginkel 2018; Vörösmarty et al. 2010). Others scrutinise the process of water governance per se. The adaptive management approach pays serious attention to the fact that social-ecological systems are unpredictable (Huitema et al. 2009). Consequently, it is how people adapt and make decisions in different situations that matter. Authors argue that multi-types of learning, experimentation, and participatory processes are pivotal and should be part of water governance (Jager et al. 2020; Reed et al. 2010). The regulation approach focuses on economic mechanisms and incentives to design water governance. This approach is often applied to the regulation of drinking water supply and wastewater management. Authors question the effectiveness of economic regulation as well as regulatory agency organisation and control mechanisms, such as performance indicators and price-setting (Bakker 2010; Pinto, Simões, and Marques 2017; Porcher and Saussier 2019).

(c) Urban governance

Cities are nodes within a more extensive (territorial) development system. As a node, they concentrate not only people, but also economic and political resources. While they depend on their hinterland, they attract people and resources. City development results from both synergies and competition with a city's hinterland or with other cities (Fujita, Krugman, and Venables 1999). Consequently, the dynamics of city development and the governance of cities are specific. Cities balance costs and gains from agglomeration. On the other hand, within each city, the need exists for coordination to shape sustainable development and to consider the fact that actors within cities both conflict with each other and coalesce to benefit from urban resources (Stone 1993). Consequently, urban governance deals with allocation of resources, including water, between a city and its hinterland and within a city. For instance, Los Angeles secures 8,895 million litres of water per day from other water basins (McDonald et al. 2014). Within Los Angeles County, water rights and water rates vary significantly, causing unequal access to water (Pincetl, Porse, and Cheng 2016).

Decentralisation is a second reason for urban governance specificity. Cities have more and more responsibilities and are in charge of implementing national policies. While theoretical governance mechanisms are not changed, the urban perimeter should be accounted for. Due to decentralisation and territorial competition, cities have developed competencies and planning strategies. It raises the question of the effect of urban management and organisation on urban sustainability. For instance, interlocal collaboration, stakeholder support, and membership in a global network that offers support (e.g., ICLEI) have a significant positive effect on implementing environmental measures (Swann 2017). More broadly, organisational environment, stakeholder participation, and urban population attributes (e.g., political preference and growth) affect a city's likelihood of adopting sustainability measures (Swann and Deslatte 2018). Those results emphasise that urban governance has a lot to do with governing fragmentation and enabling cooperation with multiple partners (Kim et al. 2020).

Studies have found non-linear relationships, which stress that with governance, context matters. Therefore, many advocate that experimenting with reforms and sharing knowledge within and across cities is essential to conduct transitions and transformative changes in urban areas with consideration given to multilevel embeddedness. Farrelly and Brown (2011) highlight the conflict between change and path-dependency in urban areas. They show that local-scale experiments are an essential source for learning but that their ability to influence the more extensive system faces inertia due to traditional urban water management. They have identified a list of 36 factors that favour or impede such experiments, encompassing agency, regulatory environment, politics, and belief in technology. As urban governance is also about establishing cooperative arrangements with rural areas, this aspect is of major importance. For instance, in the 1990s, Munich conducted an innovative reform of its drinking water supply by implementing financial support for upstream farmers to turn to bio-agriculture (Grolleau and McCann 2012). Nowadays, it is a standard scheme.

0.4 Ongoing challenges

We are experiencing global changes that challenge urban water governance sustainability and understanding. These challenges relate either to social transformation, environmental changes, or the ability to handle complex problems. These phenomena intensify stress on water resource allocation among ecosystem and people, and they require consideration of the social, hydrological, and environmental dimensions of water simultaneously both in academic analysis and in practice (Baldassarre et al. 2019; Vollmer et al. 2018).

First, with increasing urbanisation and inequalities, competition for water within an urban area, such as in Los Angeles, and with the immediate environment and rural regions is likely to intensify dramatically. In addition, inequalities increase heterogeneity and complicate finding inclusive solutions. For instance, in Rio de Janeiro, wealthy neighbourhoods with private swimming pools sit adjacent to the *favellas*, which have poor access to drinking water. Inequalities associated with urbanisation are likely to increase the complex configuration of urban areas that are still poorly addressed in the water management literature (van den Brandeler, Gupta, and Hordijk 2019). Innovation through technological change and systematic thinking to invent and build non-conventional urban water systems are needed; and the effects of systems must be anticipated to avoid misadaptations or unanticipated indirect perverse effects (Hoffmann et al. 2020; Larsen et al. 2016). Integrative and adaptive urban planning is pivotal to design and craft systems that address multidimensional constraints, e.g., topography, density, or socio-economic characteristics.

Second, environmental changes, including climate change, biodiversity loss, and ecosystem degradation, reveal the limits of our current systems and limit our ability to plan. Indeed, non-linearities and irreversibility are intrinsic to these changes, altering the predictability and understanding of social-ecological systems, which pose new challenges (Baldassarre et al. 2019; Dudgeon 2019). For instance, climate change is likely to mean more frequent and more intense droughts and floods in the same place, and urban development must respond in view of the vulnerabilities posed by these hazards (Berbel and Esteban 2019; Bolognesi 2015).

Third, those social and environmental changes require novel ways of thinking and instruments to devise coherent, legitimate, and effective urban water governance measures (Pahl-Wostl 2020); some talk of a complexity paradigm (Balland et al. 2022). To reach improved outcomes, integrated modelling must be developed to take advantage of the digital era by aiming at combining different dimensions of social-ecological systems; for instance, to set realistic and effective water tariffs (Pinto, de Carvalho, and Marques 2021). Another approach consists in exploring how governance integrates different sub-systems to improve the outcome of polycentric systems (Berardo and Lubell 2019; Trein et al. 2021). Finally, local experimentation should not be denied. It could serve as an incubator for new and effective practices (Farrelly and Brown 2011).

0.5 Urban water governance: How to take stock from the literature?

Urban water governance is multifaceted, resulting in a fragmented scholarship. This handbook seeks to put together widely accepted pieces of this fragmented knowledge. It offers a panorama of various approaches that might be combined to understand and address current issues. Indeed, while simple, some basic questions remain timely, both for academics and for practitioners: how does urban water governance work? How is it shaped? What are its impacts? These questions are important for practitioners because, as noted, there are considerable challenges ahead. They are important for academics because these simple questions reveal that urban water governance figures as a privileged field for testing theories. This handbook seeks to (1) offer a state-of-the-art perspective on these questions, taking into account the disciplinary fragmentation of the field; and (2) pave the way for future research going beyond this disciplinary fragmentation in order to favour knowledge accumulation.

The handbook is structured in presenting a progressive entry into the complexity of urban water governance, starting with technical dimensions and including more and more layers of the social phenomena. Each concept and dimension is exemplified by taking advantage of the current processes of urban water governance reform worldwide. Five aspects of urban water governance structure the handbook: (1) deliverance of urban water, (2) wastewater collection and treatment, (3) utilities regulation, (4) political-economic aspects, and (5) sustainability. Parts 1 and 2 deal with the technical means by which water services are delivered in distinguishing between traditional and urban systems as well as drinking water and sanitation (chapter 1–10). Part 3 addresses how institutions and economic regulation shape urban water governance organisation and outcome (chapter 11–15). Part 4 depicts the policy and political processes of urban water governance in considering network, power, and change (chapters 16–19). Part 5 addresses different sustainability issues of urban water governance, echoing previous chapters but in specific contexts such as slums, infrastructure management, pollution, or integrative approaches (chapters 20–25). This structure facilitates a progressive understanding of urban water governance, which relies on an iterative approach to the urban water system, focusing alternatively on the "urban water cycle" and the "water institutions."

Note

1 Information on SDG 6 is accessible from: https://www.un.org/sustainabledevelopment/water-and-sanitation/, accessed 23/12/2021.

References

Aoki, Masahiko. 2007. "Endogenizing Institutions and Institutional Changes." *Journal of Institutional Economics* 3(1): 1–31.

Bakker, Karen. 2010. *Privatizing Water: Governance Failure and the World's Urban Water Crisis*. Ithaca: Cornell University Press.

Balian, E.V., H. Segers, K. Martens, and C. Lévêque. 2008. "The Freshwater Animal Diversity Assessment: An Overview of the Results." In *Freshwater Animal Diversity Assessment, Developments in Hydrobiology*, eds. E.V. Balian, C. Lévêque, H. Segers, and K. Martens. Dordrecht: Springer Netherlands, 627–37. https://doi.org/10.1007/978-1-4020-8259-7_61.

Balland, Pierre-Alexandre et al. 2022. "The New Paradigm of Economic Complexity." *Research Policy* 51(3): 104450.

Bazilian, Morgan et al. 2011. "Considering the Energy, Water and Food Nexus: Towards an Integrated Modelling Approach." *Energy Policy* 39(12): 7896–906.

Beckie, Roger. 2005. "Fundamental Hydrologic Equations." In *Encyclopedia of Hydrological Sciences*. John Wiley & Sons, Ltd. https://onlinelibrary.wiley.com/doi/abs/10.1002/0470848944.hsa004.

Berardo, Ramiro, and Mark Lubell. 2019. "The Ecology of Games as a Theory of Polycentricity: Recent Advances and Future Challenges." *Policy Studies Journal* 47(1): 6–26.

Berbel, Julio, and Encarna Esteban. 2019. "Droughts as a Catalyst for Water Policy Change. Analysis of Spain, Australia (MDB), and California." *Global Environmental Change* 58: 101969.

Biswas, Asit K. 2004. "Integrated Water Resources Management: A Reassessment." *Water International* 29(2): 248–56.

Blöschl, Günter et al. 2019. "Twenty-Three Unsolved Problems in Hydrology (UPH) – a Community Perspective." *Hydrological Sciences Journal* 64(10): 1141–58.

Bolognesi, Thomas. 2014. "The Paradox of the Modernisation of Urban Water Systems in Europe: Intrinsic Institutional Limits for Sustainability." *Natural Resources Forum* 38(4): 270–81.

Bolognesi, Thomas. 2015. "The Water Vulnerability of Metro and Megacities: An Investigation of Structural Determinants." *Natural Resources Forum* 39(2): 123–33.

Bolognesi, Thomas. 2018. *Modernization and Urban Water Governance: Organizational Change and Sustainability in Europe*. London: Palgrave Macmillan UK. http://link.springer.com/10.1057/978-1-137-59255-2.

Bolognesi, Thomas, and Stéphane Nahrath. 2020. "Environmental Governance Dynamics: Some Micro Foundations of Macro Failures." *Ecological Economics* 170: 106555.

van den Brandeler, Francine, Joyeeta Gupta, and Michaela Hordijk. 2019. "Megacities and Rivers: Scalar Mismatches between Urban Water Management and River Basin Management." *Journal of Hydrology* 573: 1067–74.

Carlisle, Keith, and Rebecca L. Gruby. 2019. "Polycentric Systems of Governance: A Theoretical Model for the Commons." *Policy Studies Journal* 47(4): 927–52.

Commons, John Rogers. 1924. *Legal Foundations of Capitalism*. New York: MacMillan.

Cura, Robin et al. 2017. "The Old and the New: Qualifying City Systems in the World with Classical Models and New Data." *Geographical Analysis* 49(4): 363–86.

Di Baldassarre, Giuliano et al. 2019. "Sociohydrology: Scientific Challenges in Addressing the Sustainable Development Goals." *Water Resources Research*. https://agupubs.onlinelibrary.wiley.com/doi/abs/10.1029/2018WR023901.

Dudgeon, David et al. 2006. "Freshwater Biodiversity: Importance, Threats, Status and Conservation Challenges." *Biological Reviews* 81(2): 163–82.

Dudgeon, David et al. 2019. "Multiple Threats Imperil Freshwater Biodiversity in the Anthropocene." *Current Biology* 29(19): R960–67.

Farrelly, M., and R. Brown. 2011. "Rethinking Urban Water Management: Experimentation as a Way Forward?" *Global Environmental Change* 21(2): 721–32.

Flaminio, Silvia, Hervé Piégay, and Yves-François Le Lay. 2021. "To Dam or Not to Dam in an Age of Anthropocene: Insights from a Genealogy of Media Discourses." *Anthropocene* 36: 100312.

Fujita, Masahisa, Paul R. Krugman, and Anthony Venables. 1999. *The Spatial Economy: Cities, Regions, and International Trade.* Cambridge, MA: MIT Press.

Gabaix, Xavier. 1999. "Zipf's Law for Cities: An Explanation." *The Quarterly Journal of Economics* 114(3): 739–67.

Garrick, Dustin et al. 2019. "Rural Water for Thirsty Cities: A Systematic Review of Water Reallocation from Rural to Urban Regions." *Environmental Research Letters* 14(4): 043003.

Geist, Juergen. 2011. "Integrative Freshwater Ecology and Biodiversity Conservation." *Ecological Indicators* 11(6): 1507–16.

Gerber, Jean-David, Peter Knoepfel, Stéphane Nahrath, and Frédéric Varone. 2009. "Institutional Resource Regimes: Towards Sustainability through the Combination of Property-Rights Theory and Policy Analysis." *Ecological Economics* 68(3): 798–809.

Gerlak, Andrea K. et al. 2018. "Water Security: A Review of Place-Based Research." *Environmental Science & Policy* 82: 79–89.

Gleick, Peter H. 2018. "Transitions to Freshwater Sustainability." *Proceedings of the National Academy of Sciences* 115(36): 8863–71.

Grolleau, Gilles, and Laura M. J. McCann. 2012. "Designing Watershed Programs to Pay Farmers for Water Quality Services: Case Studies of Munich and New York City." *Ecological Economics* 76: 87–94.

Hering, Janet G., and Karin M. Ingold. 2012. "Water Resources Management: What Should Be Integrated?" *Science* 336(6086): 1234–35.

Health: The Freshwater Health Index'. *Science of The Total Environment* 627: 304–13.

Hoekstra, Arjen Y., Joost Buurman, and Kees C. H. van Ginkel. 2018. "Urban Water Security: A Review." *Environmental Research Letters* 13(5): 053002.

Hoffmann, Sabine et al. 2020. "A Research Agenda for the Future of Urban Water Management: Exploring the Potential of Nongrid, Small-Grid, and Hybrid Solutions." *Environmental Science & Technology*, 54(9), 5312–5322 https://doi.org/10.1021/acs.est.9b05222.

Huitema, Dave et al. 2009. "Adaptive Water Governance: Assessing the Institutional Prescriptions of Adaptive (Co-)Management from a Governance Perspective and Defining a Research Agenda." *Ecology and Society* 14(1). http://www.ecologyandsociety.org/vol14/iss1/art26/.

Huntington, Thomas G. 2006. "Evidence for Intensification of the Global Water Cycle: Review and Synthesis." *Journal of Hydrology* 319(1): 83–95.

Hussey, Karen, and Jamie Pittock. 2012. "The Energy–Water Nexus: Managing the Links between Energy and Water for a Sustainable Future." *Ecology and Society* 17(1). https://www.jstor.org/stable/26268988.

Ingold, Karin et al. 2018. "Misfit between Physical Affectedness and Regulatory Embeddedness: The Case of Drinking Water Supply along the Rhine River." *Global Environmental Change* 48: 136–50.

Jager, Nicolas W., Jens Newig, Edward Challies, and Elisa Kochskämper. 2020. "Pathways to Implementation: Evidence on How Participation in Environmental Governance Impacts on Environmental Outcomes." *Journal of Public Administration Research and Theory* 30(3): 383–99.

Kallis, Giorgos, and Richard B. Norgaard. 2010. "Coevolutionary Ecological Economics." *Ecological Economics* 69(4): 690–99.

Kim, Serena Y. et al. 2020. "Updating the Institutional Collective Action Framework." *Policy Studies Journal* (Early view). https://onlinelibrary.wiley.com/doi/abs/10.1111/psj.12392.

Knight, Jack. 1992. *Institutions and Social Conflict.* Cambridge University Press, Cambridge.

Larsen, Tove A. et al. 2016. "Emerging Solutions to the Water Challenges of an Urbanizing World." *Science.* https://www.science.org/doi/abs/10.1126/science.aad8641.

Liu, Jianguo et al. 2018. "Nexus Approaches to Global Sustainable Development." *Nature Sustainability* 1(9): 466–76.

Mahoney, James, and Kathleen Thelen, eds. 2009. *Explaining Institutional Change: Ambiguity, Agency, and Power.* Cambridge: Cambridge University Press.

McDonald, Robert I. et al. 2014. "Water on an Urban Planet: Urbanization and the Reach of Urban Water Infrastructure." *Global Environmental Change* 27: 96–105.

van de Meene, S. J., R. R. Brown, and M. A. Farrelly. 2011. "Towards Understanding Governance for Sustainable Urban Water Management." *Global Environmental Change* 21(3): 1117–27.

Nielsen, Joshua, and Megan A. Farrelly. 2019. "Conceptualising the Built Environment to Inform Sustainable Urban Transitions." *Environmental Innovation and Societal Transitions* 33: 231–48.

North, Douglass C. 2005. *Understanding the Process of Economic Change*. Princeton: Princeton University Press.

Ortega-Montoya, Claudia Y., and Arpan Johari. 2020. "Urban Ecological Footprints." In *Sustainable Cities and Communities*, ed. Walter Leal Filho et al. Cham: Springer International Publishing, 812–24. https://doi.org/10.1007/978-3-319-95717-3_76.

Ostrom, Elinor. 2005. *Understanding Institutional Diversity*. Princeton: Princeton University Press. http://www.jstor.org/stable/j.ctt7s7wm.

Ostrom, Elinor. 2009. "A General Framework for Analyzing Sustainability of Social-Ecological Systems." *Science* 325(5939): 419–22.

Ostrom, Elinor. 2010. "Beyond Markets and States: Polycentric Governance of Complex Economic Systems." *American Economic Review* 100(3): 641–72.

Pahl-Wostl, Claudia. 2015. *10 Water Governance in the Face of Global Change*. Switzerland: Springer.

Pahl-Wostl, Claudia. 2020. 'Adaptive and Sustainable Water Management: From Improved Conceptual Foundations to Transformative Change'. *International Journal of Water Resources Development* 36(2–3): 397–415.

Piégay, Hervé et al. 2020. "Remotely Sensed Rivers in the Anthropocene: State of the Art and Prospects." *Earth Surface Processes and Landforms* 45(1): 157–88.

Pierre, Jon, and B. Guy Peters. 2020. *Governance, Politics and the State*. Bloomsbury Publishing, New York.

Pincetl, Stephanie, Erik Porse, and Deborah Cheng. 2016. "Fragmented Flows: Water Supply in Los Angeles County." *Environmental Management* 58(2): 208–22.

Pinto, Francisco Silva, B. de Carvalho, and R. Cunha Marques. 2021. "Adapting Water Tariffs to Climate Change: Linking Resource Availability, Costs, Demand, and Tariff Design Flexibility." *Journal of Cleaner Production* 290: 125803.

Pinto, Francisco Silva, Pedro Simões, and Rui Cunha Marques. 2017. "Raising the Bar: The Role of Governance in Performance Assessments." *Utilities Policy* 49: 38–47.

Porcher, Simon, and Stéphane Saussier. 2019. *Facing the Challenges of Water Governance*. Cham: Springer International Publishing. https://doi.org/10.1007/978-3-319-98515-2_1.

Reckien, Diana et al. 2018. "How Are Cities Planning to Respond to Climate Change? Assessment of Local Climate Plans from 885 Cities in the EU-28." *Journal of Cleaner Production* 191: 207–19.

Reed, Mark S. et al. 2010. "What Is Social Learning?" *Ecology and Society* 15(4). https://www.jstor.org/stable/26268235.

Rees, William, and Mathis Wackernagel. 1996. "Urban Ecological Footprints: Why Cities Cannot Be Sustainable: And Why They Are a Key to Sustainability." *Environmental Impact Assessment Review* 16(4): 223–48.

Rhodes, R. A. W. 2007. "Understanding Governance: Ten Years On." *Organization Studies* 28(8): 1243–64.

Rodell, M. et al. 2018. "Emerging Trends in Global Freshwater Availability." *Nature* 557(7707): 651–59.

Roland, Gérard. 2004. "Understanding Institutional Change: Fast-Moving and Slow-Moving Institutions." *Studies in Comparative International Development* 38(4): 109–31.

Simon, Herbert A. 2000. "Bounded Rationality in Social Science: Today and Tomorrow." *Mind & Society* 1(1): 25–39.

Stone, Clarence N. 1993. "Urban Regimes and the Capacity to Govern: A Political Economy Approach." *Journal of Urban Affairs* 15(1): 1–28.

Swann, William L. 2017. "Examining the Impact of Local Collaborative Tools on Urban Sustainability Efforts: Does the Managerial Environment Matter?" *The American Review of Public Administration* 47(4): 455–68.

Swann, William L, and Aaron Deslatte. 2018. "What Do We Know about Urban Sustainability? A Research Synthesis and Nonparametric Assessment." *Urban Studies*, 56(9), 1729–1747. 0042098018779713.

Trein, Philipp et al. 2021. "Policy Coordination and Integration: A Research Agenda." *Public Administration Review* 81(5): 973–77.

UN Habitat. 2020. *World Cities Report 2020. The Value of Sustainable Urbanization*. UN Habitat. https://unhabitat.org/sites/default/files/2020/10/wcr_2020_report.pdf.

Varone, Frédéric, Stéphane Nahrath, David Aubin, and Jean-David Gerber. 2013. "Functional Regulatory Spaces." *Policy Sciences* 46(4): 311–33.

Vasquez, William F., Pallab Mozumder, Jesús Hernandez-Arce, and Robert P. Berrens. 2009. "Willingness to Pay for Safe Drinking Water: Evidence from Parral, Mexico." *Journal of Environmental Management* 90(11): 3391–400.

Vörösmarty, C. J. et al. 2010. "Global Threats to Human Water Security and River Biodiversity." *Nature* 467(7315): 555–61.

Vollmer, Derek et al. 2018. 'Integrating the Social, Hydrological and Ecological Dimensions of Freshwater

Williamson, Oliver E. 1996. *The Mechanisms of Governance*. New York: Oxford University Press.

WWF. 2018. *Living Planet Report 2018: Aiming Higher*. eds. R. E. A. Almond, and M. Grooten. Gland, Switzerland: WWF--World Wide Fund for Nature.

PART I

Technical and historical aspects of Water supply systems

1

URBAN WATER CYCLE AND SERVICES

An integrative perspective

Francisco Silva Pinto, Thomas Bolognesi, and Christopher Gasson

1.1 Introduction

A civilization may be conceived as a set of infrastructure (Yevjevich, 1992). From that perspective, the interacting systems that allow societies to achieve sustainable development in terms of social, economic, and territorial cohesion assume a major, even vital, role (European Commission, 2017). The existing urban water cycles and services are crafted by the co-evolution over time and space of such natural and anthropogenic systems (Nace, 1975).

To frame the discussion on the interactions of water-human systems(e.g., water uses and related services), it is useful to take advantage of existing heuristic/conceptual representations. Thus, one may use the hydrological cycle (i.e., in broad terms, also known as the water cycle) to represent the material flow of water on our planet, i.e., through the biosphere, atmosphere, lithosphere, and hydrosphere (Sivapalan, 2018). This overall cycle has mainly two components that define the water flow between those spheres in all its states of matter (e.g., liquid, solid, gas): (1) movement and (2) storage. Those two components define the basis of water balance throughout the planet. The resulting water availability, in terms of accessible quantity and quality, constrains the possible water uses of the human population and, thus, the inherent services. Those services include water supply, wastewater collection and management, drainage, and recreation, among others (Jenerette & Larsen, 2006). If we consider the integration of these services in the water cycle, and its application to densely populated, built-up developed areas, while keeping the cycle's conceptual structure, we can characterize an urban water cycle (UWC). Figure 1.1 schematizes a possible representation of a UWC.

The UWC, similar to its parent cycle, undergoes constant change due to, often interlinked, impacts (Pan et al., 2018), namely: (1) environmental, as climate change and ecosystem degradation; and (2) anthropogenic, as population growth, urbanization (e.g., urban development and land use characteristics), and cultural aspects. Those impacts vary considerably across the territory and their understanding is paramount for effective water resources (and somehow UWC services) management (di Baldassarre et al., 2019).

The integrated UWC management (hereafter, IUWM) strains the relationship between supply and demand for the services required to enable the multiple water uses that exist in the UWC components (as in Figure 1.1). The IUWM allows for assessment of the resulting variation of

DOI: 10.4324/9781003057574-3

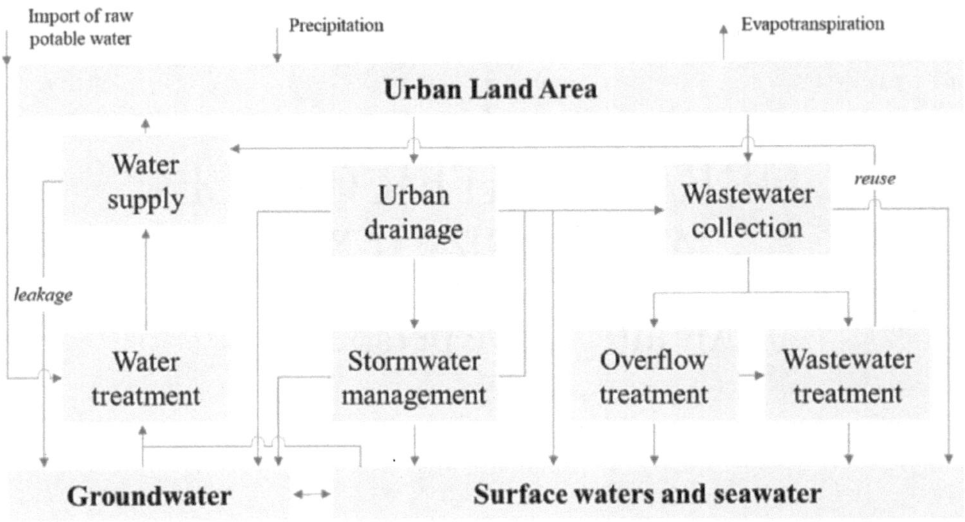

Figure 1.1 Possible schematic representation of UWC components and pathways. Source: Adapted from Marsalek et al. (2008).

water quantity and quality through the different UWC components while addressing climatic, hydrologic, ecological, land use, and engineering issues (Marsalek et al., 2008). To successfully achieve this supply/demand equilibrium, as well as suitable planning and development, identifying the interactions between the urban environment and the respective complementary neighboring areas, e.g., suburban, and rural environments (Bhaskar & Welty, 2012) is also required. In short, changing climatic, land, and human conditions often cause increased water stress that requires supply from outside the urban area or even another catchment, resulting in intensified resource movement, i.e., from source (resource abstraction) to effluent discharge into receiving waters.

In the end, the UWC services sustainability requires some questions to be clearly considered, depending on their rivalry and excludability (OECD, 2009): "who gets the service?" "when they get it?" "how (and how much) they get?" and "how (and how much) they pay for it?" Thus, for the sake of simplicity and worldwide comparison, we will focus on urban water supply and wastewater management, referring, when appropriate, to the remaining services (e.g., stormwater management and urban amenity).

As a starting point, in countries perceived as developed economies, access to safe water supply and sanitation (in which wastewater management is included) has largely been ensured following significant previous investments, mainly at the end of the last century. Nonetheless, considerable investments will still be required to rehabilitate existing infrastructure so as to bring it into conformity with stricter environmental and health regulations and to maintain service quality over time (Bo et al., 2021). On the other hand, in the remaining countries, referred to as transition or developing economies, the challenges are more daunting. A large share of the population has no access and many other residents have unsatisfactory services (Adelodun et al., 2021). An urban water crisis has persisted over decades even as the international community acknowledges these disparities and is committed to achieving milestones to counter problems such as poverty, inequality, and unequal power relationships as well as flawed water management policies that exacerbate the existing scarcity. Those goals that aim to reach sustainability are now defined as the Sustainable Development Goals (SDGs), where SDG 6 focuses on ensuring availability and sustainable management of water and sanitation for all.

This chapter highlights the different characteristics of supply and demand of UWC services across the world and how environmental and anthropogenic impacts (un)balance this equilibrium. As a result of the preceding point, it is important to understand the relationship between competing uses, existing resources, and infrastructure requirements as well as the subsequent variability in costs and prices of UWC services in different contexts (Hukka & Katko, 2015). Finally, a demand assessment, detailing and categorizing the different water uses (both consumptive and non-consumptive) and "used water" characteristics, is required to understand worldwide differences in water allocations and wastewater management tendencies and acknowledge market and merit considerations.

After this brief introduction, we provide a detailed description of urban water uses and their financing requirements, addressing "who" and "how much," as well as the main worldwide trends. Lastly, we provide some policy implications, opportunities, and concluding remarks.

1.2 Urban water cycle, systems, and services

1.2.1 Insights into the UWC systems and services

To support the worldwide demographic, economic, and social (e.g., in terms of behavior) growth, a focus has been placed on production and cost efficiency, which has favored the "take, make, consume, dispose" concept. Due to those targets, the main components of UWC systems, and the provision of related services, were addressed separately, putting an increased stress on both ends of that linear economy approach, i.e., on resources and on the environment. Their interactions were often disregarded or underestimated, which is obviously unsustainable under limited resources. As an example, the demand for water supply has reached an alarming level, providing the motivation to assess the potential of "reducing, reusing, and recycling" those resources, and thus promote IUWM.

UWC systems need to reach subsistence and, perhaps, sustainability. At this stage, IUWM becomes extremely important to achieve the former and to enable the latter. In general, those systems should aim to: (1) continuously supply safe drinking water; (2) collect and treat wastewater for disease protection and prevention of environmental harmful impacts; (3) control, collect, and transport stormwater to protect from flooding and pollution; (4) reclaim, reuse, and recycle water and nutrients.

To achieve an IUWM, the system needs to be considered as a whole and a fuller role must be given for each element/stakeholder identified as nodes (e.g., treatment plants), due to possible dependencies (e.g., treated wastewater may be mixed with "drinking water supply"). Those dependencies can reach a relevant level of complexity; nonetheless, they allow for several synergies and gains (Figure 1.2).[1]

Indeed, the improved management of outputs (e.g., wastewater and stormwater) can offer positive-sum solutions to human societies and natural ecosystems. Some examples can be linked to energy generation, water reuse, and recovery of distinct constituents (Figure 1.2), namely: (1) reclaimed water for aquifer recharge; (2) water swaps with irrigators to deliver more freshwater to urban users; (3) treatment of wastewater for liquid or dry fertilizer and soil improvement (for an extensive list of possibilities, see Rao et al., 2017). In Figure 1.2, the "materials" and "energy" flows are defined by possible node inputs (as the consumption of recovered materials like biomass and energy), and node outputs (as the generation of energy).

1.2.2 Defining urban water uses

The UNSD (1997) defines water use as the "use of water by agriculture, industry, energy production and households, including in-stream uses such as fishing, recreation, transportation and

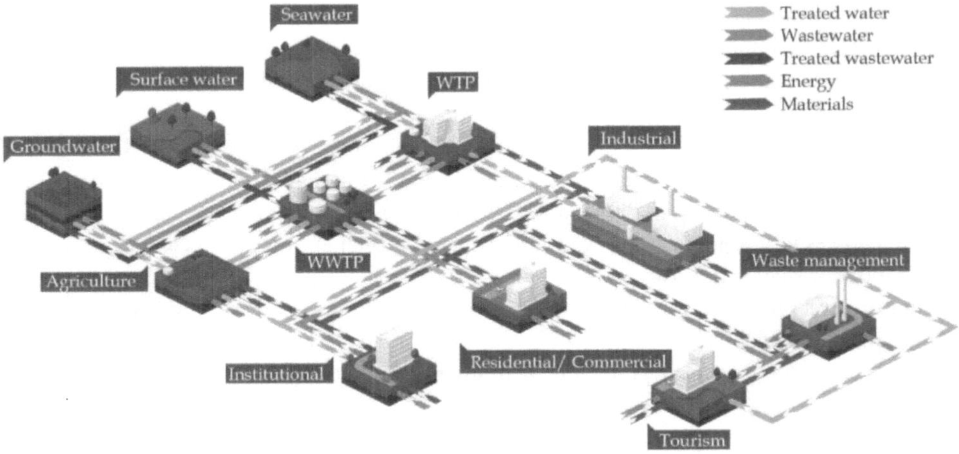

Figure 1.2 Possible general features of an integrated management process (and other related outputs).

waste disposal." Nonetheless, to clarify the concept, a differentiation of closely related terms, such as withdrawal, demand, or consumption, must be made. A withdrawal is defined as the removal of water from a water course (ground or a surface water source) for off-stream use. Demand for water is an economic concept used to describe the quantity that users are willing and able to purchase under particular circumstances. Consumption is the quantity of water withdrawn and made unavailable for reuse, including losses to evaporation or contamination (Gleick, 2003). As follows, a consumptive use is the part of withdrawn water that is "evaporated, transpired, incorporated into products or crops, consumed by humans or livestock, or otherwise removed from the immediate water environment" (Mays, 1996).

The relationship between water uses and natural hydrologic systems outlines the in-stream use, off-stream use, and return flow (Mays, 1996). The first is when water is used but not withdrawn from its course (as hydroelectric power generation, navigation, and recreation). The second is when water is withdrawn or derived from its course to supply different uses, not necessarily consumptive uses. The third is when water is released from a node of use and reaches a water source, becoming available for further use.

The water uses themselves can be defined through different categories. If we take into account Figure 1.2, the "treated water" edges flow into "water use" nodes, categorized through its broad water use features, e.g., the residential use covers water use in or around a house, including indoor (as shower, faucets, toilets) and outdoor (as watering lawns and gardens) household uses. Water uses have been aggregated in multiple ways, following economic, geographic, or engineering backgrounds. Examples are: (1) municipal water uses that cover residential, commercial, industrial, and other uses, where "other" also includes water leaks and unaccounted-for water uses (e.g., firefighting); and (2) urban, periurban, and rural water uses, a territorial typology, where each category may include several water uses (see Figure 1.2) depending on their predominance.

The definition of water uses is important in understanding the overall perspective of demand under competing uses, stressing "who gets" and enabling discussion of "when," "how much," and "how is it financed," among other important issues. Naturally, each water use has distinct impacts in terms of resource quantity and quality variations. Examples are "in-stream navigation" water use, which does not impact quantity variations but may imply quality changes through pollution (as leakages of dangerous substances), and water supply for sanitation (as toilet flush), which implies quantity and quality variations (due to excludability and rivalry of water supply and the resulting wastewater stream).

1.2.3 Measuring urban water uses

To evaluate the UWC water balance, we have to measure the different flows within the UWC (Xenochristou et al., 2020). However, systematic collection of usage data remains scarce, with strong regional disparities, leading to recurrently inadequate and incomplete data. A significant share is not measured or is not quantifiable, as the evaluation of recreational uses or ecosystem services (Bolognesi, 2018). Therefore, we will focus on water supply and wastewater management, in which two issues stand out: the first is how water uses are measured, the second is data quality. If we consider water supply and wastewater management services, utilities often use meters, e.g., in the edges of Figure 1.2 for leakage control as in district metered areas or to charge their consumers. Nonetheless, several utilities may question their use due to increased costs and resulting prices to customers (Barraqué, 2011). Thus, in several cases, water usage data are estimated rather than accurately measured as: (1) wastewater management services being charged at 90% of water supplied, or (2) at supply nodes (Figure 1.2), consumption being influenced by unknown water leaks or infiltration (mostly in wastewater flows).

Situations may exist where we have distinct types of water volumes, covering (Pinto et al., 2021): (1) production (costs, water withdrawn); (2) consumption (benefit, user consumption); and (3) billing (revenue, what utilities charge). The relation between them depends on water losses (real, apparent, and unbilled, i.e., non-revenue water) and tariff structure (e.g., guaranteed consumption).

When it comes to comparing costs and the price of water, the difficulties deepen due to decreased availability (e.g., commercial/trade secrets) or increased complexity (e.g., to consider purchasing power parities). In a seminal study, OECD (2010, p. 34) states that:

> (1) pricing, cost and other relevant data on water-related services is fundamentally local, that any aggregation or averaging exercise implies a loss of information; (2) choices regarding sampling and aggregation affect national values; and (3) extreme care should therefore be taken in proposing cross-country comparison on such variables.

Therefore, interpreting or comparing distinct studies or databases, with certain different assumptions, requires careful analysis.

Overall, the collection and interpretation of water use and related data are challenging; however, it is fast improving as several databases are emerging, fostering continuous improvement toward modernization and rationalization of water management.

1.3 Financing urban water cycle services

1.3.1 Project feasibility – no solution fits all

IUWM usually focuses on the actual/projected UWC services demand. Imbalance from supply and demand mostly covers changes in traditional water resources (e.g., surface and groundwater) availability, increases in demand, and infrastructure requirements (Grant et al., 2012). Therefore, implementing IUWM depends to a large extent on climatic, topographical, and socioeconomic conditions (among others: landlocked cities vs. coastal cities, altitude, and social inequality), which may require different governance, technical, and financing approaches.

An important feature in IUWM is its cyclic approach, where wastewater and stormwater must be considered as more than undesirable commodities that should be discharged and disposed of in a prompt and efficient fashion (Angelakis & Snyder, 2015). Several cases of their reuse can be found in the United States and in Australia, and increasingly across the world, even if stormwater reuse shows a comparatively slower growth (Global Water Intelligence, 2017).

In the end, the feasibility of IUWM projects will surely depend on possible funding structures within a variety of physical and socioeconomic circumstances.

1.3.2 Key concepts from water economics

Understanding UWC service costs were developed over time to assess existing constraints. They include the cost of forgone options, possible alternative uses (the non-accounting ones), and the costs (or benefits) imposed due to the "mutually interfering usage." From the previous statement, four concepts stand out: costs, value, price, and tariffs. A sensible definition of those concepts can be found in Table 1.1.

Drawing from the previous concepts, cost does not, in general, drive economic value, neither does price measure it, and items with no market price can still have a positive economic value. The different dimensions of value are well characterized in Smith's (1776) quote comparing diamonds and water. Now, if we compare the cost (Equation 1) and value (Equation 2) components of water, as a function of different components (adapted from Rogers et al., 2002), there is an easier assessment of its possible price.

← externalities (ext.) →

$$\text{Full cost} = f\left(\text{O \& M costs, capital charges, opportunity costs, economic ext., environmental ext.}\right) \quad (1)$$

Table 1.1 A definition for cost, value, price, and tariff

Concept	*Definition (reference)*
Cost	"O&M costs, capital costs, opportunity costs, costs of economics, environmental externalities. From a different perspective several support costs may have to be included in (or differentiated from) the classical cost outlays, as institution building, human resources development, information systems, monitoring and assessment, regulation, planning and strategy development." (Cardone & Fonseca, 2003)
Value	"Value to users of water, net benefits from return flows, net benefits from indirect use, adjustments for societal objectives and intrinsic value." (Rogers et al., 1998)
Price	"Amount set by the political and social system to ensure cost recovery, equity and sustainability. The price may or may not include subsidies. Prices for water are not determined solely by cost." (Rogers et al., 2002)
Tariff	"A tariff is the system of procedures and elements which determines a customer's total water bill (any part of that bill can be called a charge, measured in ${}^{\text{money}}/_{\text{time}}$ units or money units alone; and any unit price can be called a rate, usually measured in ${}^{\text{money}}/_{\text{volume}}$ units)." (OECD, 1999)

← Full supply →
← Full economic costs →

← Net benefits from →

$$\text{Full value} = f\left(\text{users of water, return flows, indirect uses, societal objectives, intrinsic value}\right) \quad (2)$$

← Economic value →

additional concepts are required to analyze UWC services, as their delivery achieves a distinct status from the delivery of other commodities, since those services entail vital functions and externalities. For that purpose, the reader is redirected to Berg and Tschirhart (1995) for a listing of topics addressed in the literature on utility pricing and capacity, covering concepts such as price elasticity, price discrimination, and marginal cost pricing.

Due to unique features of water-related industries (e.g., higher capital intensity and technical characteristics, institutional status, environmental importance) that are connected to the public/ private economic nature of those services and their social value, UWC services require the assessment of several sources of finance. If we distinguish the economic nature of some WSS in line with the rivalry in consumption, as well as the "excludability" in accessing the "good"/ "service" and to whom the benefits are accrued (i.e., to direct users or to a pool of beneficiaries), diversified sources of financing can be adopted (OECD, 2009). When a UWC service resembles a private good, as traditional water supply, the provider may require a payment from users (inducing the user-pay and polluter-pay principles). On the other hand, if a UWC service falls into the common pool/club/public good categories, it will be increasingly more difficult to do so, consequently reinforcing the integration of transfers and taxes. For all UWC services, a different share of those sources may be required, achieving distinct 3Ts (tariffs, taxes, transfers) policies (OECD, 2010). Nonetheless, additional sources, such as loans and bonds, must be explored due to flexibility benefits and to allow for large upfront investments normally required in the water sector (Pinto & Marques, 2017a).

1.3.3 Revenue streams

In the case of IUWM, decision-makers may have to compare water supplies besides the traditional ones, which are usually less costly. To discuss the feasibility of IUWM projects, it is useful to adopt a financing decision framework, in which, after measuring the costs and benefits to check if it is justifiable, a judgment is made whether the project is mandatory or not. If it is mandatory, the financing framework can cover customer charges below the water supply network charges, and the remaining funding can be shared by other beneficiaries. In case it is not mandatory, commercial viability must be assessed, and charges may vary between a cost-sharing model proportional to the benefits received and customer willingness to pay (WTP). The pivotal point is the requirement to achieve an appropriate mix between 3Ts (with potential external funding partners) and repayable mechanisms (Byrnes et al., 2006).

Since there are several possible funding avenues, and particularly when projects do not pay for themselves, it is of paramount importance to discuss the most appropriate financing structure and how it may vary. In fact, besides the direct beneficiaries, the wider utility customer base should also be considered, as state or national governments (Pinto & Marques, 2017b). Occasionally, funding for these projects can be charged to property developers (Lazarova et al., 2001) or to relevant industrial customers (e.g., in the Municipality of Constância, Portugal, 75%

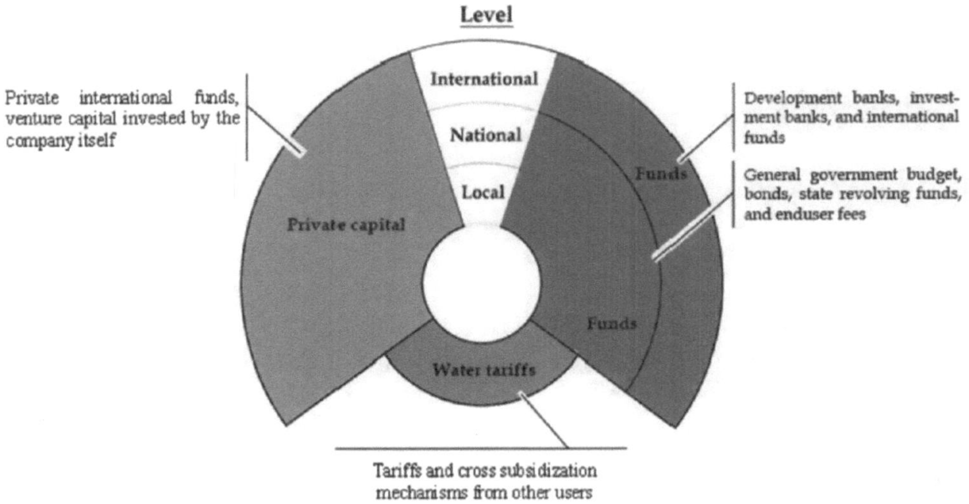

Figure 1.3 Financing sources classified according to type of funding and geographical level. Source: Adapted from de Paoli and Mattheiss (2016).

of municipal wastewater collected is treated for free at the wastewater treatment plant of a pulp [paper] production company). Thus, when deciding upon the most appropriate financing structure, different types of funding at distinct levels should be considered (see Figure 1.3).

Furthermore, and relying on the broad set of benefits brought by IUWM projects, the possibility exists to allocate the source of payment outside the direct beneficiaries. For water reuse schemes, some possibilities are:

- fees paid by developers;
- recreational fees (fishing licenses, park entrance fees);
- stormwater management fees;
- wastewater user charges; and
- donations or dues made to nonprofit groups and associations.

Finally, it is important to identify a concern for customer charges affordability, where there is a need to consider local knowledge of low-income households' expenditures, ability and WTP, although caution should be used in interpreting WTP estimates. In the absence of this information, tariff levels and structures may be based on erroneous assessments of affordability constraints that may not match WTP. In cases where tariffs are kept too low, the result may be a vicious circle of underfinanced services, lower than needed investment and maintenance, and lack of access to water services.

1.4 Urban water worldwide trends

1.4.1 Key water databases

Water-related data can be exported from different sources. Without conducting an extensive analysis, we will present important apps that maintain a continuous effort to include more granular data, higher resolution, new and aggregated data sets, and indicators from different sources.

The first is the AQUASTAT (https://www.fao.org/aquastat/), FAO-UN's global information system on water resources and their use. Established in 1960, it includes over 180 variables and indicators by country. The second is the WHO/UNICEF JMP database (washdata.org/), which, since 2000, includes estimates of progress in household drinking water, sanitation, and hygiene from data produced by national authorities. These databases are complemented by UNstat (https://unstats.un.org/) which is the core UN statistics division data portal, with a focus on the SDGs. To follow SDG's six updates, a dedicated data portal (www.sdg6data.org) draws data from UNstat and other sources.

The Organisation for Economic Co-operation and Development (OECD) also has its own water data portal (https://data.oecd.org/environment.htm#profile-Water) with data and indicators on water withdrawals and wastewater management, among others.

The Aqueduct Water Risk Atlas (https://www.wri.org/applications/aqueduct/water-risk-atlas/) from the World Resources Institute gathers data to measure water-related risks by aligning different indicators, such as physical quantity, quality, and regulatory and reputational risk categories. This app uses open-source, peer-reviewed data to help informed decision making based on best practices in water resources management and to enable sustainable growth in a water-constrained world.

The Water Peace Security partnership initiative is led by IHE Delft and draws considerably from the Aqueduct's database. The app (https://waterpeacesecurity.org/map) aims to help local stakeholders identify, understand, and address (i.e., mobilize, learn, and promote dialogue) water-related security risks.

World Bank Water Data Portal (https://wbwaterdata.org/) is the app for all water-related open data from World Bank–funded initiatives, such as the IBNET, the SIASAR, and the GFDRR, supplementing these with dozens of quality data sources, such as the OECD, the UN, WRI, WWF, and a number of governments. It contains more than 2650 data sets under three pillars (Build Resilience, Deliver Services, and Sustain Water Resources).

Our World in Data (https://ourworldindata.org/clean-water-sanitation) is an app developed at Oxford University to make the knowledge accessible and understandable. This app focuses on research and data (e.g., UNICEF/WHO JMP data and AQUASTAT) to build reports and make progress in dealing with problems related to, in this case, water resources and services.

EEA's databases (https://www.eea.europa.eu/data-and-maps/), including the Waterbase and the Water Information System for Europe (WISE) Water Framework Directive database, deal with the status and quality of Europe's surface and groundwater bodies and with the quantity of Europe's water resources and water withdrawals and discharges, among others.

GWI WaterData (https://www.gwiwaterdata.com) is a market-oriented app produced by Global Water Intelligence and covers high-value business information for the water sector.

In the end, water data analyses must still confront of comparability. Improvements have been made, but the harmonization of definitions or calculation methods between the different indicators in the existing data is (more frequently than desired) found to be lacking.

1.4.2 Water risk, sectoral perspective, and access

By using the Aqueduct Water Risk Atlas to disclose potential impacts that may threaten water quantity, quality, and accessibility and therefore constrain its sustainable use, an overall current situation or a baseline is illustrated in Figure 1.4. To reach this baseline, we consider a combination of water risk indicators selected from the Physical Risk Quantity (water stress, inter-annual and seasonal variability, groundwater table decline, riverine and coastal floods, and droughts, accounting for 70%),

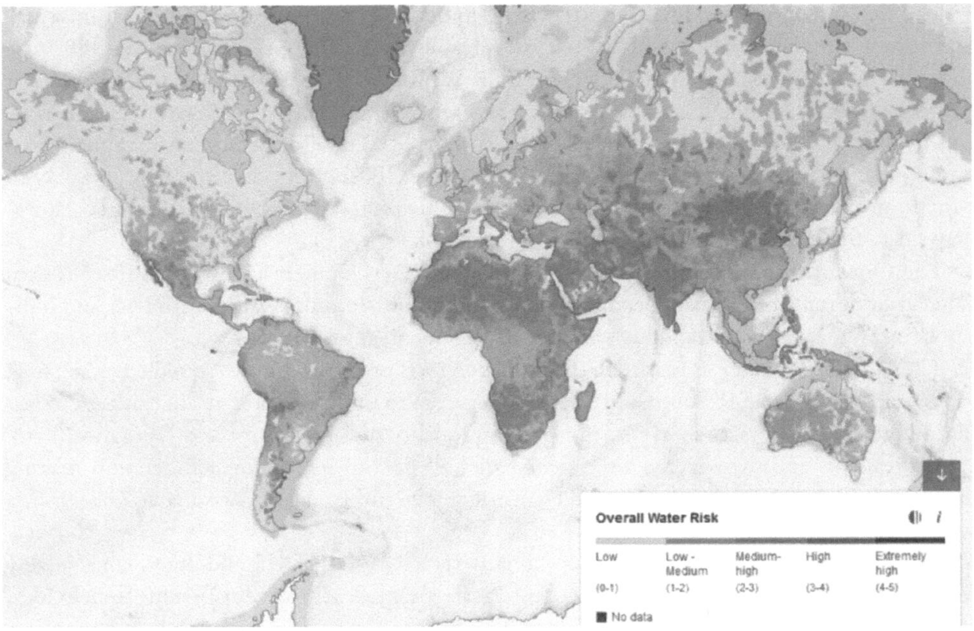

Figure 1.4 Overall water risk baseline (water quantity, water quality, and regulatory and reputational indicators). Source: WRI Aqueduct, www.wri.org/applications/aqueduct/water-risk-atlas, accessed on 05 January 2022.

Quality (untreated connected wastewater and coastal eutrophication potential, accounting for 12%), and Regulatory and Reputational Risks (unimproved/no drinking water, unimproved/no sanitation, Peak RepRisk country in "environmental, social, and governance," accounting for 18%) categories; the relative importance of those indicators was kept at default levels. For more detail on the assembly of those indicators, see Hofste et al. (2019).

For a clear understanding of how the future will unfold, we focus on water stress changes from the baseline (projections for 2040 considering societal water demand divided by available water) in Figure 1.5. We present a business-as-usual scenario[2] to promote a sensible impact.

Figure 1.5 shows that most urban areas will experience a significant increase in water stress changes, which is expected to take a toll on the overall water risk and, thus, impose constraints on urban water uses. Regarding the main off-stream water uses worldwide (i.e., agricultural, industrial, and municipal), they are measured through the quantity withdrawn due to their water demand, with significant differences around the world (Table 1.2).

The "hidden" uses through private boreholes or illegal network connections (as in informal settlements) bring an additional difficulty. Nonetheless, we can still raise awareness to potential shifts in consumption patterns, considering improved supply and sanitation services. In Figure 1.6, we draw attention to the evolution of drinking water supply coverage between urban and rural areas throughout approximately the last 20 years (2000–2020).

Regarding access to safe sanitation in urban areas (mostly wastewater management), in Figure 1.7, we highlight, for particular groups of countries, their evolution over the 2000–2020 timeframe.

1.4.3 In-depth urban water uses

To achieve SDG6, and possibly establish a lower boundary on IUWM, we need to consider the requirements to achieve access to drinking water and sanitation, to counter industrial and

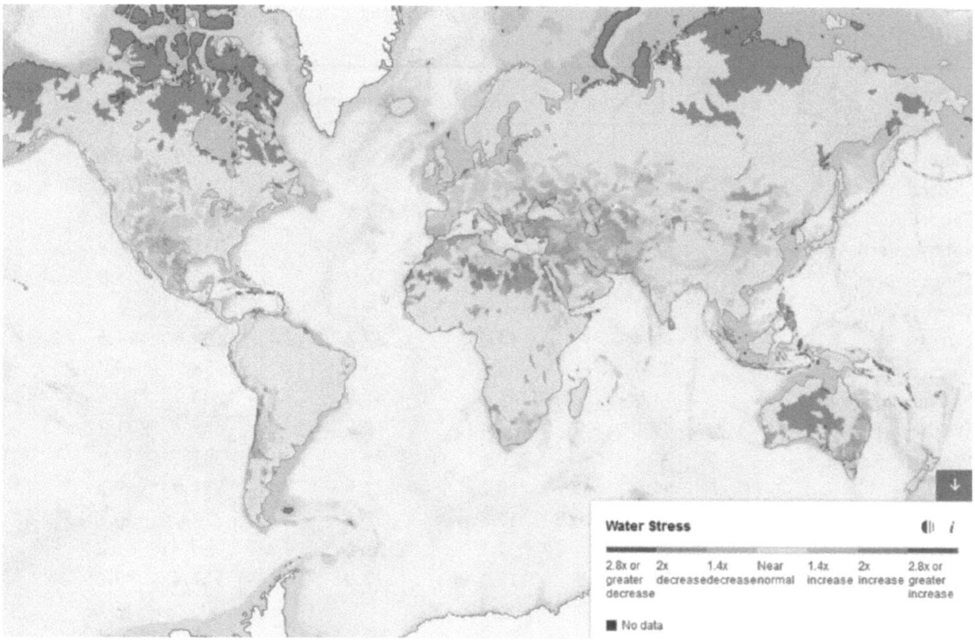

Figure 1.5 Water stress changes from baseline, timeframe 2040. Source: WRI Aqueduct, www.wri.org/applications/aqueduct/water-risk-atlas, accessed on 05 January 2022.

agricultural pollution, as well as to obtain water scarcity and improve water management. To understand what the inherent costs would entail, WRI prepared an "Achieving Abundance: Understanding the Cost of a Sustainable Water Future" data set for display on "Resource Watch." In this data set, the cost of action required to close the gap between current and desired conditions is calculated for each country, as highlighted in Figure 1.8. Major cities were also introduced and were categorized according to the "Economic Capacity Categorization of Cities" data set. It categorizes 769 cities into four groups based on: (1) their current economic strength (through their income), and (2) their projected population and economic growth (income growth relative to population growth, between 2015 and 2030). Those four categories are defined according to their high and low grades, respectively: struggling (low, low), emerging (low, high), thriving (high, high), and stabilizing cities (high, low). This categorization can help to identify cities that will likely face challenges in providing urban services in the future, possibly UWC services.

The daunting situation that several cities may face will certainly increase abstraction costs (as in withdrawals from greater distances and at greater depths). In general, they may also confront greater water and wastewater treatment costs. In such cases, Noiva et al. (2016) suggest that they must increase water use efficiency, demand management, reuse, and recycling. Still, to effectively achieve sustainable practices, utilities need to have detailed information so as to understand water use patterns and the myriad factors that influence them (for a household-level analysis, see Abu-Bakar et al., 2021). Some reports aggregate some databases to that end (Larson et al., 2021), but unfortunately, they are still too few in number and heterogeneous (di Mauro et al., 2021).

Table 1.2 Water withdrawal by sector, from sdg6data database latest data available (accessed on 05 January 2022)

Region	*Agriculture (%)*	*Industry (%)*	*Municipal (%)*	*m^3 / pop*	*Urban pop (%)*
Asia	**82.2**	**8.6**	**9.3**	**585.9**	**49.6**
Central Asia	87.7	8.4	4.0	1716.5	48.2
Eastern Asia	64.6	21.1	14.3	447.5	62.5
Southeastern Asia	85.1	6.4	8.5	771.3	48.8
Southern Asia	91.2	1.9	6.9	589.0	35.8
Western Asia	84.0	6.3	9.6	632.0	71.9
Europe	**29.4**	**43.4**	**27.2**	**409.0**	**74.5**
Eastern Europe	22.9	54.1	23.0	378.7	69.7
Northern Europe	11.9	41.1	47.0	203.9	82.1
Southern Europe	59.1	21.4	19.5	639.2	71.7
Western Europe	5.4	56.6	38.0	381.0	79.9
Oceania	**59.0**	**18.2**	**22.8**	**554.0**	**69.5**
Australia and New Zealand	60.1	17.8	22.1	714.0	86.1
Melanesia	10.7	37.1	52.2	50.3	17.2
Latin America and the Caribbean	**71.0**	**12.1**	**17.0**	**512.4**	**80.6**
Caribbean	57.6	24.8	17.6	567.7	70.3
Central America	74.8	9.3	16.0	568.1	74.5
South America	70.7	12.0	17.4	484.0	84.1
North America	**37.3**	**49.5**	**13.2**	**1317.9**	**82.2**
Sub-Saharan Africa	**75.5**	**6.6**	**17.9**	**93.0**	**40.4**
Eastern Africa	88.1	2.7	9.2	113.3	27.7
Middle Africa	45.7	17.4	36.9	22.0	49.5
Southern Africa	64.2	10.0	25.8	260.2	63.9
Western Africa	65.0	9.6	25.4	73.4	46.4
Northern Africa	**82.2**	**5.3**	**12.5**	**572.6**	**52.0**
World	**71.6**	**16.3**	**12.1**	**529.4**	**55.1**

Source: FAO AQUASTAT / UN-Water, www.sdg6data.org/tables, latest year with data is 2018 (last accessed on 5/1/2022).

1.5 Policy implications and opportunities

In general, IUWM can target several issues in following a more sustainable approach to dealing with rural poverty, water-food-energy security, health and environmental protection, climate mitigation and adaptation, and natural resources management (WWAP, 2017). In fact, the desired outcomes of healthier people, increased prosperity, more equitable societies, protected ecosystems, and resilient communities can be promoted through such key impact pathways as (UN-Water, 2014): (1) improved access to safe drinking water, sanitation and hygiene, increasing water quality and higher service standards; (2) new revenue, income, investment, cost saving, services, and human development outcomes; (3) robust and effective water governance with more effective institutions and administrative systems; (4) improved water quality and resource management within carrying capacity of ecosystems; and (5) reduced risk of water-related disasters to protect vulnerable groups and minimize economic losses.

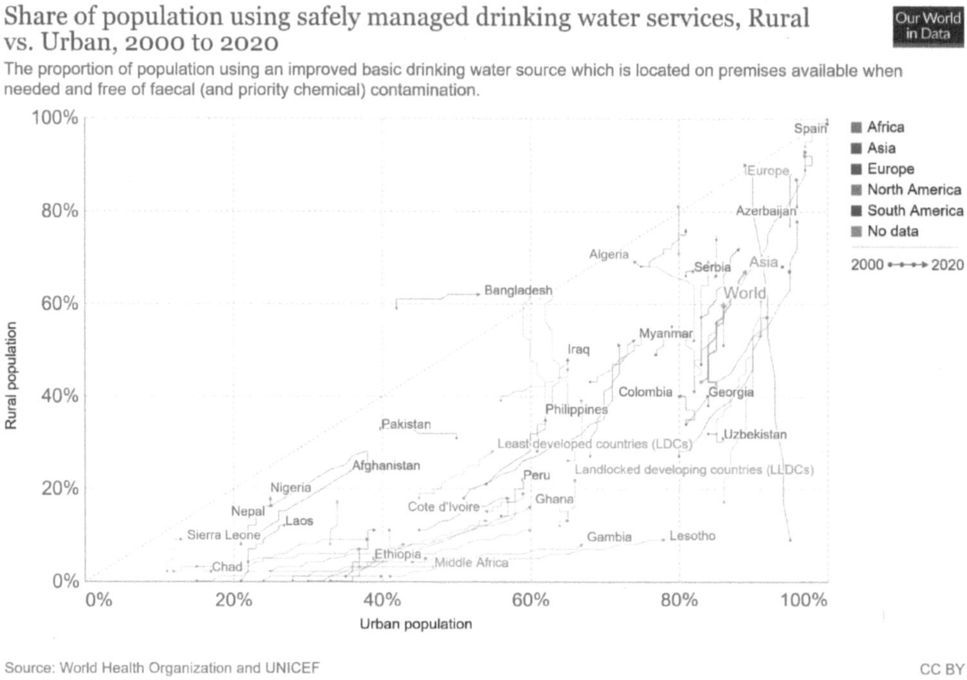

Figure 1.6 Share of population using safely managed drinking water services, rural vs. urban, 2000 to 2020. A safely managed drinking water service is one that is located on premises, available when needed, and free from contamination. Source: WHO and UNICEF, Our World in Data, https://ourworldindata.org/water-access, accessed on 05 January 2022.

For an effective IUWM, several urban and institutional design challenges can be cited. To solve those challenges, coherent participation from multiple actors is required, from water sector to strategic urban planners, environmental and economic regulators, as well as local governments (Marques et al., 2016). As seen in Section 1.4, water supply and wastewater management services are somehow already provided within contextual boundaries; however, that is not the case when you include stormwater management and other outcomes like urban amenity. For an effective delivery of those services, clear guidance in required.

Along those lines, the roles and responsibilities for appropriate IUWM need to be further clarified. In the end, if governments want to obtain stormwater management and urban amenity as outcomes of IUWM, then land and water planning have to be linked and targeted at a variety of scales, both spatial and time. The former from national to regional to local scale and the latter at the appropriate time to cover a project life cycle viewpoint (Loubet et al., 2016). Water planners, local governments, and water policy and statutory land planners need to collaborate, as a different suite of agencies are required through planning, construction, implementation, and ongoing management phases.

All services in an IUWM require consistent institutional and funding arrangements; yet, we can clearly draw a line of decreased maturity and definition within water supply, wastewater, stormwater management, and urban amenity services. The nature of those services highlights different (and increasing) funding challenges. Examples can be linked with health improvements through accessibility to improved UWC services, as quality green space, or the inherent uplift in property value due to improved livability.

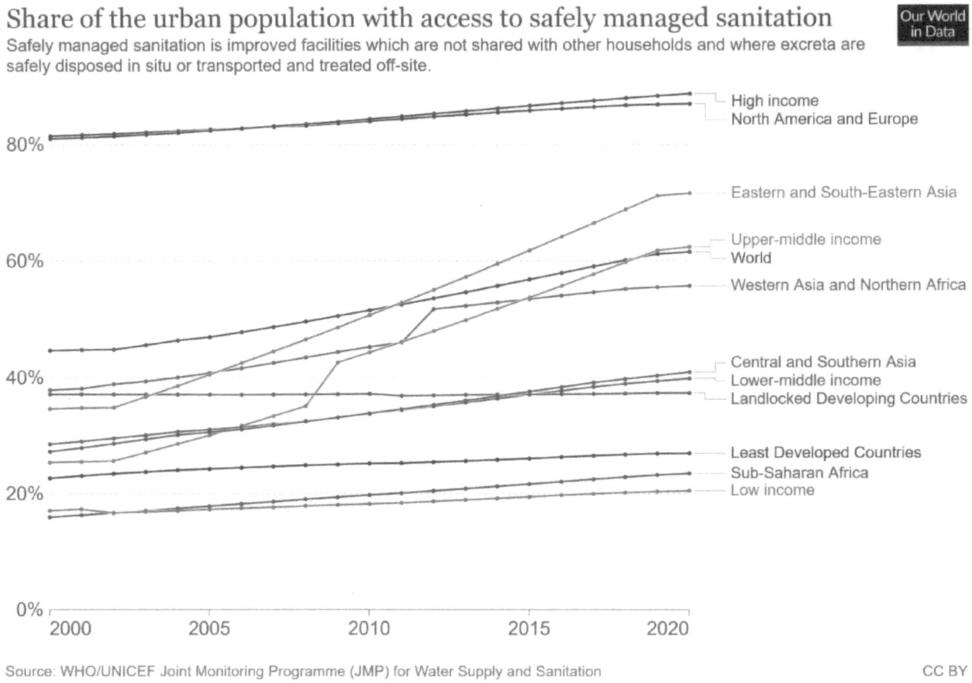

Figure 1.7 Share of the urban population with access to safely managed sanitation. Safely managed sanitation is improved household facilities where excreta are safely disposed in situ or transported and treated off-site. Source: WHO and UNICEF, JMP, Our World in Data, https://ourworldindata.org/sanitation, accessed on 05 January 2022.

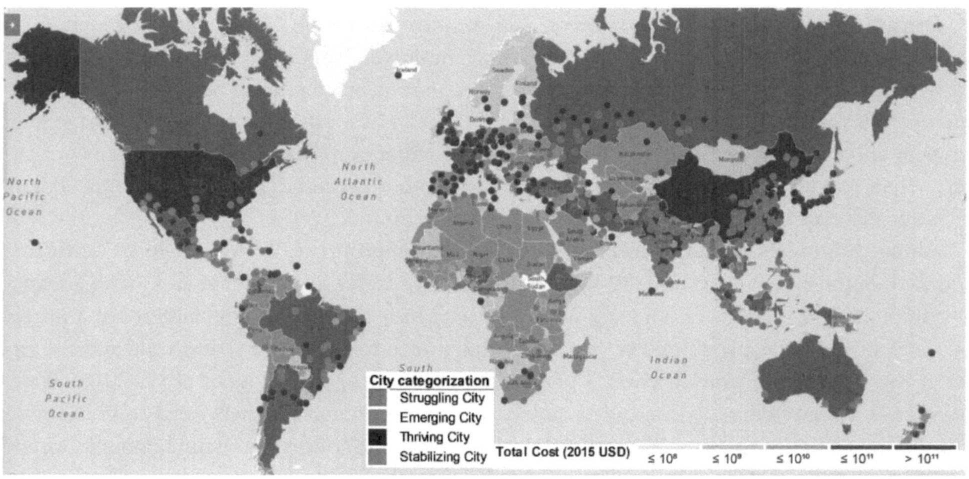

Figure 1.8 Total cost of sustainable water management (2015 USD) and city categorization. Source: WRI, www.wri.org/data/achieving-abundance-understanding-cost-sustainable-water-future-data, and Oxford Economics/WB/UN/WRI, datasets.wri.org/dataset, accessed on 05 of January 2022.

The different "rivalry" and "excludability" characteristics of UWC services render it difficult to understand who benefits and who should pay for what, hindering investment selection. This resembles a catch-22 situation, which often leads to the selection of the lowest cost option that will generally exclude the livability or the environmental options.

Those options may also be forgone in transboundary situations, where upstream vs. downstream water supplies (shift in withdrawals and discharges may lead to resource displacement) or polluting discharges (compliance among regulations and standards) may provoke conflict and hamper development efforts (Earle et al., 2015).

To reach IUWM, a long list of stakeholders must be included in the decision-making process at different levels, perhaps with rules and contracts to fulfill certain outcomes over a timeframe. In fact, these constraints may actually foster network-based governance toward sustainable synchronous regimes (Renou & Bolognesi, 2019).

1.6 Concluding remarks

We have examined the relationship among competing uses, existing resources, and infrastructure requirements, as well as the subsequent variability in costs and prices of UWC services in distinct contexts. The assessment of current knowledge reveals that there are no silver bullet solutions to balance supply and demand as well as anthropogenic and environmental sides of UWC. In short, even if water resources may be displaced elsewhere through appropriate infrastructure, they have a local nature and so do water stress issues (e.g., supply–demand imbalance).

Climatic and hydrologic conditions, infrastructure, and water demand in a given time and space are factors that shape those issues, requiring solutions tailored by location. Additionally, we are aware of the difficulties inherent in an international comparative assessment, which is why we report the parameters limiting the comparability of data (both measured and estimated), allowing an improved discussion.

The gap between research and practice remains difficult to overcome. Conflicts exist between developed scientific principles and popular thought about preferred solutions (e.g., lowest cost option). Usually, the latter carries multiple biases and noise, lacking full information and allowing for a possible continuous misapplication of, perhaps, interesting solutions.

As far as solutions go, a multitude of forces point in new directions, with a heightened emphasis on institutional advance through improved policy design and policy choice, as opposed to technological advances and further structural investments. Yet costly structures continue to be proposed and occasionally undertaken, and public interest in more dams and conveyances perseveres despite the highly harnessed water environment. For all these reasons, the potential role of engineering and management professionals remains high. To end this chapter, we would highlight in the words of Hanemann (2006):

> In short, while there clearly are some distinctive emotive and symbolic features of water that make the demand for water different, there are also some distinctive physical and economic features which make the supply of water different and more complex than that of most other goods. This fact has often been overlooked.

Notes

1 For such a system, see the case of Kalundborg Symbiosis at: www.symbiosis.dk/en/ (last accessed on 16/12/2021).

2 Scenario SSP2 RCP8.5, represents a world with stable economic development and steadily rising global carbon emissions, with CO_2 concentrations reaching ~1370 ppm by 2100 and global mean temperatures increasing by 2.6–4.8°C relative to 1986–2005 levels.

References

Abu-Bakar, H., Williams, L., & Hallett, S. H. (2021). A review of household water demand management and consumption measurement. *Journal of Cleaner Production*, *292*, 125872.

Adelodun, B., Ajibade, F. O., Ighalo, J. O., Odey, G., Ibrahim, R. G., Kareem, K.Y., Bakare, H. O., Tiamiyu, A. O., Ajibade, T. F., Abdulkadir, T. S., & others. (2021). Assessment of socioeconomic inequality based on virus-contaminated water usage in developing countries: A review. *Environmental Research*, *192*, 110309.

Angelakis, A. N., & Snyder, S. A. (2015). Wastewater treatment and reuse: Past, present, and future. *Water*, 7(9), 4887–4895.

Barraqué, B. (2011). Is individual metering socially sustainable? The case of multifamily housing in France. *Water Alternatives*, *4*(2), 223–244.

Berg, S., & Tschirhart, J. (1995). Contributions of Neoclassical Economics to Public Utility Analysis. *Land Economics*, *71*(3), 310–330.

Bhaskar, A. S., & Welty, C. (2012). Water balances along an urban-to-rural gradient of metropolitan Baltimore, 2001–2009. *Environmental & Engineering Geoscience*, *18*(1), 37–50.

Bo, Y., Zhou, F., Zhao, J., Liu, J., Liu, J., Ciais, P., Chang, J., & Chen, L. (2021). Additional surface-water deficit to meet global universal water accessibility by 2030. *Journal of Cleaner Production*, *320*, 128829.

Bolognesi, T. (2018). *Modernization and Urban Water Governance: Organizational Change and Sustainability in Europe*. Palgrave Macmillan UK.

Byrnes, J., Crase, L., & Dollery, B. (2006). Regulation versus pricing in urban water policy: The case of the Australian National Water Initiative. *Australian Journal of Agricultural and Resource Economics*, *50*(3), 437–449.

Cardone, R., & Fonseca, C. (2003). *Financing and cost recovery.* . *Thematic Overview Paper* 7. IRC International Water and Sanitation Centre.

de Paoli, G., & Mattheiss, V. (2016). *Deliverable 4.7: Cost, pricing and financing of water reuse against natural water resources.*. *DEMOWARE.*

di Baldassarre, G., Sivapalan, M., Rusca, M., Cudennec, C., Garcia, M., Kreibich, H., Konar, M., Mondino, E., Mård, J., Pande, S., & others. (2019). Sociohydrology: Scientific challenges in addressing the {Sustainable Development Goals}. *Water Resources Research*, *55*(8), 6327–6355.

di Mauro, A., Cominola, A., Castelletti, A., & di Nardo, A. (2021). Urban water consumption at multiple spatial and temporal scales. A review of existing datasets. *Water*, *13*(1), 36.

Earle, A., Cascão, A. E., Hansson, S., Jägerskog, A., Swain, A., & Öjendal, J. (2015). *Transboundary water management and the climate change debate*. Routledge.

European Commission. (2017). *My Region, My {Europe}, Our Future (Issue 7)*. European Commission.

Gleick, P. H. (2003). Water use. *Annual Review of Environment and Resources*, *28*(1), 275–314.

Global Water Intelligence. (2017). Companies and markets. In *Global Water Market 2017* (Vol. 1). Media Analytics Ltd.

Grant, S. B., Saphores, J.-D., Feldman, D. L., Hamilton, A. J., Fletcher, T. D., Cook, P. L. M., Stewardson, M., Sanders, B. F., Levin, L. A., & Ambrose, R. F. (2012). Taking the "waste" out of "wastewater" for human water security and ecosystem sustainability. *Science*, *337*(6095), 681–686.

Hanemann, W. (2006). The economic conception of water. In P. P. Rogers, M. R. Llamas, & L. Martínez-Cortina (Eds.), *Water Crisis: Myth or Reality?* (pp. 61–92). Taylor & Francis.

Hofste, R., Kuzma, S., Walker, S., E. H.Sutanudjaja, Bierkens, M. F. P., Kuijper, M. J. M., Sanchez, M. F., Beek, R. van, Wada, Y., Rodríguez, S. G., & Reig, P. (2019). *Aqueduct 3.0: Updated Decision-Relevant Global Water Risk Indicators. Technical Note.* World Resources Institute.

Hukka, J. J., & Katko, T. S. (2015). Appropriate pricing policy needed worldwide for improving water services infrastructure. *Journal-American Water Works Association*, *107*(1), E37–E46.

Jenerette, G. D., & Larsen, L. (2006). A global perspective on changing sustainable urban water supplies. *Global and Planetary Change*, *50*(3–4), 202–211.

Larson, K. B., Saulsbury, J. W., Wong, M. A. Y. L., Bellgraph, B. J., Mitchel, K. C., Reicher, D. W., & Tagestad, J. D. (2021). *Improving Discovery, Sharing, and Use of Water Data: Initial Findings and Suggested Future Work. Pacific Northwest National Laboratory.*

Lazarova, V., Levine, B., Sack, J., Cirelli, G., Jeffrey, P., Muntau, H., Salgot, M., & Brissaud, F. (2001). Role of water reuse for enhancing integrated water management in {Europe and Mediterranean} countries. *Water Science and Technology*, *43*(10), 25–33.

Loubet, P., Roux, P., Guérin-Schneider, L., & Bellon-Maurel, V. (2016). Life cycle assessment of forecasting scenarios for urban water management: A first implementation of the WaLA model on Paris suburban area. *Water Research*, *90*, 128–140.

Marques, R. C., Pinto, F. S., & Miranda, J. (2016). Redrafting Water Governance Guiding the way to improve the status quo. *Utilities Policy*, *43*, 1–3.

Marsalek, J., Cisneros, B. J., Karamouz, M., Malmquist, P.-A., Goldenfum, J. A., & Chocat, B. (2008). *Urban Water Cycle Processes and Interactions* (C. Maksimović & A. Téjada-Guibert, Eds., Vol. 2). UNESCO Publishing and Taylor & Francis.

Mays, L. W. (1996). *Handbook of Water Resources*. McGraw-Hill.

Nace, R. L. (1975). The hydrological cycle: Historical evolution of the concept. *Water International*, *1*(1), 15–21.

Noiva, K., Fernández, J. E., & Wescoat Jr, J. L. (2016). Cluster analysis of urban water supply and demand: Toward large-scale comparative sustainability planning. *Sustainable Cities and Society*, *27*, 484–496.

OECD. (1999). *Household Water Pricing in OECD Countries* (P. Herrington, Ed.). OECD Publishing.

OECD. (2009). *Managing Water for All*. OECD Publishing.

OECD. (2010). *Pricing Water Resources and Water and Sanitation Services*. OECD Publishing.

Pan, H., Deal, B., Destouni, G., Zhang, Y., & Kalantari, Z. (2018). Sociohydrology modeling for complex urban environments in support of integrated land and water resource management practices. *Land Degradation & Development*, *29*(10), 3639–3652.

Pinto, F. S., de Carvalho, B., & Marques, R. C. (2021). Adapting water tariffs to climate change: Linking resource availability, costs, demand, and tariff design flexibility. *Journal of Cleaner Production*, *290*, 125803.

Pinto, F. S., & Marques, R. C. (2017a). Desalination projects economic feasibility: A standardization of cost determinants. *Renewable and Sustainable Energy Reviews*, *78*, 904–915.

Pinto, F. S., & Marques, R. C. (2017b). New era/new solutions: The role of alternative tariff structures in water supply projects. *Water Research*, *126*, 216–231.

Rao, K. C., Otoo, M., Drechsel, P., & Hanjra, M. A. (2017). Resource recovery and reuse as an incentive for a more viable sanitation service chain. *Water Alternatives*, *10*(2), 493–512.

Renou, Y., & Bolognesi, T. (2019). Governing urban water services in Europe: Towards sustainable synchronous regimes. *Journal of Hydrology*, *573*, 994–1006.

Rogers, P., Bhatia, R., & Huber, A. (1998). *Water as a Social and Economic Good: How to put the Principle into Practice*. TAC Background Papers No. *2)*.Global Water Partnership.

Rogers, P., de Silva, R., & Bhatia, R. (2002). Water is an economic good: How to use prices to promote equity, efficiency, and sustainability. *Water Policy*, *4*(1), 1–17.

Sivapalan, M. (2018). From engineering hydrology to Earth system science: Milestones in the transformation of hydrologic science. *Hydrology and Earth System Sciences*, *22*(3), 1665–1693.

Smith, A. (1776). *An Inquiry into the Nature and Causes of the Wealth of Nations* (Vol. 1). W. Strahan and T. Cadell.

UN Water. (2014). *A Post-2015 Global Goal for Water: Synthesis of Key Findings and Recommendations from UN-Water*. UN Water.

UNSD. (1997). Glossary of Environment Statistics. In *Studies in Methods (Issue 67)*. United Nations.

WWAP. (2017). *Wastewater: The Untapped Resource*. WWAP.

Xenochristou, M., Kapelan, Z., & Hutton, C. (2020). Using smart demand-metering data and customer characteristics to investigate influence of weather on water consumption in the UK. *Journal of Water Resources Planning and Management*, *146*(2), 4019073.

Yevjevich, V. (1992). Water and civilization. *Water International*, *17*(4), 163–171.

2
TRADITIONAL SYSTEMS OF DRINKING WATER DELIVERY
Technical aspects and sources

Raziyeh Farmani and Chris Sweetapple

Water distribution systems are manmade systems that transfer water from water sources to distribution points, from where water is then distributed among different users. The idea of water supply can be traced back as early as 700 BCE, when Qanats (an underground gravity- based water transport system) that brought water to Persia were built. Romans later started constructing aqueducts (water bridges) from 312 BCE (Ormsbee 2006). One of the first major pipelines was to the Palace of Versailles, France, in 1664 (Walski 2006), and in countries such as the United Kingdom, most people had piped water supply by the early 20th century (Ofwat 2021).

Water distribution systems transfer raw water from a range of sources to treatment plants through transmission mains. The raw water sources predominantly used to supply systems with the required water include groundwater and surface water, or a combination of the two. Groundwater abstracted from aquifers is typically of good quality and requires minimal treatment, as different layers of soil act as natural filters and remove impurities in the water as it infiltrates through the soil to reach aquifer. In comparison, surface water abstracted from rivers and reservoirs requires more treatment as the water is exposed to more contaminants and pollutants. The location, availability, and quality of different types of water sources are some of the key factors in deciding which should be used for a given system. In regions where freshwater availability is limited, other resources, such as saline water from sea or ocean, rainwater, or treated wastewater, may be used to accommodate the required demand.

Water distribution systems are built to deliver water to the end-users with adequate quantity and quality under both normal and abnormal conditions. Depending on the topography of the distribution area different types of distribution, systems may be considered such as gravity-based, pump-based, or a combination of the two.

In what follows, an overview of water distribution systems and their main components is provided. This will be followed by current practice on how to design, operate, and manage these systems.

2.1 Water distribution system components

The main components of water distribution systems are pipes, pumps, storage tanks, valves, and fire hydrants. Other components may include loggers/sensors, meters, etc.

 DOI: 10.4324/9781003057574-4

2.1.1 Pipes

Water distribution systems contain different types of pipes, including:

- water transmission mains that transport treated (potable) water from storage facilities to different parts (e.g., supply pressure zones or district metered areas) of the system;
- distribution pipes, which are smaller in diameter than transmission mains and distribute potable water within local areas; and
- supply/customer connection pipes, which provide connectivity between an individual end-user property and the water distribution system.

Pipes are manufactured in different sizes and materials. The most common materials include cast or ductile iron, reinforced concrete, polyvinyl chloride (PVC), steel, etc. Different materials have their own advantages and disadvantages considering elasticity, strength, longevity, resistance to corrosion, chemical reaction properties, weight, and cost of production, transport, and installation. Cast iron pipes, for example, can withstand high pressures, but they are generally available only in shorter lengths; concrete pipes are non-corrosive, but bulky and harder to handle during installation; and PVC pipes are easy to handle and install due to their light weight, but they are more susceptible to damage. Depending on their size and material, pipes can withstand different pressures, and this should be taken into consideration when designing a distribution system.

2.1.2 Pumps

Pumps are used to aid the transport of water, including to higher elevations, by imparting energy to the flow and increasing pressure. They can be used in different parts of distribution system, including to:

- transfer raw water from source (surface water or groundwater) to treatment plant;
- discharge water under pressure to the distribution system; and
- increase pressure in the parts of the system with low pressure (booster pumps).

Different types of pumps are available, with centrifugal pumps being the most commonly used in water distribution systems. These are easy to install and operate, are low cost, and can be operated under a variety of conditions. Specific types of centrifugal pump include axial-flow, radial-flow, and mixed-flow. Radial-flow pumps provide (relatively) low capacity and high head, and, in water distribution systems, they may be used where there is an elevation difference between supply and distribution areas, to enable water to be transferred from low elevation to high elevation during low energy tariff times and used during peak demand time. Axial-flow pumps, conversely, provide high capacity but low head, and they can be used where there is relatively little difference in elevation – for example, transferring raw water from a reservoir to a treatment plant. Mixed-flow pumps deliver a moderate flow of water with medium head.

Depending on requirements of the pumping station, either single or variable speed pumps can be selected, and multiple pumps may be operated either in parallel or in series.

2.1.3 Storage tanks

In distribution systems, storage tanks are used to:

- enable fluctuating demands to be met;
- equalize operating pressure; and
- provide reserves for firefighting and emergency requirements.

Tanks play a key role when demand exceeds the capacity of the pumping station, treatment plant, or raw water supply in the distribution system. During low-demand times such as the early morning hours, water can be transferred to the tanks. The stored water is then delivered to the distribution system by gravity during peak demand hours to compensate for a shortage of water delivered from the source to the distribution system.

Tanks not only overcome issues related to limited capacity at the upstream of the distribution system, but they also allow distribution pipes and pumps to be sized for average demand rather than peak demand. This reduces complexity and issues related to overcapacity systems and provides capital savings for the water company. Tanks also allow operators to schedule the operation of pumping stations to coincide with lower energy tariff time periods. This increased flexibility in the timing of pumping has the additional advantage of increasing the useful life of the pumps by enabling more efficient operation and thus reducing wear and tear due to frequent switching on and off.

Tanks (along with pressure-reducing valves) may also be used to avoid excess pressures in low elevation parts of the system and low pressure in high elevation parts, respectively. This is important since, although provision of water with an adequate pressure is one of the main requirements of water distribution systems, excess pressure in the system could cause leakage and bursts. By introducing an additional tank, a new pressure zone is introduced to serve the customers at high elevation.

In addition to meeting demand fluctuation and providing adequate pressure in the system, tanks also provide additional storage for emergency events such as firefighting.

Some buildings may have their own storage tanks in addition to those in the distribution system. In particular, water distribution systems may not provide adequate pressure in very tall buildings, and thus these buildings will need to install their own pumps or storage facility in order to deliver water with adequate pressure to higher floors. Tanks in individual buildings may also be used in developing countries due to an imbalance between supply and demand resulting in insufficient and/or unreliable pressure in the distribution system. This is true especially in intermittent water distribution systems where water is delivered only a few hours a day or a few days a week – in such cases, those customers who can afford to invest in individual storage tanks that allow them to store as much water as possible during supply hours and then use it throughout the day.

Depending on the topography of an area, tanks could be built as a surface tank (at ground elevation) or as an elevated tank when a surface tank would not provide sufficient head. The location, elevation, and size of the tank can be considered as design variables to reduce the capital and operational costs while delivering high-quality water. For example, overcapacity tanks would result in long retention time and water stagnation at the tank, thereby resulting in deterioration in water quality. Conversely, undercapacity tanks would have a short retention time and so no associated stagnation, but they would be insufficient to provide the required supply at times of peak demand.

2.1.4 Valves

Valves are required to control or regulate flow and pressure in the distribution systems. There are different types of valve that may be used:

- Isolation valves. These are used to shut down a segment of a distribution system or isolate a pipe for maintenance or repairs. Distribution systems typically have a large number of these valves distributed uniformly throughout, so that small segments of the system can be isolated during emergency response events. Any branched parts of a distribution system contain an isolation valve, and at least two or three isolation valves would be expected at T-junctions and cross junctions, respectively.

- Pressure-reducing valves (PRVs). These maintain a constant lower pressure at the downstream of the valve than that of its upstream in order to reduce leakage and burst.
- Pressure-sustaining valves (PSVs). These maintain a constant pressure at the upstream of the valve in case downstream water demand causes low pressure at the upstream.
- Flow control valves. These regulate the flow by changing the rate of opening of the valve to deliver a pre-set flow to the downstream of the valve.
- Time-controlled valves. These, as the name implies, are controlled on a time basis in response to demand pattern during operation time.
- Throttling valves. These can be used to stop or regulate the flow.
- Air release valves. These are required at high points in the system to release air.

2.1.5 Sensors/Meters

Supervisory control and data acquisition (SCADA) is a widely distributed computerized system that remotely monitors and controls assets from a water company's control room. The types of data collected may include flow in pipes, pressure at demand nodes, water levels in tanks, and the status of valves and pumps. The collected data could be used for daily operation of a water distribution system, leakage management, emergency response planning, etc.

2.2 Water distribution system types

Depending on the location of water sources and the topography of the supply area, different types of water distribution systems can be considered, including:

- Gravity-based systems: In these systems, the source of supply is at a sufficient elevation above the distribution area that no pumping is required (Figure 2.1a).

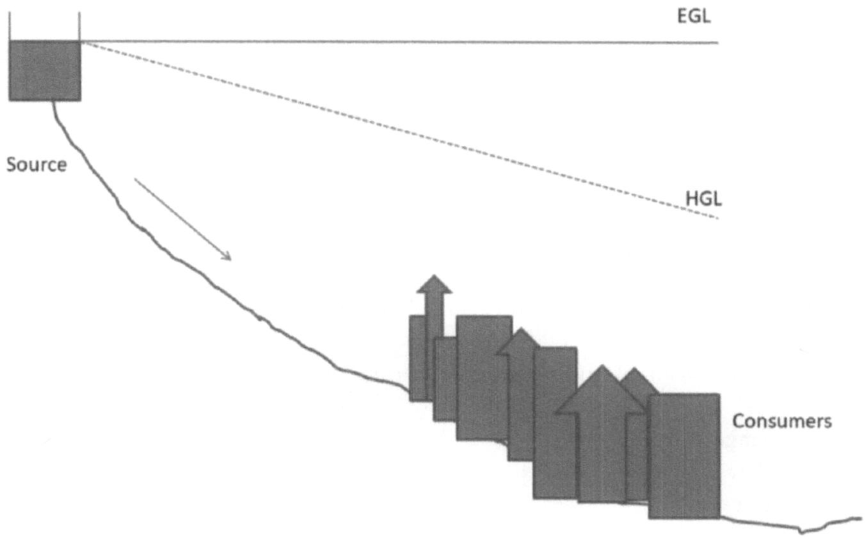

Figure 2.1a Water distribution system types (EGL – energy grade line, HGL – hydraulic grade line).

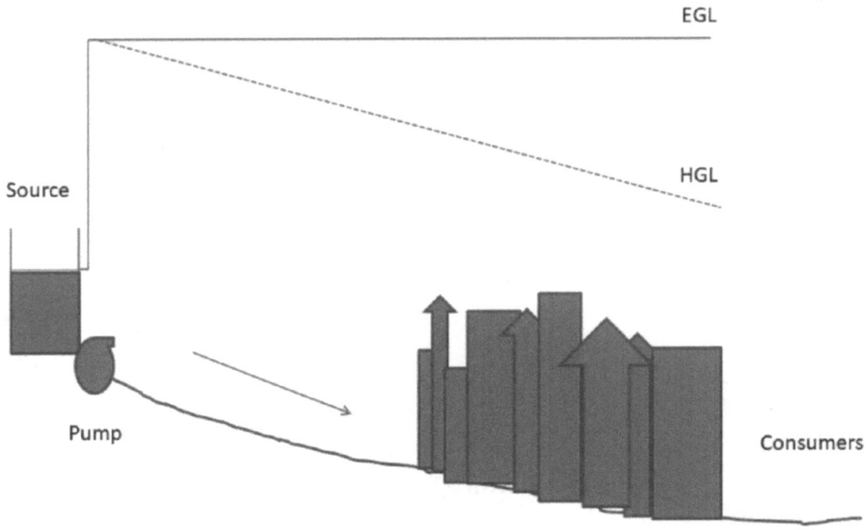

Figure 2.1b Continued.

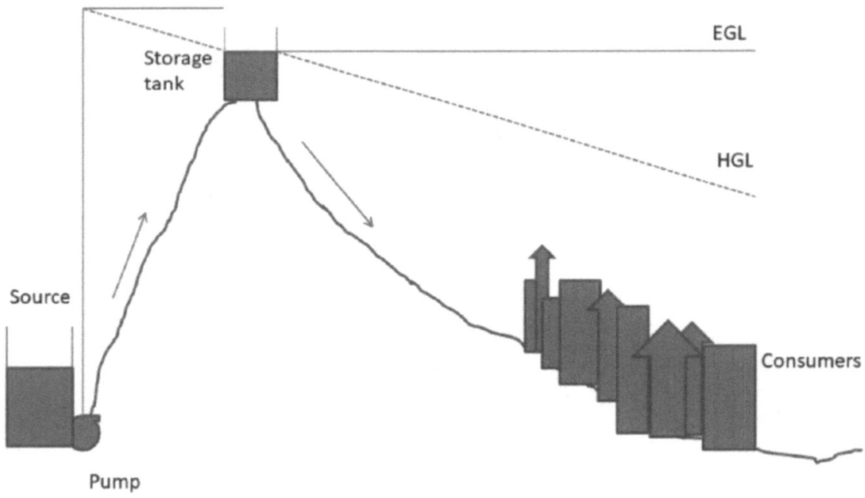

Figure 2.1c Continued.

- Pumped supply systems: In these systems, the source of supply is at an elevation lower than the distribution area or not high enough elevation to maintain minimum pressure, and thus pumping is necessary to ensure sufficient pressure is achieved (Figure 2.1b).
- Combined systems: In these systems, the source of supply is at a location that needs to be pumped to a storage tank, but onward distribution to the supply area can be achieved by gravity (Figure 2.1c).
  a. Gravity-based system
  b. Pumped-based system
  c. Combined-pumped and gravity-based system

2.3 Water distribution system layout

Water systems and land use patterns have always shaped each other (Farmani and Butler 2014). Carter et al. (2005) argue that "a fundamental contributing factor to many urban water management problems is a failure to adopt an integrated approach to water management and land use planning." The configuration of a distribution system is dictated by urban form, topography, location of the source, and location of the treatment plant.

In large cities, water distribution systems contain a combination of looped and branched (tree) subsystems. In branched parts of a system, water flows in one direction only, which can result in issues with reliability of supply. In these parts of the system, consumers are supplied through only a single connection to the rest of the system and, in the case of pipe failure, consumers will be without water supply during the repair time. Another issue with branched systems is that if there is a low demand in these dead ends, water will become stagnant and quality may deteriorate.

Looped networks, conversely, deliver a higher level of reliability and can allow the system to serve customers even under abnormal conditions (for example, when there is a failure in the system). Although looped systems are preferred over branched networks, highly looped systems may cause water quality issues due to longer travel times in the system.

2.3.1 Supply pressure zones and district metered areas

Supply pressure zones are areas or sections of a distribution system of relatively constant elevation. They are supplied by single sources and have good infrastructure connectivity with the rest of the system. Zones are used to maintain relatively constant pressures in the system over a range of ground elevations: (a) to ensure the minimum pressure requirement at demand points is satisfied, and (b) to reduce maximum pressure in order to minimize bursts and leakages in the system. This type of division is also important for enabling better control of water quality–related issues.

To comply with regulatory requirements and reduce leakage, district metered areas (DMAs), for which all water supplied can be measured, are implemented in supply zones of a distribution system. Traditionally, a DMA is a discrete part of a distribution system that usually has just a single inlet (Wright et al. 2016), and it is formed by closing isolation valves. However, in some systems, DMAs may be implemented by installing flow meters at all the pipes that connect the DMA to other parts of the distribution network rather than by closing them with valves. This is in response to the potential negative impacts of DMAs on water quality (Armand et al. 2018) and the resilience of the WDS (Diao et al. 2016). DMAs allow better assessment of water balance for leakage management and also improve pressure management.

2.4 Water distribution operation and management

End-users of a water distribution system may include residential, commercial, and industrial customers, and additional demands may be placed on it by, for example, firefighting and street cleaning. The required capacity for a water distribution system is determined based on peak demand plus fire flow. Peak demand can be assessed based on historical data, micro-component demand analysis, or smart metering data, while required fire flows may be legislated. Using scenario planning, future demand can be forecasted based on baseline year demand. The baseline and forecast demands will determine the sizing of infrastructure for new systems or expansion and rehabilitation of existing systems (Farmani et al. 2005).

2.4.1 Reliability and resilience

Traditionally, water distribution systems are designed to provide service that is reliable under predefined scenarios. The term "reliability" in water distribution systems mainly refers to the ability of the network to provide consumers with an adequate and high-quality supply under normal and abnormal conditions (Farmani et al. 2005). This necessitates provision of some additional operating capacity in the system so it will deliver the required level of service even under abnormal operating conditions. To be considered 100% reliable, a system must never fail. This reliability-based approach to water system management represents the conventional planning process, which is threat-based and mitigation-focused and relies on identification of potential threats that may then be embedded in the planning process. The system design decisions are assumed to deliver the required level of service under prevailing conditions. The design of a water distribution system will have an impact on its operation and maintenance and, in particular, systems that are designed with excess capacity to provide a higher level of reliability may have a lot of water quality issues, thereby resulting in additional operational costs (Farmani et al. 2006).

Climate change, population growth, and demographic change all pose a challenge to water management. It will become more difficult to manage water systems under highly variable future scenarios, and a reliable system will no longer be sufficient when subject to unexpected threats that exceed design conditions. Levels of service will be significantly challenged by multiple threats and the speed, magnitude, and uncertainty of future change (Butler et al. 2017). These vulnerabilities highlight the need to improve the resilience of current solutions to the potential effects of changes in the future. Systems should be designed and operated, therefore, to overcome rather than avoid failure altogether. Butler et al. (2014) defined resilience as the degree to which the system minimizes the level of service failure magnitude and duration over its design life when subject to exceptional conditions. Different system properties may contribute to the reliability or resilience of a water system, but the provision of any one property does not guarantee a certain level of service. A number of studies have concluded that reliability is necessary but not sufficient for resilience (Scholz et al. 2012; Blockley et al. 2012; Pickett et al. 2014).

Past studies that have addressed unknown or unspecified threats have typically used an approach that is qualitative (rather than quantitative) or based on system properties (rather than system performance). Butler et al. (2017), however, proposed a holistic approach to improve the response of water systems to unknown threats and build resilience to these threats, thus going beyond more traditional approaches that are typically focused on foreseeable and identifiable threats. This approach includes a "middle-based" assessment methodology for resilience that focuses on system failure modes and the resultant level of service failures, rather than on the causal threats. Since system failures modes are more easily identifiable than threats, and all threats (known or unknown) that result in the level of service failure will do so only if they also affect the system, this approach enables a comprehensive assessment of resilience considering system performance under exceptional conditions, including those that are unforeseen. Instead of focusing on specific predefined scenarios, a middle-based assessment is able to capture the effects of both probable and highly improbable system failures, without requiring knowledge of their probability (or even their possibility). Such a holistic assessment of network resilience can help to direct emergency responses and long-term capital investment more precisely, thereby improving the efficiency of planned expenditure and reducing the chance of strategic surprises arising from unassessed failure scenarios during event management. This facilitates cost-effective response planning during emergency events to overcome short-term disruptive shocks and contributes to the development of resilient water infrastructure systems to overcome long-terms stresses.

A comprehensive list of operational (e.g., optimum valve operations) and capital investment (e.g., expansion/rehabilitation of the system) intervention options could be implemented for event management in real time and for improvement of system resilience. Multiobjective optimization could be used to find alternative solutions that address the trade-off between cost and different performance indicators, including reliability and resilience.

2.4.2 Leakage and pressure management

Climate and global changes and asset deterioration are putting great pressure on water systems. Some of the challenges in the water sector include climate change risks, water efficiency, reducing abstraction, leakage, and interruption to water supplies, by improving management of incidents by modelling, forecasting, mitigating, and responding. The water industry has made great progress in responding to these challenges, as well as addressing environmental commitments, customer expectations, and regulatory requirements. However, still more needs to be done, especially on topics such as leakage, which is affecting the majority of utilities in developed and developing countries. Leakage is the loss of water from the supply network through uncontrolled actions. In addition to loss of water, it can also result in negative consequences such as:

- potential risks to public health due to the entrance of contaminants from the environment into the pipes in areas of negative pressure;
- environmental issues due to wasted energy used by pumps to deliver water that is then lost from the system and to compensate for the pressure drop;
- the amount of chemicals used to treat the water in the treatment plant;
- leaked water ending up in surface water and having a negative impact on living organisms due to high chlorine concentration.

All of these negative consequences lead to socioeconomic losses and, as such, much effort has been made to manage leakage. These efforts can be categorized as follows (Khorsandi and Farmani 2021):

- Leakage prevention. This involves making special considerations (pipe material, pressure levels) at the planning and design stage, and pressure zoning to prevent the occurrence of leaks. It could also involve maintenance of pipes before leaks happen and replacing old pipes with pipes made from better quality materials such as stainless steel.
- Leakage assessment/awareness. This is the process of determining the overall level of leakage in a distribution network, and it is usually based on water balance. This can be done by three approaches: (1) integrated/top-down approach at the company level; (2) minimum night flow/bottom-up approach at DMA level; and (3) burst and background estimation/component analysis approach at DMA level in combination with the minimum night flow.
- Leakage control. This is the process of efficiently controlling current and future leakage. Different leakage control approaches exist, such as on-line and off-line pressure rezoning; WDS optimization to reduce leakage and energy consumption by better operation of the system considering pump scheduling and demand prediction; and pressure management considering variable speed pumps and pressure-reducing valves (PRVs).
- Leakage detection and localization. Timely detection of leaks can have many benefits. It can reduce environmental impacts by reducing water losses; it can allow any required interruptions of supply to be planned, therefore reducing the impact on customers; and it could

also reduce financial consequences by reducing the level of pumping and financial losses linked to the amount of water lost. Localization of leakages has three different phases. The first phase is narrowing down the location of a leak to a specific district. For this, sensor-based leakage detection methods that use time-series data (for example, from pressure and flow sensors) to identify potential areas of the system that may have leakage are the most common methods. The sensors create big data, and by applying appropriate data analytics to them, valuable information can be obtained to make the detection and localization of leakages possible. The second phase is identifying which pipe in the district area has leakage, and the third phase is pinpointing the leakage location to a specific area as close as possible to the leak (e.g., within a radius of 50 cm). Leakage localization is important in order to reduce disruption to customers and traffic by identifying the location of the leak as close as possible and minimizing road closures, etc. There are different methods available, such as step testing, infra-red imaging/thermography, ground-penetrating radar, tracer gas, and acoustic sensing.
- Leakage repair. This is the final stage in the leakage management process.

The focus on big data is relatively new in the water industry (Shaw 2017) and the majority of asset management and operational decision making is done either independently or with limited use of the available data. Future water distribution systems will be smarter. These water systems will be more connected and operated with more data in real time to achieve maximum efficiency and effectiveness.

References

Armand H., Stoianov I., Graham N. (2018) Impact of network sectorisation on water quality management, *Journal of Hydroinformatics*, 20(2), p. 424–439.

Blockley D., Agarwal J., Godfrey P. (2012) Infrastructure resilience for high-impact low-chance risks, *Proceedings of the Institution of Civil Engineers: Civil Engineering*, 165(6), p. 13–19.

Butler D., Farmani R., Fu G., Ward S., Diao K., Astaraie-Imani M. (2014) A new approach to urban water management: Safe and SuRe, 16th Conference on Water Distribution System Analysis, WDSA 2014. Elsevier, Bari, Italy.

Butler D., Ward S., Sweetapple C., Astaraie-Imani M., Diao K., Farmani R., Fu G. (2017) Reliable, resilient and sustainable water management: The Safe & SuRe approach, *Global Challenges*, 1(1), p. 63–77.

Carter N., Kreutzwiser R.D., de Loe R.C. (2005) Closing the circle: Linking land use planning and water management at the local level, *Land Use Policy*, 22, p. 115–127

Diao K., Sweetapple C., Farmani R., Fu G., Ward S., Butler D. (2016) Global resilience of water distribution systems, *Water Research*, 106, p. 383–393.

Farmani, R., Butler D. (2014) Implications of urban form on water distribution systems performance, *Water Resources Management*, 28, p. 83–97.

Farmani R., Walters G.A., Savic D.A. (2005) Trade-off between total cost and reliability for Anytown water distribution network, *Journal of Water Resources Planning and Management*, 131(3), p. 161–171.

Farmani R., Walters G.A., Savic D.A. (2006) Evolutionary multi-objective optimisation of the design and operation of water distribution network: Total cost vs reliability vs water quality, *Journal of Hydroinformatics*, 8(3), p. 165–179.

Khorsandi P., Farmani R. (2021) Data analytics for leakage detection in water distribution networks: A systematic literature review, *Journal of Water Resources Planning and Management*, Under review.

Ofwat (2021) *Water Sector Overview, Water Services Regulation Authority*, Water sector overview - Ofwat, accessed February 2021 https://www.ofwat.gov.uk/regulated-companies/ofwat-industry-overview/.

Ormsbee L.E. (2006) The history of water distribution network analysis: The computer Age, 8th Annual Water Distribution systems Analysis Symposium, Cincinnati, OH, August 27–30.

Pickett S.T.A., McGrath B., Cadenasso M.L., Felson A.J. (2014) Ecological resilience and resilient cities, *Building Research and Information*, 42(2), p. 143–157.

Scholz R.W., Blumer Y.B., Brand F.S. (2012) Risk, vulnerability, robustness and resilience from a decision-theoretic perspective, *Journal of Risk Research*, 15(3), p. 313–330.

Shaw (2017) Understanding big data in the water industry, Water online, Understanding Big Data In The Water Industry (wateronline.com)

Walski, T.M. (2006) A history of water distribution, *Journal AWWA, American Water Works Association*, 98(3), p. 110–121.

Wright R., Abraham E., Parpas P., Stoianov I. (2016) Control of water distribution networks with dynamic DMA topology using strictly feasible sequential convex programming, *Water Resources Research*, 51(12), p. 9925–9941

3

HYBRID WATER SUPPLY SYSTEMS

Resilience and implementability

Casey Furlong, Ryan Brotchie, Peter Morison, Lindsey Brown, and Greg Finlayson

3.1 Introduction

3.1.1 Water security through water resource planning

Water resource planning focuses on ensuring water security and is one of the most important aspects of water governance. Water security is the "capacity of a population to safeguard sustainable access to adequate quantities of acceptable quality water for sustaining livelihoods, human well-being, and socio-economic development … and for preserving ecosystems in a climate of peace and political stability" (UN-Water, 2013). In this sense, water security can be considered as the tolerable level of risk for supplies not being able to meet demands (Grey et al., 2013). Water resource planners are tasked with creating projections of water demand (how much water will be needed by a city) and water supply (how much water will be available from planned and existing resources) (Xu & Singh, 2004).

To support water security objectives, it is necessary for water resource plans to be robust, considering, and responding to all plausible risks. Robust demand estimates must incorporate potential scenarios for population, housing and industry, and consumer behaviour (Dong et al., 2013). Robust supply estimates must incorporate climate change, catchment rainfall/run-off relationships, infrastructure failure, source depletion (e.g., over-extraction of aquifers), and source contamination (e.g., long-term build-up of contaminants in groundwater or the impact of bushfires) (Chiew et al., 2009). Robust water resource strategies must consider and respond to overarching societal, economic, and transitional risks, and they can be improved through consideration of unexpected "shocks" (e.g., disasters), for example, through a process of "stress testing" worst-case scenarios (Hall et al., 2019). While each city has unique local factors, which are difficult to model, some researchers have attempted to run models of water supply and demand on a global scale, including most large cities (e.g., the WaterGap Model) (McDonald et al., 2014; Flörke et al., 2018).

Water supply shortfalls are becoming increasingly common in the context of triple pressures of climate change, water demand growth (e.g., by population growth), and environmental pollution (Srinivasan et al., 2012; Vörösmarty et al., 2010). When water resource planners identify a current or approaching shortfall in supply relative to demand, it is their responsibility to identify and assess options for reducing demand and/or increasing supply (Victorian Government, 2016; LADWP, 2020).

 DOI: 10.4324/9781003057574-5

3.1.2 Hybrid water supply systems

To bolster water security, water authorities have had to look beyond traditional ground and surface waters towards alternative water sources, such as rainwater, stormwater, recycled water, and desalination[1] (Sapkota et al., 2015). Through an Integrated Water Management (IWM) approach, these alternative water sources can not only support water security, but also provide other social, environmental, and economic benefits, such as providing new sources of water for river environmental flows or protecting receiving environments from excess wastewater effluent and urban stormwater (James et al., 2015; Furlong et al., 2017). A "hybrid water supply system" is defined here as one that combines a mix of sources, including both traditional sources (surface and groundwater) and new sources (rainwater, stormwater, recycled water, and desalination).

This paper speaks specifically to the contrasting examples of two cities: Melbourne and Singapore. Melbourne's efforts towards hybridisation have focused on a combination of dams and desalination as well as a mix of non-potable (not for drinking) water substitution using recycled water and decentralised rainwater and stormwater harvesting. Singapore has also diversified its water supplies, but without opting for a decentralised non-potable approach, instead focusing on large-scale addition of stormwater and recycled water to its centralised drinking water system. While this paper focuses on these two cities, many other cities are well known for a hybrid water supply approach. For example, both Los Angeles (USA) and Perth (Australia), have a strategy that focuses on directing treated wastewater into an underground aquifer, for later treatment and addition to a centralised drinking water network (Boxall, 2019; Water Corporation, 2021).

In general, the more sources that are added to a water supply system, the more complex planning and operations are likely to become (Sapkota et al., 2015). When hybridisation is done not only for water security, but also for other social and environmental reasons, this further increases planning complexity, requiring engagement with more diverse stakeholders, processes, and drivers, such as maintaining natural waterway flow regimes (Li et al., 2017).

From purely a water supply perspective, planning becomes complex when there is a wider variety of options to compare over the long term (Rathnayaka et al., 2016). Operations become more complex when there is a wider variety of sources to choose from in the short term, as each source will have a different point of connection to the mains transfer network, a different treatment and transfer cost, and different considerations around initiating/stopping supply (Maiolo & Pantusa, 2016). For example, operators may have to compare (a) the cost of turning a desalination plant on, against (b) the disadvantage of releasing water from an upstream dam. This process seeks optimisation, whereby costs are minimised, often across a large geographic region with diverse sources and users. Hybridisation of sources makes such operational assessments exponentially more complicated in a way that is still not well understood (Sapkota et al., 2015).

Not all attempts at water supply hybridisation are successful (Furlong et al., 2017; Turner et al., 2016). Many schemes have encountered challenges during construction or operation, resulting in some schemes being decommissioned (West et al., 2015). Sometimes there is an error in the strategy and associated decision making, such as how options are compared in relation to risks and costs (Furlong et al., 2017). In other cases, difficulties emerge during implementation, owing to complexity and the involvement of various parties (Hill & Hupe, 2002).

3.1.3 Purpose of this chapter

This chapter argues that water supply option failure, or least disappointment, ultimately occurs regarding either (a) source "supply resilience" and/or (b) source "implementability." Supply resilience is used here to mean that water is reliably available for consumption, as there is the ability to adapt water supply operations to meet water needs, regardless of climatic or infrastructure conditions (e.g., droughts, storms, fires, other climate change impacts, treatment process, or pumping failures). Implementability is used here as a general term to summarise the many factors that influence the difficulty of implementing an option, including cost, regulatory, governance (e.g., roles and responsibilities for centralised or decentralised infrastructure), environmental, and community-acceptance factors.

To identify solutions offering a good trade-off between supply resilience and implementability, a certain amount of experimentation in every city is necessary and valuable. However, the process of adaptive learning through trial and error can be costly and eventually strain credibility if too many attempts are unsuccessful. Therefore, it is logical to learn as much as possible from the attempts of other cities and incorporate these lessons rather than replicating previous errors (refer to Chapter 17 in this book on Policy Transfer for further discussion).

This chapter seeks to provide guidance on hybrid water supply systems by exploring: (a) various alternative water source/end-use combinations in regard to supply resilience and implementability; (b) contrasting the experience of Singapore and Melbourne as illustrative case studies; and (c) outlining a high-level process and principles for selecting a preferred portfolio of water sources. The Ambiguity-Conflict Model of Matland (1995) is then used to provide an additional perspective on implementability that acknowledges government and community views of various policy options within the parameters of ambiguity and conflict.

3.2 Alternative water sources

A variety of water supply source and end-use combinations exist across the world. This chapter is focused on the "alternative" or "non-traditional" of these water source/end-use combinations, which have emerged through the search for new supply options. These are shown in coloured cells in Table 3.1. This section introduces the common alternative water source options that planners will consider as part of water resource planning. Each of the water sources is discussed below in relation to supply resilience and implementability.

3.2.1 *Overview of water supply resilience*

Water supply resilience as a concept includes consideration of a wide set of trends and shocks that would cause a problem for water supply, either through impacting water quantity or through quality. A key and ever-present factor in water supply resilience is the amount of dependence on rainfall. Climate change will, in many cities, reduce average annual rainfall, in terms of either average annual volume, frequency, or predictability (Vörösmarty et al., 2010; Dong et al., 2013). This means that water sources that are independent of rainfall will, in dry cities, generally be the most resilient. Resilience also considers incidents such as bushfires, earthquakes, electricity blackouts, and other technology failures.

There are varying degrees of, and ways to achieve supply resilience, and for a single option resilience can change over time. For example, a very large dam may have provided a resilient urban water supply for the entire historical record of a town, but this supply may or may not be resilient in the face of climate change. Figure 3.1 gives an initial overview of the rainfall independence of various alternative water sources, mapped against centralisation/decentralisation of infrastructure.

3.2.2 *Overview of implementability*

The degree to which an alternative water source is implementable is made up of levelised cost,[2] environmental impacts, governance, and community perceptions. Any of these issues on their

Table 3.1 Water source and end-use combinations (coloured cells indicating non-traditional source/use combinations, with traffic light index indicating a spectrum of common to rare)

	Surface water	*Ground water*	*Desalinated sea water*	*Recycled water*	*Urban storm water capture*	*Rainwater tanks*
Mains water (i.e., drinking)	Very common	Very common	Becoming common	Emerging	Rare[1]	Areas without mains supply[2]
Domestic non-potable	Commonly sourced from mains water (i.e., same source as drinking water)			Some examples		Common
Municipal non-potable					Emerging	
Industry non-potable					No known example	
Agriculture	Very common	Very common	Somewhat rare[3]	Common	Rare[4]	N/A

[1] For instance, Singapore.
[2] Warrnambool, Australia, is one example of rain tanks connecting directly to a dam.
[3] For instance, Sundrop Farms in South Australia and some areas in the Middle East.
[4] Western Water area in Victoria is one example where stormwater is captured and traded to agriculture.

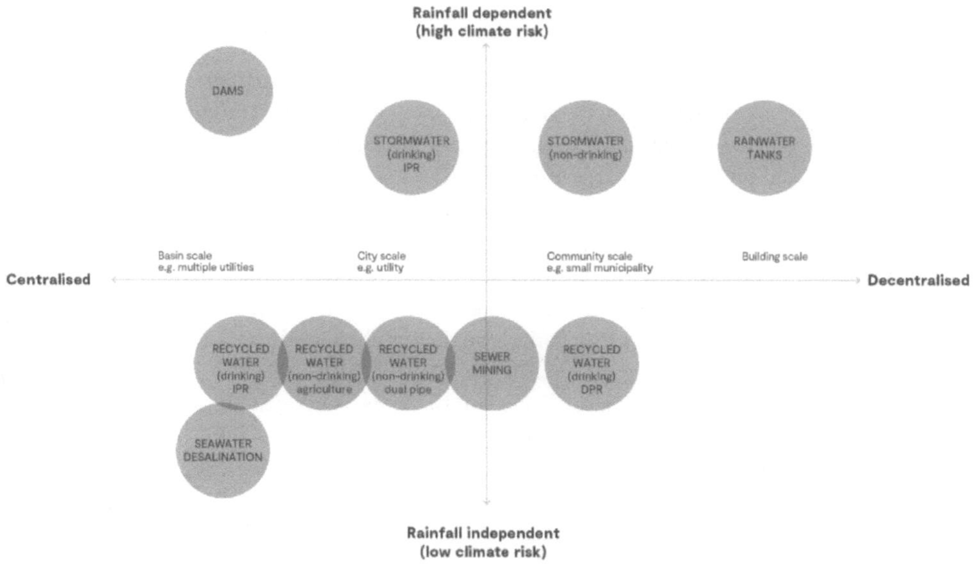

Figure 3.1 Rainfall independence of alternative water sources mapped against centralisation/decentralisation of infrastructure.

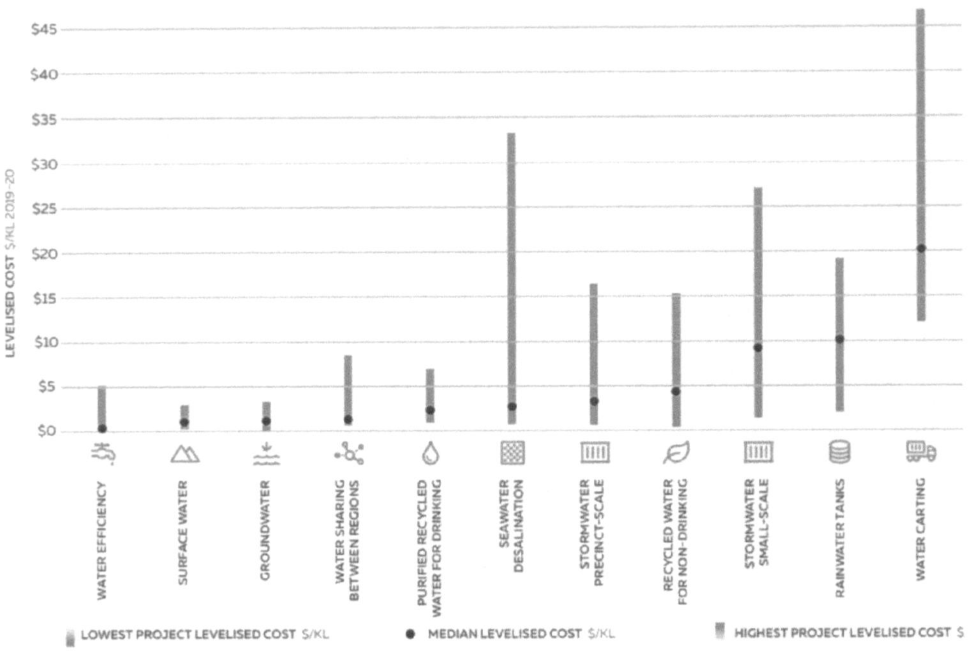

Figure 3.2 Relative levelised cost of various water sources in Australia, based on survey of~400 schemes (Source: WSAA, 2020).

own can be enough to prevent an option from being attempted or cause an option to fail or disappoint when it is attempted. Cost and environmental impacts have been recognised and considered in water supply augmentation decisions for many decades (Dee et al., 1973). The relative costs of alternative water sources in Australia are shown in Figure 3.2.[3]

Another key aspect of water source option implementability is the amount of water that can be produced per site, how many sites are needed, and what this means for governance, particularly stakeholder and community participation. The relative water yields from alternative water sources per project in Australia are shown in Figure 3.3.

Supply solutions that create large amounts of water per site, such as desalination plants, mean that fewer sites are needed (e.g., 1–3 projects per city). These large projects can be managed in a centralised manner, operated by utilities or equivalent specialists. For larger, more centralised projects, it is possible to predict, with a relatively higher level of confidence, the amount of water they will produce.

If a supply solution creates only a small amount of water per site, then many individual projects may be needed, spread across a city. These decentralised pieces of infrastructure may need to be managed by residents, businesses, park managers, etc. Decentralised supply infrastructure can create many environmental and social opportunities, such as managing pollution closer to its source, keeping water in the landscape, and protecting waterways. However, from a water supply implementability perspective, the more that power is devolved and distributed, the lower the confidence is in the yield produced, especially in the long term (discussed further in the Melbourne case study below).

It is important to note that "implementability" also changes with time. New concepts are hard to implement, due to inexperience in all aspects: design, fulfilment of regulatory requirements, costs, risks, environmental management, community capacity/views, and appraisal of

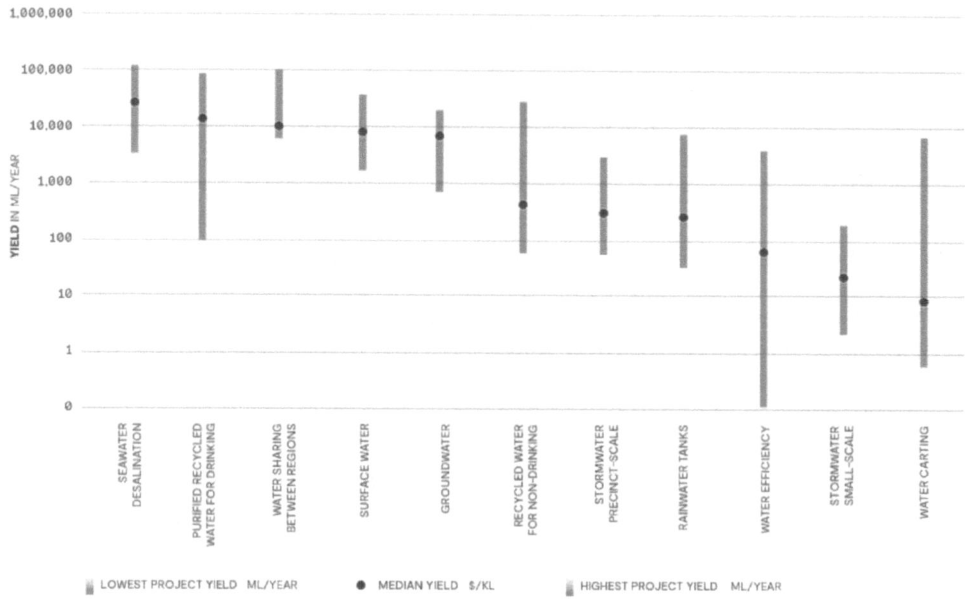

Figure 3.3 Relative yield of various water sources in Australia, based on survey of~400 schemes (Source: WSAA, 2020).

costs and benefits. As a concept becomes more familiar, the difficulty of implementation reduces. Implementation can be seen as having an "activation energy," whereby the amount of energy required reduces each time a project is implemented in a given location. For example, desalination is increasingly being seen as a standard option now, but it was at one point considered experimental and innovative (Tal, 2011). For decentralised infrastructure, it is also possible to drive lasting change in community culture and capacity in a way that increases confidence that infrastructure will be appropriately managed. For example, the Netherlands is recognised for its high community interest and capacity in water management (Furlong et al., 2018).

3.2.3 *Desalination*

3.2.3.1 Desalination source and supply resilience

Desalination projects remove salt from seawater, and then generally supply treated water into the mains network for drinking and all other uses (Ghaffour et al., 2013). Technologies for desalination include thermal, such as Multi-Stage Flash Distillation, and Membrane, such as Reverse Osmosis or Electrodialysis. Reverse Osmosis has come to be the most common technology (Pinto & Marques, 2017). There are some examples of using desalination for agriculture (e.g., tomato growing in greenhouses), but this is less common (Hitchin, 2014). Desalination is, for obvious reasons, more suitable when demands are near seawater. Construction and operational costs, for pipes and pumps, increase with distance from the coast. Desalination is entirely independent from rainfall, as the source water comes from the sea and is always present.

3.2.3.2 Desalination implementability

Desalination can in many instances be relatively simple for governments and water utilities to deliver. Governments have significant experience with the delivery of large projects, from experiences with transport, health, and other sectors. Funding partners are sought, often from the private sector, and contracts are arranged between public and private entities. Once a suitable site and connection to the existing water transfer system are found, and environmental concerns are addressed, desalination can be a relatively straightforward process. As it is connected to the existing reticulation network, both new and existing suburbs can rely on the generated water.

Desalination has a reputation for being very expensive. These projects typically involve substantial upfront capital investment, and therefore they often involve public-private partnership (PPP) arrangements. However, a recent Australian study found that they are generally approximately the same levelised cost as potable recycled water projects, and cheaper per unit volume than more decentralised alternative water sources, such as small-scale stormwater, rainwater, and non-potable recycled water (WSAA, 2020). Some commentators hold the view that PPP arrangements result in a higher cost to the community than would be incurred by governments spending their own money (Potter, 2020), but many governments see a benefit in having less debt on a balance sheet and in sharing risks with a private partner.

There are also some concerns around desalination in relation to broader social and environmental considerations. There are some concerns around the impact on aquatic ecosystems from the return of higher salinity brine, although some studies (e.g., at Carlsbad, USA) have found no negative impact on biological outcomes despite higher salinity (Lykkebo Petersen et al., 2019), thus highlighting the importance of a site-specific Environmental Impact Assessment (Lattemann & Höpner, 2008).

A larger concern is generally the amount of greenhouse gas emissions to create the required energy. The emissions can be significant if the electricity is generated through burning fossil fuels. Energy usage has led some commentators to view desalination overall as not environmentally sustainable. However, it is possible to fully offset the energy required using renewable energy (e.g., this was done for the Victorian Desalination Project near Melbourne).

3.2.4 Recycled water

3.2.4.1 Recycled water source and supply resilience

Recycled water is a term used to describe the capture and reuse of wastewater (human/household and industrial sewage).[4] Recycled water has been widely used as a source of water for agriculture since the Bronze Age, as distribution of treated effluent onto farm paddocks has been a common means of effluent disposal as well as land fertilization (Angelakis et al., 2018).

In recent decades, new approaches to recycled water have emerged. There are some recent examples of non-potable recycled water use within cities for public open space, industrial uses, and in some cases "dual pipe" or "third pipe" connections to residential/commercial buildings for toilets, laundry, and gardens (Chen et al., 2013; Turner et al., 2016). Recycled water is increasingly being considered for centralised treatment and reticulation through mains networks for drinking (Furlong et al., 2019). Recycled water is generally captured/produced at wastewater treatment plants (WWTPs), but it is sometimes captured through a "sewer mine," which extracts wastewater directly from sewer pipes, far away and upstream from the WWTP, to generate recycled water closer to demands.

Like desalination, recycled water can be considered significantly independent from rainfall, as the source water is generally constant. Although unlike desalination, recycled water cannot be the only water source for a city, in a closed loop, because there would be some losses from the system. Therefore, some external supplementation from other sources, such as rainfall dependant sources or desalination, will always be needed.

3.2.4.2 Purified recycled water for drinking purposes – implementability

Purified recycled water for drinking is most often in the form of "indirect potable reuse" schemes, a term that means that an environmental buffer is used. Los Angeles and Perth use groundwater aquifers as a buffer, whereas San Diego and Singapore supplement surface water dams (Furlong et al., 2019). As indirect potable reuse requires a suitable environmental buffer (e.g., large dam/waterway/aquifer) before connecting to the mains network, the amount of infrastructure required can vary between cities. For example, if the WWTP is located far from a suitable buffer, then an expensive pipeline construction traversing existing suburbs may be necessary. In other cities, where the WWTP is located adjacent to a suitable dam, waterway, or aquifer, the associated pipe/pump infrastructure costs will be minimised.

Indirect potable reuse is like desalination in that it involves large-scale advanced treatment, and it is approximately a similar cost, generally slightly cheaper than desalination (WSAA, 2020). It is unlike desalination in two respects: it requires more rigorous testing and monitoring for safety; and there is typically a significant community concern around its suitability for drinking. For these reasons, these projects can take up to a decade to implement successfully (City of San Diego Public Utilities Water and Wastewater, 2015). Scientifically it is clear that advanced treatment can make wastewater safe to drink, and this is evidenced by the fact that there is already treated effluent present in drinking water supplies across the world (known as "de facto reuse") (Rice & Westerhoff, 2014).

Communities generally require significant consultation and engagement before they are confident that recycled water is safe to drink. Many politicians hold the view that their voters will never accept it, and so they are nervous to introduce it. However, a review of the available research indicates that populations are responsive to engagement and education campaigns, such as Singapore, Los Angeles, and San Diego (Furlong et al., 2019).

3.2.4.3 Recycled water for non-drinking purposes – implementability

Recycled water for non-drinking purposes is very different from desalination and indirect potable reuse, as it is not added to existing water mains, but rather requires an entirely new and separate distribution system. These can be separated into two general categories: (a) systems that supply a few specific locations, such as industrial, agricultural, or public open space demands; and (b) systems that supply a large number of properties across a suburb, known as "dual pipe" or "third pipe"[5] connections to houses.

Industrial, agricultural, and public open spaces schemes often require context-specific contractual arrangements with end-users (e.g., separate pricing structures), and independent transfer infrastructure (i.e., pipes, pumps). A key issue is whether users are obligated to pay for and/or use water even when they don't need it, a key variable in financial feasibility of such schemes (Furlong et al., 2017). There is an example in Australia where farmers are negotiating with a water utility to consider accepting recycled water onto their properties, even when they do not want/need it, in return for financial incentives (e.g., a lower price). This saves the water utility from having to dispose of wastewater effluent elsewhere. Agricultural and industrial schemes can in some instances provide an affordable source of supply, where wastewater effluent of sufficient quality is located nearby to demands.

"Dual pipe" residential non-potable recycled water schemes generally involve supplying recycled water for toilets, laundry, and private garden demands (i.e., not drinking, cooking, or bathing). From an infrastructure perspective, dual pipe schemes are more complicated than agricultural or open space irrigation. They require a more extensive separate reticulation system (to connect to a larger number of streets/properties) and a separate in-building plumbing system to be constructed in residential areas. These systems can be expensive to retrofit (tearing up and replacing street surfaces to place pipes), and so they are more likely to be implemented in new suburbs where pipes and plumbing systems can be placed more cheaply.

Urban planning controls have been used in some instances to require developers to pay for the dual pipe reticulation and plumbing networks in new developments. There are examples in the USA where the main recycled water reticulation pipe is built first, and then any buildings that are knocked down and rebuilt nearby to the recycled water main are required to connect (e.g., Austin, TX Code of Ordinances § 6-4-11 requires new commercial developments or redevelopments within 250 feet of a reclaimed water main to connect for irrigation, cooling, and other significant non-potable water uses).

Dual pipe schemes have been attempted in the USA and Australia to mixed results. On the positive side, some water utilities believe that this can be a cost-effective means of disposing of wastewater effluent, when environmental regulators discourage discharge to waterways. There have been some issues, including demands being overestimated because irrigated private gardens have shrunk (Melbourne, Australia) and other cases of developers being opposed to the cost imposition of duplicated plumbing (Newcastle, Australia). On a volumetric basis, dual pipe schemes are relatively expensive, but they are justified where they avoid other costs (e.g., there is nowhere else to dispose of the effluent and the government does not allow potable reuse) (WSAA, 2020).

Often these schemes are justified based on avoided costs to other systems. However, there are risks associated with assuming an avoided cost on the basis that another piece of infrastructure does not need to be built. Sometimes others decide to build it anyway, nullifying the identified benefit. There was a scheme in Gold Coast, Australia, where dual pipe was justified to avoid a new ocean outfall (for discharging effluent). However, the uptake and use of recycled water were less than predicted, the outfall was built anyway, and the recycled water scheme was turned off despite capital investment in construction (West et al., 2015).

3.2.5 Stormwater

3.2.5.1 Stormwater source and supply resilience

Stormwater capture involves the capture of rainwater after it touches the ground within a city. This stormwater can be collected through wetlands or storage tanks and used for non-drinking purposes, such as the irrigation of public open space, or it can be collected into a larger basin, dam, or aquifer for the purposes of treating to potable standards and injection into the drinking water network. It is generally the lack of a suitably sized storage (e.g., dam or aquifer) within an urban stormwater catchment, which is the biggest barrier to large-scale capture and reuse.

Stormwater and rainwater harvesting are dependent on rainfall; if there is no rain, then there is no water supply. However, they are not as vulnerable to drought as surface waters (rivers and dams) because of the catchment "rainfall/runoff relationship." Rainfall/runoff relationships refer to the amount of runoff that is generated by a certain amount of rainfall. Permeable earth (i.e., in rural/forested areas) must be saturated before water begins to flow over its surface. Impervious surfaces such as roads and roofs create more runoff, faster. This means that, for a dam in a permeable catchment, a small reduction in rainfall can often result in a larger reduction in inflows to dams (DELWP, 2016). Due to water immediately running off hard surfaces, a rainwater/stormwater system will generally collect water in months when many dams and rivers run dry.

3.2.5.2 Stormwater for drinking purposes – implementability

Singapore has implemented what may be the largest intentional stormwater capture project in the world, by barraging Marina Bay, removing seawater, and thus converting the bay into a man-made dam, to provide stormwater for drinking purposes (Hui et al., 2010). There is one small active stormwater to potable scheme in Australia (Orange City Council, n.d.). There are some examples where stormwater is mixed with other sources as part of a potable supply, e.g., Monterey (USA) (Monterey One Water, n.d.). When there is a very large storage available, such as a dam or an aquifer, stormwater implementation can be similar to desalination and indirect potable reuse, in that it is one large centrally controlled project.

Stormwater harvesting for any end-use is easier to achieve in new suburbs because a stormwater collection system can be purpose built to maximise flow, storage, and site for treatment plant (and potential for developers to support funding). However, the outcomes in Singapore show that, if there is a suitably sized storage, it is possible to retrofit urban catchments and waterways so that they increase stormwater quality (refer to Singapore case study for more detail). In Australia large-scale stormwater harvesting can be almost cost competitive with desalination (WSAA, 2020).

3.2.5.3 Stormwater for non-drinking purposes – implementability

In Australia the most common stormwater capture scheme involves very small-scale decentralised interception of stormwater, for irrigation of public open space (James et al., 2015; Ferguson et al., 2013). These schemes are typically operated by municipal open space managers. While the volumetric cost of these schemes is high, they are promoted for benefits they provide to waterways, by removing excess stormwater associated with the proliferation of impervious surfaces. This creates significantly more complexity for planning and governance, as more stakeholders need to be involved to identify the best locations for decentralised stormwater harvesting across a large geographical area (Melbourne Water, 2018). These small schemes typically collect water in above-ground wetlands/lakes/basins or underground tanks. There is one larger scheme in Salisbury (Australia) that injects harvested stormwater into an aquifer, for later use through a non-potable network (City of Salisbury, n.d.).

As already stated, these schemes are cheaper in new suburbs because a stormwater collection system can be purpose built (and potential for developers to support funding), but small non-potable schemes are also commonly built in existing neighbourhoods, and they can simply divert water from existing underground stormwater drains or other stormwater assets. In general, small-scale non-potable reuse of stormwater is expensive, roughly three times the median cost of desalination per volume (WSAA, 2020).

3.2.6 Rainwater

Rainwater harvesting involves the capture of rainwater before it touches the ground, by connecting a storage directly to a building roof. Harvesting through domestic tanks for non-potable uses is common for detached dwellings (i.e., stand-alone houses) in some countries and rare in others (Campisano et al., 2017). In cities this water is commonly used for gardening, and it is sometimes plumbed into houses for toilets and laundry demands. There is one example in Melbourne (Aqua Revo), of rainwater being treated with chlorine and supplied to hot-water systems for bathing purposes (South East Water, 2019). There is one other example in Australia (Warrnambool, Victoria), of multiple residential roofs being directly connected to an existing nearby dam (via pipes so the water connects from roof to dam without touching the ground) (Wannon Water, 2018).

Rainwater harvesting is typically conducted by individual households, and it may be required or encouraged by water authorities or municipalities, either for water security, flooding, or waterway health reasons. In Melbourne, the building code requires that each new home install either a rainwater tank or a solar hot-water system. Some municipalities require new buildings to install rainwater tanks to meet stormwater quality targets (to reduce flooding and waterway damage). Amsterdam, for example, has an urban planning regulation for flood-related reasons that requires new buildings to be "water neutral" for rainfall of less than 60 mm (in 48 hours) by retaining water on-lot (Furlong et al., 2018).

From an implementation perspective, it is typically straightforward to require new and renovated buildings to install rainwater tanks. These systems are relatively easy to install, but they require strong planning controls or incentives for builders/residents to do so. Rainwater reuse in Australia has been found to be relatively expensive on a volumetric basis, compared to other sources (WSAA, 2020).

The fundamental implementability issue with rainwater tanks relates to their ongoing maintenance and operation. There is a possibility that a large proportion of tanks are not maintained or operated correctly, as there is not enough data on this (Moglia et al., 2015). For this reason, it is very difficult to be confident about how much yield a network of tanks is really contributing towards a city's water security. Also, as rainwater is not always available, there always needs to be a centralised back-up, reducing the potential for any avoided system costs. Real-time "Smart" monitoring systems could eventually resolve this issue and increase confidence in supply yield.

3.3 Ambiguity and conflict

One useful lens to view implementability through is the Ambiguity-Conflict Model (Matland, 1995). The model adopts two variables – ambiguity and conflict – to produce a typology of the form of implementation of a particular policy, in this case relating to potential new water sources. Following a matrix pattern (see Figure 3.4), Matland's model identifies four types of

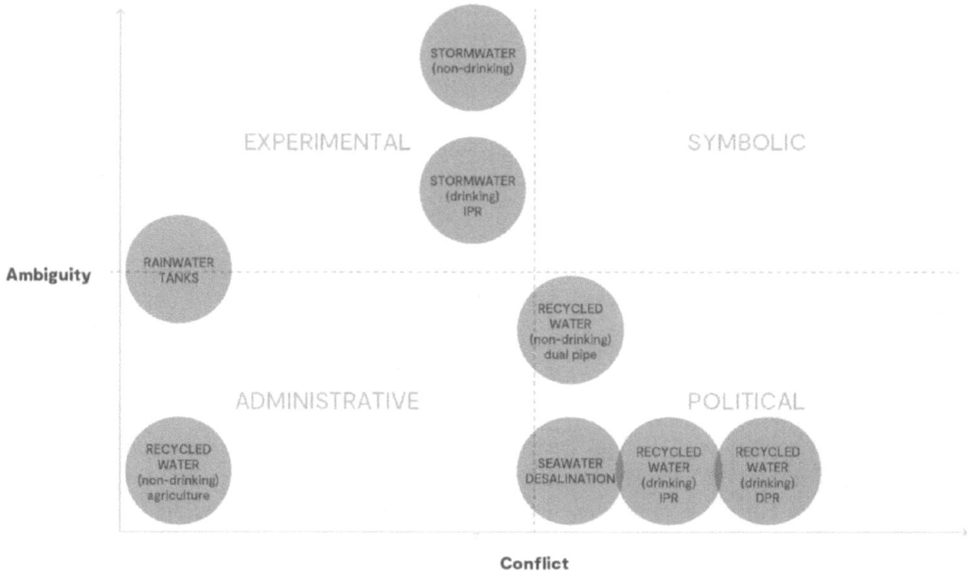

Figure 3.4 Relative ambiguity and conflict of alternative water sources, using the Ambiguity-Conflict Model of Implementation (Source: Matland, 1995).

implementation that relate to the combination and degree of ambiguity and conflict observed within the physical, social, and political context of the water system.

For the purposes of this paper, conflict is used in the traditional sense of the word (disagreement or opposition). Ambiguity is used here to mean both uncertainty around roles and responsibilities, as well as uncertainty around how much water will result from implementation. For example a water utility can know with a high degree of confidence how much water will be produced from a desalination plant but has a lower degree of confidence regarding the use of household rainwater tanks and their impact on water security.

"Administrative implementation" is said to occur in conditions of low ambiguity and low conflict. These conditions are ideal for policy implementation that may be replicable and ongoing. For example, there may be a process for Expressions of Interest, and costs, associated with farmers nearby to WWTPs seeking recycled water.

"Political implementation" is said to occur where high conflict but low ambiguity exists. Disagreements are likely to occur between parties affected by its implementation (e.g., whether the community accepts drinking recycled water). The implementability of the policy relies on protagonists' ability to influence. Water supply technologies may start here, such as desalination or indirect potable reuse, but they may over time move towards administrative.

"Experimental implementation" involves low conflict but high ambiguity. The ambiguity presents in cases where it is not clear who will identify, pay for, operate, and/or regulate a particular water source, how effective they will be, and how much water source will ultimately be created and used. This causes variability of implementation from site to site (e.g., who makes sure that rainwater tanks on private properties are operating correctly), thus creating a series of policy experiments rather than replicates. Over time, after many experiments, it may be possible to move technologies from experimental to administrative. Rainwater tanks could be considered as on the border between the two, as they are administrative to require, but experimental to control over time (in the absence of "smart" monitoring and control systems).

"Symbolic implementation" involves both high conflict and ambiguity, resulting in differing interpretations of what a policy is intending to achieve and how it is supposed to work. The strength of its coalitions at a local level typically dictates the policy's implementation outcome. Due to the challenges associated with non-potable reuse (wastewater or stormwater), such as the cost of duplicating reticulation networks, planners need to ensure that these projects are done for a clear reason and with a clear logic around which areas should or should not be targeted. Without a clear logic, these schemes can fall into the Symbolic category. The options that fall into the Symbolic category are likely to be ones that water security planners have less confidence in, as ambiguity and conflict will both create barriers to scheme implementation, operation, and monitoring.

This is shown visually in Figure 3.4.

This model can be used to understand the implementability of alternative water sources. Administrative implementation is generally favoured by government. For example, in places where desalination has been successfully implemented, its choice has been associated with low ambiguity (governments are confident they will work), and moderately low conflict. However, this is context-dependent, and there are situations where conflict has been high, such as in Melbourne after the desalination plant was built, when extensive rainfall replenished dam supplies, and the government faced criticism for the desalination decision. As dam storages eventually reduced, the conflict waned.

The next preference of governments is typically political implementation. For example, recycled water for drinking is low ambiguity in terms of yield, but typically generates higher conflict in relation to public health concerns. Experimental projects, such as rainwater, are typically less favoured. These are associated with lower conflict but higher ambiguity, leaving governments with limited confidence they will fulfil water security targets.

3.3.1 Singapore case study

Singapore is world renowned for its efforts to achieve water security. To achieve greater political independence from Malaysia, the city-state adopted a multipronged strategy. A first element of this strategy was desalination, building four desalination plants. The next element in this strategy was NEWater, which treats wastewater to a very high quality for industrial uses and for supplementing drinking water. A third element was "city as a catchment," converting much of the city, including Marina Bay, into a network of catchments and man-made dams, for the purposes of collecting stormwater for drinking. This included rehabilitating the city's waterways to improve the quality of the stormwater that enters the dams (Jensen & Nair, 2019). This combination of approaches provides the city with greater supply resilience.

All these solutions have been resilient and implementable. They are resilient in the sense that they are significantly rainfall independent as well as diversified. They are implementable in the sense that they are large-scale schemes operating under centralised governance, resulting in higher likelihood of effective maintenance and operation as well as lower cost.

On the ambiguity-conflict matrix, these projects are generally "political," with low ambiguity and high conflict. Even stormwater harvesting, which on a smaller scale is somewhat ambiguous, was converted into having lower ambiguity and higher conflict by implementation at such a large scale. Singapore did an exceptional job at community education (if not consultation) to increase the public willingness to drink recycled water, thus reducing the associated conflict with this choice (Furlong et al., 2019).

The potential for environmental complications from the capping of a saltwater bay and conversion to a constructed freshwater dam in some other countries such as Australia may have resulted in a level of conflict (i.e., environmental concern) that could not have been overcome. At the time it was not clear how many of the 139 fish species from 57 families that inhabited Marina Bay and the connected river systems would be able to survive the conversion from saltwater to freshwater, and the limitations of fish passage (Hui et al., 2010).

However, the authors do not wish to give the impression that Singapore generally shows disregard for environmental outcomes, as this project example should be considered in the context of the significant investment and improvement in waterways and ecosystems across Singapore and positive biodiversity outcomes (Ng & Corlett, 2011). "Active, Beautiful, Clean Waters Programme" or "ABC Waters" for short, was the name given to the initiative of cleaning up and beautifying waterways, which not only improved water quality prior to harvesting, but also created a wide array of social, environmental, and biodiversity benefits through the city (Public Utilities Board, n.d.).

3.3.2 Melbourne case study

Melbourne is world renowned for some aspects of Integrated Water Management, particularly stormwater quality management in growth areas. This is done through a process of Water Sensitive Urban Design embedded in urban planning provisions, to meet minimum standards for stormwater quality treatment (Potter & RossRakesh, 2007). There are requirements for Green Infrastructure, such as wetlands, raingardens, and swales in all new developments (Ferguson et al., 2013).

In relation to water security, most water supply still comes from surface water dams. The next largest contribution to water security comes from the construction of a large desalination plant (150 GL/year able to supply a third of the city's demand). There have been significant efforts towards other alternative water sources as well.

At present, it is likely that rainwater is the next most significant source of water. Rainwater harvesting has had a strong uptake, with 27% of homes having rainwater tanks for outdoor demands, and 8% of homes having tanks plumbed into houses for indoor toilet and laundry demands as well (Australian Bureau of Statistics, 2011). However, little is known about the proportion of tanks that are maintained and used effectively, and the available evidence is from limited surveys and inspections of a relatively small number of tanks (Moglia et al., 2015).

Stormwater harvesting for open space is expanding in popularity, but these schemes are very small, with an average yield per scheme of 28 ML/year (WSAA, 2020). So, despite many schemes, stormwater does not add up to a notable proportion of Melbourne's supply (Melbourne Water, 2017). In 2012 there were estimated to be 108 schemes (Ferguson et al., 2013). If, in 2021, there are currently 200 schemes of the average size operating effectively,[6] this would add up to 1% of Melbourne's 450 GL/year demand.

Recycled water for agriculture at one point achieved a 20% target (i.e., 20% of wastewater was recycled), but, after drought ended in 2010, recycled water demand dropped significantly (Low et al., 2015). Recycled water for non-potable urban uses is increasing slowly, through mandated dual pipe networks in new developments. These will eventually make up a notable proportion of Melbourne's supply (perhaps 5% of the city's future demands), but it will be decades before this point is reached.

On the ambiguity-conflict matrix, many of Melbourne's attempts towards alternative water are "experimental," with high ambiguity and low conflict. Non-potable recycled water supplied to homes through dual-pipe, small-scale stormwater harvesting and residential rainwater harvesting are all examples of things that populations generally support (i.e., low conflict), but they are ambiguous in regard to the amount of water security that will eventually be provided. Due to their smaller scale, they are also more complex in regard to governance (e.g., operation, monitoring, and cost-sharing).

3.4 Navigating a diversity of water source options

3.4.1 Two perspectives on supply resilience

Under the emerging trend towards hybridisation of water supplies, through the introduction of alternative water sources, it is necessary to consider a portfolio of supplies rather than each supply independently. For example, a low-cost low-reliability supply may be complemented by a higher-cost higher-reliability supply.

Modern Portfolio Theory (MPT) has been widely adopted by investment portfolio managers to develop investment strategies involving combinations of assets that maximise return while reducing the overall risk (Markowitz, 1952). Rather than considering the risk of individual investments, risk is managed across the whole portfolio. MPT has been applied to water systems in considering the various supply options and seeking out the most efficient diversification of infrastructure (Shin & Park, 2018). This allows factors such as cost and risks to be accounted for in the determination of a portfolio of water source options.

Water resource planners seek to develop an optimised water security portfolio that is resilient. Here we discuss two (of many) different perspectives on how water security should be optimised. The first is the view that greater ***diversity and decentralisation*** results in greater supply resilience, a view representative of efforts to achieve Integrated Water Management in Melbourne. The second is that supply resilience does require ***diversity, but does not require decentralisation***, a view more representative of the Singapore approach.

Proponents of water supply ***diversity and decentralisation*** highlight that increased uptake of decentralised alternative water source projects reduce reliance on centralised water supply systems. This also reduces the likelihood of all water systems failing simultaneously. For example, rainwater tanks, recycled water, and stormwater harvesting systems will, to varying degrees, have water available sometimes even when centralised systems are undergoing water restrictions (e.g., under extreme drought).

This decentralisation mindset is reflected in the popular phrase "Fit-for-purpose," which suggests that treatment costs can be avoided by providing lower quality water for non-drinking purposes, rather than the same water for all uses. Proponents of these systems also emphasise the social and environmental benefits of decentralised systems, such as reducing pollution to waterways, and the amenity and cooling benefits from capturing water in the landscape.

Proponents of ***diversity without decentralisation***, would agree that a mix of sources is needed. A mix of low-energy, gravity systems (e.g., dams), and high-energy systems (e.g., desalination/recycled water) may allow for water to be supplied even in the context of disasters (e.g., major bushfires or total power outages). There is a clear benefit in having multiple treatment plants, attached to different sources, as a way of introducing redundancy into a system, e.g., to allow for occasional mechanical failures or source contaminations.

These proponents would argue that, if looking purely through a water security lens, decentralised systems are typically outperformed by centralised solutions. Rainwater tanks in cities, and stormwater harvesting systems, are usually more expensive than desalination or indirect potable reuse (WSAA, 2020). They are also reliant on rainfall, meaning that they often don't have water when it's needed most (e.g., the driest 10–20% of the time). Non-potable dual pipe systems often run into problems, such as costs being underestimated, or demands being overestimated.

From this perspective, there is more value in focusing on city-scale interventions, such as a combination of desalination, indirect potable reuse, and only doing non-potable recycled water/stormwater/rainwater in specific locations where it is proposed for other benefits or avoided costs (e.g., to improve waterway health or dispose of wastewater effluent).

Which approach is better for supply resilience and implementability, in each city, will depend on local context. Depending on the city there will be a different (a) level of tolerance for financial and environmental impacts, (b) attitude around which water sources are "acceptable" for drinking, and a (c) level of capacity for centralised (e.g., at treatment plants) and decentralised management (e.g., in individual homes).

It is also important to note that local context can change over time. For example, a supply option that is "high conflict," such as drinking recycled water, can be moved to "low conflict" through a concerted effort to meaningfully engage with and educate the community (e.g., community education around indirect potable reuse in Singapore).

3.4.2 Processes and principles for navigating uncertainty

It is common for cities with water security concerns to develop formal water resource plans on a regular basis (LADWP, 2020; Victorian Government, 2016). When shortfalls emerge, which cannot be overcome by demand-side measures alone, it is necessary to identify and compare new supply sources.

Planners use a few common techniques to assist decisions between different new water supply options, including: Cost Effectiveness Analysis (CEA), Multi-criteria Analysis (MCA), and

Cost Benefit Analysis (CBA). CEA can be used to identify the cheapest source of water, without considering any broader economic, social, or environmental implications, and it can be done through summing capital and operating costs over a project life (e.g., 30 years), and dividing it by the amount of water produced, creating a "levelised cost" (WSAA, 2020). MCA can be used to compliment CEA with a mix of mostly qualitative assessments of broader implications (Dodgson et al., 2009). More complex and holistic CBA involves quantification of various externalities and avoided costs, including economic (e.g., house prices), social (e.g., community health), and environmental (e.g., reduced pollution) (Young & Loomis, 2014). These techniques are generally effective at considering multiple options at a fixed point in time, e.g., "what will be the best option in three years."

Water resource plans should look far into the future (e.g., 50 years), which means there is always significant uncertainty. Best practice water resources planning should start with an assessment of all options independent of timing. As options can be complementary or mutually exclusive, it is important to consider not only the next augmentation, but also a preferred progression of augmentations over an extended period. This can be done for an ultimate time horizon or multiple time slices. Some options may look unattractive in the short term, but attractive in the long term, and so it is important to identify a pathway that does not preclude the option being undertaken in the future.

Emerging practice is the use of an "adaptive pathways" planning approach, which considers the staged implementation of portfolios through time, under a range of plausible future scenarios. This includes flexibility to switch between options or move actions forward or back in time as needed. This approach typically results in adaptive pathway diagrams. To give a simple example, it may be decided that a desalination plant is not currently needed, but that land in the best location needs to be purchased now, to keep the option open for the future.

Community engagement in long-term water planning is also important to allow the end-users of the water to make informed decisions about where they want their water to come from, and what price they are willing to pay. To give another simple example, indirect potable reuse may be marginally cheaper than desalination, and non-potable use of stormwater for parks may be significantly more expensive. The community may choose desalination due to public health concerns with indirect potable reuse, or they may choose stormwater harvesting in the interest of protecting waterways. There is no single "best" supply source, and so in a democratic country it is valid to allow the people to decide what is best, so long as the community is sufficiently informed with information on safety, taste/odour, cost, and other data.

3.5 Conclusion

In most cities a variety of water source/use combinations can be chosen. The concepts outlined in this chapter, "supply resilience" and "implementability," can be used to summarise the key metrics against which decisions are made.

It is generally accepted that the supply resilience of a water supply system can be increased through diversity, reducing the likelihood that all sources will simultaneously lose ability to supply. There are mixed views around whether decentralisation of water supplies is also needed to increase overall water supply resilience, but such schemes often provide other environmental and social benefits. In practice, the benefits of diversity will be present in most cities, while the benefits of decentralisation will depend on the local context.

Implementability of options relates to cost, environmental impacts, community views, and governance of construction, maintenance, and operation. Depending on the city's culture there will be a different level of initial tolerance for financial, environmental, taste/source changes and impacts, and expectation for water reliability. Matland's ambiguity-conflict matrix presents a useful way for categorising different water supply options for a given city and for tracking changes over time. For example, some options will start as either political (high conflict) or experimental (high ambiguity), but they may over time become administrative (low conflict and ambiguity).

There is a plethora of techniques, including levelised costing, MCA, and non-market CBA, that can assist in the comparison of options against each other. The use of adaptive pathways planning can help with the identification of preferred portfolios of options over time (i.e., how technologies should be combined together).

In all cases, significant benefits are available from learning from the experiences of other forerunner cities in advance, to adopt their successful strategies, and to avoid any potential pitfalls. Therefore, the importance of water planners conducting research into other cities before choosing their own water security strategy cannot be overstated.

Notes

1 In this chapter the word "desalination" on its own is always used to refer to desalination of sea water, rather than desalination of brackish groundwater.
2 Total lifecycle cost (including all capital and operating expenses) divided by the amount of water expected to be produced.
3 Figures 3.2 and 3.3 provide an overview of the different water supply options in terms of cost and yield. Figure 3.2 is replicated from, and Figure 3.3 is derived from, WSAA (2020). The definitions and descriptions of each of the supply options particularly relevant to this chapter, are described in the following sections on "Desalination," "Purified recycled water for drinking," "Recycled water for non-drinking," "Stormwater for drinking," and "Stormwater for non-drinking." There are two minor differences between how sources are described in this chapter and how they are described in WSAA (2020). WSAA (2020) does not include "Stormwater for drinking" as a sub-category, and divides "Stormwater for non-drinking" into two categories, "precinct-scale" and "small-scale" (the exact distinction between these is not clearly defined, but Figure 3.3 shows the difference in yield/scale).
4 Some cities have separate wastewater and stormwater pipes, while others have combined sewers. Any stormwater that enters a sewer becomes wastewater. Any reuse of wastewater is referred to here as recycled water.
5 These two terms have the same meaning. "Dual pipe" refers to the fact that there are two water supply pipes entering the home. "Third pipe" refers to the fact that there are two supply pipes, and one sewerage pipe, resulting in three pipe connections to each home.
6 An approximate estimate by the authors.

Bibliography

Angelakis, A., Asano, T., Bahri, A., Jimenez, B., & Tchobanoglous, G. (2018). Water reuse: From ancient to modern times and the future. *Frontiers in Environmental Science*, 6, p. 26.

Australian Bureau of Statistics, 2011. *Household Water and Energy Use*, Victoria, October 2011. [Online] Available at: https://www.abs.gov.au/ausstats/abs@.nsf/Lookup/4602.2Chapter400October%202011#:~:text=In%20Balance%20of%20Victoria%2C%2048,rainwater%20tank%20(figure%203.2) [Accessed 28 Feb 2021].

Boxall, B., 2019. L.A.'s ambitious goal: Recycle all of the city's sewage into drinkable water. *Los Angeles Times*, 22 February.

Campisano, A., Butler, D., Ward, S., Burns, M.J., Friedler, E., DeBusk, K., Fisher-Jeffes, L.N., Ghisi, E., Rahman, A., Furumai, H., & Han, M., 2017. Urban rainwater harvesting systems: Research, implementation and future perspectives. *Water Research*, 115, pp. 195–209.

Chen, Z., Ngo, H.H. & Guo, W., 2013. A critical review on the end uses of recycled water. *Critical Reviews in Environmental Science and Technology*, 43(14), pp. 1446–1516.

Chiew, F., Teng, J., Vaze, J., Post, D., Perraud, J., Kirino, D., & Viney, N. (2009). Estimating climate change impact on runoff across southeast Australia: Method, results, and implications of the modeling method. *Water Resources Research*, 45(10), w10414.

City of Salisbury, n.d. *Salisbury Stormwater Harvesting*. [Online] Available at: https://www.salisbury.sa.gov.au/services/environment-and-sustainability/wetlands-and-water/current-and-future-projects/salisbury-stormwater-harvesting [Accessed 30 Nov 2021].

City of San Diego Public Utilities Water and Wastewater, 2015. *San Diego's Surface Water Augmentation Projects*, San Diego: City of San Diego.

Dee, N., Baker, J., Drobny, N., Duke, K., Whitman, I., & Fahringer, D. (1973). An environmental evaluation system for water resource planning. *Water Resources Research*, 9(3), pp. 523–535.

DELWP, 2016. *Guidelines for Assessing the Impact of Climate Change on Water Supplies in Victoria*, Melbourne: Victorian Government.

Dodgson, J., Spackman, M., Pearman, A. & Phillips, L., 2009. *Multi-criteria Analysis: A Manual*, London: Department for Communities and Local Government.

Dong, C., Schoups, G. & van de Giesen, N., 2013. Scenario development for water resource planning and management: A review. *Technological Forecasting and Social Change*, 80(4), pp. 749–761.

Ferguson, B.C., Brown, R.R. & Frantzeskaki, N., 2013. The enabling institutional context for integrated water management: Lessons from Melbourne. *Water Research*, 47, pp. 7300–7314.

Flörke, M., Schneider, C. & McDonald, R., 2018. Water competition between cities and agriculture driven by climate change and urban growth. *Nature Sustainability*, 1(1), pp. 51–58.

Furlong, C., De Silva, S., Gan, K., Guthrie, L., & Considine, R., 2017. Risk management, financial evaluation and funding for wastewater and stormwater reuse projects. *Journal of Environmental Management*, 191, pp. 83–95.

Furlong, C., Jegatheesan, J., Currell, M., Iyer-Raniga, U., Khan, T., & Ball, A., 2019. Is the global public willing to drink recycled water? A review for researchers and practitioners. *Utilities Policy*, 56, pp. 53–61.

Furlong, C., Uittenbroek, C., Gulsrud, N., Termes-Rife, M., Dodson, J., & Skinner, R., 2018. *Understanding the Role of the Water Sector in Urban Liveability and Greening Interventions: Case Studies on Barcelona, Rotterdam, Amsterdam, Copenhagen and Melbourne,* Melbourne: RMIT University, https://www. waterra. com. au/_r6994/media/system/attrib/file/1643/Furlong% 20et% 20al 202018. (2018): Case study report.

Ghaffour, N., Missimer, T.M. & Amy, G.L., 2013. Technical review and evaluation of the economics of water desalination: Current and future challenges for better water supply sustainability. *Desalination*, 309, pp. 197–207.

Grey, D., Garrick, D., Blackmore, D., Kelman, J., Muller, M., & Sadoff, C., 2013. Water security in one blue planet: Twenty-first century. *Philosophical Transactions of the Royal Society A*, 371, p. 20120406.

Hall, J., Borgomeo, A., Bruce, A., Di Mauro, M., & Mortazavi-Naeini, M., 2019. Resilience of water resource systems: Lessons from England. *Water Security*, 8, p. 100052.

Hill, M. & Hupe, P., 2002. *Implementing Public Policy: Governance in Theory and in Practice*. London: Sage.

Hitchin, P., 2014. Greening the desert [seawater greenhouse]. *Engineering & Technology*, 9(6), pp. 82–85.

Hui, T., Low, M. & Peng, K., 2010. Fishes of the Marina Basin, Singapore, before the erection of the Marina Barrage. *The Raffles Bulletin of Zoology*, 58(1), pp. 137–144.

James, E., Breen, P. & Browne, D., 2015. *Stormwater Reuse to Mitigate Impacts from Increased Runoff Frequency and Volume.* 9th International Water Sensitive Urban Design Conference, Barton.

Jensen, O. & Nair, S., 2019. Integrated urban water management and water security: A comparison of Singapore and Hong Kong. *Water*, 11(4), p. 785.

LADWP, 2020. *Urban Water Management Plan*. [Online] Available at: https://www.ladwp.com/ladwp/faces/ladwp/aboutus/a-water/a-w-sourcesofsupply/ [Accessed 12 September 2020].

Lattemann, S. & Höpner, T., 2008. Environmental impact and impact assessment of seawater desalination. *Desalination*, 220(1–3), pp. 1–15.

Li, C., Fletcher, T., Duncan, H., & Burns, M., 2017. Can stormwater control measures restore altered urban flow regimes at the catchment scale?. *Journal of Hydrology*, 549, pp. 631–653.

Low, K., Grant, S., Hamilton, A., Gan, K., Saphores, J., Arora, M., & Feldman, D., 2015. Fighting drought with innovation: Melbourne's response to the millennium drought in Southeast Australia. *WIREs Water*, 2, pp. 315–328.

Lykkebo Petersen, K., Heck, N., Reguero, B., Potts, D., Hovagimian, A., & Paytan, A., 2019. Biological and physical effects of brine discharge from the Carlsbad desalination plant and implications for future desalination plant constructions. *Water*, 11(2), p. 208.

Maiolo, M. & Pantusa, D., 2016. An optimization procedure for the sustainable management of water resources. *Water Science and Technology: Water Supply*, 16(1), pp. 61–69.

Markowitz, H., 1952. Portfolio selection. *Journal of Finance*, 7(1), pp. 77–91.

Matland, R., 1995. Synthesizing the implementation literature: The ambiguity-conflict model of policy implementation. *Journal of Public Administration Research and Theory*, 5(2), pp. 145–174.

McDonald, R.I., Weber, K., Padowski, J., Flörke, M., Schneider, C., Green, P.A., Gleeson, T., Eckman, S., Lehner, B., Balk, D., & Boucher, T., 2014. Water on an urban planet: Urbanization and the reach of urban water infrastructure. *Global Environmental Change*, 27, pp. 96–105.

Melbourne Water, 2017. *Melbourne Water Systems Strategy*, Melbourne: Melbourne Water.

Melbourne Water, 2018. *Healthy Waterways Strategy*, Melbourne: Melbourne Water.

Moglia, M. et al., 2015. *Survey of Savings and Conditions of Rainwater Tanks*, Melbourne: CSIRO.

Monterey One Water, n.d. *Source Waters*. [Online] Available at: https://www.montereyonewater.org/272/Source-Waters [Accessed 30 Nov 2021].

Ng, P. & Corlett, R., 2011. Biodiversity in Singapore: An overview. In: P. Ng, R. Corlett & H. Tan, eds. *Singapore Biodiversity: An Encyclopedia of the Natural Environment and Sustainable Development*. Editions Didier Millet, pp. 18–26.

Orange City Council, n.d. *Blackmans Swamp Creek Stormwater Harvesting Scheme*. [Online] Available at: https://www.orange.nsw.gov.au/water/stormwater/ [Accessed 30 Nov 2021].

Pinto, F. & Marques, R., 2017. Desalination projects economic feasibility: A standardization of cost determinants. *Renewable and Sustainable Energy Reviews*, 78, pp. 904–915.

Potter, J., 2020. In Australia, COVID-19 has exposed a Litany of privatization disasters. *Jacobin*, 23, p. 8. https://www.jacobinmag.com/2020/08/australia-victoria-coronavirus-privatization.

Potter, M. & RossRakesh, S., 2007. Implementing water sensitive urban design through regulation. *Rainwater and Urban Design 2007*, 1, p. 886–893.

Public Utilities Board, n.d. *Active, Beautiful, Clean Waters Programme*. [Online] Available at: pub.gov.sg/abcwaters/about [Accessed 28 Nov 2021].

Rathnayaka, K., Malano, H., & Arora, M., 2016. Assessment of sustainability of urban water supply and demand management options: A comprehensive approach. *Water*, 8(12), p. 595.

Rice, J. & Westerhoff, P., 2014. Spatial and temporal variation in de facto wastewater reuse in drinking water systems across the USA. *Environmental Science & Technology*, 49(2), pp. 982–989.

Sapkota, M., Arora, M., Malano, H., Moglia, M., Sharma, A., George, G., & Pamminger, F., 2015. An overview of hybrid water supply systems in the context of urban water management: Challenges and opportunities. *Water*, 7(1), pp. 153–174.

Shin, S. & Park, H., 2018. Achieving cost-efficient diversification of water infrastructure system against uncertainty using modern portfolio theory. *Journal of Hydroinformatics*, 20(3), pp. 739–750.

South East Water, 2019. *Aquarevo: A New Way of Living*. [Online] Available at: southeastwater.com.au/CurrentProjects/Projects/Pages/Aquarevo.aspx [Accessed 12 September 2020].

Srinivasan, S.V., Lambin, E., Gorelick, S., Thompson, B., & Rozelle, S., 2012. The nature and causes of the global water crisis: Syndromes from a meta-analysis of coupled human-water studies. *Water Resources Research*, 48(10).

Tal, A., 2011. The desalination debate: Lessons learned thus far. *Environment: Science and Policy for Sustainable Development*, 53(5), pp. 34–48.

Turner, A., Mukheibir, P., Mitchell, C., Chong, J., Retamal, M., Murta, J., Carrard, N., & Delaney, C., 2016. Recycled water–lessons from Australia on dealing with risk and uncertainty. *Water Practice and Technology*, 11(1), pp. 127–138.

UN-Water, 2013. *Water Security & the Global Water Agenda: A UN-Water Analytical Brief*, Hamilton, Ontario: United Nations University.

Victorian Government, 2016. *Water for Victoria*, Melbourne: Victorian Government.

Vörösmarty, C.J., McIntyre, P.B., Gessner, M.O., Dudgeon, D., Prusevich, A., Green, P., Glidden, S., Bunn, S.E., Sullivan, C.A., Liermann, C.R., & Davies, P.M., 2010. Global threats to human water security and river biodiversity. *Nature*, 467(7315), pp. 555–561.

Wannon Water, 2018. *Warrnambool Roof Water Harvesting Initiative*. [Online] Available at: wannonwater.com.au/news-projects/projects/warrnambool-roof-water-harvesting-initiative.aspx) [Accessed 12 September 2020].

Water Corporation, 2021. *Groundwater Replenishment*. [Online] Available at: https://www.watercorporation.com.au/Our-water/Groundwater/Groundwater-replenishment [Accessed 30 Nov 2021].

West, C., Kenway, S., & Yuan, Z., 2015. *Risks to the Long-term Viability of Residential non-potable Water Schemes: A Review*, Melbourne: Cooperative Research Centre for Water Sensitive Cities.

WSAA, 2020. *All Options on the Table: Urban Water Supply Options for Australia*, Melbourne: Water Services Association of Australia.

Xu, C. & Singh, V., 2004. Review on regional water resources assessment models under stationary and changing climate. *Water Resources Management*, 612, pp. 591–612.

Young, R. & Loomis, J., 2014. *Determining the Economic Value of Water: Concepts and Methods*. 2nd ed., New York: RFF Press.

4
URBAN WATER SUPPLY AND LIFE CYCLE ASSESSMENT

Zepon Tarpani and Gallego Schmid

4.1 Introduction

The adequate governance of water in many cities around the world is constantly being overlooked and, as a result, several of them currently experience all sorts of issues derived from excessive impermeabilisation of riverbanks, lack of stormwater handling, and freshwater pollution. These issues should be addressed accordingly for the sustainable growth of urban centres (Arden and Jawitz 2019). Moreover, the availability of freshwater in proximity to several cities has been decreasing as a consequence of climate change, environmental neglect, and rapid urbanisation, experienced mostly in developing countries (Larsen et al. 2016; Rodell et al. 2018). These issues end up impairing water security in many regions, and they will soon require cities to ameliorate their urban water governance in order to guarantee sufficient drinking water for their populations (Maurya et al. 2020). Besides conscientisation for rational and efficient water use, improving urban water governance can be achieved by updating the current infrastructure, implementing the use of new water sources, or administering better treatment and distribution management practices (Silva et al. 2020; Gallego et al. 2008). However, their impacts should be evaluated and fully interpreted with the aim of decreasing associated environmental burdens and improving their sustainability. For this purpose, life cycle assessment (LCA) is an increasingly adopted methodology.

The aim of an LCA is to compile all the inputs and outputs of resources, materials, energy, emissions, and waste associated with a specific product, system, or service during its whole life cycle and, then, estimate, based on the data, a diverse set of potential environmental impacts. In the context of urban water supply, the role of an LCA is to enable their evaluation based not only on performance indicators or benchmarking (Vilanova et al. 2015), but also on the quantification and interpretation of the use of resources and pollution shifts occurring during the entire life cycle of the processes and fluxes associated with it. This is of great importance given the highly interconnected and environmentally fragile world we live in today. From the LCA perspective, the evaluation of environmental impacts of urban water supply involves aspects associated with the life cycle of electricity generation for water abstraction, treatment, pumping, and distribution; energy and heavy machinery for site preparation and pipes installation; concrete for the construction of dams and treatment plants; steel and other metals for equipment, pipes, and pumping stations; chemical products used for water treatment; transportation of repair parts and chemicals products to the treatment plant; and recycling of water treatment equipment, etc.

 DOI: 10.4324/9781003057574-6

The correct execution of an LCA enables a more complete picture of the environmental impacts associated with urban water supply. This is because the consideration of a set of potential life cycle environmental impacts and identification of their "hotspots" ease the decision making about, for instance, what is the best water source to choose or which treatment technique and distribution practice to adopt. An LCA is structured in four phases (ISO 2006). The first is the goal and scope definition, whereby the main purposes, system boundaries, assumptions, and functional unit of the study are reported and justified. This is followed by the inventory analysis, when the processes and flows within the system boundaries are compiled, described, and quantified in relation to the functional unit. The next phase is the life cycle impact assessment, performed with the use of software to model and calculate the potential environmental impact categories according to the chosen impact methodology. The impacts can be estimated and communicated either as midpoint categories (such as climate change potential, ozone depletion potential, eutrophication potential, and particulate matter formation) or as endpoint categories (such as human health, ecosystem quality, and resource depletion), and they can optionally include normalisation, grouping, and weighting. The last phase is interpretation, although this is continuous throughout an LCA since it requires constant verification and reassessment.

In this chapter the reader first finds an overview of the most important aspects regarding the application of an LCA in the evaluation of urban water supplies. Thereafter, the chapter explores some LCA studies about conventional urban water supply systems (treated freshwater delivered to consumers by centralised systems) and hybrid ones (those considering decentralised and alternative water sources). Finally, the chapter concludes with the main findings regarding the use of an LCA for the environmental performance of urban water supplies.

4.2 The application of LCA to urban water supply

The first scientific study applying an LCA approach to evaluate the environmental impacts of urban water supply dates back to the late 1990s. Since then, the application of the LCA has constantly evolved. This is the direct result of more complete databases, developments in life cycle impact methodologies, and software tools for process modelling that became available to LCA practitioners over the years. However, several methodological inconsistencies and gaps can still be identified in the literature about the topic, and, in due course, they should be filled to produce a more satisfactory interpretation of the environmental burdens associated with urban water supplies. As previously mentioned in the introduction, the LCA methodology is structured in four phases according to the International Standards Organization (ISO 2006): goal and scope definition; inventory analysis; life cycle impact assessment; and interpretation. These are discussed next, in the context of the evaluation of urban water supply.

4.3 Goal and scope definition

The goal and scope definition is the first step of an LCA, and straight away it must provide key information about the study (e.g., its purposes, choices, assumptions, completeness, and ability to be representative), the system boundaries, and the functional unit. In the case of a conventional centralised urban water supply, the system boundaries refer to freshwater (surface and groundwater) abstraction, treatment, and pumping to the distribution network, as well as the maintenance required during their life cycles. It is common practice to depict these stages with further details in a process flow diagram (see example in Figure 4.1). Nonetheless, it is not uncommon that some stages or infrastructure parts are disregarded in LCA studies, albeit this can result in significantly underestimating the impacts in some categories (Igos et al. 2014).

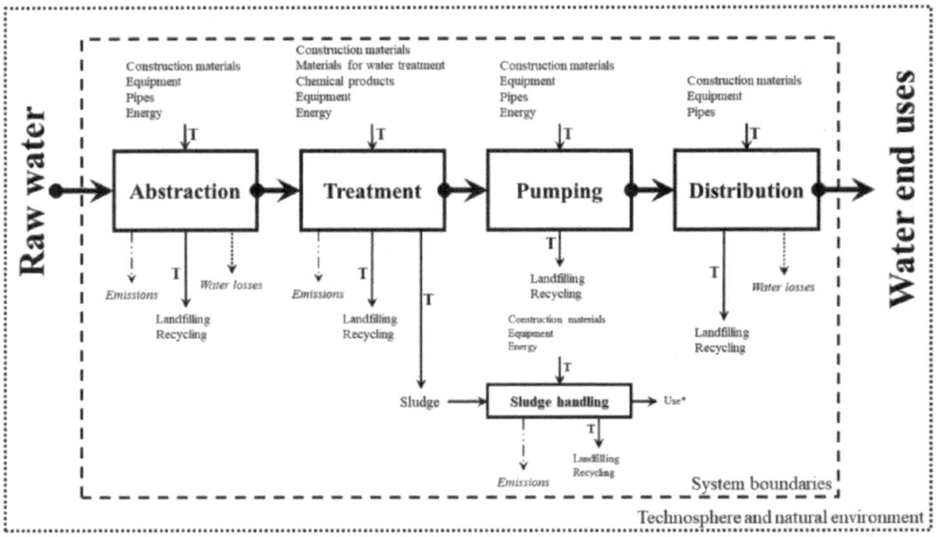

T : transport. *Optional

Figure 4.1 General example of process flow diagram to describe conventional urban water supply in an LCA.

The functional unit is the quantitative basis on which the system is evaluated, and the environmental impacts are estimated. The most common units in the LCA of urban water supplies are the provision of 1 m^3 to the consumer, 1 capita/year, and 1 city/year (Loubet et al. 2014). Although it is reasonable to estimate the impacts per unit of volume delivered to consumers, this can pose difficulties with interpretation in some cases since the initial and final water characteristics are often different, and the per capita consumption can vary widely. These are briefly commented on next.

Even though potable water standards can be considered to have no significant variation, the quality of the raw water can be quite different. Consequently, the level of treatment necessary to achieve a potable standard (i.e., the treatment of good quality groundwater compared to polluted surface water or seawater desalination) also varies (Skuse et al. 2020; Flores-Alsina et al. 2010). Therefore, disclosure of the initial and final characteristics of the water being considered in the study is recommended. Another reason for this is that interpretation problems can arise when performing the LCA of water supplies augmented with non-potable water. For instance, cities enhancing their conventional urban water supply by blending treated wastewater (i.e., indirect potable reuse) do not necessarily need to meet drinking standards for the treated wastewater being reused (Lahnsteiner et al. 2018). Hence, it is important to carefully evaluate and describe these specificities during the goal and scope definition.

In relation to per capita water consumption, literature reports values of 14-538 L/day (IWA 2018). This wide range indicates that it is an important factor to consider when comparing the environmental performance of urban water supplies, especially between cities that have distinct socioeconomic and environmental backgrounds. This is especially true for studies considering alternative techniques in hybrid supply systems, such as wastewater reuse and rainwater harvesting (Sitzenfrei et al. 2017). Therefore, this information should be taken into consideration and reported during this phase, so that the interpretation becomes clearer for those consulting the study.

4.4 Inventory analysis

In this phase the inputs and outputs in each stage within the system boundaries are described and quantified according to the functional unit. A good starting point is to create an inventory table following the processes flow diagram, showing the final values corresponding to the functional unit for the foreground processes (i.e., the relevant energy, resources, materials, emissions, and waste flows gauged by the user – see Figure 4.2). When data are not readily available (e.g., measurement is not possible or not publicly reported), they have to be estimated from literature values or using technical calculations. In all cases, the origins and logic behind the final results should be thoroughly discussed. Therefore, the main task here is to create the most complete and reliable inventory as possible to properly depict the urban water supply under study.

The infrastructure necessary for urban water supply can be highly variable. This can be due to, for example, distance between the raw water source and the final consumer, local topography, amount of water to be distributed, and level of treatment necessary to achieve a potable standard. Another topic requiring closer consideration is water losses in the distribution network (AL-Washali et al. 2020), which can be responsible for a significant share of the environmental impacts (Del Borghi et al. 2013; Pillot et al. 2016) once it is directly related to the functional unit. Additionally, accurate electricity consumption is more than often the main task during this phase. Values from several locations worldwide suggest the energy intensity of these systems, conventional and hybrid ones, are usually in the range 0.20–4.90 kWh/m^3 (Wakeel et al. 2016).

The values included in the inventory are then used for modelling, when the foreground is connected to the corresponding background process using specific software, such as Gabi, openLCA, and SimaPro (see Figure 4.2). The background processes are previously compiled data sets of inputs and outputs of resources, materials, energy, emissions, and waste over the life cycle of several thousand units or system processes. They represent a wide range of industries and services, including electricity generation, building materials, equipment, and chemical products used in the infrastructure, operation, and maintenance of urban water supply. The

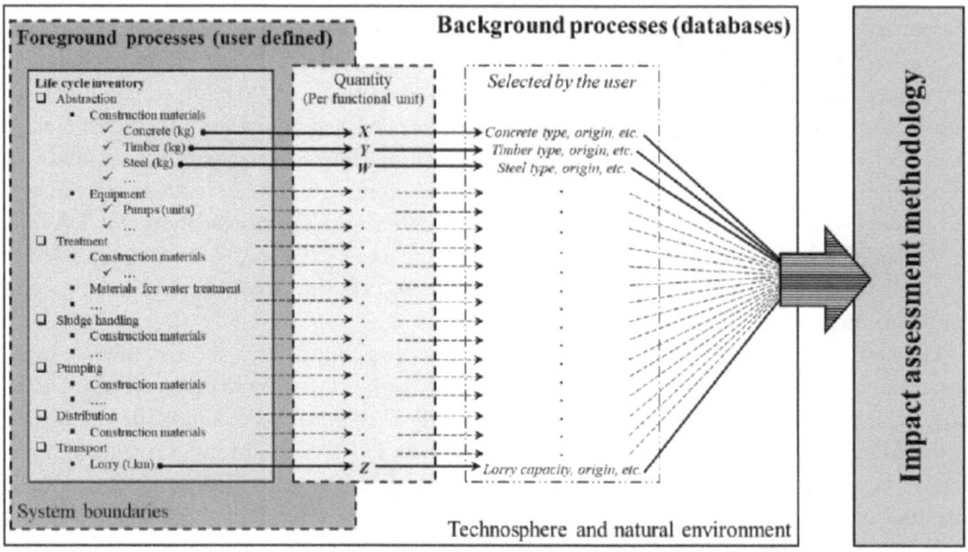

Figure 4.2 Scheme with the use of foreground and background processes for impact assessment of conventional urban water supply in an LCA.

most common database used for the LCA of urban water supplies is Ecoinvent (Wernet et al. 2016). These days they are indispensable for estimating life cycle environmental impacts because they provide a more complete and standardised extension of the foreground processes, contributing to the completeness, ability to be representative, and reliability of LCA studies (Meron et al. 2016). Since there is a large number of processes to be chosen from by the user, the selected ones should correspond to the system under evaluation as much as possible. For instance, the electricity mix from the database should correspond to the location of the urban water supply infrastructure and the period within which the data were measured (Meron et al. 2016).

4.5 Impact assessment and interpretation

The aim of the impact assessment phase is to estimate, based on the information gathered for the inventory, the potential environmental impacts associated with the urban water supply. This is performed by software, such as Gabi, openLCA, and SimaPro. The most used impact assessment methodologies for the evaluation of urban water supplies are Eco-Indicator 99, CML, and ReCiPe (Loubet et al. 2014). An overview of these and other impact assessment methodologies can be found in the handbook by the Joint Research Centre Institute (JCR 2010). The most common midpoint impacts evaluated in LCA studies of urban water supplies are climate change, eutrophication, acidification, and ecotoxicity potentials. In relation to the first, Meron et al. (2016) reported values in the range of 0.16–3.4 kg CO_2 eq./1 m^3 of water supplied, with a mean of 0.84 kg CO_2 eq./1 m^3 and a median of 0.57 kg CO_2 eq./1 m^3. Discussion of endpoint categories, normalized results, and single-score indicators is also common in the scientific literature about the topic (Loubet et al. 2014; Byrne et al. 2017). This multitude of combinations among methodologies and discussion of impact categories often hampers interpretation by readers, which can impair literature comparison. LCA practitioners should therefore be aware of this when carrying out their studies.

The interpretation phase should include a discussion about the results of each impact category and details about their "hotspots" (i.e., the main contributors to that impact), check inconsistencies found during the study, and provide recommendations according to the goal and scope definition. It is important that this phase is discussed as a coherent, non-redundant, and logical follow-up from the main topics the study seeks to clarify. A guide containing more information about LCA results interpretation has been published by the Joint Research Centre (Zampori et al. 2016). Nevertheless, mixed results among impact categories are common in LCAs, and they should be more rigorously evaluated before a certain alternative or practice is recommended or options ranked. For this purpose, multicriteria decision analysis (MCDA) can be applied for improving decision making (Zanghelini et al. 2018). Additionally, uncertainty and sensitivity analysis are important tasks to be taken into consideration for interpretation and communication of LCA results.

Uncertainty refers to random or systematic errors (e.g., modelling choices, measurement inaccuracies, or lack of scientific knowledge) relative to the system under study. Thus, uncertainty analysis refers to procedures that aim to quantify how these uncertainties are transferred to the LCA results. The most applied method for this in LCAs is the Monte Carlo technique. Sensitivity analysis is performed to obtain additional information to aid interpretation, as it quantifies the contribution of model parameters in the LCA result. It can be done locally, by varying a specific input at a time and checking its effect on the output result, or globally, by checking the contribution of each input during the propagation of several of them on the output result variance (Igos et al. 2019).

4.6 The LCA of urban water supplies

In this section a brief review is given of some relevant studies in the scientific literature about the LCA of urban water supplies. Initially, there are studies evaluating conventional urban water supply systems (centralised systems producing potable water only from freshwater sources), followed by others evaluating hybrid urban water supply systems (producing potable and non-potable water from seawater, wastewater, and rainwater). They are summarized in Table 4.1.

4.6.1 LCA of conventional water supply

As far as the authors are aware, Tillman, Svingby, and Lundström (1998) was the first study in the scientific literature to consider an LCA to evaluate the environmental performance of an urban water supply. However, the authors only briefly mentioned water treatment and distribution during the study since the focus was on the wastewater and sludge treatment stages of two Swedish municipalities. As a pioneer in LCAs, the work is mostly exploratory in terms of boundary choices and potential impacts derived from different choices for wastewater and sludge management. Almost a decade later, Lassaux et al. (2007) evaluated the impacts of supplying water to the Walloon Region (Belgium), besides wastewater treatment and its discharge to the environment. The water was abstracted from surface and groundwater sources, and the authors provided information based on local data during inventory analysis. The authors considered the use of the Eco-Indicator 99 methodology and eight of its impact categories as well as including a sensitivity analysis on groundwater production infrastructure, diameter of sewerage pipes, and use of another impact assessment methodology (the CML). The authors concluded that wastewater treatment and its discharge stages generate higher impacts than the urban water supply and, therefore, it should be the focus of eventual interventions.

The authors in Lemos et al. (2013) evaluated the urban water supply of Aveiro (Portugal). The raw water was sourced from a river located 25 km from the city, and groundwater. Water losses during the distribution stage were 38%. For the impact assessment, the authors considered eight ReCiPe 2008 midpoint impacts and its single endpoint indicator. Sensitivity analyses were carried out on several processes, including water loss, electricity consumption, and use of chemical products. The authors found that electricity consumption was the main hotspot and discussed scenarios for the improvement of its environmental performance, such as decreasing water loss and the use of an electricity mix less dependent on fossil fuels. Additionally, the authors concluded that the results could not be extrapolated to other locations since it was found that the electricity consumption for water abstraction and treatment was much greater than those commonly found in literature, possibly due to its low system efficiency and local topography. Slagstad and Brattebø (2014) performed an LCA of the urban water supply in Trondheim (Norway). The water supplying the city is abstracted from a lake, treated, and distributed to consumers. The study provides a detailed description of the system, and the authors estimated that 32% of the water leaving the treatment plant was lost by the time it reached consumers. The impacts were assessed with eight normalized midpoint impact categories from the ReCiPe 2008 methodology, and a sensitivity analysis on the electricity mix was carried out. The main conclusion was that the climate change impact of the water supply was insignificant compared to the annual CO_2-eq emissions from a person in Europe.

Barjoveanu et al. (2014) performed an LCA of the urban water supply and the wastewater treatment system of Iasi city (Romania). The water was abstracted from surface water located several kilometres away from the city and from groundwater. The electricity consumption of the system in peak hours reached 0.35 kWh/m^3, and water loss during distribution was 41%. The

Table 4.1 Studies performing the LCA of conventional and hybrid urban water supplies

Reference	*Country*	*Goal of the study*	*Functional unit*	*Main conclusions*
Conventional systems				
(Tillman et al. 1998)	Sweden	To evaluate the environmental consequences of changes in the centralised wastewater treatments of Hamburgsund and Bergsjön.	Treatment of the wastewater from one person equivalent for one year.	It is beneficial to enlarge the system boundaries to evaluate wastewater and sludge treatments.
(Lassaux et al. 2007)	Belgium	To determine the environmental impact of using one cubic metre of water in the Walloon Region.	One cubic metre of water at the consumer tap.	The stages that contribute most to environmental impacts are water discharge, wastewater treatment plant operation, and the sewer system.
(Lemos et al. 2013)	Portugal	To evaluate environmental impacts of the urban water system of Aveiro.	One cubic metre of potable water at the point of consumption.	The environmental impacts of the urban water supply are dependent on the local geography and cannot be extrapolated. Electricity consumption is the main hotspot.
(Slagstad and Brattebø 2014)	Norway	To examine the environmental impacts of operating the water and wastewater system in Trondheim.	The one-year provision of water, and collection, transportation, and treatment of wastewater (including stormwater).	The climate change potential of water supply is minor compared to the annual contribution of a person in Europe to this impact. Electricity consumption is a hotspot, contributing strongly to freshwater eutrophication.
(Barjoveanu et al. 2014)	Romania	To analyse the entire water services system in Iasi City and demonstrate the usefulness of the LCA to support water resources management.	One cubic metre of tap water delivered in the city.	The water supply generates more impacts than the wastewater treatment system because of its high energy demand and water loss in the distribution system. LCA is useful to analyse the environmental performance of water supplies.

(*Continued*)

Table 4.1 (Continued)

Reference	*Country*	*Goal of the study*	*Functional unit*	*Main conclusions*
(Jeong et al. 2015)	United States	To conduct the LCA of the centralised water system of the city of Atlanta.	One cubic metre of water distributed to the point of use.	The construction of the infrastructure is a significant contributor to the environmental impacts.
(García-Sánchez and Güereca 2019)	Mexico	To assess the environmental and social impacts of the water system in Mexico City and provide new perspectives for its sustainability.	One cubic metre of water for user consumption.	Electricity consumption for water treatment is the main contributor to climate change impact.
(Xue et al. 2019)	United States	To enable utility managers to make better decisions and highlight the importance of integrating water and wastewater management in the Greater Cincinnati region.	One cubic metre of treated and distributed meeting or exceeding National Primary Drinking Water Regulation.	Operation and maintenance are responsible for most of the energy consumption and climate change potential from water abstraction to its discharge. Infrastructure shows little contribution to impacts.
Hybrid systems				
(Lundie et al. 2004)	Australia	To examine Sydney Water's operation and its environmental impacts in the year 2021.	The provision of water supply and sewerage services in the year 2021.	Desalination increases the climate change potential of the system even when adding a little amount of water. An LCA is useful to assess financial, social, and local environmental issues of Sydney Water's operation in the future.
(Pasqualino et al. 2011)	Spain	To assess the operation of wastewater treatment in Catalunya and establish the environmental impacts of wastewater reuse.	One cubic metre of wastewater entering the wastewater treatment plant.	Reclaiming wastewater can be beneficial for non-potable uses. Potable wastewater reuse with tertiary treatment does not result in significant environmental improvement.

(*Continued*)

Table 4.1 (Continued)

Reference	*Country*	*Goal of the study*	*Functional unit*	*Main conclusions*
(Amores et al. 2013)	Spain	To use LCA to carry out the environmental impacts of the current system and of two scenarios for urban water cycle in Tarragona.	One cubic metre of potable water supplied to the consumers.	Energy consumption is the main hotspot, mostly for water distribution. Wastewater reuse scenario has similar impacts to the current water supply. Desalination should be used only in extreme drought scenarios.
(Lane et al. 2015)	Australia	To study the current water and wastewater services of the Golden Coast region using an LCA, with focus on the uncertainties.	To provide water supply and wastewater management services during a one-year period to the urban population in the Gold Coast region.	The wastewater treatment system contributes more to environmental impacts than the water supply. To further diversify water sources substantially increase the environmental impacts of the current water supply.
(Hsien et al. 2019)	Singapore	To inform stakeholders of the environmental impacts of NEWater and tap water delivered to consumers in Singapore.	One cubic metre of water (NEWater or tap water) delivered to the consumer.	The NEWater supply has a higher impact in three of the eight impact categories. The current water supply has a higher impact in the other five. Desalination promotes a disproportionately higher impact.
(Tarpani et al. 2021)	Brazil	To perform the LCA of three alternative techniques to substitute the use of drinking water from the distribution network in the north of Florianopolis.	To increase the availability of drinking water in the local distribution network by 1,000 cubic metres.	Indirect potable wastewater reuse alternatives had most of the lowest impacts in nine of the 15 categories assessed, and desalination most of the highest. An electricity mix other than Brazilian markedly increases the impacts from seawater desalination. Storage tanks are responsible for most of the life cycle impacts in rainwater harvesting.

authors considered ten normalised midpoint categories from the CML 2000 methodology, and a single score indicator from Ecological Scarcity 2006. Their conclusion was that for all categories, except those related to water pollution in the two impact methodologies, the water supply system had the highest impacts when compared to the wastewater treatment system. This was attributed mostly to the water losses during distribution. A year later, Jeong et al. (2015) performed an LCA of the urban water supply and wastewater treatment systems of the city of Atlanta (USA). The raw water was sourced from a river adjacent to the city and treated by coagulation/flocculation and filtration before distribution (with water loss at this stage being 13%). The authors provided a detailed inventory analysis, based on public information from local authorities. For the impact assessment, ten midpoint impacts from the TRACI v2.1 methodology were used. The main finding was that the infrastructure construction was a significant contributor to environmental impacts, although the life cycle data sets used for steel and cast-iron production were taken from Europe since they were not available for the United States at the time.

More recently, García-Sánchez and Güereca (2019) used an LCA to evaluate the environmental impacts of the urban water supply in Mexico City (Mexico), which is mostly sourced from groundwater, with the rest originating from surface water. The authors estimate that about 40% of the water was lost during distribution and they did not consider any infrastructure for the system. The impact assessment was performed with seven midpoint impacts from the ReCiPe v1.12 methodology, and uncertainty analysis was carried out based on the Monte Carlo approach. The main conclusion was that the environmental impacts of pumping water from source to its treatment was the main hotspot due to the high electricity consumption and the national electricity mix being heavily based on fossil fuel. Lastly, Xue et al. (2019) used two different methodologies to estimate the life cycle environmental impacts of the urban water supply and wastewater treatment of the greater Cincinnati region (USA). The water was sourced from the Ohio River, and 19% of it was lost during distribution. The data for the system were derived from the local water utility, and the authors considered the TRACI v2.0 methodology for the impact assessment as well as the ReCiPe 2008 for metal and fossil fuel depletion potentials, and performed sensitivity analyses on several input parameters, including the electricity mix. The results indicated that energy consumption for water distribution was mainly responsible for the impacts, and that infrastructure had little contribution except for metal depletion potential.

4.6.2 The LCA of hybrid water supplies

The first LCA in scientific literature considering an alternative and decentralised option for urban water supply was Lundie et al. (2004). The study created scenarios to represent the water supply situation for Sydney (Australia) in the year 2021, including demand management initiatives, higher efficiency pumps, and desalination. In relation to the latter, the construction and operation of a reverse osmosis seawater plant was chosen to supply 6% of the city's water consumption. The environmental impacts were assessed using seven midpoint indicators, including climate change, eutrophication, and human toxicity potentials. The results showed that if desalination was implemented in the city, the environmental burdens of supplying water would increase substantially. A few years later, Pasqualino et al. (2011) evaluated several options for wastewater reuse aiming to decrease the environmental burdens of supplying water to the region of Catalunya (Spain), albeit most of the study was focused on the wastewater treatment stage. The authors considered nine midpoint categories from the CML 2000 methodology to analyse the environmental impacts of four scenarios: treated wastewater discharged into the environment, secondary treated wastewater used for brine dilution, tertiary treated wastewater used for potable uses, and tertiary treatment wastewater for potable use instead of desalinated seawater.

The authors concluded that reusing wastewater for non-potable uses was environmentally beneficial, whilst for potable uses it was not, and desalinated water should only be used for potable uses due to its high energy consumption.

Still in the Catalunya region, Amores et al. (2013) performed an LCA of the urban water supply of Tarragona. The authors used actual operational data from the city of 145,000 inhabitants provided by internal reports, interviews, and previous studies in the region. In addition, the authors evaluated two scenarios: wastewater reuse for agricultural purposes after tertiary treatment, and its combination with reverse osmosis seawater desalination providing 25% of the water consumption. Seven midpoint categories from the CML 2001 methodology were used for the environmental impact assessment. The main findings were that electricity for water abstraction and distribution were major hotspots, and wastewater reuse can increase water availability in drought situations without increasing environmental impacts; desalination should only be used as a last resource. Lane et al. (2015) assessed the environmental impacts of increasing the adoption of residential rainwater harvesting, indirect potable wastewater reuse, and seawater desalination in the urban water supply in the Gold Coast region (Australia). The scenario built to evaluate this was rainwater harvesting systems corresponding to 15%, desalination to 29%, and wastewater reuse to 10% of the water being supplied to the region, and the remaining amount from the reservoir. The inventory was created from compiled information of locally or regionally measured data with available empirical and literature values, and the impact assessment was made with 14 midpoint categories from the ReCiPe 2008 methodology. The results from the LCA study suggested that diversifying the water supply in the region would substantially increase the environmental impacts of the system.

Recently, Hsien et al. (2019) evaluated the water supply of Singapore, which encompasses several subsystems supplying water to households and industries. It includes different wastewater reuse options for non-potable reuse in industries, and for augmenting the main supply system composed of surface water (from river and reservoirs) and seawater desalination. The water loss in the distribution system was 5%, and the desalination technology considered was reverse osmosis. The authors evaluated the water supply using eight midpoint impacts from the ReCiPe 2008 methodology and considered sensitivity analysis for different combinations of water sources. The results showed that although desalinated water corresponded to about a fifth of the water consumption, it was responsible for one-third of the climate change potential of the system. The wastewater reuse scheme for non-potable uses in industries had the highest impacts in three of the eight impact categories and the main water supply in five of them. Finally, Tarpani et al. (2021) performed an LCA study comparing three alternatives for increasing the urban water supply in Florianopolis (southern Brazil). The options were seawater desalination by reverse osmosis, an indirect potable wastewater reuse scheme by upflow anaerobic sludge blanket digestion (UASB), oxidation ditch and ozonation followed by soil aquifer treatment (SAT), and five different types of rainwater harvesting systems. The authors considered 15 midpoint impacts from the ReCiPe 2016 methodology for result interpretation and included a parametric analysis on key aspects of each alternative. Additionally, a sensitivity analysis was undertaken on the climate change potential from the emissions of CH_4 and N_2O of the wastewater treatment stage in the wastewater reuse alternative, and of the electricity mix. The results indicated that desalination and wastewater reuse had electricity consumption as the main contributor to impacts, whilst for rainwater harvesting it was storage tanks. The climate change potential of wastewater reuse showed highly variable results, from 0.7 to 1.3 kg CO_2-eq/m^3. Overall, however, this option showed the lowest impacts in nine categories.

More information about alternative techniques that can be used for hybrid water supply can be found in the literature dedicated to the topic. For instance, for desalination and LCA, see Tarpani et al. (2019), Aziz and Hanafiah (2020), and Lee and Jepson (2021); for wastewater reuse and LCA, see Tarpani and Azapagic (2018), Gallego Schmid and Tarpani (2019), Corominas et al. (2020), and Risch et al. (2021); for rainwater harvesting and LCA, see Zanni et al. (2019) and Ghimire et al. (2017).

4.7 Conclusions

Many LCA studies about conventional urban water supplies have been published in the last two decades. Moreover, alternative techniques that can be adopted for augmenting urban drinking water availability such as desalination, advanced treatments for wastewater reuse, and rainwater harvesting are also being increasingly evaluated using LCA. This is of great importance given the negative consequences of climate change, freshwater deterioration, and the rapid urbanisation already experienced in some cities, especially those in the developing south. The findings from these LCAs assist researchers and stakeholders to understand and minimise the environmental impacts associated with urban water supply.

LCA studies on conventional urban water supplies suggest they usually have a climate change potential below 0.60 kg CO_2-eq. per 1 m^3 delivered to customers. Variations occur mostly due to differences in local geography, infrastructure requirements, and water losses during distribution, as well as the electricity mix of the region. Estimates for other impact categories are more complex to define due to a diversity of frameworks adopted by LCA practitioners, including system boundaries, assumptions, databases, and impact assessment methodologies. A more complete description of the life cycle inventory and more sound discussion of environmental impacts, including uncertainty analysis on the most influential parameters (such as electricity consumption and water losses) are welcome. These are important aspects that need to be discussed and eventually fulfilled by LCA practitioners in the future to enable a homogenous and robust evaluation of the life cycle environmental performance of urban water supply.

A trend has been observed over recent years in the application of LCAs from conventional to hybrid urban water supplies. This suggests preoccupation not only about urban water availability, but also in relation to the environmental consequences of the adoption of alternative water sources. However, the use of LCAs to evaluate alternative and decentralised techniques for water supply must be carefully conducted, taking into consideration the quality of the final effluent and other details that can impair the correct interpretation by LCA practitioners and stakeholders. More specifically, efforts towards promoting an easier comparison in terms of their purposes, assumptions, life cycle inventory description, and impact assessment methodologies are necessary when evaluating these hybrid systems. These would greatly improve the results interpretation and decision-making processes by stakeholders and promote a more resilient and sustainable urban water supply.

References

AL-Washali, T., Sharma, S., Lupoja, R., AL-Nozaily, F., Haidera, M. and Kennedy, M. (2020). Assessment of water losses in distribution networks: Methods, applications, uncertainties, and implications in intermittent supply. *Resources, Conservation and Recycling*, 152(September 2019), p.104515.

Amores, M.J., Meneses, M., Pasqualino, J., Antón, A. and Castells, F. (2013). Environmental assessment of urban water cycle on Mediterranean conditions by LCA approach. *Journal of Cleaner Production*, 43, pp.84–92.

Arden, S. and Jawitz, J.W. (2019). The evolution of urban water systems: Societal needs, institutional complexities, and resource costs. *Urban Water Journal*, 16(2), pp.92–102.

Aziz, N.I.H.A. and Hanafiah, M.M. (2020). Application of life cycle assessment for desalination: Progress, challenges and future directions. *Environmental Pollution*, 268, 115948.

Barjoveanu, G., Comandaru, I.M., Rodriguez-Garcia, G., Hospido, A. and Teodosiu, C. (2014). Evaluation of water services system through LCA. A case study for Iasi City, Romania. *International Journal of Life Cycle Assessment*, 19(2), pp.449–462.

Del Borghi, A., Strazza, C., Gallo, M., Messineo, S. and Naso, M. (2013). Water supply and sustainability: Life cycle assessment of water collection, treatment and distribution service. *International Journal of Life Cycle Assessment*, 18(5), pp.1158–1168.

Byrne, D.M., Lohman, H.A.C., Cook, S.M., Peters, G.M. and Guest, J.S. (2017). Life cycle assessment (LCA) of urban water infrastructure: Emerging approaches to balance objectives and inform comprehensive decision-making. *Environmental Science: Water Research and Technology*, 3(6), pp.1002–1014.

Corominas, L., Byrne, D., Guest, J.S., Hospido, A., Roux, P., Shaw, A. and Short, M.D. (2020). The application of life cycle assessment (LCA) to wastewater treatment: A best practice guide and critical review. *Water Research*, 184, p.116058.

Flores-Alsina, X., Gallego, A., Feijoo, G. and Rodriguez-Roda, I. (2010). Multiple-objective evaluation of wastewater treatment plant control alternatives. *Journal of Environmental Management*, 91(5), pp.1193–1201.

Gallego-Schmid, A. and Tarpani, R.R.Z. (2019). Life cycle assessment of wastewater treatment in developing countries: A review. *Water Research*, 153, pp.63–79.

Gallego, A., Hospido, A., Moreira, M.T. and Feijoo, G. (2008). Environmental performance of wastewater treatment plants for small populations. *Resources, Conservation and Recycling*, 52(6), pp.931–940.

García-Sánchez, M. and Güereca, L.P. (2019). Environmental and social life cycle assessment of urban water systems: The case of Mexico City. *Science of the Total Environment*, 693, p.133464.

Ghimire, S.R., Johnston, J.M., Ingwersen, W.W. and Sojka, S. (2017). Life cycle assessment of a commercial rainwater harvesting system compared with a municipal water supply system. *Journal of Cleaner Production*, 151, pp.74–86.

Hsien, C., Choong Low, J.S., Chan Fuchen, S. and Han, T.W. (2019). Life cycle assessment of water supply in Singapore: A water-scarce urban city with multiple water sources. *Resources, Conservation and Recycling*, 151(March), p.104476.

Igos, E., Benetto, E., Meyer, R., Baustert, P. and Othoniel, B. (2019). How to treat uncertainties in life cycle assessment studies?. *International Journal of Life Cycle Assessment*, 24(4), pp.794–807.

Igos, E., Dalle, A., Tiruta-Barna, L., Benetto, E., Baudin, I. and Mery, Y. (2014). Life cycle assessment of water treatment: What is the contribution of infrastructure and operation at unit process level?. *Journal of Cleaner Production*, 65, pp.424–431.

ISO. (2006). *Life Cycle Assessment: Principles and Framework*. ISO 14040:2006, p.20. Geneva: International Organization for Standardization.

IWA. (2018). *International Statistics for Water Services 2018*. IWA, England. Available from: www.waterstatistics.org.

JCR. (2010). *Analysing of Existing Environmental Impact Assessment Methodologies for Use in Life Cycle Assessment*. JCR, Ispra, Italy.

Jeong, H., Minne, E. and Crittenden, J.C. (2015). Life cycle assessment of the City of Atlanta, Georgia's centralized water system. *International Journal of Life Cycle Assessment*, 20(6), pp.880–891.

Lahnsteiner, J., van Rensburg, P. and Esterhuizen, J. (2018). Direct potable reuse: A feasible water management option. *Journal of Water Reuse and Desalination*, 8(1), pp.14–28.

Lane, J.L., de Haas, D.W. and Lant, P.A. (2015). The diverse environmental burden of city-scale urban water systems. *Water Research*, 81, pp.398–415.

Larsen, T.A., Hoffmann, S., Luthi, C., Truffer, B. and Maurer, M. (2016). Emerging solutions to the water challenges of an urbanizing world. *Science*, 2016, pp.928–933.

Lassaux, S., Renzoni, R. and Germain, A. (2007). LCA case studies life cycle assessment of water from the pumping station to the wastewater treatment plant. *International Journal*, 12(2), pp.118–126.

Lee, K. and Jepson, W. (2021). Environmental impact of desalination: A systematic review of life cycle assessment. *Desalination*, 509(March), p.115066.

Lemos, D., Dias, A.C., Gabarrell, X. and Arroja, L. (2013). Environmental assessment of an urban water system. *Journal of Cleaner Production*, 54, pp.157–165.

Loubet, P., Roux, P., Loiseau, E. and Bellon-Maurel, V. (2014). Life cycle assessments of urban water systems: A comparative analysis of selected peer-reviewed literature. *Water Research*, 67(0), pp.187–202.

Lundie, S., Peters, G.M. and Beavis, P.C. (2004). Life cycle assessment for sustainable metropolitan water systems planning. *Environmental Science & Technology*, 38(13), pp.3465–3473.

Maurya, S.P., Singh, P.K., Ohri, A. and Singh, R. (2020). Identification of indicators for sustainable urban water development planning. *Ecological Indicators*, 108(August 2019), p.105691.

Meron, N., Blass, V., Garb, Y., Kahane, Y. and Thoma, G. (2016). Why going beyond standard LCI databases is important: Lessons from a meta-analysis of potable water supply system LCAs. *International Journal of Life Cycle Assessment*, 21(8), pp.1134–1147.

Pasqualino, J.C., Meneses, M. and Castells, F. (2011). Life cycle assessment of urban wastewater reclamation and reuse alternatives. *Journal of Industrial Ecology*, 15(1), pp.49–63.

Pillot, J., Catel, L., Renaud, E., Augeard, B. and Roux, P. (2016). Up to what point is loss reduction environmentally friendly?: The LCA of loss reduction scenarios in drinking water networks. *Water Research*, 104, pp.231–241.

Risch, E., Boutin, C. and Roux, P. (2021). Applying life cycle assessment to assess the environmental performance of decentralised versus centralised wastewater systems. *Water Research*, 196, p.116991.

Rodell, M., Famiglietti, J.S., Wiese, D.N., Reager, J.T., Beaudoing, H.K., Landerer, F.W. and Lo, M.-H. (2018). Emerging trends in global freshwater availability. *Nature*, 557, pp.651–659.

Silva, J. da, Fernandes, V., Limont, M., Dziedzic, M., Andreoli, C.V. and Rauen, W.B. (2020). Water sustainability assessment from the perspective of sustainable development capitals: Conceptual model and index based on literature review. *Journal of Environmental Management*, 254(September 2019), p.109750.

Sitzenfrei, R., Zischg, J., Sitzmann, M. and Bach, P.M. (2017). Impact of hybridwater supply on the centralised water system. *Water*, 9(11), 855.

Skuse, C., Gallego-Schmid, A., Azapagic, A. and Gorgojo, P. (2020). Can emerging membrane-based desalination technologies replace reverse osmosis?. *Desalination*, 2020, p.114844.

Slagstad, H. and Brattebø, H. (2014). Life cycle assessment of the water and wastewater system in Trondheim, Norway: A case study. *Urban Water Journal*, 11(4), pp.323–334.

Tarpani, R.R.Z. and Azapagic, A. (2018). Life cycle environmental impacts of advanced wastewater treatment techniques for removal of pharmaceuticals and personal care products (PPCPs). *Journal of Environmental Management*, 215, pp.258–272.

Tarpani, R.R.Z., Lapolli, F.R., Lobo Recio, M.Á. and Gallego-Schmid, A. (2021). Comparative life cycle assessment of three alternative techniques for increasing potable water supply in cities in the Global South. *Journal of Cleaner Production*, 290, p.125871.

Tarpani, R.R.Z., Miralles-Cuevas, S., Gallego-Schmid, A., Cabrera-Reina, A. and Cornejo-Ponce, L. (2019). Environmental assessment of sustainable energy options for multi-effect distillation of brackish water in isolated communities. *Journal of Cleaner Production*, 213, pp.1371–1379.

Tillman, A.-M., Svingby, M. and Lundström, H. (1998). Life cycle assessment of municipal waste water systems. *International Journal of Life Cycle Assessment*, 3(3), pp.145–157.

Vilanova, M.R.N., Magalhaes Filho, P. and Balestieri, J.A.P. (2015). Performance measurement and indicators for water supply management: Review and international cases. *Renewable and Sustainable Energy Reviews*, 43, pp.1–12.

Wakeel, M., Chen, B., Hayat, T., Alsaedi, A. and Ahmad, B. (2016). Energy consumption for water use cycles in different countries: A review. *Applied Energy*, 178(19), pp.868–885.

Wernet, G., Bauer, C., Steubing, B., Reinhard, J., Moreno-Ruiz, E. and Weidema, B. (2016). The ecoinvent database version 3 (part I): Overview and methodology. *International Journal of Life Cycle Assessment*, 21(9), pp.1218–1230.

Xue, X., Cashman, S., Gaglione, A., Mosley, J., Weiss, L., Ma, X.C., Cashdollar, J. and Garland, J. (2019). Holistic analysis of urban water systems in the Greater Cincinnati region: (1) life cycle assessment and cost implications. *Water Research X*, 2, p.100015.

Zampori, L., Saouter, E., Schau, E., Cristobal, J., Castellani, V. and Sala, S. (2016). *Guide for Interpreting Life Cycle Assessment Result*, EUR 28266 EN, Luxembourg: Joint Research Centre.

Zanghelini, G.M., Cherubini, E. and Soares, S.R. (2018). How multi-criteria decision analysis (MCDA) is aiding life cycle assessment (LCA) in results interpretation. *Journal of Cleaner Production*, 172, pp.609–622.

Zanni, S., Cipolla, S.S., Fusco, E. di, Lenci, A., Altobelli, M., Currado, A., Maglionico, M. and Bonoli, A. (2019). Modeling for sustainability: Life cycle assessment application to evaluate environmental performance of water recycling solutions at the dwelling level. *Sustainable Production and Consumption*, 17, pp.47–61.

5

MODELLING URBAN WATER INFRASTRUCTURE RENEWAL

Yves Le Gat

5.1 Introduction

The service provided to users by a drinking water utility physically relies upon a water supply network (WSN). This infrastructure is mainly composed of water production plants, reservoirs, and pipes, i.e., water mains and service connections as well as their appurtenances (various types of valves, fire hydrants, metering devices). According to Vitse et al. (2012), pipes represented around 92% of the asset replacement values for French WSNs in 2009, whereas production plants and reservoirs were, respectively, 6.6% and 1.4% worth. In this chapter, attention is focused on sole pipe assets owing to the utmost importance of their economic value.

Due to gradual changes in land use, demography, modes of water consumption, and sanitary regulations as well as infrastructure ageing (deterioration and obsolescence), a WSN needs each year to undergo the renewal of some of its segments; a noticeable proportion of annual renewals is, nevertheless, carried out also for operational reasons pertaining to the practical organisation of the worksites or to take into account opportunities to coordinate with third-party infrastructures, especially relating to road rehabilitation. Knowledge is needed concerning WSN infrastructure asset management (IAM), which consists in understanding how, and in which proportions, these various causes of asset renewal combine with each other, and how the renewal policy impacts the performance level of the service provided to users. These questions are examined throughout this chapter, considering technical phenomena from the standpoint of a disciplined representation that can be mathematically formalised.

Operating a complex technical object such as a WSN, and managing its assets, requires an adapted representation that facilitates predicting and simulating, at various timescales, its behaviour in the matter of hydraulics, water quality and temperature, and asset condition (leakage, breakage, internal wall roughness, clogging). A WSN is then commonly represented as a graph, i.e., a set of links, that stands for network segments, connected with each other by nodes, which stand most often for valves and other pipe appurtenances. Network segments are generally elementary pipes joined with each other, homogeneous with respect to their diameter, material, and date of commissioning; their length is highly variable, ranging from a few metres to several hundred metres. The WSN of a typical medium-sized European city comprises several thousands of segments, which can then be considered as individuals forming a statistical population. Probability theory and statistical methods provide a natural theoretical framework, within which engineering sciences and social and human sciences can be articulated for modelling WSN

 DOI: 10.4324/9781003057574-7

segments renewals and for designing methods and technical tools for supporting decisions for short-term renewal work programming, mid-term budget planning, and long-term IAM strategy.

The scientific and technical literature devoted to IAM utilises the central concept of "lifetime" as well as derivative concepts such as "design life," "physical life," "economic life," "service life," or "useful life" that relate to various consequences of asset ageing, but most often without clearly defining them. As a first approach, some definitions inspired by AWWA (2018) can be usefully mentioned:

- According to pipe manufacturer specifications, the design life is the period of time for which an asset is expected to function without rehabilitation.
- An asset reaches the end of its physical life when its structural deterioration involves unacceptable risks of failure.
- An asset reaches the end of its economic life when it ceases to be the lowest cost alternative to satisfy a specified level of performance or service.
- The obsolescence of an asset marks the end of its service life, mainly due to changes in its operational context (e.g., capacity insufficient with respect to the growth of water demand) or in-service standards (e.g., water quality standards).
- The useful life is the minimum of the design, physical, economic, and service lives.

The first section of this chapter will supplement these definitions by addressing the practical assessment of the lifetime of an asset or a group of assets.

Whether lifetime is explicitly considered as an intrinsic property of pipe assets or a result of operational decisions, it provides a comprehensive key for the various modelling approaches proposed in the literature since the end of the 1970s, an overview of which constitutes the subject of the second section. These approaches aim to help WSN management decisions, relating to pipe renewals, with respect to two complementary objectives: (i) assessing the budget needs in the medium term (typically around ten years ahead); and (ii) programming renewal works in the short term (typically between one and three years ahead). The relationship between these objectives and, namely, their consistency provides a second comprehensive key for the proposed modelling approaches.

The third section will revisit the question of pipe renewal decisions in offering a long-term simulation approach that takes into account the various operational determinants of pipe renewals in a probabilistic manner and ensures coherence with budget planning and alignment with strategic service performance objectives.

The last and concluding section will consider questions and future research needs regarding pipe renewal modelling.

5.2 The *lifetime* concept

Lifetime, defined as the time elapsed between the commissioning and decommissioning dates of an asset, is the basic variable considered for renewal modelling. Used for technical objects, a semantic field peculiar to living beings appears prima facie as somewhat curious; nevertheless, it is justified by the theoretical roots of reliability engineering, which developed in the 1950s (cf. Bhamare et al. 2007) by borrowing from biostatistics some of its concepts and mathematics, especially those designed for survival analysis (cf. Andersen and Keiding 2005).

In the literature devoted to pipe renewal modelling since the mid-1990s, lifetime is assessed either as:

- a determined value, most often provided by expert judgement;
- a determined value computed by economic optimisation; or
- a random value characterised by its probability distribution.

This assessment is most often performed at the scale of asset groups, considered technologically homogeneous and, then, likely to follow similar deterioration trajectories.

5.2.1 Expert judgement

Pipe manufacturers generally provide technical recommendations with regard to field implementation of their products, with a range of *design lifetime* values, according to more or less severe conditions of exploitation and the environment of the assets. Engineers working for water utilities are also sometimes able to provide ranges of *useful life* duration values based on their technical experience; these expert-based lifetime values are often considered as better adapted to the local conditions of water utilities. Representative examples can be found in technical reports, such as AWWA (2012), or in articles, such as Amaral et al. (2016).

5.2.2 Economic optimisation

Within the classical paradigm of cost-benefit analysis, the seminal paper of Shamir and Howard (1979) presents the basic principles for estimating the optimal economic lifetime of a given pipeline. To that end, the following assumptions are made:

- the pipeline will undergo $N(t) = N(0)\exp(At)$ breaks in the future year t, with $t = 0$ the present year, and A a breakage coefficient;
- the cost for repairing a break is C_B;
- the cost for replacing the pipe is C_R;
- the discount rate i, accounting for both interest and inflation rates, allows transforming future costs into their present values;

where $N(0)$ is supposedly observed and parameters A, C_B and C_R depend on pipe characteristics (material, diameter, length, *etc.*). The total cost, denoted C_T, for maintaining the pipe between years 0 and t and replacing it at year t is then:

$$C_T(t) = \frac{C_R}{(1+i)^t} + \sum_{u=0}^{u=t} \frac{N(0)\exp(Au)C_B}{(1+i)^u}$$

and an approximate calculation gives the optimal replacement year:

$$t^\star = \frac{1}{A}\ln\left(\frac{\ln(1+i)C_R}{N(0)C_B}\right)$$

Despite its apparent simplicity, the epistemological importance of Shamir and Howard (1979) has to be stressed, as it is the basic rationale of life cycle cost analysis (LCCA) approaches that are presented in Section 5.4.3.

5.2.3 Lifetime distribution

Both preceeding sections relate to what the pipe lifetime *should* be, but they do not account for the dispersion of actual lifetimes observed in water utility databases, for which a probabilistic point of view must be adopted. This leads modelling the lifetime distribution using either parametric or non-parametric estimates.

5.2.3.1 Parametric Herz distribution

The paper of Herz (1996) underlines the evidence of the lifetime variability within a given pipe group; it proposes then to characterise the lifetime with a probability distribution instead of a single value. The lifetime is then considered as a random variable T, with possible values $t\in[0,\infty]$. An original analytic form, relating to the since then called *Herz distribution*, for the survival function $S(t)=\Pr\{T>t\}$ is proposed, which depends on three parameters $\theta=(\eta,\gamma,\tau)$:

$$\begin{cases} S_\theta(t) = 1 \quad \forall t \in [0,\tau[\\ S_\theta(t) = \dfrac{\eta+1}{\eta+\exp(\gamma(t-\tau))} \quad \forall t \in [\tau,+\infty[\end{cases} \tag{1}$$

While parameters (η,γ) define the shape and aging rate of the survival, parameter τ (expressed in years) stands for the duration of the beginning of the lifetime (generally around ten years) during which no decommissioning is supposed to occur; after the initial plateau, the survival curve (i.e., graph of function $S_\theta(t)$) has a decreasing S-shape. The initial plateau assumption is an approximation of how pipe renewals actually happen; however, some, albeit very few, renewals may occur at the very beginning of the lifetime for land management reasons (e.g., in-depth rehabilitation of the road that involves relocating the pipes buried underneath), and then independently of the pipe condition. Figure 5.1 illustrates the survival curve of the Herz distribution.

5.2.3.2 Parametric Weibull distribution

The alternative analytic form of the Weibull distribution (initially designed in Weibull 1951) is proposed by Le Gat et al. (2013). For the use of Weibull distribution in the field of reliability and biostatistics, readers can refer to Kalbfleisch and Prentice (2002). The Weibull survival function depends on two parameters $\theta = (\lambda,\delta)$, which define, respectively, its scale and shape:

$$S_\theta(t) = \exp\left(-(t/\lambda)^\delta\right) \quad \forall t \in [0,+\infty[\tag{2}$$

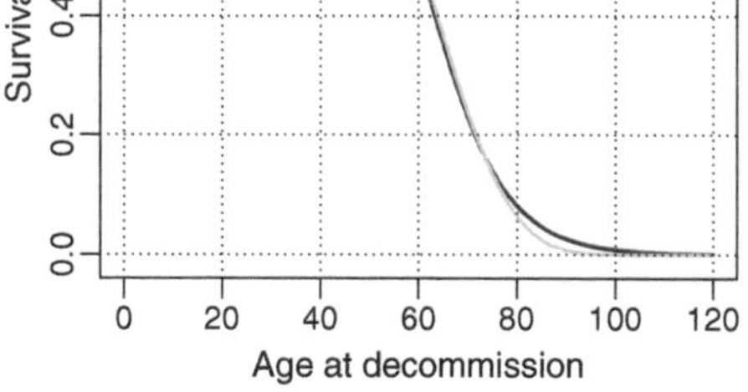

Figure 5.1 Examples of Herz (with parameters τ = 10 years, η = 400, γ = 0.12 years^{-1}) and Weibull (with parameters λ = 65.07 years, δ = 4.86) survival curves.

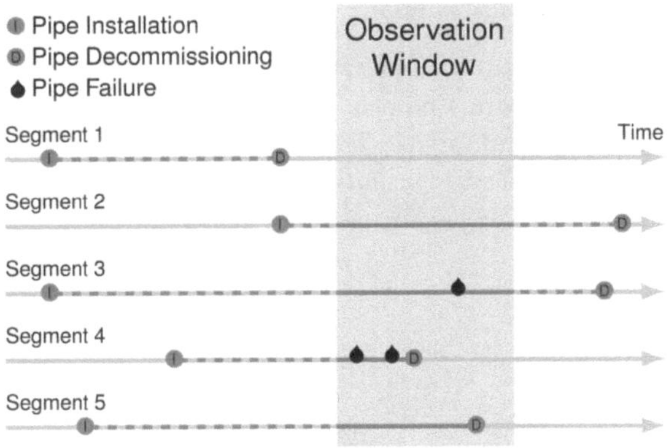

Figure 5.2 Actual water utility data collection and possible observation bias due to left truncation and right censoring.

For $\delta > 1$, $S_\theta(t)$ is also markedly S-shaped, but has stricto sensu no plateau at its beginning since, in contrast to the Herz model, very early renewals are statistically considered possible. Figure 5.1 illustrates the survival curve of the Weibull distribution; notice that the two Weibull parameters can be chosen so that the survival curve is rather close to the Herz one.

5.2.3.3 Maximum likelihood estimation

The question of estimating Herz and Weibull distribution parameters with actual lifetime values, and assessing the goodness of fit of obtained parametric distributions, is investigated by Le Gat et al. (2013). Estimating distribution parameter $\boldsymbol{\theta}$ requires building the so-called *Likelihood function* $L(\theta)$ (*cf.* Andersen and Keiding 2005 and its comprehensive bibliography), calculated with a sample of lifetime values, observed within the same observation window, as illustrated by Figure 5.2. This observation can be formalised as a set of n triplets $\{(a_i b_i c_i) i \in \{1,\ldots,n\}\}$ where:

- a_i stands for the age of segment i at the beginning of the observation window;
- b_i stands either for the age of segment i at the end of the observation window if this segment is still in service at that moment (cases of segments 2 and 3 in Figure 5.2) or for the age of the segment at its decommissioning, if decommissioning occurs within the observation window (cases of segments 4 and 5 in Figure 5.2);
- c_i is a *right-censoring* binary variable that takes the value 1 if segment i is still in service at the end of the observation window (right-censored lifetime observation), 0 otherwise.

Considering any parametric survival function, such as defined by (1) or (2), the likelihood function is defined as:

$$L(\boldsymbol{\theta}) = \prod_{i=1}^{n} \left(-\frac{dS_\theta}{dt}(b_i) \right)^{1-c_i} S_\theta(b_i)^{c_i}$$

As available observation windows are often rather short, and when the use of the pipe material considered was stopped some years before the observation window (the case, e.g., of grey cast iron pipes, the use of which stopped between 1965 and 1970, with observation data after 2000), the observed minimum age $a = \min_i(a_i)$ may not be null; this means that nothing can be directly inferred from the observations about the behaviour of $S_\theta(t)$ for ages $t\in[0,a]$. This leads to considering the *conditional* survival function $S(t \mid a) = \Pr\{T>t \mid T\geq a\}$ instead of $S(t)$; as mentioned by Le Gat et al. (2013), considering a parametric survival model like (2) allows at least estimating $S_\theta(a)$.

In Figure. 5.2, segment 1 is not observed due to its decommissioning before the observation window, whereas segment 3, installed in the same year, is observed. This is an example of left truncation: some observations are missing due to the design of the observation plan itself. This issue is addressed in subsection 5.2.3.5.

5.2.3.4 Non-parametric distribution estimate

Using the same notations as in subsection 5.2.3.3, the non-parametric estimate of the lifetime survival function is a right-continuous left-limited non-increasing step-function that undergoes a jump down at each distinctly observed decommissioning age $s{:}b_i = s, c_i = 0$. Using the celebrated result of Kaplan and Meier (1958), the so-called *Kaplan-Meier* non-parametric estimate is defined as:

$$S_{\mathrm{KM}}(t \mid a) = \prod_{s\in[a,t]} \left(1 - \frac{d_s}{Y_s}\right)$$

with:

$$d_s = \#\{i : b_i = s, c_i = 0\}$$

$$Y_s = \#\{i : s \in [a_i, b_i]\}$$

The variance of $S_{\mathrm{KM}}(t \mid a)$ can be estimated by the Greenwood's formula:

$$\widehat{\mathrm{Var}}S_{\mathrm{KM}}(t \mid a) = S_{\mathrm{KM}}(t \mid a)^2 \sum_{s\in[a,t]} \frac{d_s}{Y_s(Y_s - d_s)}$$

Provided the sample size is large enough for each observed decommissioning time, a 95% confidence interval is given by:

$$S_{\mathrm{KM}}(t \mid a)^{\exp\left(\frac{\pm 1.96\widehat{\mathrm{Var}}S_{\mathrm{KM}}(t \mid a)^{1/2}}{S_{\mathrm{KM}}(t|a)\ln S_{\mathrm{KM}}(t|a)}\right)}$$

The Kaplan-Meier survival curve estimate, with its 95% confidence band, is illustrated in Figure 5.3, relating to a sample of 82.3 km of grey cast iron pipes commissioned between 1952 and 1971, and which decommissioning process is observed between 2007 and 2018.

The Kaplan-Meier survival curve, accompanied by its 95% confidence band, can be graphically compared to any parametric survival curve. A reasonably good fit allows placing a degree of trust in the relevance of the parametric decommissioning model.

5.2.3.5 The selective survival issue

As mentioned in the introductory section, a pipe renewal decision depends partly on the risk of repeated breaks, but also on other factors, either linked to the condition of the pipe (degradation

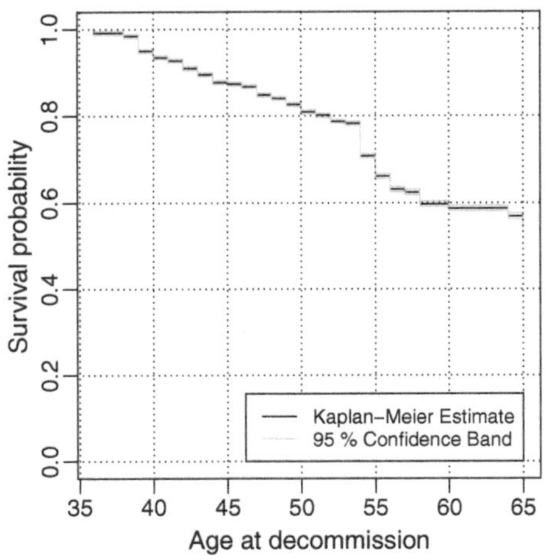

Figure 5.3 Example of Kaplan-Meier survival curve estimate, with 95% confidence band.

of water quality, leakage) or independent of the pipe condition (pipe relocation dictated by a land management operation, pipe resizing to face demographic development, coordination with roadworks, organisation of the renewal site of adjacent pipes, etc.). This raises two questions when attempting to assess pipe survival:

- Is it possible to estimate the survival probability as a combination of a breakage-dependent survival and a survival accounting for other renewal factors?
- How to take into account the left truncation of data, illustrated in segments 1 and 3 in Figure 5.2, given it cannot be excluded that some pipes (due to their renewal before the observation window) are unobserved partly because they repeatedly broke?

The second question refers to the *selective survival* phenomenon, according to which, among pipes belonging to the same cohort (i.e., of the same type and commissioning epoch), those surviving until the observation window can be considered somewhat more robust than those unobserved. This implies that the left truncation of data does not purely randomly take place; rather, it is *informative*, i.e., linked to the breakage and renewal processes of interest. It will be seen in what follows that answering the second question facilitates answering the first.

Pipe breakage and renewal processes are to be considered interdependent and, thus, to be modelled in a coupled manner. This is the reason why the textbook of Le Gat (2016) revisits the failure model, called *Linear Extension of the Yule Process* (LEYP) (formerly published in Le Gat [2014]), in the following way. The LEYP model represents the failure process (pipe breakage) by a counting process $N(t)$, which is a mapping of R_+ into N illustrated in Figure 5.4. The LEYP model postulates that the intensity function of $N(t)$ depends on pipe age t, characteristics $\boldsymbol{Z}$ (length, material, diameter, number of service connections, service pressure, etc.), and number of past breaks $N(t-)$ (that happened in $[0,t]$). To account for selective survival, the likelihood function that permits estimating the LEYP parameters explicitly considers that observable pipes are those that survived until the observation window (see segments 2, 3, and 4 in Figure 5.2); this involves introducing another counting process, denoted $R(t)$, relating to pipe renewal. The renewal process intensity is also postulated to depend on pipe age, characteristics, and past breaks.

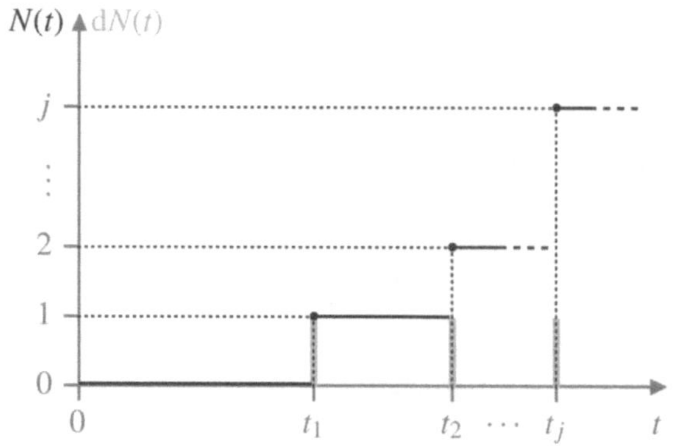

Figure 5.4 Counting Process $N(t)$ - differential $dN(t) = 1$ at each break time t_j, $dN(t) = 0$ elsewhere.

This gives the *Linear Extension of the Yule Process with selective survival* (LEYP2s) mathematically formalised as:

$$\forall t \in R_+,$$
$$\begin{cases} N(0) = 0 \\ \mathrm{E}\left(\mathrm{d}N(t) \mid N(t-);\boldsymbol{Z}\right) = \left(1 + \alpha N(t-)\right)\lambda(t;\boldsymbol{Z})\mathrm{d}t \\ R(0) = 0 \\ \mathrm{E}\left(\mathrm{d}R(t) \mid N(t-);\boldsymbol{Z}\right) = \left(\psi(t;\boldsymbol{Z}) + \phi N(t-)\right)\mathrm{d}t \\ \end{cases}$$
$$\text{with}: \alpha > 0, \quad \lambda(t) \geqslant 0, \quad \psi(t) \geqslant 0, \quad \phi > 0$$

In the LEYP2s setup, scalar parameter α accounts for the tendency of breaks to involve a rather low proportion of the WSN segments in an aggregated manner, while scalar parameter ϕ reflects the propensity of the water utility manager to prevent repeated service disruptions undergone by the same users.

The pipe survival $S_\theta(t)$ is shown in Le Gat (2016) to be the product of $S_C(t;\theta)$ relating to renewal factors independent of the breakage process, and $S_S(t;\theta)$ relating to renewal for repeated breaks reason:

$$S_C(t;\boldsymbol{\theta}) = \exp\left(-\int_0^t \psi(u)\mathrm{d}u\right)$$

$$S_S(t;\boldsymbol{\theta}) = \left(\mu(t) - v(t)\right)^{-1/\alpha}$$

$$\text{with}: \mu(t) = \exp\left(\alpha\int_0^t \lambda(u)\mathrm{d}u\right), v(t) = \int_0^t \mathrm{e}^{\phi(u-t)}\mathrm{d}\mu(u)$$

It is nevertheless to be noticed that:

- this theoretical development does not permit distinguishing within the constrained survival the relative proportions relating to mandatory renewal and to opportunity renewal;
- opportunity renewal can depend, at least partly, on the failure process, when the water utility manager can choose from among road rehabilitation operations those that can be turned to good account for implementing the coordinated renewal of pipes in bad condition.

5.3 A literature survey of pipe renewal modelling approaches

Before giving an overview of the main pipe renewal modelling approaches, the governance context in which decisions relating to pipe renewal are made must be specified. Following Burn et al. (2010) and Alegre et al. (2014) three main decision levels are to be distinguished:

- in the long term, i.e., 20 years or much more, strategic objectives are defined, mainly concerning the quality of the service provided to users;
- to meet these objectives, tactical decisions concerning the investment budget to be spent annually, over a five to ten years period, must be made;
- in the short term, i.e., for the next year (or next two or three years), a set of operational rules has to be fixed concerning the allocation of the annual investment budget to priority renewal sites.

These long-, medium-, and short-term decisions involve governance debates among water utility stakeholders, as illustrated in Figure 5.5. Much research has been carried out since the 1980s

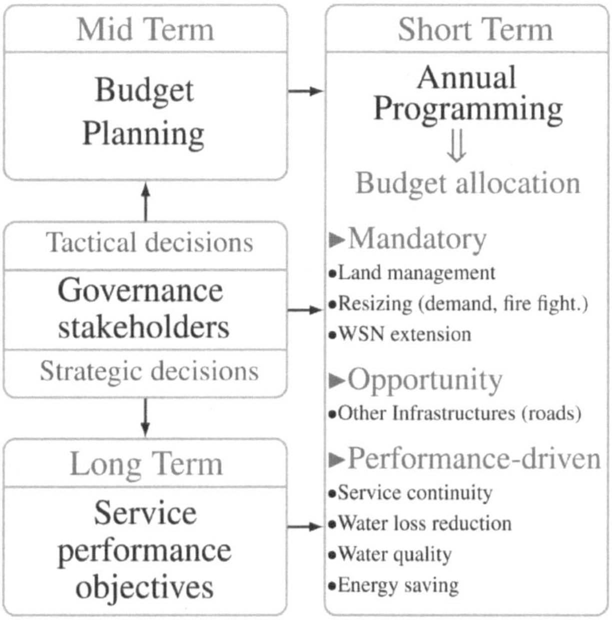

Figure 5.5 Strategic, tactical, and operational decision levels, and temporalities.

to develop modelling tools to aid decision-makers concerning mid-term investment budget planning and short-term priority works programming.

Four broad families of approaches can be distinguished according to their objectives and how they handle the lifetime concept:

- determined lifetime used to shift ahead the historical commissioning pattern;
- lifetime distribution used to shift ahead and spread the historical commissioning pattern;
- life cycle cost analysis (coupling economic optimisation and deterioration modelling);
- multicriteria prioritising.

In the following sections, some approaches consider the lifetime as an intrinsic property of pipes, while others assume lifetime to ensue from WSN management rules. This constitutes a pivotal epistemological choice that is not always explicit.

5.3.1 Simple shift

This method, initially proposed by Australian consultant Haydn Reynolds in the 1980s (according to AWWA 2001), is implemented in seven steps:

- gathering WSN segments into groups, generally with respect to their material, class of diameter, more or less severe conditions of environment and exploitation, etc.;
- collecting, for each asset group, expert judgements as to the maximal service lifetime during which such assets can be expected to provide a required level of service quality (in matter of continuity, water loss, service pressure, water quality);
- reconstructing for each asset group the commissioning history (pipe length commissioned by year);
- assessing for each asset group the present installation cost by unit length;
- shifting each group commissioning history towards the future by the expected service lifetime;
- multiplying for each group and each future year the estimated pipe length and unit installation cost; and
- cumulating over the groups for each future year the estimated renewal cost.

The simple shift method is illustrated in Figure 5.6. When represented graphically, the future annual renewal needs estimated for the whole WSN figure follow a snake-shaped curve, popularised under the somewhat humorous designation of *Nessie Curve*. As testified by two American Water Works Association (AWWA) reports, AWWA (2001) and AWWA (2012), this method enjoyed some success in the 2000s and 2010s with WSN managers in North America. It is also to be noticed that the Nessie Curve method reproduces mechanically, albeit a bit crudely, the historical installation pattern (the Nessie method and assumptions are also subjected to criticism in, e.g., Baird and Wagner 2020); the conclusions of AWWA (2001) and AWWA (2012) were expressed in a rather alarmist manner and then were instrumentalised by pipe manufacturing and installing companies to advocate a massive increase in their market segment (see, e.g., McWaneDuctile 2021; PlasticsNews 2021). The same happened in France with respect to the technical report Cador (2002) in which future investments in French WSN renewal were projected in a Nessie-like manner.

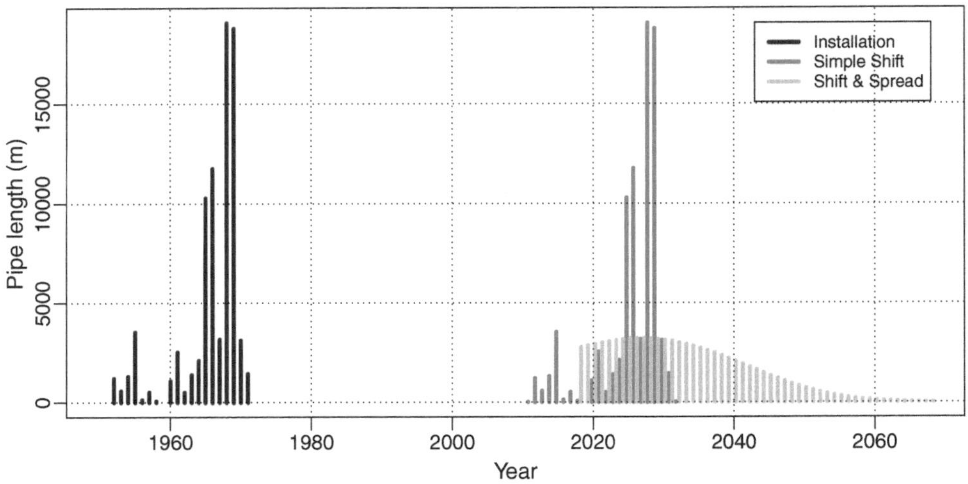

Figure 5.6 Estimating renewal needs of 82.3 km of grey cast iron pipes commissioned between 1952 and 1971, using both methods of simple shift (by 60 years) and shift and spread (by Herz distribution with parameters $\tau = 10$ years, $\eta = 400$, $\gamma = 0.12$ years^{-1}, which involve a 60 years median lifetime).

An approach, epistemologically close to Nessie, is presented by Alegre et al. (2014), consisting in assessing the global economic value of an infrastructure (water or sewer network, and considering all components, not only pipes), in calculating an *Infrastructure Value Index* (IVI) in the following way:

$$IVI(t) = \frac{\sum_{i=1}^{n} C_{Ri}(t)\left(U_i - a_i(t)\right) / U_i}{\sum_{i=1}^{n} C_{Ri}(t)}$$

with n number of infrastructure components, i component index, t a given year, $C_{Ri}(t)$ replacement cost of component i at t, u_i expected useful life duration of component i, and $a_i(t)$ age of component i at t. $IVI(t)$ is then a cost-weighted metric.

An infrastructure is considered as *mature* at year t when the value of its $IVI(t)$ is around 0.5.

Current research presented in Amaral et al. (2016) modulates the lifetimes u_i, initially assessed by utility experts, by taking into account the quality of the service supplied within the functional unit (district metered area in the case of WSN) to which the component i belongs; the service quality is then assessed by a sum of weighted metrics relating to continuity of water supply and water losses. The IVI approach is presented as aiming, first, at creating awareness among utility managers and then encouraging them to improve the asset database so as to later enable the use of more sophisticated IAM tools. IVI calculation is implemented within the open-source IAM system AWARE-P, presented in Alegre et al. (2013).

In these approaches, pipe lifetime is considered as an intrinsic property of technical assets, that ensues from their industrial design, their operational and environmental conditions of exploitation, and the level of service quality supposed to be expected by users. The use of the lifetime concept in IAM is considered misleading by Renaud et al. (2014), who advocate clearly distinguishing the annual renewal work programming and the mid-term smoothing of renewal budget planning. Moreover, this lifetime vision has led, among people involved in WSN management, to the near paradigmatic idea according to which an annual renewal rate r amounts to a $1/r$ years lifetime. This line of reasoning is erroneous:

- either because systematically renewing pipes that reach a given fixed lifetime is practically unfeasible and would financially lead to annual budget needs that may noticeably vary from year to year, whereas water tariffs have to smoothly evolve;
- or since interpreting the above mentioned renewal rate r as a statistical parameter is equivalent to consider pipe lifetime as exponentially distributed with mean $1/r$ years, which is far from being supported by the observation of actual lifetime data (see subsection 5.3.1).

5.3.2 Shift and spread

As presented in Herz (1996), when assessing in the present year the projection of replacement needs of a given cohort of assets over future years, it is far more realistic to consider a distribution of the useful lifetime instead of a single central value, as in subsection 5.3.1. The term "cohort," used in Herz (1996) in analogy with its sense in demography, means a set of assets of the same commissioning year, material, and diameter class.

The calculation of the fraction of the pipe length of a given cohort that will need to be replaced in a given future year is based on the conditional survival:

$$S(t \mid t_0) = \Pr\{T > t \mid T \geqslant t_0\}$$

$$= S(t) \,/\, S(t_0)$$

i.e. the probability that an asset survives beyond the age t (with $t \geq t_0$) given it was still in service at age t_0.

The replacement of pipes of group j commissioned in year c and still in service in present year d_0, the length of which is $L_{Cj}(c;d_0)$, will be spread over future years $d \geq d_0$ by fractions $S_j(d-c \mid d_0-c) - S_j(d+1-c \mid d_0-c)$. Given asset group j was commissioned over the years c_{Bj} to c_{Ej}, its length $L_{Rj}(d)$ needing replacement in year d is then:

$$L_{Rj}(d) = \sum_{c=c_{Bj}}^{c_{Ej}} L_{Cj}(c;d_0)\big(S_j(d-c \mid d_0-c) - S_j(d+1-c \mid d_0-c)\big)$$

The shift and spread method is illustrated in Figure 5.6; renewal needs are distributed over the years 2018 (year of inventory) to 2068, and they appear to be smoothed, despite the installation pattern roughness.

This cohort survival analysis method is currently implemented in the KANEW software, presented by Herz and Lipkow (2003), which is the long-term planning tool delivered by the 5th European Framework Programme CARE-W, presented by Saegrov (2005). This methodology was also latter used in Sweden by Malm et al. (2013), and it was subjected to further developments in the software PiReM, presented by Fuchs-Hanusch et al. (2008). In these systems, the implemented survival function is that of Herz distribution; the parameter values are often fixed according to the judgement of utility experts, sometimes with additional help from breakage analysis, when such failure history is available and linked to WSN segment identifiers. The cohort survival analysis has also been implemented by Pericault et al. (2017) in studying the impact of coordination of rehabilitation works among water, sewer, and road infrastructures on the lifetime distributions of water and sewer pipes.

5.3.3 Life cycle cost analysis

Detailed recommendations for using LCCA to help manage municipal infrastructures (not limited to piping systems) are presented by Rahman and Vanier (2004), making a reference to current Northern American standards relating to transportation and building infrastructures.

More specifically in the domain of WSN management, early applications of LCCA to pipe renewal, such as Shamir and Howard (1979), aim at calculating for each pipeline of the WSN a renewal year, optimal in an economic sense. More broadly, Kleiner et al. (2001) compare management trajectories that may consist in alternative rehabilitation operations (i.e., replace or reline) implemented at various ages. Shamir and Howard (1979) consider breakage as the sole ageing manifestation of pipe ageing, whereas Kleiner et al. (2001) account also for the loss of hydraulic capacity, modelled as an age-dependent increase in the roughness of the internal pipe wall, according to the work of Walski et al. (1986).

This research reveals that the practical use of such LCCA-based management aids refers neither to the assessment of mid-term renewal needs nor to the prioritisation of pipes to be included in annual renewal programmes. This distinction between tactical mid-term and operational short-term decisions constitutes an important forward step due to the European research project CARE-W (within 5th Framework Programme) presented by Saegrov (2005).

Later works based on LCCA, as the Australian PARMS system presented in Burn et al. (2003), clearly devote LCCA to planing renewal of budget needs in the midterm, in the so-called PARMS-PLANNING module, whereas a specific module, called PARMS-PRIORITY, presented in Moglia et al. (2006), addresses the question of annual renewal work programming.

In PARMS-PLANNING, optimal pipe lifetime is not considered as an a priori technical intrisic property of assets, but rather as resulting from management rules; such rules consist, e.g., in deciding to replace a pipe when its predicted annual number of breaks meets a given threshold (1, 2, or 3 breaks/year), which is to be chosen by LCCA. In this paradigm, it is clearly stated that lifetime can be extended according to the level of service quality that users can accept.

Some researchers use LCCA to address the effect of operational measures on pipe lifetime extension or on pipe replacement cost-effectiveness. Installing cathodic protection on pipelines buried in corrosive soils can extend their lifetime, according to Rajani and Kleiner (2004). Three operational measures are particularly studied by Moglia et al. (2006):

- Service pressure reduction is efficient for lowering the pipe failure rate, which enables deferring the pipe replacement.
- Installing shut-off valves on pipelines that serve numerous customers reduces the number of customers impacted by a breakage, thus reducing failure costs and possibly allowing deferring pipe replacement.
- Because of high setup costs, organising pipe renewal at the level of clusters of adjacent pipes instead of single pipes is economically more efficient.

The work of Pardo and Valdes-Abellan (2019) is a rare exception, where LCCA is not breakage driven but considers, instead, an ageing effect represented by an increase in pipe leakage; such an ageing model is nevertheless not yet sufficiently developed to be applied other than at the aggregated level of a district metered area completely homogeneous with respect to pipe age, which strongly limits possible applications.

An innovative idea is proposed by Alvisi and Franchini (2009), who consider ageing as represented by a breakage model (with costs associated to breaks) and a leakage model coupled with a hydraulic model; this construct is used to optimise in the midterm the annual budget allocation between pipe replacements and leakage detection operations.

It is to be noticed that all research works devoted to support renewal budget planning rely either:

- on the consideration of a technical or useful lifetime, the mean value (see subsection 5.3.1) or the distribution (see subsection 5.3.2) of which is assessed by water utility expert judgement; or
- on pipe aging-driven economic calculations.

This observation is important and is further discussed in Section 5.4, with respect to the coherence with multicriteria decision making for prioritising annual renewal works.

5.3.4 Short-term multicriteria decision aid

Once an annual renewal budget is decided, using, e.g., one of the methods presented in subsections 5.3.1, 5.3.2, or 5.3.3, its allocation between prioritary network segments has to be translated into an annual programme of works. An abundant literature has been devoted, mainly since the 1990s, to the non-trivial question of hierarchising pipelines according to some criteria. Criteria relating to pipe breakage were historically the first to be considered (cf. Shamir and Howard 1979), and they still feature prominently in the various hierarchisation methods in use among water utility managers, as evidenced by the references cited further in this section; service and traffic disruptions caused by pipe breaks and their repairs are, indeed, major and evident manifestations ofWSN deterioration that impact both water utility and users. PARMS-PRIORITY module mentioned in previous subsection 5.3.3 is an example where pipe renewal priorities in the short term are based on risk management principles in a pipe breakage-driven manner; the term *risk* means here the probability of a failure event by a valuation of breakage impacts, especially by taking into account externalities, such as negative effects of pipe breaks on users (service interruption, basement flooding, traffic and business disruption, etc.). Other examples of pipe breakage-driven short-term renewal priorities assessment are the I-WARP system presented by Kleiner and Rajani (2009), and the short-term rehabilitation module of PiReM presented by Fuchs-Hanusch et al. (2008), who emphasise consideration of social costs.

Other criteria have also been considered since the beginning of the 2000s relating to the waste of water resources due to leakage (cf. Alvisi and Franchini 2009), waste of energy due to increased pipe roughness (cf. Kleiner et al. 2001; Francisque et al. 2014), degradation of water quality due to internal corrosion of metallic pipes (cf. Kleiner et al. 2001), coordination with other infrastructure, mainly roadworks or sewers (cf. Nafi and Kleiner 2010; Pericault et al. 2017), and critical size of renewal site (cluster-replacement)-seeking economies of scale (cf. Moglia et al. 2006; Nafi and Kleiner 2010).

To rank the WSN segments according to their renewal priority, several multicriteria decision-making (MCDM) methods have been proposed.

The Weighted Sum Method (WSM) consists in choosing a set of criteria (it is supposed here that the priority level of the renewal increases with the value of the criterion), assigning a weight to each criterion, multiplying the criterion value by its weight, summing the criteria, and ranking pipes by their weighted sums. WSM is considered the simplest method; examples of implementation in WSN management are given by Roberts and Broadbent (2011), Tuhovcak et al. (2014), and Marzouk et al. (2015). The crucial step of assigning weight values is carried out according to the judgements of utility experts. WSM has nevertheless an advantage of being computationally light as a single criterion method.

Proposed in 1977 by Thomas L. Saaty, the Analytic Hierarchy Process (AHP), presented, e.g., in Saaty (1987), has been rather successful with WSN managers, according to Tscheikner-Gratl et al. (2017). The main difference with WSM lies in weighting the chosen criteria by performing systematic pairwise comparisons formalised by a square matrix of the same dimension as the number of criteria considered, diagonalising this matrix, retrieving the first eigenvector (corresponding to the highest eigenvalue), and normalising the vector by dividing the components by their sum. Here again, the pairwise comparison of the criteria is carried out according to the judgements of utility experts. The considered alternatives (i.e., WSN segments candidates for renewal) have then to be ranked for each criterion; concerning criteria the value of which cannot be expressed on a quantitative scale, the same systematic pairwise comparisons, followed by matrix computations, are performed as for weighting the criteria. In this way, the alternatives are ranked for each criterion, and, using the criteria weights, they can then be globally ranked according to their weighted sums. An example in WSN management can be found in Francisque et al. (2014). An advantage of the AHP method is put forward by Saaty and Vargas (2012), which relates to its robustness towards the possible lack of consistence of the judgements of several experts facing the same pairwise comparisons of criteria; the method is even able to quantify the departure from complete consistence. A basic criticism is, however, stressed by Belton and Gear (1982), who exhibit a very simple example in which three alternatives are ranked according to three criteria in a certain order, which happens to be reversed for two of these alternatives if a fourth one is introduced in the process. The AHP pairwise comparison matrix may be computationally difficult to implement when the number of alternatives is high.

A third MCDM method called ELECTRE, for "ÉLimination et Choix Traduisant la Réalité" (in French), i.e., "ELimination and Choice Expressing the REality," is based on the concept of outranking (cf. Figueira et al. 2005), according to which an alternative outranks another if it is at least as good on the whole set of criteria ("good" means here "to be preferred"). Several versions of ELECTRE have been developed, adapted to the three main types of decision making and cited in chronological order:

- "choosing" among a set of alternatives the one that seems to offer the most acceptable trade-off between possibly conflicting decision criteria;
- "ranking," possibly with *ex aequo*, the alternatives from the best to the worst, again in terms of acceptable trade-off;
- "sorting" the alternatives into categories.

In this third type of decision making, the so-called ELECTRE TRI method is well suited to decision problems where numerous alternatives have to be classified in a small number of ordered categories, as it is the case for WSN segments to be affected to classes reflecting the urgency of their renewal. ELECTRE TRI has been followed by Le Gauffre et al. (2004) for the annual rehabilitation planning module of the CARE-W research project presented by Saegrov (2005). This method entails also weighting criteria, according to the judgement of water utility experts; each criterion is then considered as an elector expressing preferences for the alternatives, and whose vote is worth its weight. On each criterion scale, three thresholds have to be chosen that express either a veto, an indifference, or a preference. In contrast to the WSM and AHP methods, no compensation between criteria is possible. As a consequence, two alternatives may happen to be incomparable. Examples of positive feedback from the practical use of ELECTRE TRI by water utility managers are reported in Large (2016). It is also to be noticed that, in contrast to choosing and ranking decision problems, this sorting has the computational advantage of not involving a systematic pairwise comparison of the possible alternatives.

The diversity of available MCDM methods, based on different mathematical setups as well as on considerations borrowed from cognitive sciences (in the case of AHP and ELECTRE), leads Triantaphyllou and Mann (1989) to formulate a "decision-making paradox": "What decision-making method should be used to choose the best decision-making method?"; using concurrently, e.g., WSM and AHP on simple basic examples consisting of three alternatives ranked according to three criteria, they show that these methods are not mutually consistent; ratherm they tend to behave so when the number of alternatives increases. Such comparisons of MCDM methods, carried out in actual pipe prioritisation case studies by Tlili and Nafi (2012) and Tscheikner-Gratl et al. (2017), also suggest the possibility of inconsistencies and call for the concurrent use of different MCDM methods.

Beyond these issues, the relevance itself of a pipe prioritisation process that relies on judgements by water utility experts is questionable with regard to its ability to ensure that, in the long term, the strategic objectives in securing service quality can be met. This basic question, as well as a similar one relating to the assessment of renewal budget needs in the midterm, is the subject of the next section.

5.4 The probabilistic long-term simulation approach

When studying the various decision aid tools at the disposal of water utility managers, in considering both midterm renewal budget planning and annual renewal works programming, Large et al. (2015) acknowledge two critical facts:

- Annual renewal work programming faces a huge complexity and only a fraction of renewal works are then performance oriented, i.e., aligned on strategic objectives in matter of the quality of the service delivered to users.
- Methods to assess renewal budget needs in the midterm are either based on expert judgements that reflect a strong path dependency (partly justified by the necessary stability of water tariffs) or on rather simplistic economic optimisation that addresses the sole consideration of performance-driven renewal, generally relating to pipe breakage reduction.

As a consequence, Large (2016) concludes that coherence between operational and budgetary decisions is not ensured, and their combined implementation cannot guarantee that strategic objectives can be met in the long term.

5.4.1 Necessity of a long-term vision

For economical reasons, the annual renewal rate cannot exceed 1% or 1.5% of the WSN length, and the impact on the service performance of a management policy cannot therefore be assessed on an annual basis. The necessity to assess this impact over several years militates then in favour of adopting a long-term vision.

Additionally, the aggregated nature of pipe breaks, both spatially and temporally (see Goulter et al. 1993; Kleiner et al. 2001; Burn et al. 2003; Tlili and Nafi 2012; Le Gat 2014), implies that short-term hierarchisation of pipes according to their risk of failure strongly relies on consideration of observed past breaks. Computations that consider the development of the breakage process and attempt to assess the effect of renewal works over several future years, should therefore account for the "information gain" made up of the observation of breaks during the lifetime of the pipe (for a similar idea in sewer network management, see Taillandier et al. [2020]).

A method currently under development is reported by Rulleau et al. (2020) that consists in carrying out long-term simulations, in a way partially inspired by Alvisi and Franchini (2009) but proceeding probabilistically. The probabilistic approach is well adapted to a prospective framework. It has the following justifications:

- The simulation of the information gain that guides performance-driven operational decisions cannot proceed deterministically by considering the average development of failures.
- The precise location of future land management operations and road rehabilitation is unknown beyond some years ahead, and the exposure of WSN segments to these events is thus to be simulated in probability.

5.4.2 Long-term simulation set-up

A set of basic modelling hypotheses is proposed:

- The pipe breakage process develops according to LEYP models (generally by pipe material), the parameters of which have been calibrated beforehand using a breakage history available in the information system of the water utility and, if possible, corrected against the selective survival bias through the use of a LEYP2s model, provided a pipe decommissioning history is also available.
- The renewal costs are modelled (either by a function or by an abacus) for each WSN segment, depending on the characteristics (material, diameter, length, urban vs. rural environment, etc.) of both decommissioned and replacement pipes; a factor K is also considered that modifies the renewal cost in case of coordinated work; this factor is constant among segments with a view towards simplification, but it can be defined as a function of pipe characteristics without difficulty.
- According to its hydraulic importance, each WSN segment is characterised by its hydraulic criticality (see Ayala Cabrera et al. 2017; Marlim et al. 2019) that measures the impact of its performance failure on the other WSN segments, weighted by their sensitivity to this impact (in particular on critical service users) as well as the potential impact on the built and natural environment.

Long-term simulation conceptually pertains to a what-if logic and is then based on a set of parameters that can be tuned to investigate asset management trajectories:

- a sequence of annual renewal budgets over a range of years considered by the long-term simulation;
- an allocation key that allows distributing the annual budget among mandatory, opportunity, and performance-oriented renewal works;

- a coordination level parameter that translates the ability of the utility manager to choose among the road works which ones concern problematic WSN segments that deserve to be renewed.

5.4.3 Long-term simulation algorithm

The proposed long-term simulation algorithm is based on the exploration of the possible trajectories for each WSN segment, by iteratively drawing, from one year to the next, a probability tree diagram, which is illustrated in Figure 5.7.

Estimating for each year *a* probabilities p_{aj} in Figure 5.7 involves the following steps:

- ranking WSN segments by failure risk (i.e., product of segment criticality and expected number of failures in year $a + 1$ given the observed failure history, using the LEYP model);
- determining the observed failure history for each segment by generating for year *a* a random number of failures given the observed failure history, using the LEYP model;
- allocating the renewal budget of year *a* between budget shares, denoted B_M, B_O, and B_P, respectively, dedicated to mandatory, opportunity, and performance-oriented works in a probabilistic way;
- aggregating the trajectories that involve the segment renewal in year *a* (illustrated by ascending arrows in Figure 5.7), and combining the resulting renewal probabilities into the p_{aj}'s.

Notice that aggregating some trajectories in the above last step mitigates the combinatorial inflation of possible trajectories, thus making the computations feasible in a reasonable time with an ordinary personal computer.

The calculations for a given year *a* that lead to evaluating the probabilities p_{aj} are presented throughout the next three subsections (5.4.4, 5.4.5, and 5.4.6), where the reference to the year is nevertheless dropped in order to lighten the notations.

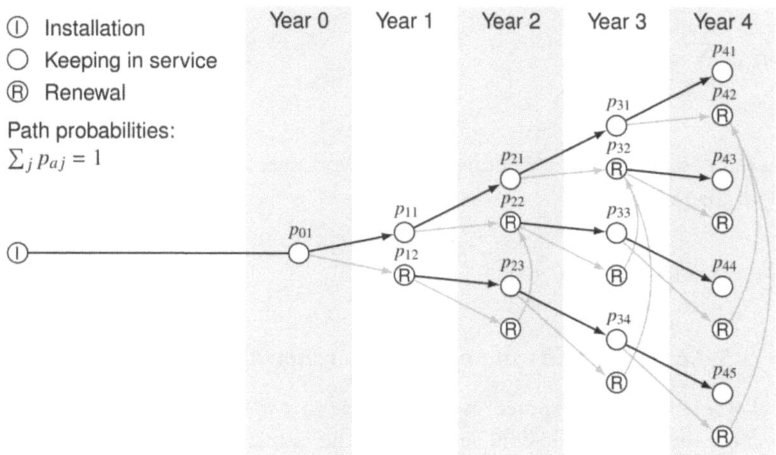

Figure 5.7 Exploring the possible trajectories for a given WSN segment.

5.4.4 Probability of mandatory renewal

Several reasons may lead to a mandatory renewal of some WSN segments:

- large-scale land management operations or in-depth road rehabilitation require very often relocating pipes;
- urban development and demographic changes involve pipe resizing or abandoning some segments and installing new ones elsewhere;
- sanitary regulations for renewed pipes that degrade water quality (e.g., emission of chlorine vinyle monomers by PVC pipes manufactured before 1980);
- customer reluctance to use coloured water due to corroded metallic pipes that gives rise to a suspected renewal of segments;
- elimination programme of segments the maintenance of which is problematic (e.g., high cost overrun of asbestos-cement pipe repair);
- planned gradual improvement of WSN configuration, aiming at easier operation or management (see, e.g.Gilbert et al. 2017 concerning leakage control).

Estimating for each segment its probability τ_{Mi} to undergo in a given year a mandatory renewal for any of the above reasons necessitates assessing the possible transformations of the WSN under given prospective hypotheses, validated by hydraulic modelling, as in Rulleau et al. (2020).

Denoting KC_{Ri} the renewal cost of WSN segment i (accounting for a possible effect of coordination on the cost via the K factor), the expected value of the annual budget share B_M consumed by mandatory renewal is:

$$\mathrm{E}B_{\mathrm{M}} = \sum_i KC_{\mathrm{R}i}\tau_{\mathrm{M}i}$$

5.4.5 Probability of performance-oriented renewal

The probability of performance-oriented renewal is calculated so as to consume the budget share B_p. The WSN segments are ranked to that end by decreasing the expected risk of failure, and the first n_p segments are selected for performance-oriented renewal. The rank n_p is such that:

$$\sum_{i=1}^{n\mathrm{p}} \mathrm{E}C_{\mathrm{R}i} \leqslant B_{\mathrm{P}} < \sum_{i=1}^{n\mathrm{p}+1} \mathrm{E}C_{\mathrm{R}i}$$

where expected renewal costs depend on whether renewal works are also mandatory or liable to be coordinated with road renovation:

$$\mathrm{E}C_{\mathrm{R}i} = \left[\left(\tau_{\mathrm{M}i} + \tau_{\mathrm{O}i}\right)K + \left(1 - \tau_{\mathrm{M}i} - \tau_{\mathrm{O}i}\right)\right]C_{\mathrm{R}i}$$

5.4.6 Probability of opportunity renewal

The probability τ_{oi} of segment i to be exposed in a given year to a road renovation that offers the opportunity to renew the concerned WSN segment can be estimated knowing the annual road renovation rate and its possible geographical variations. This has to be hypothesised over the mid- and long terms, according to prospective scenarios. As the annual road renovation rate is most often noticeably higher than the annual pipe renewal rate, the water utility manager can face one of two opposing situations:

- the annual road renovation programme is either known sufficiently in advance, so that, coordinated with WSN segment renewal works, it can be selected in a performance-oriented way; or
- or this performance-oriented selection is not possible and coordinated WSN segment renewal works are randomly decided, regardless of pipe deterioration.

These opposite cases can be characterised by an "anticipated knowledge rate" τ_A, the value of which is worth 1 in the first case of "complete anticipation" of the road renovation programme, 0 otherwise.

To go further into this presentation, two Bernoulli random variables (BRV) can be introduced:

- the BRV V_{oi} is a binary variable with value $V_{oi} = 1$ if segment i is exposed to a road renovation that offers an opportunity for pipe renewal, $V_{oi} = 0$ otherwise; its probability is $\Pr\{V_{oi}=1\} = \tau_{oi}$;
- the BRV R_i is a binary variable with value $R_i = 1$ when the WSN segment i is renewed in the given year.

For pipes with ranks higher than n_p, the following random sum of costs is considered:

$$\forall k \in \{n_P + 1,\ldots,n\}, \quad S_k = \sum_{i=n_P+1}^{k} KC_{Ri}V_{Oi}$$

where n is the number of WSN segments. The expectation of S_k is then:

$$ES_k = \sum_{i=n_P+1}^{k} KC_{Ri}\tau_{Oi}$$

and its variance:

$$\mathrm{Var}S_k = \sum_{i=n_P+1}^{k} K^2C_{Ri}^2\tau_{Oi}\left(1-\tau_{Oi}\right)$$

The probability of opportunity renewal with anticipated knowledge of the road work programme is then:

$$\Pr\left\{R_k = 1 \mid V_{Ok} = 1, \tau_A = 1\right\} = \Pr\left\{S_k \leqslant B_O\right\} \simeq \Phi\left(\frac{B_O - ES_k}{\sqrt{\mathrm{Var}S_k}}\right)$$

where a Gaussian approximation is considered, and Φ() denotes the standard normal cumulative distribution function.

In the case of non-anticipated knowledge, this probabilty is merely uniform:

$$\Pr\{R_k = 1 \mid V_{Ok} = 1, \tau_A = 0\} = \frac{B_O}{ES_n}$$

And in the case of partially anticipated knowledge of the road work programme (i.e., $\tau_A \in [0,1]$):

$$\Pr\{R_k = 1 \mid V_{Ok} = 1, \tau_A\} \simeq \tau_A \Phi\left(\frac{B_O - ES_k}{\sqrt{\mathrm{Var}S_k}}\right) + (1-\tau_A)\frac{B_O}{ES_n}$$

5.4.7 Combining probabilities

The renewal probability of segment i is then:

either $\Pr\{R_i = 1 \mid i \leqslant n_P\} = 1$

or $\Pr\{R_i = 1 \mid i > n_P\} = \tau_{Mi} + \tau_{Oi}\left(\tau_A \Phi\left(\frac{B_O - ES_k}{\sqrt{\mathrm{Var}S_k}}\right) + (1-\tau_A)\frac{B_O}{ES_n}\right)$

5.4.8 Annual syntheses

After the algorithm has reached the horizon year of simulation, annual syntheses can be calculated at the WSN level, relating to means indicators, such as annual renewal rates, and performance indicators, such as annual failure rates.

5.5 Conclusion

The literature overview carried out in Sections 5.2 and 5.3 leads to a consideration of lifetime as the result of asset management rules rather than as an intrinsic property of pipes, and to represent it as a probability distribution, characterised by a wide dispersion rather than as a single central value. The expert judgement-based methods for renewal modelling are likely to be essentially driven by experience and then reinforced by the path dependence of management decisions. The proposed economic-based methods introduce a degree of rationality in the decision-making process, but they fail to clearly distinguish short-term works programming and midterm budget planning, consider the pipe failure development in a deterministic way even in the long term, and ignore the importance of information gain for operational decisions. Short-term works programming faces a huge diversity of methods and variants, and it ultimately depends on the degree of confidence placed on the judgements of experts as to the relative importance of hierachisation criteria and their alignment with strategic service performance objectives.

The long-term simulation approach presented in Section 5.4 intends to take into account the operational constraints (mandatory and coordinated works) the annual programming is subjected to, reducing the complexity of multicriteria choice to budget shares consumption (namely considering the ratio B_p/B_o), and adopting a probabilistic point of view. This approach is designed to check whether a given annual renewal budget, coupled with a given allocation key, facilitates meeting long-term objectives concerning service quality; due to its what-if logic, a solution acceptable for governance stakeholders is to be found by trial and error. Whether an optimal solution could be searched by embedding the long-term simulation algorithm within an operational research procedure is an open question.

The long-term simulation approach can be improved regarding the following issues:

- So far, the approach takes into account the sole performance dimension relating to pipe breakage, and it needs then to be extended to pipe leakage and internal pipewall roughness (hydraulic capacity reduction).
- Integrating the performance dimension relating to pipe leakage; considers also the renewal of service connections, which are generally considered as responsible for a large proportion of the WSN water losses.
- The renewal probabilities presented in subsections 5.4.4, 5.4.5, and 5.4.6 need to be redefined at the level of clusters of segments instead of single segments, to take into account the practical requirement of organising renewal sites of sufficient size.
- Coupling with hydraulic modelling within a prospective framework is promising, as shown by Rulleau et al. (2020).

Another pivotal issue pertains to the credibility of simulation results. They aim indeed at facilitating governance debates by ensuring a mediation between governance stakeholders, who represent diverse scientific and technical cultures, as well as various sensitivities towards socio-economical and environmental stakes. This leads to questioning both:

- the scientific validity of models and simulation tools; and
- the relation to "reality" of tools based on numerical data and algorithms designed to convert data into "information."

These questions cannot be legitimately ignored, and answers are to be sought in the field of philosophy of science; the work of French philosopher Gilbert Simondon and, particularly, Simondon (1958), who promotes the development of a technical culture within human societies, deserves to be seriously considered in this respect.

The scientific validity of models and simulations is a pivotal epistemological question that calls for at least a minimal answer according to which the validity can be debated only within a sound theoretical framework. The one chosen here is the axiomatic theory of probability.

Concerning the ontological nature of information, an in-depth discussion can be found in Logan (2012) and in Hui (2012). A complex socio-technical system such as a water utility draws its materiality from the physical infrastructure, social relationships, and information. Research works should thus assess the cost for water utilities to improve their information system (data, models, staff skills), i.e., increase their "informational capital," and the gain they can expect in return in the long term regarding the efficiency of their IAM policy.

References

Alegre, H., Coelho, S.T., Covas, D., Almeida, M.C., and Cardoso, M.A. (2013). A utility-tailored methodology for integrated asset management of urban water infrastructure. *Water Science and Technology: Water Supply*, 13(6), 1444–1451.

Alegre, H., Vitorino, D., and Coelho, S.T. (2014). Infrastructure value index: A powerful modelling tool for combined long-term planning of linear and vertical assets. *Procedia Engineering*, 89, 1428–1436.

Alvisi, R. and Franchini, J.S. (2009). Multiobjective optimization of rehabilitation and leakage detection scheduling in water distribution systems. *Journal of Water Resources Planning and Management*, 135(6), 426–439.

Amaral, R., Alegre, H., and Matos, J.S. (2016). A service-oriented approach to assessing the infrastructure value index. *Water Science and Technology*, 74(2), 542–548.

Andersen, P.K. and Keiding, N. (2005). Survival analysis, overview. In *Encyclopedia of Biostatistics*. John Wiley & Sons.

AWWA. (2001). *Dawn of the Replacement Era: Reinvesting in Drinking Water Infrastructure*. Technical report, American Water Works Association.

AWWA. (2012). *Buried no Longer: Confronting America's Water Infrastructure Challenge*. Technical report, American Water Works Association.

AWWA. (2018). *Awwa Asset Management: Definitions Guidebook, Version 1.0*. Technical report, American Water Works Association.

Ayala Cabrera, D., Piller, O., Herrera, M., Parisini, F., and Deuerlein, J. (2017). Analysis of water distribution networks in a context of mechanical failures. In Congreso de Métodos Numéricos en Ingeniería (CMN 2017) (pp. 1589–1604), July 2017, Valence, Spain.

Baird, G.M. and Wagner, T.B. (2020). The high cost of using awwa's buried no longer pipe service life table for capital budgeting. In Pipelines 2020, August 9–12, 2020, San Antonio, TX.

Belton, V. and Gear, T. (1982). On a short-coming of saaty's method of analytic hierarchies. *Omega*, 11(3), 228–230.

Bhamare, S.S., Yadav, O.P., and Rathore, A. (2007). Evolution of reliability engineering discipline over the last six decades: A comprehensive review. *International Journal of Reliability and Safety*, 1(4), 377–410.

Burn, S., Tucker, S., Rahilly, M., Davis, P., Jarrett, R., and Po, M. (2003). Asset planning for water reticulation systems the PARMS model. *Water Science and Technology: Water Supply*, 3(1–2), 55–62.

Burn, S., Marlow, D., and Tran, D. (2010). Modelling asset lifetimes and their role in asset management. *Journal of Water Supply: Research and Technology–AQUA*, 59(6–7), 362–377.

Cador, J.M. (2002). *Le renouvellement du patrimoine en canalisations d'eau potable en france (in French). the renewal of drinking water pipelines in france*. Technical report, GEOPHEN Université de Caen-Basse Normandie.

Figueira, J., Mousseau, V., and Roy, B. (2005). *Multiple Criteria Decision Analysis: State of the Art Surveys, Chapter 4: Electre Methods* (pp. 133–162). Springer.

Francisque, A., Shahriar, A., Islam, N., Betrie, G., Siddiqui, R.B., Tesfamariam, S., and Sadiq, R. (2014). A decision support tool for water mains renewal for small to medium sized utilities: A risk index approach. *Journal of Water Supply: Research and Technology–AQUA*, 63(4), 281–302.

Fuchs-Hanusch, D., Gangl, G., Kornberger, B., Kölbl, J., Hofrichter, J., and Kainz, H. (2008). PiReM, pipe rehabilitation management developing a decision support system for rehabilitation planning of water mains. *Water Practice and Technology*, 3(1), 9.

Gilbert, D., Abraham, E., Montalvo, I., and Piller, O. (2017). Iterative multistage method for a large water network sectorization into dmas under multiple design objectives. *Journal of Water Resources Planning and Management*, 143(11), 27.

Goulter, I., Davidson, J., and Jacobs, P. (1993). Predicting water-main breakage rates. *Journal of Water Resources Planning and Management*, 119(4), 419–436.

Herz, R. (1996). Ageing processes and rehabilitation needs of drinking water distribution networks. *Journal of Water Supply: Research and Technology–AQUA*, 45(5), 221–231.

Herz, R. and Lipkow, A. (2003). Ageing processes and rehabilitation needs of drinking water distribution networks. *Water Science and Technology: Water Supply*, 3(1), 35–42.

Hui, Y. (2012). What is a digital object? *Metaphilosophy*, 43(4), 380–395.

Kalbfleisch, J.D. and Prentice, R.L. (2002). *The Statistical Analysis of Failure Time Data, Second Edition*. John Wiley & Sons.

Kaplan, E.L. and Meier, P. (1958). Non-parametric estimation from incomplete observations. *Journal of the American Statistical Association*, 53(282), 457–481.

Kleiner, Y. and Rajani, B.B. (2009). I-WARP: Individual water main renewal planner. In Proceedings of the Tenth International Conference on Computing and Control for the Water Industry, CCWI 2009, Integrating Water Systems, 1–3 September 2009, Shefleld, UK.

Kleiner, Y., Adams, B.J., and Rogers, J.S. (2001). Water distribution network renewal planning. *Journal of Computing in Civil Engineering*, 15(1), 15–26.

Large, A. (2016). *Une meilleure gestion patrimoniale des réseaux d'eau potable : le modèle de prévision du renouvellement à long terme OPTIMEAU* (in French). *A Better Water Network Asset Management: The Long Term Renewal Prediction Model OPTIMEAU*. PhD thesis, I2M, University of Bordeaux, France.

Large, A., Le Gat, Y., Elachachi, S.M., Renaud, E., Breysse, D., and Tomasian, M. (2015). Improved modelling of 'long-term' future performance of drinking water pipes. *Journal of Water Supply: Research and Technology–AQUA*, 64(4), 404–414.

Le Gat, Y. (2014). Extending the Yule process to model recurrent pipe failures in water supply networks. *Urban Water Journal*, 11(8), 617–630.

Le Gat, Y. (2016). *Recurrent Event Modeling Based on the Yule Process: Application to Water Network Asset Management. Mathematics and Statistics Series (Vol. 2): Mathematical Models and Method in Reliability Set.* ISTE-Wiley.

Le Gat, Y., Kropp, I., and Poulton, M. (2013). Is the service life of water distribution pipelines linked to their failure rate? *Water Science and Technology: Water Supply*, 13(2), 386–393.

Le Gauffre, P., Laffréchine, K., Baur, R., Poinard, D., Haidar, H., and Schiatti, M. (2004). Aide multicritère aux décisions de réhabilitation d'un réseau d'eau potable (in French). multicriteria decision aid for drinking water network rehabilitation. In XXIIièmes Rencontres Universitaires de Génie Civil 2004, Ville & Génie Civil, France.

Logan, R.K. (2012). What is information?: Why is it relativistic and what is its relationship to materiality, meaning and organization. *Information*, 3, 68–91.

Malm, A., Svensson, G., Bäckman, H., and Morrison, G.M. (2013). Prediction of water and wastewater networks rehabilitation based current age and material distribution. *Water Science & Technology: Water Supply*, 13(2), 227–237.

Marlim, M.S., Jeong, G., and Kang, D. (2019). Identification of critical pipes using a criticality index in water distribution networks. *Applied Sciences*, 9(19), 14.

Marzouk, M., Abdel Hamid, S., and El-Said, M. (2015). A methodology for prioritizing water mains rehabilitation in egypt. *Housing and Building National Research Center Journal*, 11, 114–128.

McWaneDuctile. (2021). Buried no longer confronting america's water infrastructure challenge. https://www.mcwaneductile.com/blog/buried-no-longer-confronting-america-s-water-infrastructure-challenge/. Accessed: 2021-01-23.

Moglia, M., Burn, S., and Meddings, S. (2006). Decision support system for water pipeline renewal prioritisation. *Journal of Information Technology in Construction*, 11, 237–256.

Nafi, A. and Kleiner, Y. (2010). Scheduling renewal of water pipes while considering adjacency of infrastructure works and economies of scale. *Journal of Water Resources Planning and Management*, 136(5), 519–530.

Pardo, M.A. and Valdes-Abellan, J. (2019). Pipe replacement by age only, how misleading could it be? *Water Supply*, 19(3), 846–854.

Pericault, Y., Bruaset, S., Ugarelli, R., Saegrov, S., Viklander, M., and Hedström, A. (2017). Coordinated long term planning of sewer and water mains rehabilitation. In 2017 IWA LESAM Conference Proceedings, Trondheim, Norway.

PlasticsNews. (2021). Saving water with plastics. https://www.plasticsnews.com/article/20150728/NEWS/150729901/saving-water-with-plastics/,. Accessed: 2021-01-23.

Rahman, S. and Vanier, D.J. (2004). Life cycle cost analysis as a decision support tool for managing municipal infrastructure. In CIB 2004 Triennial Congress (pp. 1–12), Toronto, Ontario, May 2–9, 2004.

Rajani, B. and Kleiner, Y. (2004). Alternative strategies for pipeline maintenance/renewal. In AWWA 2004 Annual Conference (pp. 1–16), Orlando, Florida, June 13–17, 2004.

Renaud, E., Bremond, B., and Le Gat, Y. (2014). Water pipes: Why 'lifetime' is not an adequate concept on which to base pipe renewal strategies. *Water Practice & Technology*, 9(3), 307–315.

Roberts, M. and Broadbent, C. (2011). Effective use of infrastructure funds: Prioritizing pipe rehabilitation projects. In *NC Currents, North Carolina Section of the American Water Works Association* (pp. 84–87), Summer 2011.

Rulleau, B., Salles, D., Gilbert, D., Le Gat, Y., Renaud, E., Bernard, P., Gremmel, J., Giard, A., Assouan, E., de Grissac, B., Eisenbeis, P., Husson, A., Rambonilaza, T., and Stricker, A.-E. (2020). Crafting futures together: Scenarios for water infrastructure asset management in a context of global change. *Water Supply*, 20(8), 3052–3067.

Saaty, L.G. (1987). The analytic hierarchy process: What it is and how it is used. *Mathematical modelling*, 9(3–5), 161–176.

Saaty, T.L. and Vargas, L.G. (2012). *Models, Methods, Concepts & Applications of the Analytic Hierarchy Process*. Springer.

Saegrov, S. (2005). *CARE-W Computer Aided Rehabilitation for Water Networks*. IWA Publishing.

Shamir, U. and Howard, C.D.D. (1979). An analytic approach to scheduling pipe replacement. *Management and Operations, Journal AWWA*, 71(5), 248–258.

Simondon, G. (1958). *Du mode d'existence des objets techniques (in French). About the Existence Mode of Technical Objects*. Aubier-Montaigne.

Taillandier, F., Elachachi, S.M., and Bennabi, A. (2020). A decision-support framework to manage a sewer system considering uncertainties. *Urban Water Journal*, 17(4), 344–355.

Tlili, Y. and Nafi, A. (2012). A practical decision scheme for the prioritization of water pipe replacement. *Water Science & Technology: Water Supply*, 12(6), 895–917.

Triantaphyllou, E. and Mann, S.H. (1989). An examination of the effectiveness of multi-dimensional decision-making methods: A decision-making paradox. *Decision Support Systems*, 5, 303–312.

Tscheikner-Gratl, F., Egger, P., Rauch, W., and Kleidorfer, M. (2017). Comparison of multi-criteria decision support methods for integrated rehabilitation prioritization. *Water*, 9(68), 1–28.

Tuhovcak, L., Taus, M., and Mika, P. (2014). Indirect condition assessment of water mains. *Procedia Engineering*, 70, 1669–1678.

Vitse, L., De Fremery, G., and Golla, G. (2012). Etude de calcul de la récupération des coûts des services liés à l'utilisation de l'eau pour les bassins hydrographiques français en application de la directive cadre sur l'eau (in French). *Cost Recovery Study for Water Services in French Major River Basins in Application of the Water Framework Directive*. Technical report, OIEau, Ernst & Young.

Walski, T.M., Sharp, W.W., and Shields, F.D. (1986). Predicting internal roughness in water mains. Technical report, USAEWES Environmental Laboratory.

Weibull, W. (1951). A statistical distribution function of wide applicability. *Journal of Applied Mechanics*, 18(3), 293–297.

6

TERRITORIES AND TECHNOLOGIES

History and current trends of their interaction in urban water services

Bernard Barraqué

6.1 Introduction

Urban water services are now increasingly confronted with new challenges questioning their sustainability. Environmental issues and climate change require more efficient use of energy and water resources, for the sake of biodiversity protection; and the rise of "water poverty" triggers renewed interest in social sustainability. Full cost recovery is broadly supported by economists: in their view, public subsidies should be coming to an end, although at the very moment when water services need to invest to renew their assets. Now can all these challenges be addressed at the same time? A new articulation of urban water services with natural water resources is needed.

This chapter sets the historical framework of the evolution of urban water services and their governance. It looks at various countries and sets out a description of their past technical and territorial evolutions within a context of institutional path dependencies. A sociohistorical comparison is made of urban water services that was initiated almost 30 years ago by a research lab that focused on innovation in urban networks (Dupuy 1984). Each country develops a specific organisation of its water services, depending upon the role of the state and its relations with local authorities or upon geography and jurisdictional aspects (Guerin-Schneider et al. 2002). However, a historical analysis of the evolution of water services in developed countries, where the water industry has reached a certain maturity, can provide an opportunity to reveal common trends in governance evolution. Thus, this chapter reviews the functional and institutional paths of well-established urban water services observed in western Europe, Australia, and the United States of America. The chapter mainly focuses on drinking water services, but examples are cited in wastewater services as well, as there is a clear interaction between them in most cases.

Four major issues emerge in this historical review of urban water sector governance: (i) the evolution of private sector participation in urban water services and the recurring public vs. private debate; (ii) the emergence of two opposite but complementary trends in territorial scales of infrastructure management and stakeholders' participation, from upscaling to downscaling; (iii) the potential alternative to these territorial reorganisations offered by various degrees of integration/unbundling of local public services; and (iv) the increasing consideration of the

DOI: 10.4324/9781003057574-8

relationship between urban water services and water resources, as illustrated by the "Water Framework Directive" set by European legislation (WFD 2000/60/EC): it requires that water policies derive from the aquatic environment's recovery, i.e., improving *territorial* management before looking for *technological* solutions.

Much existing literature focuses on the public vs. private debate (Bakker 2003). We argue that granting importance to this antagonism may result in overlooking other dimensions of the water services constitution. To give just one example, the debated disvestiture and full privatisation of water services in England and Wales in the 1990s tend to obscure another important structural move: a progressive centralisation of water services at the scale of supposedly "optimal" water resources management, the ten regional water authorities (Saunders 1983), followed by increased centralisation and lower interest in water users' participation after the privatisation of water supply and sanitation (WSS) services.

The four themes of the evolution should be considered together, to better show the mutual influence of geographic, socioeconomic, and political contexts on the scaling and technical-financial arrangements of WSS services in various countries (Barraqué 1992; Pezon 2006). In addition, resource management institutions and urban water services institutions should be studied separately, as their scales, regulations, and economics differ at least in developed countries, and therefore discussions about privatisation of water resources and of urban water services should be held separately too (Barraqué 2003).[1] It is all the more important that recent evolutions, in particular with the European Union WFD, tend to rearticulate the categories of services and those of resources, for the sake of reducing the ever-increasing costs incurred by "end-of-pipe" technologies.

The four issues of water services governance received different answers within the group of countries selected here. However, together they offer a heuristic approach to encapsulate the past and current trends in urban water services governance. Due to length constraints, we will focus on the upscaling/downscaling and the integration/unbundling issues; for a deeper discussion of the private/public and water resources/services issues, please see, respectively, Barraqué (2009) and Barraqué et al. (2008).

6.2 A wide array of public–private arrangements

In the 1990s, the World Bank promoted private sector involvement aimed at improving urban infrastructure in developing countries, including WSS services. In less than a decade, many Third World and "alter" NGOs organised strong opposition to it, and their arguments against the social and political impacts of private sector participation on WSS services resulted in some cities choosing to terminate these contracts and to return to public procurement.

However, history teaches that most countries developed their water services through a varying mix of public and private participation in the financial, technical, and institutional aspects, resulting in a wide range of hybrid governance models. A review of public–private partnerships in WSS might then focus on the implications of this hybridisation, rather than remaining in polarised arguments supporting either public or private models.

We can mention here that only England adopted full privatisation while the rest of Europe offers a wide array of management formulas: on the public management side, direct labour by a municipal department, with no separate budget (less and less frequent); direct provision via an autonomous public service with separate budget; mixed economy utilities with varying degrees of private capital; formal private utility with 100% public capital, called SPL in France (*société publique locale*), but more frequent and older in the Netherlands or Germany. On the delegation side, by order of growing responsibility of private partners (and also of contract duration), simple management contract (*gérance*), management contract with bonus (called *régie intéressée* in French), lease (*affermage*), and concession contracts.

6.3 Upscaling/downscaling WSS management

In all the countries we studied for this chapter, water services were initially run within the competence of the municipalities – or were organised at the scale of hamlets in rural areas, as is still the case in low-density areas of Ireland (group water schemes: Brady and Gray 2013), Denmark, northwestern Spain, and Portugal. To face the challenges met in their development from the early 19th century, WSS services often changed territorial scales – due to resource problems, technology innovations, and also to the evolution of the local authorities and of their legal responsibilities. The general trend went towards larger territories to achieve economies of scale and a better management of sanitary or financial risks. But in some cases, decentralisation and downscaling took place to solve specific management issues. In terms of the sustainability of water services, the debate on the appropriate territorial scale is as important as the private vs. public management issue. For what specific reasons, and how, did they move to present-day management territories?

6.3.1 Upscaling: Looking for improved performance of water services

In France, a large number of water services – above 12,000 entities for 64 million inhabitants in 2009 – is a legacy of the extreme fragmentation of the 36,000 communes (Pezon and Canneva 2009). Intermunicipal cooperation for some public services has emerged as a practical alternative to municipal reforms. Single-purpose joint boards came into existence as early as 1890; multipurpose joint boards in 1959. The 1999 Chevènement law created new kinds of multimunicipal boards, today called EPCI (*établissements publics de coopération intercommunale*): one for the metropolitan areas (responsible for WSS), one for rural areas, and one for cities (both could choose to provide these services, or not). It has led to a relative concentration of water services (Pezon and Canneva 2009).

With the most recent laws (MAPTAM 2014; NOTRe 2015), another step took place with the creation of *métropoles* for cities above 200 000 inhabitants and the obligation for all EPCI to become the authorities responsible for water supply. This reform met some resistance because the new territories did not match previous service delegation or direct supply areas. But within a few years, the total number of utilities could be divided by between 7 and 10.

In contrast, the rationalisation of wastewater treatment in the Paris region took place much earlier. In the beginning, it was formed within the former Seine *département*.[2] The breakup of this large county into four smaller ones in 1968 resulted in the creation in 1970 of a common board called SIAAP (Syndicat Interdépartemental d'Assainissement de l'Agglomération Parisienne) in charge of treating the wastewater from most of the metropolis (almost 9 million inhabitants). The upscaling of wastewater treatment gave economies of scale, which allowed SIAAP to invest in sophisticated technology and in measures for protecting the environment. Conversely, the sewer systems stayed within the competence of the communes, while rainwater drains and large collectors are managed by the counties. This is typical of needed multilevel governance for megapolises, and it illustrates how the concentration of water and sanitation services and the mutualisation of infrastructure can be justified on economic or technical grounds.

The move towards the larger scale, with the creation of multi- or single-purpose joint boards, has taken place in all the reviewed countries. In the Netherlands, today only 10 water companies are supplying water to 17 million inhabitants. Their territories partly match those of the Dutch provinces in size, and they are run by mixed boards of provinces and communes (Loijenga 2009). On the other hand, sewerage remains a municipal competence (350 municipalities now, under consolidation process). But sewage treatment was entrusted in the 1960s to the waterboards (*Waterschappen*). These drainage and flood protection institutions were created in the Middle Ages and have intensely clustered in the last decades: there were more than 2,600 waterboards in 1945, only 24 waterboards in 2010 (Kuks 2010), and 21 today. This concentration process went along with a decrease in the number of sewage works: between 1980 and 2010, their number went down from 505 to 390. The overall rationale for concentration – based on a voluntary process – is the growing conscience of flood risks (the 1953 flood killed 1,830 people); water pollution issues also became more acute after the 1950s (Kuks 2010). To face financing challenges, the water boards had to make significant investments. Mergers into larger entities made them a credible partner of the government to run all the major dyke systems in the country, as well as to return space for river expansion under swells.

In Germany, despite a reduction of the water footprint, the collapse of water consumption in the eastern *Länder* after unification was an early warning of WSS services' potential unsustainable evolution, as the income was generated by water bills based on metering volumes purchased. But this did not prevent the federal government from shutting down the former 15 *Volkswasserbetriebe* (supralocal water bureaucracies) and returning WSS services to municipalities. In turn, these were soon involved in larger joint boards, and resulting water utilities even make capital exchanges with other services that operate at the regional level, like typically electricity, thus pursuing the logic of multiservice integration of the *Stadtwerke*.[3]

In England and Wales, a similar concentration process of WSS services took place after World War II. The adoption of the Water Act in 1973 led to a complete regionalisation of these services. At the beginning of the 1970s there were 198 water entities, of which 101 were joint boards, 64 were municipal services, and 33 were small private companies. The Water Act merged them into 10 regional water authorities (RWAs), the borders of which matched hydrological basins (Saunders 1983). RWAs were put in charge of the whole drinking water service, wastewater treatment, and resource management. Only local sewer systems remained within the competence of the municipalities. This regionalisation illustrates how technical matters were thought to have precedence over political or administrative considerations (Saunders 1983). Actually, the creation of the RWAs was the final outcome of a political process started in the 1940s: it eroded the local authorities' responsibilities and power over public services, and it makred a decisive step towards technical and industrialised management of water and sanitation services. The privatisation of WSS services in 1989 did not really impact the territorial organisation of the water services. It only resulted in a few more mergers of "water-only" companies and of planning and resource management authorities (from 10 to 8).

With the 1994 Galli law, Italy attempted to adopt the British model. The Italian water services needed huge investments for wastewater collection and treatment so as to comply with the Urban Waste Water Treatment Directive (UWWTD) (Massarutto 2011). But the banks were unwilling to lend money to a multitude of small municipal entities. The concentration of 6,600 existing water services into larger entities, and their opening to private sector participation, appeared as a solution for getting cheaper loans. Originally WSS services were supposed to be gathered in river catchments called *ambiti territoriali ottimali* (ATOs), but, in most cases, the regions, which were implementing the reorganisation, finally decided to merge water and wastewater services together at the administrative scale of the provinces (except in Sardinia and Puglia, where WSS were already regionalised). Thus, some 91 ATOs were created to manage WSS services in order to produce economies of scale and of scope. Given the historical power of Italian mayors regarding all local public services – one used to talk about "the mayor's water" in Italy – this top-down reform met strong opposition among the local elected representatives. Fifteen years after the reform, 19 ATOs had not yet appointed their water service operator (Mangano 2010), and now commercial rationalisation with upscaling is increasingly threatened by the success of an anti-privatisation referendum held in 2011, with a popular movement to keep the small municipal services.

6.3.2 Consolidation of water production: A strategic path to deal with water scarcity

Local authorities' implication in the development of WSS services was a "low noise" success in Europe. Today it seems that centralisation of water resources abstraction and potabilisation at a larger territorial scale could help overcome some difficulties in water supply. In Germany, for example, infrastructure for water production in areas with low rainfall but major urbanisation has been managed at a supra-local scale since the early 20th century: in 1912 in Baden-Württemberg the imperial government set up a regional public enterprise, the "Staatliche Landeswasserversorgung," to produce and transport drinking water to almost 100 different communes via a 750 km system.[4] In 2007, the 100 largest water operators delivered half the total volume of drinking water produced in Germany, illustrating the concentration (Kraemer et al. 2007).

Upscaling as a solution to water scarcity has also been adopted in the United States. In southern California, the risk of water shortage led utilities around Los Angeles to build a large aqueduct to bring water from the Colorado River. A "wholesaler," the Metropolitan Water District of Southern California (MWD), was put in charge to produce and transport water to 26 water distribution districts, supplying 19 million inhabitants. A similar organisation was developed in Boston but for different reasons: in 1985 the Environmental Protection Agency (EPA) required better treatment of wastewater. Boston decided to create the Massachusetts Water Resources Authority (MWRA), a regional institution responsible for the treatment of effluent and the production of drinking water for 61 communities. Today, more than 2.5 million people in Boston and its suburbs get water from the MWRA. While member municipalities distribute water and collect effluent on their territories, water and sewage works and interceptors are managed regionally.

As for Australia, after the droughts of the early 2000s, the federal government and water operators made huge investments in infrastructure to improve water security. Desalination and water recycling plants flourished, whereas local governments tried to rationalise and interconnect their networks. Queensland, where 2.8 million inhabitants live around the fast-growing city of Brisbane, decided to centralise water production. The South East Queensland Water Grid, initiated by the state government, is one of the largest Australian infrastructure projects.[5] AU$9,000 million was spent on an integrated water production system to secure water supply: a large system connects drinking water plants and pipes, which allow transfers from existing or projected water sources.[6] According to the designers, this system has the advantage of relying on water sources that depend in different ways upon the local climate – dams or rainwater storage vs. desalination or recycling plants. Centralised management also allows better allocation of water resources between different users, including the aquatic environment itself. Subsequently, major water supply systems were reorganised in such a way as to separate production and transport, on the one hand, and distribution, on the other.

In the end, the result of both concentration and centralisation processes is the rise of multilevel governance, whereby municipalities and their joint boards are not the only institutions implied in WSS services. The new political issue is whether to completely eliminate the role of local authorities, which have long carried the democratic and welfare ideals of public services, whilst remaining close to citizens. In Portugal, some water experts claim that even the water distribution and sewage collection should be operated at supra-municipal level. But in Australia, local stakeholders, who were once put aside, are now coming back to take part in the management of WSS services. First, local authorities are producing recycled water for their parks and sports grounds. Then citizens, individually or as part of local communities, are progressively turning into co-producers of the water service at the infra-municipal scale since rain harvesting has become compulsory for new buildings. To what extent do these decentralised systems reveal a common trend in the functional evolution of water services in other developed countries?

6.3.3 Downscaling and decentralised systems: Next step towards sustainability?

In contrast with European cities, which developed centralised water supply systems in the 19th century, hamlets and scattered settlements in rural areas have always relied on self-sufficient water systems, traditionally managed at the neighbourhood scale. In France, where the communes are already very small entities, these decentralised infra-municipal systems are quasi absent. But in Denmark, Lower Saxony, rainy countries like Ireland, the north of Portugal and northwestern Spain, this kind of neighbourhood association in charge of the local water supply is still common. Moreover, while they seemed bound to disappear in the short term (given industrialisation of water services), they are now being reconsidered. Indeed, the emergence of new sustainable urban settlements encourages the development of decentralised water and wastewater systems at neighbourhood or even block level, which are considered closer to natural water cycles and benefit from lower renewal costs. In Germany there are "eco-neighborhoods" experiments where grey water is recycled for non-potable uses, with a drastic reduction of consumption.

In Magdeburg, the local *Stadtwerk* plans on combining upscaling and downscaling solutions: a supra-municipal enterprise could produce bulk freshwater for the whole region, while decentralised systems at neighbourhood or building block scales could make it drinkable and deliver it to the end consumers. This would minimise the time drinking water spends in the pipes. Many cities from the former East Germany are facing the same situation, so some researchers are questioning the traditional centralised water supply system and are now thinking about adapting the actual network to shrinking demand via well-designed alternative systems (Moss 2008; NetWORKS 2008).

In France, despite governmental incentives towards sustainable urban projects, decentralised water systems are still seen as technical innovations and remain scarce. But we could very well imagine development of these systems in the near future. It would be consistent with the 5 million septic tanks still functioning on French territory! In addition, due to the high connection rate and the low population density, the cumulated length of water supply pipes exceeds 1 million km. The water services community is worried that systems' maintenance in the countryside will be impossible or overcostly, leading to reintroducing of on-site water services. The general issue at stake is the risk that these solutions reintroduce strong differences in the quality of service provided to various social or territorial groups. Graham and Marvin (2001) call attention to this potential "splintering urbanism," which might bring rich countries' water services close to the fragmented situation existing in developing ones. Arizona offers a striking example of this fragmentation (Megdal 2012): there are very large public utilities (e.g., in Phoenix and Tucson), but a significant fraction of the population depends on private community systems, sometimes with no or tiny networks and very different degrees of technical sophistication. Individualism and parochialism make development of public services with on-site technology unlikely. Yet there are projects combining robust small-scale systems with sophisticated information and communication technologies (ICT) systems to allow for remote control and reduced O&M costs.[7] In the end, the issue is how to separate the notion of public service and the trust it generates from the historical preferred technology of centralised systems in urban areas.

Territorial scales of WSS services management are moving in opposite but not incompatible directions, combining upscaling and downscaling, concentration, and decentralisation. The municipal territory is no longer the only or most relevant level to run WSS services because of water shortages, lack of money to renew old and centralised assets, a drop in water consumption, among others. The examples given in this part of the chapter show that the evolution of a territorial scale of management often goes along with questioning the integrated water service production chain. For instance, there may be a separation between water production at the regional scale and water distribution at the municipal or infra-municipal scale. Can disintegration and/or unbundling be part of the solution to some water services issues?

6.4 Water services' integration vs. unbundling

WSS management encompasses water resource abstraction and protection, drinking water production and distribution, wastewater and stormwater collection and treatment, and ending with disposal of treated wastewater into the environment. These different steps have been either integrated or unbundled, depending on the time or place considered. So, in addition to the question of geographic scaling of water services, one can consider the degrees of integration of the WSS services according to the following concepts:

1. vertical integration (or concentration): how far are the various tasks of water management, planning, and operation integrated together or separated (e.g., from water resources to the consumer, and back to the resource);
2. horizontal integration related to other sectors (single water service or multiactivity service);
3. structural integration (or autonomy): to what extent is the water service autonomous and independent from other public administrations (in terms of legal status, budget, accounting, etc.).

In this section, vertical integration issues will be left aside, since what would be described is largely redundant with either upscaling-downscaling, or public vs. private issues. Unbundling, in particular, was and is advocated by neoliberal economists willing to open as many segments of the water industry to market competition as possible/feasible.

However, one can recall the complex WSS organization in greater Paris: securing surface water availability even in summer droughts is achieved by a regional institution operating four upstream reservoirs. The city of Paris and several joint boards in the suburbs potabilise and distribute the water, either directly or through delegation contracts. During the mandate of Mayor Jacques Chirac, there even was a split between a mix economy company (SAGEP), in charge of producing drinking water from surface and groundwater and conveying it to the capital, and subsidiaries of Veolia (right bank of the Seine) and Suez (left bank), in charge of inner system maintenance (leaks control), distribution to customers, and metering and billing. And concerning wastewater, the above section showed the three territorial levels implied: communal (sewers) county (stormwater and interceptors), and intercounty (sewage treatment). Drinking water, sewage collection, and sewage treatment are separated in the Netherlands.

At the other extreme, all the tasks are fully integrated in New York City, from the protection of freshwater in the Catskill Mountains and upper Delaware River through payments of environmental services to farmers through its conveyance with a 250-km aqueduct to the city, distribution, and billing, plus incentives to conserve water, down to wastewater collection and treatment. Integration of abstraction and transfer of water resources, together with in-city supply is done by some large cities (San Francisco, Oakland, São Paulo, Rome, Madrid, Lisbon) while others depend on a regional company to tap and transfer the resource (above-mentioned Landwasserversorgung, Marseille, Barcelona, Acquedotto Pugliese …).

Usually, the sanitation sector remains separate from the drinking water even though funded now through water bills. It is a consequence of historical roots of water supply as a commercial activity while sewer services were often linked with drainage and street maintenance, and they were imposed on urbanites, publicly managed, and funded by taxes. But in some countries, WSS is becoming a single unified sector under a unified management (e.g., the privatised regional water companies in England and Wales, the *consorzi idrici integrati* in Italy, the integrated water utilities in Australia before 1994). Most of these institutional evolutions aim at improving the efficiency of the water and wastewater sectors, either separately or together. But it typically results in the need for coordination of activities/institutions, i.e., multilevel governance.

6.4.1 Horizontal integration vs. single purpose

Water supply and sanitation can be organised as single-purpose activities, or else brought under a broader multiutility company. In the beginning, water supply was initiated by private companies, with single-purpose contracts. The know-how about systems of sealed pipes under pressure with no leaks was partly borrowed from the coal gas industry, and, in fact, in most European countries water and gas engineers were gathered in the same associations. But in England, despite the return of both services under direct management after the historical change in Birmingham by Mayor Joseph Chamberlain, there was no movement to integrate both services in a single operation. In France, both activities remained separate, and in fact the general philosophy was to keep all services separate to avoid giving too much power to their operators against elected authorities: "*un seul service ne donne pas beaucoup de pouvoir.*"

In Germany, conversely, water supplies were at first separate from the main municipal organisations and were later transformed into private companies owned by local authorities. As early as the interwar period (1920s–1930s) they were frequently merged into multiutility services called *Queverbund* or *Stadtwerk*, which included water, gas, electricity, district heating, transport (but less frequently wastewater and solid waste). This reorganisation allowed water services to access a broader range of financing resources (public loans, financial markets) and operate financial and operational cross-subsidies among different local public utilities (Herdt 2009).

This horizontal integration of water services into multi-utilities can also be found to a lesser degree in Austria, Switzerland, and Italy (with the well-known example of Rome's ACEA, responsible for the water supply chain, sewage treatment, and energy supply). Copenhagen also offers an example of the complex mix that can emerge: while the suburbs have opted for joint boards, the central city has an integrated water and energy company. This company is now a regional power supply company, which still provides water to the central city but also additional volumes to suburban water supplies (Barbier and Michon 2010).

6.4.2 Water services: Financial autonomy or dependence?

A core factor in determining whether WSS services are integrated or unbundled is how their financial relations with other services and with the budgets of local authorities are arranged. The century-long transformation of water services into commercial utilities has turned them into autonomous entities with their own financial power and status. But the evolution was uneven. As mentioned above, wastewater services were mostly developed by local city governments along with drainage, linked with street maintenance and funded by taxes. For public health reasons they were usually imposed as administrative services rather than as commercial operations. Conversely, drinking water services were often created by private companies, and, apart from the United Kingdom (and Ireland), their costs were covered by bills; even after water services were reclaimed by municipalities as public entities, their operating costs were frequently charged to customers. However, major amounts of public money were injected to extend water services to all inhabitants (Pezon 2000). Progressively, however, water utilities took advantage of metering and billing to move closer to full internal cost pricing; and beyond, an increase in tariffs is now supported by a general move to protect water resources and reduce their water footprint (Walker 2009).

In several countries, wastewater services are increasingly considered an amenity, just like drinking water, and their costs are incorporated into water bills. This has led to sharp price increases (doubling or more over time) and can cause secondary sustainability problems – like large consumers exiting the public utility, creating an imbalance in the water services budget (e.g., in Wallonia); or like the "water poor" phenomenon. Some consumer NGOs even wrongly perceived the price increase as due to unchecked profits of private operators and gave credit to remunicipalisation projects. Yet there is little chance for WSS to be reintegrated into the general budgets of local or regional authorities, and their financial autonomy will probably remain, even after remunicipalisation, and even in Italy!

In France, the levies of the *agences de l'eau* are directly superimposed on water bills, and the available money partly returns to WSS authorities' projects that reduce pollution discharges or abstractions. The two levies (pollution and abstraction) could be considered as proxies for environmental costs and resource costs; while the mutualisation of the levies allows provision of cheap money to local authorities. Thus, the *agences* can be considered not only as mutual savings banks, but also as participating in the general commodification of WSS services.

The financial autonomy of WSS services was indeed achieved in France and the United Kingdom through various forms of privatisation or delegation to private companies, which have always operated water services commercially. But it is also the case in other continental European countries, where cities or regional governments have broadly reorganised their WSS services but have kept them as public utilities. It is less the case in Mediterranean countries, where climate variability and periods of authoritarian governments have led to regional/national bulk water production and transport paid by public money. Wastewater services' investments are still subsidised, so that water prices paid by consumers remain low, and the push towards full-cost pricing and commercialisation is all the more resisted by local populations.

WSS services end up being organised as a complex array of formulas involving direct public control, or financial autonomy, delegation to the private sector, publicly owned private companies, multi-utility companies, and the like, all aimed at interfacing with other policies in an efficient and coordinated way. Economies of scale and scope seem to be the main driver for vertical or horizontal integration. And indeed, upscaling is often adopted for the sake of averaging out the financial needs and the growing risks due to climate change.

This is clearly the case in France, where extreme fragmentation became unsustainable in the face of European directives. Even the *agences de l'eau* refused to subsidise separately the small projects multiplying in rural areas, and they preferred to pool their aid at county level. Before the territorial reforms of 2014–2015, there was a trend in some *départements* to get involved in consolidation processes (Barbier and Michon 2010), but it did not take the shape of a separation between water and wastewater plants operations at the supralocal level, and systems operation at the local level, as in Portugal for instance. There is a "reverse case" in the Vendée (western France), where all the rural municipalities plus coastal resorts[8] gathered into a single distribution system but left the water production units in the hands of each local utility. Mutualising the water distribution harmonised the financing of a county-wide network and secured the provision of water on the coast in summer. Supralocal management of drinking water production was not needed.

In any case the evolutions remain slow, and the frequent preservation of unbundled and efficient water services, tend to suggest that no "one solution fits all."

6.5 Interactions between WSS and water resources

In previous sections the change in management scales, or integration/unbundling of WSS services, was analysed chiefly in terms of economic rationalisation. We presented a wide array of institutional arrangements and governance, but hardly took into account water resource allocation issues. Indeed, the merit of sanitary engineering, with the invention of water potabilisation and sewage treatment, was to liberate cities and their services from water resources–related constraints, particularly in densely populated temperate areas where water quality was the chief problem.

However, this technology-based model is increasingly being questioned. First, making drinking water production systems overly complex results in the following loop: criteria → monitoring → chemical treatment → new dangerous substances → more criteria. To avoid the costly race for quality, the water industry (e.g., in Germany) prefers to look for naturally clean water. Second, large cities impact water resources not only with wastewater, but also with urban runoff, which is polluted and can create problems like local flooding and the choking of rivers. Land-use-based management and urban hydrology (peak runoff storage) might not only help to reduce the reliance upon infrastructure, but also turn a nuisance into a resource (groundwater recharge with stormwater). In any case, the WFD proposes to derive all water policy from the primary task to recover the aquatic environment quality, and, since urban water is the one that gives the greatest value to the use of the resource, it is particularly concerned with this recovery.

The economic dimension of this issue is covered by article 9 of the WFD, which recommends that water users should pay as close as possible to the full cost of the services rendered by water, including resource and environmental costs, on top of O&M and investment depreciation. It can be assumed that resource or opportunity cost relates to a potential rivalry for quantities of water, while the environmental cost relates to water quality problems. To illustrate this, one could say that if there is not enough water for its users in a given territory, economists would recommend allocating it primarily to the user giving it its highest value (the water services), but that other users should be compensated (which is difficult to assess); and if a city discharges in a river pollution loads that impair downstream uses, it should either pay damages to them or alternatively invest in sewage works in such a way as to reduce its discharge to a point considered acceptable by downstream users. Altogether this could add up to a much higher price than the full internal cost, depending on geographic conditions.

Two issues are raised when implementing this article 9: Should the full cost be covered by the tariff or, following the OECD's suggestion, could it be allocated between the 3 Ts, tariffs taxes. and transfers? And if yes, which territorial institutions would be the best fit to manage the budgets generated from the 3 Ts (Barraqué 2020)?

Full-cost recovery according to article 9 of the WFD, even in the richest European states, should not necessarily be done through the sole tariff increases. If it is a service fee paid in water bills, then the money should be returned to improving the WSS public services. But part of water policy does not correspond to a consumer service and should then be covered through taxes. And wherever water is managed as a common pool resource, then the institution in charge should bring its users in a community following the 8 Ostromian rules for a sustainable management, and, in particular, allocating the cost of the policy equitably between categories of users. It could give way to payments for environmental services between water resource users, which does correspond to the third T of OECD transfers.

6.6 Conclusion

This chapter has aimed to illustrate the complexity of the many arrangements found in cities of several developed countries to improve WSS sustainability. But perhaps the most important is that such arrangements are now beyond the public vs. private debate that emerged in the 1990s, opposing the World Bank and neoliberal economists to alter-globalist movements. In particular, the institutional diversity of water services in Europe resulted not only in a vast array of public–private arrangements, but also the choices have usually been combined with other issues: local level provision vs. concentration/centralisation at upper scales, integration vs. unbundling, with diverse new links between WSS services and water resources management.

In the end, privatisation and water marketing remain quite limited in the countries reviewed in this chapter (apart from California and the Murray-Darling Basin in Australia). But this does not mean water is to be considered a pure public good to be managed by administrations either. Water as a resource is still frequently considered a common pool resource, resulting in the survival of community-type institutions, functioning under equitable principles. Conversely, water as a service grew out of the other form of impure public goods, defined by R. Musgrave and P. Samuelson: club goods. As an invention of the Enlightenment, the club replaces community self-obligation and equity rules with new rules based on freedom and equality among its members.[9] Public water supplies are a special kind of club, since they are not closed but potentially open to all, and, in addition, they must follow the rules of public service: universality, continuity, equality, and adaptability. They are also supposed to be funded by their users through some form of consumerisation (i.e., billing). This makes a serious difference in developing countries where water services are not yet complete: lack of financial capital delays the universalisation of good WSS services. For all those who are ill- or not connected, their relationship to water is to a resource, not to a public service, and, in most cultures, access to water resources for domestic uses should be free. This attitude explains the frequent low level of social acceptance to pay for water services, which, in turn, prevents improvements to them. In developed countries, a great deal of complexity, and sometimes misunderstanding, results from the coexistence of the two forms of impure public goods (common pool and clubs) in the articulation of water resources and WSS policies. Alter-globalist NGOs seeking a drastic improvement of WSS in developing countries unfortunately often fail to see how different the situation is in developed nations in terms of, first, cost coverage by tariffs. Additionally, national governments, even in federal countries, still play a role in the regulation of water resources policies and, more recently, in the protection of nature and of water as natural capital. This may lead some citizens to claim water should be a pure public good (e.g., repeatedly by the French Communist Party).

As in other public policies, a need is growing for coordination between various types of institutions related to water services – territorial communities, club goods, market institutions, and the state – resulting in better coordinated multilevel governance. Still, how to achieve such coordination is not obvious, as these policy mixes also call for a change in government culture. The failure of the socialist model, as it was developed in Soviet regimes, parallels the questioning of the national statism policies supported earlier by the World Bank, in which economic development is based upon the mobilisation of water resources in multipurpose projects. But the alternative is not just markets. Water management, in particular, illustrates the notion of multilevel governance. Here lies the opportunity for a water culture that would make room for all different kinds of water institutions, and the capacity to combine them, as a new form of social capital. Indeed, the need is clear to discuss which part of the articulation between water services and water resources could be best managed by common-pool resource institutions run under participative democracy.

Notes

1 In France and even more in Germany, supporters of remunicipalisation of water services wrongly argue that water, being a common good, cannot be delivered by a private company; and yet once returned in direct public procurement, water services remain utilities and go on being financed by water bills. While water resources can be common pool, water services are in fact particular club goods subject to the rules of public service (e.g., universal access, continuity, equal treatment).

2 The Seine county gathered together the four present-day *départements* – Paris, Hauts-de-Seine, Seine-Saint-Denis, and Val-de-Marne – until it was reformed in 1968.

3 In Magdeburg, the capital city of Sachsen-Anhalt, where the population has dropped by 20% since reunification, the area supplied by the *Stadtwerk* is steadily losing consumers, so that by 2050 there will be only 540,000 people compared to 800,000 today. The technical impacts of the subsequent decrease in water consumption are already serious. In some neighbourhoods, water stays for more than 25 days in the pipes before being used (Herdt 2009)! The solution is increased consolidation plus participation of the regional electricity company in the capital of the *Stadtwerk*.

4 However, after World War II, this state enterprise was typically decentralised into a joint board gathering together the served municipalities.

5 See https://www.seqwater.com.au/

6 Adjusted for inflation (2008 → 2021) the investment would reach AU$11,560 million, or US$8,260 million.

7 For example, the work by Yoram Cohen on decentralised systems in the Institute of Environment and Sustainability at UCLA.

8 In the beginning, the three largest cities did not join: at the time they were under delegation contracts and their private operators could offer a cheaper water price.

9 Indeed, citizens are not obliged to subscribe to the public water service, and in some countries, there is, in fact, no obligation to extend the service to all citizens (such as hamlets far from the network). The tariff is the same for the entire category, and those who do not pay are temporarily disconnected. Until recently, few WSS operators considered the issue of the "water poor" since water was affordable to virtually all the population. Major price increases have given rise to a new issue that brings the developed countries closer to developing nations, and these increases have created an unprecedented level of complexity.

References

Bakker, K. J. (2003). From public to private to … mutual? Restructuring water supply governance in England and Wales. *Geoforum*, *34*(3), 359–374.

Barbier, R. & S. Michon (2010). Gestion de l'eau destinée à la consommation humaine: enquête sur l'action des départements. Retrieved April 23, 2012 from http://aquadep.irstea.fr/wp-content/uploads/2012/04/Aquadep-Synthese-Enquete-CG.pdf.

Barraqué, B. (1992). Water management in Europe: beyond the privatization debate. *Flux*, 7, 7–26. DOI 10.3406/flux.1992.1165.

Barraqué, B. (2003). Past and future sustainability of water policies in Europe. *Natural Resources Forum*, *27*, 200–211.

Barraqué B. (2009). Managing potable water supplies. Chapter 10 in Ferrier, R. C. & Jenkins, A. (eds.), *Handbook of catchment management* (pp 235–252). Wiley Interscience, Sept. 2009.

Barraqué B. (2020). Full cost recovery of water services and the 3 T's of OECD. *Utilities Policy*, *62*, 100981, 8 p. https://www.sciencedirect.com/science/article/pii/S0957178719303340

Barraqué, B., Formiga Johnsson, R. M., & Nogueira de Paiva Britto, A. L. (2008). The development of water services and their interaction with water resources in European and Brazilian cities. *Hydrology and Earth System Sciences, 12*(4), 1153–1164.

Brady J. & Gray N. F. (2013). Analysis of water pricing in Ireland and recommendations towards a more efficient water sector. *Water Policy, 15*(3), 435–457.

Dupuy G. (ed.) (1984). *Réseaux techniques urbains: Les annales de la recherche urbaine* (pp. 23–24). Dunod (ISBN/ISSN 0180930).

Graham, S. & Marvin, S. (2001). *Splintering urbanism. Networked infrastructures, technological mobilities and the urban condition*. Routledge.

Guerin Schneider, L., Nakhla, M., Grand d'Esnon, A., & Baudot, B. (2002). Gestion et organisation des services publics d'eau en Europe. Cahiers de recherche 19. Paris: Ecole des mines de Paris, Centre de gestion scientifique.

Herdt H. (2009). *Städtische Werke Magdeburg*. Retrieved February 13, 2012, from http://eau3e.hypotheses.org/files/2009/11/ATHENS_Helmut_Herdt.pdf

Kraemer, R. A., Pielen, B. & De Roo, C. (2007). Regulation of water supply in Germany. *CESifo DICE Report, 5*, 21–26.

Kuks, S. (2010, November). *Water management by the regional water authorities in the Netherlands*. Retrieved February 13, 2012, from http://eau3e.hypotheses.org/files/2010/11/Athens_Kuks_NL_10.pdf

Loijenga, H. (2009, November). *Water Governance in the Netherlands*. Retrieved February 13, 2012, from http://eau3e.hypotheses.org/files/2009/11/ATHENS_Henk_Loijenga.pdf.

Mangano, A. (2010, November). *Water services in Italy*. Retrieved February 13, 2012, from http://eau3e.hypotheses.org/fi les/2010/11/ATHENS_Mangano_10.pdf

Massarutto, A. (2011). Urban water reform in Italy: A live bomb behind outward unanimity. In Barraqué, B. (Ed), *Urban water conflicts* (pp. 247–268). Taylor & Francis.

Megdal, Sharon B. (2012) The role of the public and private sectors in water provision in Arizona, USA. *Water International, 37*(2), Routledge, 156–168.

Moss, T. (2008). 'Cold spots' of urban infrastructure: shrinking processes in Eastern Germany and the modern infrastructural ideal. *International Journal of Urban and Regional Research, 32*, 436–451.

NetWORKS (2008 September). Minute. In Minute of the International Workshop held in Berlin on "Sustainable Urban Water Infrastructure: Possibilities of Adaptation and Transformation".

Pezon, Ch. (2000). *Le service d'eau potable en France de 1850 à 1995* (441 pp). Presses du Centre de Recherche en Economie et Management.

Pezon Ch. (2006). *Intercommunalité et durabilité des services d'eau et d'assainissement. Etudes de cas français, italiens, portugais* (136 pp). Report for the Programme Politiques Territoriales et Développement Durable, MEDD.

Pezon, Ch. & Canneva, G. (2009). Petites communes et opérateurs privés: généalogie du modèle français de gestion des services d'eau potable. *Espaces et Sociétés, Usages et régulations des eaux urbaines, 4*, 21–38.

Saunders, P. (1983). *The "Regional State": A review of the literature and agenda for research*. Urban and Regional Studies paper, University of Sussex.

Walker, A. (2009). *The independent review of charging for household water and sewerage services*. Final Report. Published by the Department for Environment, Food and Rural Affairs. Retrieved April 23, 2012 from http://www.defra.gov.uk/publications/2011/12/06/walker-review/

PART II

Technical and historical aspects of wastewater systems

7

CONVENTIONAL SYSTEMS FOR URBAN SANITATION AND WASTEWATER MANAGEMENT IN MIDDLE- AND HIGH-INCOME COUNTRIES

Jenifer R. McConville

7.1 Introduction

Sanitation refers to the provision of infrastructure and services for the safe management of human excreta and wastewater. Lack of safe management of excreta and wastewater leads to disease and environmental degradation. The World Health Organization defines safely managed sanitation as the use of improved facilities (e.g., designed to hygienically separate excreta from human contact), which are not shared with other households, and where excreta are safely disposed in situ or transported and treated off-site. However, provision of safely managed sanitation remains a significant challenge around the world, even within the urban context. Only 47% of the global urban population in 2017 had access to safely managed sanitation (WHO/UNICEF Joint Monitoring Programme, 2021). During the last decades, the number of people accessing safe sanitation has steadily increased from 1.7 billion people in 2000 to 3.4 billion people in 2017 (ibid.). However, due to population increases, over 4 billion people still lack access to safe sanitation. On average, high-income countries safely treat about 70% of the generated wastewater, while in middle-income countries only 28% to 38% is treated (Sato et al., 2013). In low-income countries, data are often unavailable, but average treatment rates are less than 10% (ibid.). Thus, urban sanitation systems are still in a process of development.

The conventional sanitation system that has evolved in middle- and high-income countries is a waterborne system in which domestic and industrial waste is flushed away from the point of generation and generally treated off-site. Waterborne sanitation has a long history with continuous progress being made in methods for collection, transport, and treatment (refer to Chapter 12 for a history of technology change in wastewater management). While conventional systems all manage waterborne waste flows, i.e., wastewater, there are variations in how these systems are designed and the technologies that are used. Factors that affect the design of urban sanitation systems include urban density, land use, typology, and the composition of the wastewater. For example, housing density and geological typology will affect how collection pipes or sewers can

DOI: 10.4324/9781003057574-10

be laid and, thus, the design of the collection system. The composition of the wastewater, including the source of the wastewater, will dictate what treatment technology is appropriate. Finally, the proximity and characteristics of the discharge site for the treated water and solids dictate the choice of treatment technologies and possibilities for reuse and/or disposal.

This chapter provides an overview of conventional urban sanitation and wastewater management systems in middle- and high-income countries. It explains the composition of wastewater that is managed in these systems, including key parameters to be controlled in treatment. Next, it covers the sanitation service chain, including an overview of conventional technologies used in urban sanitation. Finally, it discusses the selection and design of sustainable urban sanitation systems.

7.2 Wastewater composition and the key parameters to be controlled

To properly design sanitation systems, it is important to understand what is being managed and why. Urban sanitation systems typically manage a number of different flows of wastewater within the urban area. Wastewater is the collective term applied to the used water and other liquid waste fractions, including excreta, generated by a community. There are two main sources of wastewater – domestic and industrial. Domestic wastewater includes human excreta (urine and faeces) and water that has been used for washing and cleaning, referred to as greywater. Industrial wastewater is the discarded water from industrial processing of, e.g., food products, textiles, paper, metals, or chemicals. Depending on the industry, the qualities of industrial wastewater can be vastly different. Depending on the design of the urban sanitation and drainage systems, stormwater may also be managed with the wastewater.

Sanitation systems play a critical role in protecting health and the environment. Wastewater is contaminated with a variety of microbial pathogens and other organic and inorganic pollutants, which can contribute to environmental and public health problems if they are not controlled through wastewater treatment. Designing and implementing appropriate wastewater treatment systems requires a basic understanding of wastewater composition and the key parameters to be controlled. Concentrations of pathogens, nutrients, and solids can vary depending on the nutrition habits, sanitation practices, and socioeconomic situation of the population. The majority of pathogen contaminants are found in faecal matter, while a significant portion of the nutrient loading is found in urine.

There are five main categories of contaminants in urban sanitation systems (Table 7.1). These contaminants primarily: (i) affect human health through disease transmission (pathogens and parasites), (ii) cause eutrophication and degradation of water bodies (suspended solids, organic matter, and nutrients), or (iii) result in human or eco-toxicity (heavy metals and emerging pollutants). Traditionally, sanitation systems were first designed to remove suspended solids and pathogens. More recently, attention has focused on the removal of nutrients, primarily phosphorus and sometimes nitrogen, as well as the reduction of heavy metals. A growing number of emerging pollutants and chemicals end up in our wastewater systems, including pharmaceuticals, oxidants, organics chemicals, PFAS, and micro-plastics. The impacts of these chemicals on humans or the environment are often unknown, especially when there are multiple different compounds present (cocktail effect). However, research has increasingly linked many of these compounds to toxic effects on both aquatic and terrestrial ecosystems (Tran et al., 2018). Indeed, a number of conventions and legislation monitor and control these chemicals, such as the global Stockholm Convention on Persistent Organic Pollutants (POPs) and the European REACH legislation on chemicals. Removal of these chemicals from wastewater is difficult, and current technology is insufficient, although development is advancing rapidly. Source control measures

Table 7.1 The main categories of contaminants in urban domestic wastewater, including their primary impacts, sources, and options for removal. Note since industrial wastewaters are so variable, they are not included as potential sources in this list

Contaminant	*Impact*	*Primary source*	*Removal options*
Pathogens & parasites	Disease transmission	Excreta (primarily faeces)	Storage/detention time Disinfection (UV, ozone)
Suspended solids & organic matter	Eutrophication of waterbodies	Excreta, food particles, fats & oil, soil particles (primarily from greywater), stormwater	Primary treatment (sedimentation) Secondary treatment (biological degradation)
Nutrients	Eutrophication of waterbodies	Faeces, urine, food particles	Tertiary treatment
Heavy metals	Toxicity	Greywater, excreta, stormwater	Sedimentation, chemical precipitation, ion-exchange, bio-sorption. Source control
Emerging pollutants (pharmaceuticals, organic compounds, micro-plastics, etc.)	Toxicity	Depends on pollutant and how it is used	Oxidation, ion-exchange, membranes, activated carbon. Source control

to avoid contaminants from entering sanitation systems have been the most widely applied approach so far to reduce heavy metals and emerging pollutants.

The primary source of pathogens and parasites is human faeces. Human faeces contain an abundance of bacteria and viruses. One gram of faeces can contain 10 million viruses, 1,000,000 bacteria, 1,000 parasite cysts, and 100 parasite eggs. These microscopic organisms cause a variety of diseases that are responsible for over 13 million deaths a year. Coliform bacteria are the most abundant faecal microorganism, comprising 20–33% of the total weight of faeces, and they are therefore commonly used as indicators for the biological strength of the wastewater. Containment and treatment of faeces is the most effective way to reduce the spread of disease. Faeces can also contaminate other wastewater fractions with pathogens, such as cross-contamination of urine in a urine-diverting toilet. Similarly, greywater may contain traces of pathogens, e.g., from washing diapers or vomit. It is also possible for pathogens that originate from food waste to enter the sanitation system. Thus, all wastewater treatment should include the removal of pathogens.

Management of eutrophication impacts is more complex since it involves the management of suspended solids, organic matter, and nutrients. The environmental impacts of wastewater are closely linked to the chemical composition, specifically the concentrations of nitrogen (N), phosphorus (P), and dissolved organic matter, i.e., biochemical oxygen demand (BOD). Nitrogen and phosphorus are key nutrients for plants and are therefore the main causes of eutrophication. BOD is a measure of the amount of dissolved oxygen that will be consumed during microbial decomposition of the organic matter in the wastewater; it is also measured as chemical oxygen demand (COD). Discharging wastewater with high levels of N, P, and BOD/COD into surface waters can result in eutrophication and a corresponding increase in algae and plant growth. The subsequent high organic decomposition rates in eutrophic waters lead to oxygen depletion that endangers fish and other aquatic animals, ultimately resulting in fish kills and water quality degradation. This degradation of water quality and loss of aquatic life can have significant consequences for the natural environment, the biodiversity, and the communities that depend on it for their economic livelihoods.

The organic matter entering the wastewater system comes from faeces, anal cleansing products (e.g., toilet paper), food waste, and dirt from washing. The fraction of solids to liquid will determine possibilities for transport of the wastewater and treatment rates at the wastewater treatment plants. Solid concentrations are highly variable depending on water usage, anal cleansing practices and toilet design, and the collection system (e.g., leakages and additional inputs from storm or groundwater).

The majority of nutrients (e.g., nitrogen, phosphorous, potassium, sulphur) found in domestic wastewater come from excreta, particularly urine. Depending on diet, urine from one person during one year (approximately 300 to 550 L) contains on average approximately 3 to 4 kg of nitrogen (N), 0.3 kg of phosphorus (P), and 0.7 kg of potassium (K) (Rose et al., 2015). In contrast, the faeces from one person for one year contain on average about 0.6 kg of N, 0.2 kg of P, and 0.3 kg of K (ibid.). Greywater generally has lower nutrient concentrations, but higher concentrations of solids and organic matter due to the presence of food wastes. Approximately 40% of the BOD loading from domestic wastewater comes from greywater, but only 20% of the phosphorus and 10% of the nitrogen (Kujawa-Roeleveld and Zeeman, 2006). However, the use of detergents can contribute 10% to 30% of the total P loading to the wastewater system depending on the location (WHO, 2006). Many countries have now banned the use of P in detergents as a way to reduce nutrient emissions. The volumes of greywater produced are also much larger than the volumes of excreta produced, accounting for approximately 65% of the wastewater produced in households with flush toilets. Thus, separation of the excreta from the greywater can significantly reduce nutrient loading in the wastewater to be treated. Removal of suspended solids will still be necessary to reduce eutrophication, but such treatment is simpler and less energy intensive than nutrient removal.

Wastewater can contain contaminants that are toxic to humans, animals, and ecosystems if they are not removed. These toxic compounds fall into two general categories: heavy metals and emerging pollutants that include a wide variety of chemicals. The most common toxic metals that end up in wastewater are arsenic (As), lead (Pb), mercury (Hg), cadmium (Cd), chromium (Cr), copper (Cu), nickel (Ni), silver (Ag), and zinc (Zn). The sources of heavy metals vary depending on the metal. For example, Cd and Ni are emitted from stainless steel and can thus end up in washing water and excreta (Sörme and Lagerkvist, 2002). Other metals are in our drinking water, such as As, Cu, Ni, and Zn. Zinc, copper, and cadmium are also common in food. The plumbing system can also contribute to metals such as Cu, Zn, and Pb. Stormwater runoff from roofs, roads, and buildings can also be a significant source of heavy metals. In particular, runoff from roads contains heavy metals from fuel exhaust and wear from brakes and tyres (Gupta, 2020). A study in Sweden found that heavy metal from excreta contributes less than 10% of total metal loading from households, with the exception of zinc (45% from excreta) and mercury (50% from excreta) (Jönsson et al., 2005). The most dominant emission source for mercury is amalgam in teeth (Sörme and Lagerkvist, 2002).

A growing number of organic chemicals are in use in society today, many of which can enter our sanitation systems. The database of the European Chemical Agency contains 23,117 unique substances, few of which are regularly monitored in the environment. Organic pollutants that are frequently found in wastewater include dyes, petroleum, surfactants, pesticides, polycyclic aromatic hydrocarbons (PAHs), polychlorinated biphenyls (PCBs), and pharmaceuticals. These contaminants can enter the sanitation system through (1) industrial effluent from production processes (e.g., manufacturing of active ingredients, production, and packaging), (2) release via human excreta after ingestion (e.g., medications or pesticide residues on food), (3) residues released to greywater through washing (e.g., dyes, surfactants, and pesticides), and (4) disposal of the leftover medicine and other wastes into the sanitation systems (e.g., flushing of drugs, medicines, and other products down the toilet). Many of these substances are designed to be persistent in the environment and are thus difficult to break down and remove. The toxicity effects of many of these substances are unclear. There are currently 211 substances classified as very high concern by the European Union. The greater concern however is the potential cocktail effect of mixing multiple substances in the environment. The complex interactions from the multitude of chemicals that we are releasing from society into our drinking water sources and the environment are just beginning to be understood.

Another important characteristic to consider for design purposes is the total volume of wastewater to be managed. The production rate of wastewater within a community is closely linked to water usage. On average 60% to 90% of per capita water consumption is collected as wastewater, depending on how much water is used for lawn and garden irrigation (Metcalf and Eddy, 2003). However, the actual amount of wastewater that arrives at the wastewater treatment plant also depends on the amount of water that leaks into or out of the collection systems. Inflows or filtration into collection systems may come from direct sources, such as stormwater, or indirect sources, such as defective pipes, pipe connection, and leaky manholes. Combined sewer systems that are designed to collect stormwater as well as wastewater will increase the amount of water needing treatment. High groundwater tables and cement mortar or brick joints in the collection system will increase inflows to the system. There is a wide variation in the estimated contribution of infiltration to total flows, with estimates ranging from 0 % in newly replaced sewer networks to 80% in wet weather. Conversely, low water tables and leaky collection systems can lead to exfiltration with wastewater leaking into the ground. Again, there is a variation for exfiltration from sewers, with leakages of 0–13% of dry weather flow being reported (Rutsch et al., 2006). The potential for infiltration or exfiltration needs to be considered on a case-by-case basis. Design of wastewater treatment systems should consider the area serviced by the collection systems, age and material composition of the collection system, variability of the groundwater table, and frequency of rain events when estimating flowrates and concentration of contaminants.

7.2.1 Sanitation service chain

In an urban context, safe management of excreta and wastewater means that the flows described above must be transported and treated off-site. Thus, urban sanitation will require a service chain that manages sanitation products from the point of generation through treatment and reuse/disposal (Figure 7.1). This section presents the dominant technologies that are used in middle- and high-income countries for each step. The text below highlights the different components of the service chain and the various technologies that are commonly used. It also discusses the advantages and disadvantages of these technologies.

7.2.2 Capture

The water flush toilet has become the golden standard in the majority of high- and middle-income countries. While there are drawbacks to this toilet, managers of urban water systems need to understand why this technology has been so successful. Modern cistern flush toilets use between 3 and 6 litres per flush, while older models from before the 1990s use 12 or more litres per flush. Considering that the average person flushes the toilet 5-6 times a day, means that we

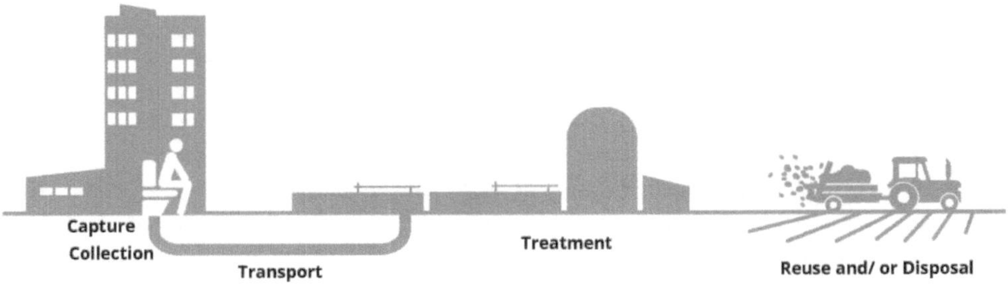

Figure 7.1 The sanitation service chain including capture, collection, transport, treatment, and reuse and/ or disposal of wastewater. (Illustration: Annika Nordin, SLU)

are flushing around 20–30 litres of water per person each day just to flush the toilet. Given global water shortages and pollution pressure on our water bodies, using so much water to transport our excreta (1–2 L per person/day), may seem like a bad idea. However, the flush toilet provides a user experience that is hard to beat. The flush water transports the excreta out of the household simply by pushing a button. Porcelain or other finishes on the toilet, in combination with water left in the bowl, make cleaning very easy. There is little smell and users do not have to see or handle the excreta. While dry toilets have been tried in urban areas, it is difficult to compete with the flush toilet experience in terms of cleanliness and convenience. That said, there are urine-diverting and ultra, low-flush toilets available today that use less than 1 litre per flush. Vacuum toilets that use 0.6 L of water or less are also gaining in popularity. Keep in mind that the acceptability of a sanitation system will largely hinge on how users experience the toilet. It will need to provide the same qualities of comfort, convenience, and cleanliness that are achieved today by the flush toilet.

In addition to toilets, greywater is captured from sinks, showers, baths, and washing machines through drains and pipes, typically merging all household pipes together into a single pipe exiting the building. In a majority of cases, the blackwater (excreta and flush water) from the toilet joins the greywater in the household piping system so that a mixed wastewater fraction leaves the building in a single pipe. In the urban area, industrial wastewater is collected from drains, sinks, and catchments before being sent to collection systems along with domestic wastewater from households (Figure 7.2).

The conventional urban sanitation system is structured around the capture of mixed wastewater from multiple sources. However, the increasing trend in the sanitation sector towards seeing wastewater as a resource rather than a waste, has led to suggestions, and in some cases

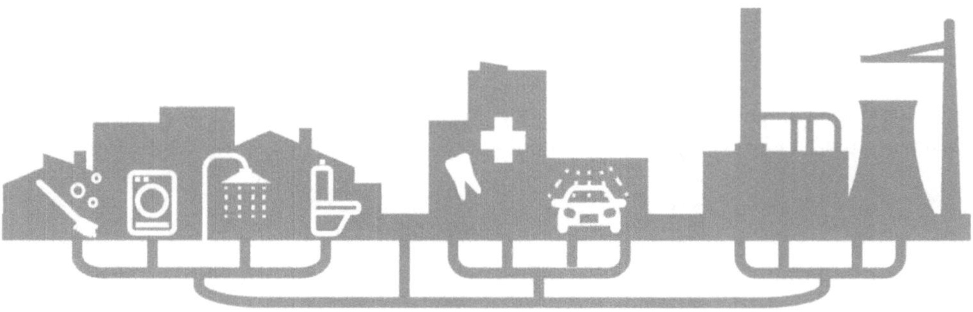

Figure 7.2 Capture of wastewater flows from different sources in the urban area, including urine and faeces in the toilet, greywater from domestic washing, and industrial wastewater. (Illustration: Annika Nordin, SLU).

implementation, of separate collection systems for different wastewater fractions. Separate greywater collection can facilitate heat and water recovery, as it is not mixed with cold flush water nor the high pathogen loading from blackwater. Separate collection of urine and/or faeces can facilitate nutrient removal and recovery as the majority of nutrients are in excreta. Separate collection of excreta can also make treatment of pathogens and pharmaceuticals easier since the volumes requiring advanced treatment will be only a fraction of the total wastewater volumes.

7.2.3 Collection and transport

Once the wastewater from the toilet and other appliances has been captured in the household, it is either led from the household directly into a collection system for transport to a treatment plant or first sent to an on-site storage unit. On-site store units include sealed pits or septic tanks that temporarily collect wastewater before further transportation to treatment. In dense urban areas, infiltration of wastewater is difficult and not recommended due to risks associated with groundwater contamination. However, sealed on-site storage units can reduce the loading of solids, oils, and grease to the collection system, thus allowing for simplified designs and maintenance of the collection system. In addition, on-site septic tanks and improved septic tanks, such as anaerobic filters or anaerobic baffled reactors (ABR), provide biological treatment, thus reducing loading on the treatment plant. The accumulated sludge in these units needs to be regularly removed and taken for further treatment. In some parts of the world, it is common for urban households to have on-site storage units prior to the collection system (e.g., Asia and South America), while in other parts of the world it is not (e.g., Europe and North America).

In high- and middle-income countries, collection and transportation of wastewater is primarily done using piped sewer systems. There are a number of ways to design a sewer system, including having separate pipes for different wastewater fractions or a single system for mixed wastewater (most common). There are also a variety of different types of sewer networks, including conventional gravity, simplified, solids-free, and pressurized sewers (Figure 7.3).

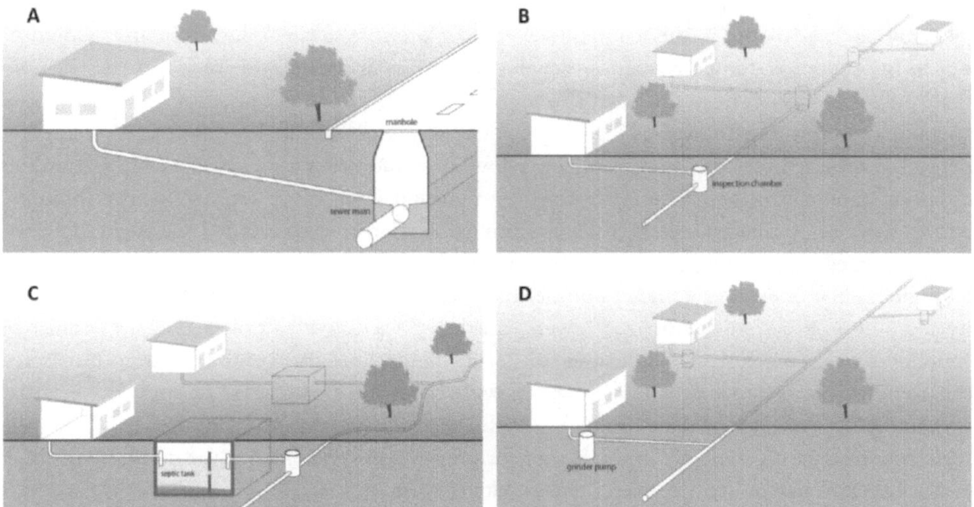

Figure 7.3 Examples of collection and transportation systems for urban wastewater: (A) gravity sewers, (B) simplified sewers, (C) solids-free sewers and (D) pressurized sewers. (Illustration adapted from Tilley et al., 2014).

While little data are available regarding the type of sewers implemented, it may be assumed that a majority are gravity sewers. Gravity sewers are designed so that the wastewater flows via gravity downhill to a pumping station or treatment plant. Typically, the collection system is divided into branching networks that collect wastewater to the lowest local point, from which it is pumped to larger mains that cumulate at the wastewater treatment plant. Gravity sewers are designed with a minimum velocity in the pipes so that they are self-cleaning. Thus, they require a minimum amount of water and downhill gradient. In many cities, stormwater from urban runoff is also collected in the sewer network. A major advantage of the gravity sewers is that they can be designed to minimize the need for energy inputs for the transport of wastewater. However, to get the gravity flow, parts of the network may require deep excavation. Thus, construction costs for gravity sewers can be quite high. The gravity sewer system also requires large volumes of water to prevent deposition of solids and blockages. Depending on the terrain and expanse of the urban area being served by the sewer network, a number of pumping stations will be required, adding to operational costs.

Simplified versions of gravity sewers have been developed to reduce costs, especially in areas that are difficult and costly to excavate and/or with fluctuating water usage. Simplified sewers, also known as small-bore or condominial sewers, use small-diameter pipes that can be constructed at a flatter gradient than gravity sewers. An inceptor tank at each household connection is often used in this model to capture solids and grease, thus reducing blockages and easing maintenance. A similar system is a solids-free sewer in which a small-diameter pipe network is used to transport effluent water from septic tanks or other on-site containment systems at the household level. In general, these alternative systems are cheaper to install, but more expensive to maintain since they are more prone to blockages. Similar to conventional gravity sewers, simplified sewers pose a risk for wastewater leaking from the pipes into the groundwater, or conversely groundwater infiltrating into the pipes, increasing loading at the treatment plant. Leakages in any of these systems are difficult to identify.

Other alternatives to gravity sewers use pressurized pipes. Low-pressure sewers can also be installed when the terrain is variable, difficult to excavate, or if the location is below the mainline sewer elevation. These systems have a grinder pump installed at each household connection that "pushes" the wastewater to the mainline sewer. Low-pressure sewers can thus be installed without deep excavation. Similarly, vacuum sewer systems use air pressure to create a partial vacuum that "pulls" wastewater from the point of generation to a central vacuum station. An advantage of vacuum and pressurized sewers is that they operate well with minimal water in the system, thus being the most appropriate options for ultra-low flush and water-saving toilets. They also offer environmental advantages since infiltration from groundwater into these systems is minimized, and, in the case of vacuum sewers, leakages from exfiltration are also avoided. Construction costs for pressurized systems are generally lower than gravity systems. However, maintenance costs can be higher. They often have a higher number of pumps than gravity systems and rely on electricity for pump operation. Pressurized sewers can thus be sensitive to power outages. Similar to simplified sewer, pressurized system use smaller diameter pipes that can be prone to blockages, thus requiring more maintenance. However, with improved fault-finder sensors and automation in the network, operational challenges with alternative sewers can be minimized. Life cycle replacement costs are expected to be higher for pressurized systems since the components have lower life expectancies than conventional gravity systems.

In reality, the sewer networks in many cities may be combinations of these different sewer types. In many older cities, the existing gravity sewers are reaching the end of their life expectancy and will need replacement. Expansion and renewal of urban sewer systems offer new opportunities for replacement and upgrading of our systems for the collection and transport of excreta and wastewater. New transportation systems may explore a combination of both pipe and vehicular transport. Previous designs for collection systems focused on adapting them to the physical constraints of the terrain. Future designs should also consider how to minimize leakages, especially in environmental sensitive areas. Opportunities to design collection systems that optimize water-saving, energy, and nutrient recovery should also be considered. In the future, we will see cities with a variety of wastewater collection networks, perhaps designed as decentralized subsystems, adapted to different parts of the city (see Chapter 8).

7.2.4 Treatment

Conventional wastewater treatment uses physical, biological, and chemical processes to clean the incoming wastewater (Figure 7.4). The design and processes used in treatment have developed over time in response to increasingly strict treatment standards. Treatment plants focused initially on removing suspended solids and, to some degree, pathogens. These treatment processes are referred to as primary treatment. Secondary treatment processes developed in the 1960s in response to the need for further reduction of dissolved organic matter. As our understanding of the eutrophication impacts from nutrient loading increased so did the need for increased nutrient removal. Tertiary treatment processes emerged for phosphorous and nitrogen removal in the 1970s with wide-scale implementation in the 1990s. The majority of wastewater treatment plants in high-income countries today have at least secondary, if not tertiary treatment. Note, however, that the level of treatment varies between countries. In some European countries, less than 50% of urban wastewater treatment passes through tertiary treatment, while in others it is nearly 100% (European Environment Agency, 2020). Wastewater treatment standards in many countries are guided by effluent discharge points, with stricter treatment standards applied to discharge in smaller and/or sensitive water bodies. Traditionally, it was assumed that the greater dilution factors of rivers and oceans would offset the need for treatment, thus allowing for lower treatment of wastewater discharged into these bodies. However, evidence is growing that aquatic environments, including oceans, have a limited capacity to absorb the nutrient, metal, and organic contaminants released in our wastewater discharges (Stark et al., 2016; Steffen et al., 2015).

Primary treatment aims to remove suspended solids and other large solid contaminants from the wastewater. Physical processes capture and sediment solids. Typically, the incoming wastewater flows through screens to capture any large objects that may have been flushed or washed into the collection system, e.g., rags, plastic bags, diapers, or large debris. The solids will need to

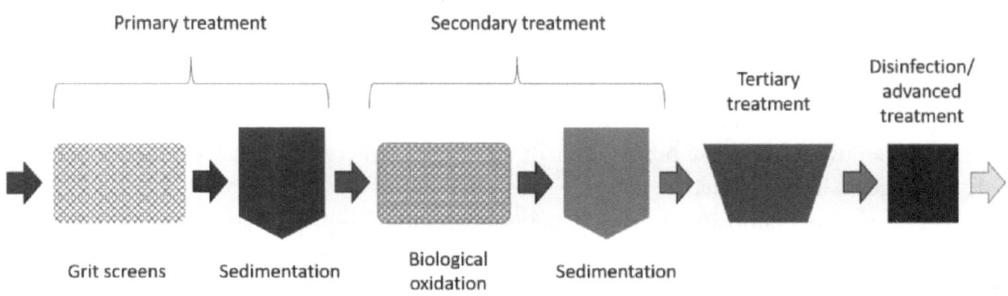

Figure 7.4 Typical configuration of a conventional wastewater treatment plant. It includes primary treatment for solids removal, secondary treatment to reduce organic loading, tertiary treatment for nutrient removal, and possibly a disinfection step for pathogen removal and/or advanced treatment for chemical contaminants.

be regularly removed from the screens, which can be done manually or mechanically. Following the screens, a grit chamber allows for the settling of heavy solids such as sand/gravel, seeds, coffee grounds, or larger food particles. Grit chambers will also need to be cleaned regularly, again using manual or mechanical means. Grit chambers are particularly important when the collection system also brings stormwater to the treatment plant, since the runoff from streets and fields may contain a lot of sand and gravel. Finally, the grit chamber is generally followed by a primary sedimentation tank or clarifier. Finer solids settle to the bottom of this tank and are removed mechanically from the bottom of the tank. Floating debris, such as grease and small plastics, can also be removed using surface skimming devices. Primary treatment is relatively simple to design, does not need advanced mechanical parts or energy inputs, or lots of space. However, primary treatment removes only large suspended solids, thus potential negative impacts on health, eutrophication, and toxicity still exist. Nevertheless, primary treatment is often a prerequisite for efficient functioning of subsequent treatment steps.

Secondary treatment aims to remove dissolved organic matter, thus reducing the potential for eutrophication. Biological processes in which microorganisms are cultured to consume the organic matter are the primary means used to achieve this. There are three main methods used for secondary treatment: biofiltration, aeration, and oxidation. In biofiltration, the wastewater is passed slowly over or through a filter material, such as a trickling filter, contact filter, or rotating biological contactor discs. Microorganisms grow on the surface of the filter material and consume the dissolved organic matter as it flows past. The most common aeration process is activated sludge treatment. Wastewater is mixed with a solution of microorganisms (generally obtained from internal recycling of sludge within the plant) in an aerated basin. Compressed air is injected into the basins to keep them aerated. Microorganisms, particularly bacteria, thrive in the aerated water, actively consuming organics. There are many variations of the activated sludge process, including oxidation ditches, sequencing batch reactors, and surface-aerated basins. Secondary sedimentation clarifiers are needed after biofiltration and aeration units to remove the microorganisms that have been grown. Oxidation ponds are also used for secondary treatment. In oxidation ponds, algae are cultured to produce the oxygen needed for the bacteria that consume the organics. Mechanical aerators may also be used to supplement the algae, thus reducing the size requirements of the ponds. Ponds are generally shallow and the penetration of sunshine has a sanitizing effect on pathogens, meaning a higher pathogen removal rate. Similar to biofilters and activated sludge, oxidation will require settling afterward to remove bacteria and filtration or chemical precipitation to remove algae.

Secondary treatment steps generally require skilled personnel for operation and maintenance. The biological processes are sensitive to changing conditions, such as incoming chemical spikes or large water flushes when the system is connected to stormwater collection. Operators need to be able to manage internal flows to optimize treatment. Secondary treatment units are associated with high capital costs for construction and high operating costs due to required energy inputs for mechanical parts and particularly for aeration units. Remember, secondary treatment is primarily designed to remove BOD and not nutrients. Although some nutrient reduction does occur, adjustments to the process and/or additional tertiary treatment steps are needed to reduce nutrients in the effluent.

Tertiary treatment aims to remove nutrients, primarily phosphorus and nitrogen. Chemical precipitation of phosphorus is an increasingly common practice, as is nitrification-denitrification for nitrogen removal (Ramasahayam et al., 2014). Effluent polishing using granular filter materials is also used to increase BOD and pathogen removal. This may be followed by disinfection using ultraviolet (UV) radiation, ozonation, or chlorination for pathogen destruction. However, both chlorination and ozonation can form toxic disinfection by-products, especially when there are high amounts of organic matter.

In addition to disinfection and tertiary treatment for microbial and nutrient removal, awareness is growing of the need for specialized treatment for contaminants of emerging concern, e.g., medicines, cleaning products, flame retards, and nanoparticles. Current technologies to remove these pollutants include carbon adsorption, oxidation processes, ozonation, UV radiation, and membrane filtration (Krzeminski et al., 2019; Rizzo et al., 2020). However, in 2016, only 6% of the big cities in Europe had more stringent treatment for chemicals other than nutrients (European Environment Agency, 2020).

The treatment methods described above are for conventional systems with mixed wastewater from domestic sources since that is currently the dominant means of treating urban wastewater. It is worth noting, however, that human excreta accounts for 60% of the organic matter, 90% of the nitrogen, and 80% of the phosphorus loading at the wastewater treatment plants, while only contributing 2% of the volumetric flow. Human excreta also contains a majority of the pharmaceutical residues. Separate collection and treatment of excreta would greatly reduce the amount of contaminants to be removed at the treatment plant.

7.2.5 Reuse and disposal

The final step in the sanitation service chain is the reuse or disposal of treated products. In the case of conventional systems for urban sanitation in high- and middle-income countries, the end-products are generally sewage sludge and effluent water. The water is typically released into local waterways, although in some cases it is used for groundwater recharge. Reuse of treated wastewater is not a common practice, although this is changing. In Europe, just over 2% of the treated urban wastewater effluent is currently reused (European Commission, n.d.). The European Commission is developing rules and incentives to stimulate and facilitate water reuse. Countries in water-stressed areas have come further regarding water reuse. For example, Singapore uses treated wastewater for direct non-potable use and indirect potable use (through injection in reservoirs) to meet over 40% of its total water demand.

Reuse of organic matter and nutrients from sanitation systems is more common. Approximately half of the sludge generated is applied to land and the rest landfilled or incinerated. In the European Union, roughly 47% is used in agriculture as a soil conditioner and 9% is used in compost and other land applications (Bianchini et al., 2016). In the United States, approximately 50% of the sludge (called Biosolids) is applied on land (30% in agriculture), while the remainder is disposed of in landfills (22%) and incineration (16%) (United States Environmental Protection Agency, 2019).

A movement is growing toward resource recovery from sanitation systems, including nutrients, energy, and water (Figure 7.5). An increasing number of technologies are available for nutrient recovery from wastewater streams. Several of these are already operating commercially, including precipitation and extraction of phosphorus in struvite and ammonia stripping for nitrogen recovery. Extraction of energy in the form of heat exchangers or biogas production is also commonly practised at conventional wastewater treatment plants. Global water shortages are also pushing for increased water reuse. We will likely see more resource recovery and reuse from sanitation systems in the future.

7.3 Selection and design of sustainable urban sanitation systems

The principles that guide the selection and design of urban sanitation systems have changed over time. Many of the existing sanitation systems in the urban areas of high- and middle-income

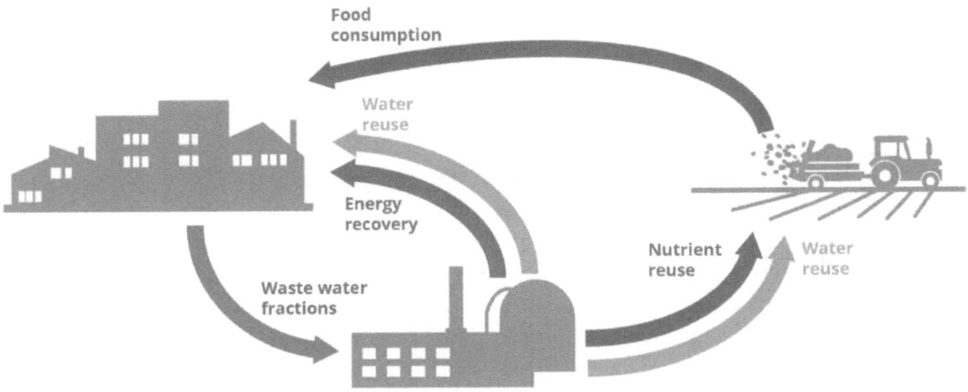

Figure 7.5 Reuse potential of nutrient, water and energy resources in conventional sanitation systems. (Illustration: Annika Nordin, SLU).

countries are systems that have evolved over a period of 50–100 years. During this time, society has changed and the demands on the sanitation system have changed, too. Originally designed for sanitary reasons, urban sanitation systems have been upgraded and retrofitted to provide environmental protection. Today, a holistic definition has emerged of what a sustainable sanitation system should achieve. Sanitation systems are seen as a critical part of achieving sustainable development, as recognized by the inclusion of sanitation in the Sustainable Development Goal 6 – Clean water and sanitation for all. A sustainable sanitation system is one that protects and promotes human health, does not contribute to environmental degradation or depletion of the natural resource base, and is technically and institutionally appropriate, economically viable, and socially acceptable (Andersson et al., 2016). Based on this definition, a number of indicators have been developed to measure the sustainability of our sanitation systems (Table 7.2). Note that these indicators cover a range of social, economic, and environmental issues.

Designing sustainable urban sanitation systems requires recognition of the complexities within the service chain and how each step may impact the sustainability performance of the entire system. Qualitative aspects of sustainability mean that each system must be adapted to the local context –not only the physical environment, but also the social and institutional structures of the people who will use and manage the system. This requires an understanding of the interests, needs, and capacity of the local stakeholders. Planning and decision-making processes should be adapted to include a systematic and inclusive understanding of the sanitation system itself and its potential sustainability in the local context.

The traditional method of planning for sanitation was a top-down process of written rules and procedures, often with a budgetary focus. Well-meaning officials set targets and designed options, usually with little participation from the beneficiaries. In contrast, the strategic planning frameworks that have emerged in the past 50 years focus on more participatory, bottom-up methodologies. Planners solicit the participation of a variety of stakeholders in a democratic planning process to gain understanding of the social, cultural, and institutional aspects that can affect design. These planning processes integrate technical and social aspects, and they apply sustainability thinking to multi-stakeholder, participatory processes. These participatory and holistic methods for sanitation planning are believed to increase the potential for sustainability through better understanding and management of the potential impacts/risks associated with different system choices, improved social acceptability, as well as capacity development of local stakeholders for continued operation and maintenance of the systems. In addition, systematic

Table 7.2 Indicators used to assess the sustainability of an urban sanitation system, organized according to the sustainability dimensions highlighted in the definition of sustainable sanitation, i.e., health, environment, economic, social, institutional, and technical (adapted from Cossio et al., 2020)

Dim.	*Indicator*	*Description*	*Suggested units*
Health	Risk of disease/ toxicity	Risk of exposure to hazardous substances and/or pathogens	DALYs Qualitative
	Quality of effluent and sludge	Pollutants discharged into water and toxic compounds discharged into soil	mg/L kg/p.e.-year
Environment	Energy use	Energy used per volume unit of treated wastewater or inhabitant	kW/p.e. kWh/m^3
	Global warming potential	Emissions of greenhouse gases (GHG) into the atmosphere	kg CO_2eq./p.e.-year kg CO_2eq./m^3
	Eutrophication potential	Reduction of environmental impact on the recipient water body due to release of organic matter and nutrients	mg/L kg/p.e.-year
	Land use	Land area required for the wastewater treatment facility	m^2/p.e.
	Potential recycling	Reuse of treated wastewater: nutrients (N, P) and energy	g/p.e.-year %
Economic	Investment costs	Cost of construction and installation of the WWTP	\$/m^3 \$/p.e.-year
	Operating and maintenance costs	Operating costs per volume unit of wastewater treated	\$/m^3 \$/p.e.-year
Social & institutional	Public acceptance	Opinion of the local population affected by the plant	Qualitative
	Aesthetics	Measured level of nuisance deriving from, e.g., odor, noise, visual impact, insects, and other pests	Qualitative
	Institutional capacity	Availability of knowledge, skills, organization structures, etc. to efficiently plan, implement, operate, monitor, and maintain the system	Qualitative
Technical	Reliability	Infrastructure or mechanical reliability; resilience; security; ability to endure shock loads and/or seasonal effects; potential for overflow	Qualitative
	Complexity of construction and O&M	Ease of construction, complexity of plant construction and system installation; complexity of operation and maintenance; professional skills required for operation and maintenance	Qualitative

and inclusive planning processes are intended to help decision-makers in selecting appropriate technology to satisfy the functional requirement of the various stakeholders. Planning sanitation should thus aim to build: (1) a common vision between the city administrations and the water and waste utilities, (2) cross-sectoral cooperation, (3) strong leadership, and (4) a focus on sustainable innovation and adaption (Lennartsson et al., 2019).

The conventional urban sanitation system is not a static or uniform system. Indeed, the sanitation sector will continue to adapt to pressures from urbanization, changing consumption patterns, changes in land use, water availability, and environmental concerns. Growing urban populations continue to challenge our ability to safely collect and treat a growing volume of wastewater. At the same time, treatment standards are increasingly strict as awareness grows of how our management or mismanagement of wastewater affects the health of humans, waterways, ecosystems, and the global climate. Future development of urban sanitation systems must hold fast to the principles of protection of health and the environment. We need to expand sanitation services to all urban inhabitants. Our technologies need to adapt to the emergence of new chemicals and continue to improve treatment levels. At the same time, a clear shift in the sanitation sector is occurring from seeing sanitation as an end-of-pipe problem to recognizing that the sector has a role to play in several of the global challenges facing society. Sanitation systems can recover resources for use in agriculture, energy, transportation, and industrial sectors. The provision of equitable sanitation services allows more children (particularly girls) to attend school. Energy recovery from sanitation can mitigate climate change. Future developments will be guided by a comprehensive understanding of sustainability and a desire for win-win solutions. Future sanitation systems will be designed not only to reduce negative impacts on health and the environment, but also to also create net positive societal impacts.

References

Andersson, K., Rosemarin, A., Lamizana, B., Kvarnström, E., McConville, J., Seidu, R., Dickin, S., & Trimmer, C. (2016). *Sanitation, Wastewater Management and Sustainability: from Waste Disposal to Resource Recovery*. United Nations Environmental Programme and Stockholm Environment Institute, Nairobi and Stockholm.

Bianchini, A., Bonfiglioli, L., Pellegrini, M., & Saccani, C. (2016). Sewage sludge management in Europe: a critical analysis of data quality. *Int. J. Environ. Waste Manag.* 18, 226. https://doi.org/10.1504/ijewm.2016.10001645

Cossio, C., Norrman, J., Mcconville, J., & Mercado, A. (2020). Indicators for sustainability assessment of small-scale wastewater treatment plants in low and lower-middle income countries. *Environ. Sustain. Indic.* 6, 100028. https://doi.org/10.1016/j.indic.2020.100028

European Commission. (n.d.). Water reuse [WWW Document]. URL https://ec.europa.eu/environment/water/reuse.htm (accessed 2021-02-18).

European Environment Agency. (2020). Urban waste water treatment in Europe [WWW Document]. Indic. Assess. WAT 005. URL https://www.eea.europa.eu/data-and-maps/indicators/urban-waste-water-treatment/urban-waste-water-treatment-assessment-5 (accessed 2.18.21).

Gupta, V. (2020). Vehicle-generated heavy metal pollution in an urban environment and its distribution into various environmental components. In *Environmental Concerns and Sustainable Development* (pp. 113–127). Springer. https://doi.org/10.1007/978-981-13-5889-0_5

Jönsson, H., Baky, A., Jeppsson, U., Hellström, D., & Kärrman, E. (2005). *Composition of Urine, Faeces , Greywater and Biowaste for Utilisation in the URWARE Model, Urban Water Report*. Gothenburg, Sweden.

Krzeminski, P., Tomei, M. C., Karaolia, P., Langenhoff, A., Almeida, C. M. R., Felis, E., Gritten, F., Andersen, H. R., Fernandes, T., Manaia, C. M., Rizzo, L., & Fatta-Kassinos, D. (2019). Performance of secondary wastewater treatment methods for the removal of contaminants of emerging concern implicated in crop uptake and antibiotic resistance spread: A review. In *Science of the Total Environment* (Vol. 648, pp. 1052–1081). Elsevier B.V. https://doi.org/10.1016/j.scitotenv.2018.08.130

Kujawa-Roeleveld, K., & Zeeman, G. (2006). Anaerobic treatment in decentralised and source-separation-based sanitation concepts. *Rev. Environ. Sci. Biotechnol.* 5, 115–139. https://doi.org/10.1007/s11157-005-5789-9

Lennartsson, M., Mcconville, J., Kvarnström, E., Hagman, M., & Kjerstadius, H. (2019). Investments in innovative urban sanitation: Decision-making processes in Sweden. *Water Altern.* 12, 588–608.

Metcalf & Eddy. (2003). *Wastewater Engineering: Treatment and Reuse* (Metcalf & Eddy Inc., G. Tchobanoglous, F. Burton, & H. D. Stensel, Eds.). McGraw-Hill Education.

Ramasahayam, S. K., Guzman, L., Gunawan, G., & Viswanathan, T. (2014). A comprehensive review of phosphorus removal technologies and processes. *J Macromol Sci, Part A* 51(6), 538–545. https://doi.org/10.1080/10601325.2014.906271

Rizzo, L., Gernjak, W., Krzeminski, P., Malato, S., McArdell, C. S., Perez, J. A. S., Schaar, H., & Fatta-Kassinos, D. (2020). Best available technologies and treatment trains to address current challenges in urban wastewater reuse for irrigation of crops in EU countries. In *Science of the Total Environment* (Vol. 710, p. 136312). Elsevier B.V. https://doi.org/10.1016/j.scitotenv.2019.136312

Rose, C., Parker, A., Jefferson, B., & Cartmell, E. (2015). The characterization of feces and urine: A review of the literature to inform advanced treatment technology. *Crit. Rev. Environ. Sci. Technol.* 45, 1827–1879. https://doi.org/10.1080/10643389.2014.1000761

Rutsch, M., Rieckermann, J., & Krebs, P. (2006). Quantification of sewer leakage: A review. In *Water Science and Technology* (Vol. 54, Issues 6–7, pp. 135–144). IWA Publishing. https://doi.org/10.2166/wst.2006.616

Sato, T., Qadir, M., Yamamoto, S., Endo, T., & Zahoor, A. (2013). Global, regional, and country level need for data on wastewater generation, treatment, and use. *Agric. Water Manag.* 130, 1–13. https://doi.org/10.1016/j.agwat.2013.08.007

Sörme, L., & Lagerkvist, R. (2002). Sources of heavy metals in urban wastewater in Stockholm. *Sci. Total Environ.* 298, 131–145. https://doi.org/10.1016/S0048-9697(02)00197-3

Stark, J. S., Bridgen, P., Dunshea, G., Galton-Fenzi, B., Hunter, J., Johnstone, G., King, C., Leeming, R., Palmer, A., Smith, J., Snape, I., Stark, S., & Riddle, M. (2016). Dispersal and dilution of wastewater from an ocean outfall at Davis Station, Antarctica, and resulting environmental contamination. *Chemosphere* 152, 142–157. https://doi.org/10.1016/j.chemosphere.2016.02.053

Steffen, W., Richardson, K., Rockström, J., Cornell, S. E., Fetzer, I., Bennett, E. M., Biggs, R., Carpenter, S. R., Vries, W. de, Wit, C. A. de, Folke, C., Gerten, D., Heinke, J., Mace, G. M., Persson, L. M., Ramanathan, V., Reyers, B., & Sörlin, S. (2015). Planetary boundaries: Guiding human development on a changing planet. *Science* 347, 1259855. https://doi.org/10.1126/science.1259855

Tilley, E., Ulrich, L., Lüthi, C., Reymond, Ph., Schertenleib, R., & Zurbrügg, C. (2014). *Compendium of Sanitation Systems and Technologies*. 2nd Revised Edition. Swiss Federal Institute of Aquatic Science and Technology (Eawag). Dübendorf, Switzerland.

Tran, N. H., Reinhard, M., & Gin, K. Y. H. (2018). Occurrence and fate of emerging contaminants in municipal wastewater treatment plants from different geographical regions-a review. In *Water Research* (Vol. 133, pp. 182–207). Elsevier Ltd. https://doi.org/10.1016/j.watres.2017.12.029

United States Environmental Protection Agency. (2019). Basic information about biosolids [WWW Document]. URL https://www.epa.gov/biosolids/basic-information-about-biosolids#uses (accessed 2.8.21).

WHO. (2006). *Guidelines for the Safe Use of Wastewater, Excreta and Greywater. Volume 3: Wastewater and Excreta Use in Aquaculture.* World Health Organization. Available at: www.who.int

WHO/UNICEF Joint Monitoring Programme. (2021). WASH data [WWW Document]. Database. URL https://washdata.org/data/household#!/dashboard/new (accessed 2.18.21).

8

SANITATION SYSTEMS

Are hybrid systems sustainable or does winner takes all?

Max Maurer

8.1 Introduction

The quest for the optimal system configuration to provide an inherently distributed service is not new but still very present in the academic literature. Services, such as wastewater treatment, electricity, heat provision, and many more, can be produced locally with short distribution paths or centrally by relying on an extensive distribution network. Recent examples in the literature are electricity, hydrogen distribution, building heating, manure treatment, computing, chlorine production, and biomass gasification (see also overviews in Eggimann et al. 2015; Dahlgren et al. 2013).

Wastewater management as a service has some particularities that distinguish it from other services. In terms of mass, it constitutes more than 90% of a city's waste flux (Brands 2014) and requires considerable effort to be transported around. Due to wastewater's rapid and unpleasant degradation, the preferred transport mechanism is underground gravity sewers. This means that – different from water supply or electricity networks – the local topography strongly impacts the network. Both characteristics make it difficult to compare it directly with approaches for electricity or heating.

For wastewater management, the centralized versus decentralized debate can be found in the literature of the 1970s as a consequence of the US Clean Water Act. The main goal was to identify decision criteria for the optimal degree of connectivity mainly based on economic criteria (Downing 1969; Dajani and Gemmell 1971; Adams et al. 1972). More recent work was done by Wang (2014), Tchobanoglous and Leverenz (2013), Maurer (2009), and Fane and Fane (2005). Much of these approaches focus strongly on finding a financial optimum for the treatment plant(s) only. The only exception is Roefs et al. (2017), who tries to model simplified but entire wastewater systems to draw some basic conclusions. All this work assumes that treatment performance is more or less independent of plant size.

From a technical point of view, the size-independent performance is well documented. Ignoring the large body of literature for low-income countries (e.g. Gutterer et al. 2009; Sasse 1998), the literature on high-end systems shows nicely that small plants can achieve treatment performances similar to those of large facilities (Geenens and Thoeye 2000; Abegglen et al. 2008; Barca et al. 2014; Mažeikienė and Grubliauskas 2020). Also, in terms of organic micropollutant elimination, they show similar behaviour (Abegglen et al. 2009; Hube and Wu 2021), indicating

 DOI: 10.4324/9781003057574-11

that the biological processes are more or less size independent. However, empirical evidence shows that the overall system performance is not what one would expect from the technical performance achievable. Kaminsky and Javernick-Will 2013 attributes this not to the technical hardware but "*software is more likely to be the root cause of system failure*" – defining software as institutional factors ("*e.g., knowledge, institutions, education*").

Consequently, decentralised systems are generally viewed as a stopgap solution where sewers cannot be built or are too expensive. This is surprising as from a global perspective, sewers networks also have some substantial disadvantages, such as very long lead and construction times, large spare capacity due to long planning horizons, high upfront investment costs, low flexibility to adapt to unforeseen changes, high water demand, and great efforts for resource recovery (Roefs et al. 2017; Larsen et al. 2016; Brands 2014). These disadvantages could be overcome or at least softened by adding decentralized systems to the engineering portfolio and utilising both systems' approaches to provide wastewater services. However, such hybrid solutions seem not to be very common around the world.

This chapter examines why hybrid wastewater systems – as one potential example of a complementary system alternative - are not more common and which barriers and research gaps might play a crucial role. A very technical view is taken on this issue and the focus is on wastewater management only. Many relevant aspects (governance, organisational aspects, institutional, socioeconomic setting, etc.) are be ignored or only hinted at, trusting that other authors will complement these gaps with more authority.

8.2 Centralized, modular, and hybrid sanitation systems (definitions)

The literature is plagued with terminology to distinguish sanitation system alternatives from the "traditional" centralised system. This ranges from distributed and on-site to non-sewered systems, primarily relying on an intuitive understanding of the proposed underlying system layout. In the following, some definitions are given that we find helpful for discussing the main concepts in this article:

Degree of centralization (DC) or *connection rate* is defined as one minus the ratio of sinks to sources in a given catchment, weighted by the amount of wastewater per sink and source. A DC = 0 indicates that every source has its own sink, and a DC = 1 means that the sink is outside the catchment area (Eggimann et al. 2015). We generally equate a source with a single building and a sink as a wastewater treatment plant.

Centralized sanitation systems have a high DC. Consequently, these are network- and transport-dominated systems where the wastewater is transported to a few large wastewater treatment facilities.

Modular sanitation systems have a very low DC and therefore shorter and fragmented sewer networks. Compared with the number of sources, they have a relatively large number of treatment plants. This definition does not make a statement about the size of the treatment plants. However, it can be expected that in a modular system, the units are substantially smaller than in a comparable centralised system. The term "modular" is probably closest related to the term "distributed." We deliberately chose this term to distinguish it from (a) "on-site," which is equal to a DC of zero, and (b) from the term "decentralised wastewater treatment system" or DEWATS, which is frequently associated with wastewater management in low-income environments (e.g., Gutterer et al. 2009; Sasse 1998) or in the United States predominantly with septic tanks or any small treatment plants with less than a few thousand population equivalents but still a high DC, e.g., US-EPA (2005) or McCray et al. (2009).

Hybrid sanitation systems are a combination of centralized and modular systems. Part of the catchment has a very high DC and part of the system has a very low DC. The overall DC of the catchment might be anywhere larger than 0 but smaller than 1.

8.3 Economies of scale of wastewater systems

Empirical evidence shows that sewer networks show strong path dependencies (Fam et al. 2009; Wolf and Störmer 2010). Besides the arguments of the institutional regime and high sunk costs, there is also a strong element of (apparent) economies of scale leading to a natural monopoly situation. This means that the benefits of adapting the existing centralized regime are neither trivial nor obvious. The case for treatment plants is quite well documented over a wide range of capacities, e.g., see compilation in Maurer 2009. A doubling of the capacity corresponds roughly to a 20% decrease in investments per population equivalent and about 15% in operating expenses.

The evidence for economies of scale in sewer networks is much shakier. Although the investments into the network far dominate the installation costs (Maurer 2013), little is known about their cost behaviour in terms of size and spatial extent. Nauges and Van den Berg (2013) for example, concluded from an analysis of national data that "*the cost structure of the water and wastewater sector varies significantly between countries and within countries, and over time,*" finding evidence for economies of scale in some but not in other countries.

Our own work found a not surprising impact of settlement structure on the scale economies (Maurer et al. 2010). Larger cities have more expensive networks due to larger pipes, but higher population densities might distribute these higher costs over more users. Looking at data from Switzerland, cost and settlement structure hint at very small economies of scale for combined systems (Maurer et al. 2013). A doubling in inhabitants corresponded with a decrease in 9% in replacement value for the entire network. However, the analysis was averaged over whole cities and towns and ignored heterogeneity within the settlement structure. This averaging is also the case in other investigations in the literature, e.g., Nauges and Van den Berg (2013). We see in Figure 8.1 that heterogeneity can have a significant impact on local costs and, therefore, economies of scale. Thus, a more "individual" approach for specific catchments provides better insights into the viability of hybrid systems.

8.4 Connect or not to connect: The optimal degree of centralization

A key challenge for a given wastewater management catchment is which houses should be connected to a sewer network and how extended these networks should be. This spatial explicit problem needs to consider topography – as most sewer networks are gravity driven. In Eggimann et al. (2015) we developed such a model to explore the cost for different connection rates utilizing a heuristic routing algorithm. As a first step, this tries to connect a house with its neighbours and compares the resulting cost with the option of having a local treatment plant. If the difference exceeds a threshold, then the houses are connected. In the next step, it moves on to the next house and the algorithm is repeated. If there are no cost-effective connections anymore, then a treatment plant is positioned at the lowest point of the sewer network. Changes in the threshold can be used to favour or disadvantage connections over treatment, which allows exploring all the possible degrees of connections. The algorithm also provides freedom to align sewers with the street network or other relevant features.

The approach assumes that the treatment performance is independent of plant size (e.g., biological active, not septic tanks). The cost functions are based on realistic data on sewer construction costs (depending on depth, diameter, and if it is under a street or not) and investment

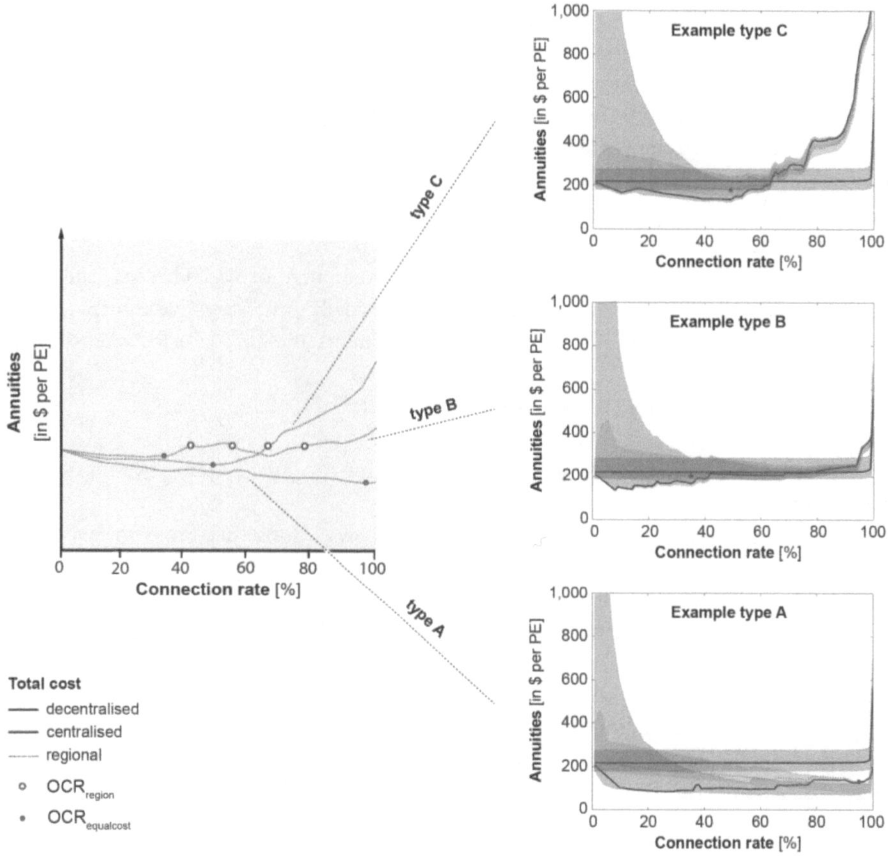

Figure 8.1 Cost curve configuration typology with examples. For each example type, standard parameter runs (thick line) and scenario uncertainties are provided. The "Annuities" are annualised OPEX and CAPEX (Eggimann et al., 2016) and used with permission by the creator.

costs for treatment plants (depending on their capacity). The operation and maintenance costs also include spatial explicit travel and transport costs for inspection and sludge removal. Details can be found in Eggimann et al. (2015). Although the absolute costs are based on Swiss data and most probably change substantially from country to country, we still assume that the impact of the relative costs between sewers and treatment plants are still providing generic answers.

Eggimann et al. (2016) used the model to explore the total costs for the entire range of DC - from purely modular systems (a DC of zero) to fully centralized systems (a DC of one) – for 25 real and mostly rural catchments. From the results, three types of cost behaviour emerged, summarized in Figure 8.1.

Catchment type A shows clear economies of scale, where a fully centralized approach constitutes a natural monopoly. The other clear case is catchment type C, where the costs show a pronounced global minimum. The settlement structure and the topography strongly impact this result (e.g., a compact town centre with more loosely settled surroundings or houses in a hilly or mountainous area). It is also worth noting that for very high connection rates the costs for the decentralized part of the system increases strongly, due to the high transportation costs for the low density of the remaining small-scale treatment plants.

Type C is closest to the idea of a hybrid system and the analysis shows that there are circumstances where from a purely economic point of view, hybrid systems are a viable end point. Changes in the cost structure might affect at what DC the minimum might be, but the fundamental characteristics of a hybrid system remain. A second interesting result was that the global minima did not coincide with the point of equal costs, where an individual house could connect or then go modular at cost parity. This means that the incentive structure matters. Letting the homeowner decide (and assuming a *homo economicus*) will produce a DC that is not necessarily the overall cost optimum for the entire catchment.

Catchment type B shows a very flat cost curve over a large range of DC. This indicates that small changes in costs can influence the optimal DC substantially. An alternative interpretation is that there is no clear optimal DC for such type B catchments and that hybrid systems are a cost-effective solution.

The following conclusion can be drawn from these results:

- Under the investigated (Swiss-)cost structure for operating modular systems, hybrid systems can be cost effective.
- The internal heterogeneity of a settlement structure plays a substantial role in favouring hybrid systems.
- The mix within the hybrid system (optimal DC) depends on the institutional framing. The catchment-wide cost minimum might be at a different point than the individual cost optimum. Therefore, it matters if a catchment-wide entity (benevolent dictator) or the homeowner decides if a house should be connected to the grid or not.
- Many total cost curves turned out to be very flat over wide ranges of DC, with substantial uncertainties. Small changes in the cost structure can have drastic impacts on DC.
- It is very likely that with the increased production of units, the costs for modular systems will go down. For example, Mayor 2020 identified a plant size-independent learning effect of 12–30% for desalination technologies, depending on the technology. This will most likely substantially affect the extent of modular systems that become cost effective but most likely will not change the fundamental cost curve behaviour.

These modelling exercises, illustrative as they are, have some substantial restrictions in answering the question of whether hybrid systems are sustainable. First, they assume green-field conditions and ignore existing infrastructure. Second, it is a very static snapshot of the situation and does not consider changes in population or costs or technology. Third, it considers only (average) costs as the main deciding factor and ignores any environmental, institutional and organizational aspects. In the following sections, these issues are addressed in more detail.

8.5 Path dependency of centralised systems

The next critical question worth focusing on is the dynamic stability of existing systems over time. Empirical evidence indicates that sewer networks show strong path dependencies (Fam et al. 2009; Wolf and Störmer 2010). Therefore, introducing or maintaining hybrid systems might not be viable, even if they would be the best and cheapest option. Here we will focus exemplary on two regime stabilizing aspects: the perceived synergies between stormwater and wastewater drainage and cost calculation methods. We will ignore all the other relevant institutional and governance aspects, as this would go beyond the scope of this article. However, it is essential to acknowledge that these – from an engineering point of view – "soft" aspects play a crucial role in stabilising the path dependency of a dominant sociotechnical system.

8.5.1 Stormwater and wastewater

The option of using combined sewers makes things more complicated. They promise the alluring prospect of combining two essential urban water services with one single infrastructure and therefore utilizing synergies. As with any hydrological catchment, a city must provide ways for stormwater to drain. The transport capacity of a corresponding pipe network is proportional to the drained area and typical rain intensities. In many "wet" climate zones, the capacities needed to drain stormwater dwarf the wastewater flow – typically by two or more orders of magnitude. Consequently, wastewater can be easily added to the drainage infrastructure often without needing any additional capacity.

Such combined sewer systems create a path dependency of their own. Compared with stormwater drainage, combined systems have very particular technical requirements. They are generally closed conduits (compared with the option of open stormwater drains), with higher demand for water tightness (to prevent wastewater leaking) and deeper construction depths to accommodate sufficient flow velocities and minimise pumping needs. Moreover, for water pollution protection the combined sewers cannot just drain into the next surface water body but need to lead to a treatment plant. In this respect, they are more similar to foul sewers but with substantially more transport capacity. However, not all stormwater can be treated in the wastewater treatment plant, and wastewater is discharged from the system during rain events – also known as combined sewer overflows. These require substantial additional efforts and infrastructures (e.g., overflow tanks) to minimize their environmental impact.

Although this is a simplified view of the two fundamental urban drainage design approaches, these examples highlight an important point. Combined sewer systems are not just stormwater systems with wastewater but distinct technological systems with specific adaptation needs. Correspondingly, combined sewer systems have their own lock-in logic and breaking the path dependency has most likely its particular challenges compared to separate systems. To our knowledge, this aspect is not explored in sufficient detail in the literature.

In terms of hybrid systems, stormwater management is a significant step ahead of sewage. Science and practice provide many more decentralized engineering alternatives to mitigate the challenges and opportunities stormwater management has for cities. We will touch on this in the next section about transition management.

8.5.2 Cost calculation methodology

A crucial element stabilizing the strong path dependency of sewer networks are the economic methods used to assess the cost effectiveness of system adaptation. Current costing and financing approaches are suited for evaluating incremental system adaptation but not for fundamental system changes and therefore tend to stabilize the path dependency of the current system. Here we would like to mention two key points: Insufficient life costing methodologies and the cross-subsidizing effect of tariffs.

The extremely long lifespan of pipe networks (typically in the range of 30 – 100 years) and the corresponding long planning horizon pose a "wicked problem" (Rittel and Webber 1973) for proper life cycle costing methodologies mainly because of four reasons:

i. Long-term discount rates are highly sensitive for life cycling costing, but unfortunately very subjective. For example, Weitzman (2007) points out that we are a lot less sure about what interest rate should be used for very long-term discounting than is commonly acknowledged. High discount rates favour delayed investments, while low discount rates enable much more present spending.

ii. Even small population changes over the long planning horizon lead to substantial overcapacities as typically these infrastructures are designed to deal with peak demand. Consequently, cost methodologies focus on the cost of providing peak demand. Examples in Maurer (2009) show that the so-created unused capacities lead to substantially higher per capita costs compared with no growth situation or more flexible alternatives. For example, modular or hybrid systems could be built as needed and the overall investment pattern would adapt more closely to the actual future demand and therefore could be cheaper, especially under high discount rate regimes. It should be noted that this is also true for negative growth rates, where modular systems have the distinct benefit of shorter life spans.
iii. Prognosis uncertainty over several decades of planning is horrendous and therefore usually not considered systematically in cost calculations. The effect of this is nicely shown in de Neufville and Scholtes (2011) for different engineering fields. For wastewater systems, it can be demonstrated that uncertainties increase the probability of much higher specific costs over the lifetime of the considered infrastructure (Hug et al. 2010). Again, the flexibility of hybrid solutions could provide a cost advantage due to the demand-driven investment pattern. This is especially relevant in areas with very high growth rates, where two factors – population change and high uncertainties – come together. Ignoring uncertainties generally favours static approaches.
iv. Decisions are generally made for one specific project or measure and without consideration of the long-term lock-in effects. This favours distinctively incremental changes and, therefore, path dependency, as they can rely on the already existing sunk cost infrastructure. Compared to this, options that aim for more radical changes will fare worse as they often require a higher initial investment, not just in infrastructure, but also governance and organizational structures. See also the call for long-term planning instrument in section 8.6 to help to overcome this specific disadvantage.

Another stabilizing factor for the centralized regime is the financing of the system over tariffs. Most tariff schemes apply the same rules to everybody in a catchment and do not distinguish the specific costs needed to connect an individual house. This leads to a situation in which the connection of unfavourable neighbourhoods to the grid is subsidised by the rest of the catchment. This averaging effect of tariff schemes decreases the incentive to find cheaper alternative solutions and stabilises the path dependency of the current system.

8.6 Transition planning

Interestingly, in stormwater management, hybrid systems are widely accepted and actively implemented in practice. Several key concepts were developed since the 1990s, such as SUDS, LID, WSUD, or Sponge City (see for a good overview, Fletcher et al. 2015) that rely heavily on modular systems to strengthen the resilience and performance of urban stormwater drainage. This offers an opportunity to learn from the successes and failures to establish hybrid systems and transfer some lessons learnt to wastewater.

In a detailed investigation of the modular stormwater systems in Melbourne (Kuller et al. 2018a), the authors concluded that

> opportunistic WSUD (water sensitive urban design considering stormwater hybrid systems) planning leads to unintentional outcomes that fail to capitalize on the full potential of WSUD benefits. [...] integrated planning for WSUD are encouraging and

> have the potential to significantly improve the outcomes of water quality, flood safety and amenity for urban communities.

This exemplary highlights the need for suitable planning tools that are capable of transcending the complexity of technological options and to capitalize on the additional benefits a hybrid system can provide over a centralized approach.

This emphasis on planning is also one of the main conclusions of an investigation into the feasibility of modular systems (Särkilahti et al. 2017). The authors identified a series of drivers, barriers, and enablers for system change, and they concluded:

> The results indicate that sustainability transition can be facilitated through impartial urban planning that allows the early participation of actors and improved communications. Additionally, studying the impact of alternative solutions and city guidance according to environmental policy may enhance transition.

This highlights the importance of urban water infrastructure planning as a key element for transition management. From the evidence presented above, we can conclude that a suitable planning approach should minimally fulfil the following four key requirements: (i) representing very long planning horizons, (ii) the need to be spatially explicit, (iii) consideration of a wide range of technological development options that includes modular systems, and (iv) the need to consider a sufficiently broad range of goals and benefits (Kuller et al. 2018b; Brands 2014).

In the literature, several of these elements can be found for water infrastructure planning in different combinations. To our knowledge, only two modelling approaches can be found in the literature that cover all the factors mentioned above. One is DAnCE4Water (Dynamic Adaptation for enabling City Evolution for Water, Rauch et al. 2017; Urich et al. 2013) and the other is SinOptikom (Baron et al. 2017; Baron et al. 2015). Both simulate, in detail, the transition of water infrastructures for an entire catchment over long periods considering deep uncertainty and multi-criteria assessment algorithms to evaluate numerous options to meet targets at a variety of spatial scales. Unfortunately, DAnCE4Water is mainly focused on stormwater management, and SinOptikom disappeared into obscurity after the end of the project.

An excellent indication of the importance of a wide range of goals for infrastructure transition can be found in Lienert et al. (2015) and Zheng et al. (2016). The authors present a procedure that combines long-term scenario planning with a sophisticated multi-criteria decision analysis (MCDA) approach. One of the key results was the finding that most stakeholders highly valued performance, resource protection, and intergenerational equity while cost and social acceptance were only of minor importance (Zheng and Lienert 2018). This result is valid only for the analysed Swiss cases. Nevertheless, the authors could show for realistic decision cases that stakeholders value modular systems as high or higher than centralized approaches (Beutler et al. 2021). However, the presented approach does not support the modelling of the spatial explicit transition of the infrastructure.

8.7 Conclusions

The final verdict on whether hybrid systems for wastewater can be sustainable in urban catchments is still out. The evidence indicates that there are good reasons to believe that, once established, they would improve sustainability under certain circumstances. The main strength of modular systems is the high flexibility in accommodating demand change and planning uncertainty, providing locally adapted solutions in low-density areas and avoiding significant upfront investments into sewer networks (Larsen et al. 2016). Fortifying centralized systems with these benefits enables overcoming some of the critical weaknesses of the one-size-fits-all centralised systems approach – especially in cities with large heterogeneities.

However, centralised systems show very strong path dependencies that need to be overcome for hybrid systems to be implemented in the first place. Besides a wide range of institutional and organisational barriers (Särkilahti et al. 2017), the current economic and planning tools also need to be improved and adapted. Much of the existing engineering tools for infrastructure management, such as costing or long-term planning, work very well for incremental adaptation of the dominant centralised system, but they have essential weaknesses in terms of integrating fundamentally novel technological approaches.

From an engineering perspective, the following research and development gaps still need to be closed for hybrid wastewater systems to have widespread dissemination:

- The development of long-term transition planning tools that are spatially explicit and can consider a wide range of modular technologies. The main challenge is identifying the proper degree of model resolution and accuracy as such models can get very demanding very fast (Bach et al. 2014; Rauch et al. 2017). The goal of such models is not to predict the future in all detail but to facilitate short-term decision making by assessing the long-term consequences of the decisions.
- There is also an urgent need to find ways to consider cross-sectoral integration appropriately, such as energy, stormwater, and water supply, to understand the broader synergies that hybrid systems can provide (Trapp et al. 2017).
- Stronger integration of multi-criteria decision analysis (MCDA) or any other structured decision-making procedure into the planning is essential to consider the broader range of goals and benefits hybrid systems can provide.
- Engineering economic methodologies need to be developed so that they can include planning uncertainties and adaptability. Real option approaches (de Neufville and Scholtes 2011) might be the first step. However, due to the difficulty of quantifying unsatisfying performance in monetary terms, real options are not readily applicable in wastewater.

This is not to say that there is no research need in the realm of engineering "hardware." However, compared with the needs to support regime change on a system level, these are relatively minor.

References

Abegglen, C., Joss, A., McArdell, C.S., Fink, G., Schlusener, M.P., Ternes, T.A. and Siegrist, H. (2009) The fate of selected micropollutants in a single-house MBR. *Water Research* 43(7), 2036–2046, https://doi.org/10.1016/j.watres.2009.02.005.

Abegglen, C., Ospelt, M. and Siegrist, H. (2008) Biological nutrient removal in a small-scale MBR treating household wastewater. *Water Research* 42(1–2), 338–346, https://doi.org/10.1016/j.watres.2007.07.020.

Adams, B.J., Dajani, J.S. and Gemmell, R.S. (1972) On the centralization of wastewater treatment facilities1. *JAWRA Journal of the American Water Resources Association* 8(4), 669–678, https://doi.org/10.1111/j.1752-1688.1972.tb05208.x.

Bach, P.M., Rauch, W., Mikkelsen, P.S., McCarthy, D.T. and Deletic, A. (2014) A critical review of integrated urban water modelling: Urban drainage and beyond. *Environmental Modelling & Software* 54, 88–107, https://doi.org/10.1016/j.envsoft.2013.12.018.

Barca, C., Meyer, D., Liira, M., Drissen, P., Comeau, Y., Andrès, Y. and Chazarenc, F. (2014) Steel slag filters to upgrade phosphorus removal in small wastewater treatment plants: Removal mechanisms and performance. *Ecological Engineering* 68, 214–222, https://doi.org/10.1016/j.ecoleng.2014.03.065.

Baron, S., Dilly, T.C., Kaufmann Alves, I. and Schmitt, T.G. (2017) *Transformation of Urban Drainage Systems in Rural Areas*, IWA, Prague, Chech Republic.

Baron, S., Hoek, J., Kaufmann Alves, I. and Herz, S. (2015) Comprehensive scenario management of sustainable spatial planning and urban water services. *Water Science and Technology* 73(5), 1041–1051, https://doi.org/10.2166/wst.2015.578.

Beutler, P., Larsen, T., Maurer, M., Staufer, P. and Lienert, J. (2021) Value-based MCDA helps set research priorities for non-grid wastewater management. in preparation.

Brands, E. (2014) Prospects and challenges for sustainable sanitation in developed nations: a critical review. *Environmental Reviews* 22(4), 346–363, https://doi.org/10.1139/er-2013-0082.

Dahlgren, E., Göçmen, C., Lackner, K. and van Ryzin, G. (2013) Small modular infrastructure. *The Engineering Economist* 58(4), 231–264, https://doi.org/10.1080/0013791x.2013.825038.

Dajani, J.S. and Gemmell, R.S. (1971) *Economics of Wastewater Collection Networks*. Research Report No. 43. Water Resources Center, Illinois.

de Neufville, R. and Scholtes, S. (2011) *Flexibility in Engineering Design*, MIT Press, Cambridge, MA.

Downing, P.B. (1969) *The Economics of Urban Sewage Disposal*, Praeger, New York .

Eggimann, S., Truffer, B. and Maurer, M. (2015) To connect or not to connect? Modelling the optimal degree of centralisation for wastewater infrastructures. *Water Research* 84, 218–231, https://doi.org/10.1016/j.watres.2015.07.004.

Eggimann, S., Truffer, B. and Maurer, M. (2016) The cost of hybrid waste water systems: A systematic framework for specifying minimum cost-connection rates. *Water Research* 103, 472–484, https://doi.org/10.1016/j.watres.2016.07.062.

Fam, D., Lopes, A., Willetts, J. and Mitchell, C. (2009) The challenge of system change: An historical analysis of Sydney's sewer systems. *Design Philosophy Papers* 7(3), 195–208, https://doi.org/10.2752/144871309X13968682695316.

Fane, A.G. and Fane, S.A. (2005) The role of membrane technology in sustainable decentralized wastewater systems. *Water Science and Technology* 51(10), 317–325, https://doi.org/10.2166/wst.2005.0381.

Fletcher, T.D., Shuster, W., Hunt, W.F., Ashley, R., Butler, D., Arthur, S., Trowsdale, S., Barraud, S., Semadeni-Davies, A., Bertrand-Krajewski, J.L., Mikkelsen, P.S., Rivard, G., Uhl, M., Dagenais, D. and Viklander, M. (2015) SUDS, LID, BMPs, WSUD and more - The evolution and application of terminology surrounding urban drainage. *Urban Water Journal* 12(7), 525–542, https://doi.org/10.1080/1573062x.2014.916314.

Geenens, D. and Thoeye, C. (2000) Cost-efficiency and performance of individual and small-scale treatment plants. *Water Science and Technology* 41(1), 21–28.

Gutterer, B., Sasse, L., Panzerbieter, T. and Reckerzügel, T. (2009) *Decentralised Wastewater Treatment Systems (DEWATS) and Sanitation in Developing Countries: A Practical Guide Water, Engineering and Development Centre (WEDC)*, Loughborough University, UK, in association with Bremen Overseas Research (BORDA), Germany.

Hube, S. and Wu, B. (2021) Mitigation of emerging pollutants and pathogens in decentralized wastewater treatment processes: A review. *Science of the Total Environment* 779, 18, https://doi.org/10.1016/j.scitotenv.2021.146545.

Hug, T., Dominguez, D. and Maurer, M. (2010) The cost of uncertainty and the value of flexibility in water and wastewater infrastructure planning. *Conference proceedings from Water Environment Federation WEF: Cities of the Future 2010/Urban River Restoration*, 487–500. Boston, MA, USA

Kaminsky, J. and Javernick-Will, A. (2013) Contested factors for sustainability: Construction and management of household on-site wastewater treatment systems. *Journal of Construction Engineering and Management* 139(12), 9, https://doi.org/10.1061/(asce)co.1943-7862.0000757.

Kuller, M., Bach, P.M., Ramirez-Lovering, D. and Deletic, A. (2018a) What drives the location choice for water sensitive infrastructure in Melbourne, Australia? *Landscape and Urban Planning* 175, 92–101, https://doi.org/10.1016/j.landurbplan.2018.03.018.

Kuller, M., Farrelly, M., Deletic, A. and Bach, P.M. (2018b) Building effective Planning Support Systems for green urban water infrastructure: Practitioners' perceptions. *Environmental Science & Policy* 89, 153–162, https://doi.org/10.1016/j.envsci.2018.06.011.

Larsen, T.A., Hoffmann, S., Lüthi, C., Truffer, B. and Maurer, M. (2016) Emerging solutions to the water challenges of an urbanizing world. *Science* 352(6288), 928–933, https://doi.org/10.1126/science.aad8641.

Lienert, J., Scholten, L., Egger, C. and Maurer, M. (2015) Structured decision-making for sustainable water infrastructure planning and four future scenarios. *EURO Journal on Decision Processes* 3, 107–140, https://doi.org/10.1007/s40070-014-0030-0.

Maurer, M. (2009) Specific net present value: an improved method for assessing modularisation costs in water services with growing demand. *Water Research* 43(8), 2121–2130, https://doi.org/10.1016/j.watres.2009.02.008.

Maurer, M. (2013) *Source Separation and Decentralization for Wastewater Management.* Larsen, T.A., Udert, K.M. and Lienert, J. (eds), pp. 85–100, IWA Publishing, London.

Maurer, M., Scheidegger, A. and Herlyn, A. (2013) Quantifying costs and lengths of urban drainage systems with a simple static sewer infrastructure model. *Urban Water Journal* 10(4), 268–280, https://doi.org/10.1080/1573062X.2012.731072.

Maurer, M., Wolfram, M. and Herlyn, A. (2010) Factors affecting economies of scale in combined sewer systems. *Water Science and Technology* 62(1), 36–41, https://doi.org/10.2166/wst.2010.241.

Mayor, B. (2020) Unraveling the Historical Economies of Scale and Learning Effects for Desalination Technologies. *Water Resources Research* 56(2):e2019WR025841, https://doi.org/10.1029/2019wr025841.

Mažeikienė, A. and Grubliauskas, R. (2020) Biotechnological wastewater treatment in small-scale wastewater treatment plants. *Journal of Cleaner Production*, 279, 123750. https://doi.org/10.1016/j.jclepro.2020.123750.

McCray, J.E., Geza, M., Murray, K.E., Poeter, E.P. and Morgan, D. (2009) *Modeling Onsite Wastewater Systems at the Watershed Scale: A User's Guide*, IWA Publishing, London.

Nauges, C. and Van den Berg, C. (2013) *How "Natural" Are Natural Monopolies In The Water Supply And Sewerage Sector? Case Studies From Developing And Transition Economies*, Worldbank, Washington, D.C., https://doi.org/10.1596/1813-9450-4137.

Rauch, W., Urich, C., Bach, P.M., Rogers, B.C., de Haan, F.J., Brown, R.R., Mair, M., McCarthy, D.T., Kleidorfer, M., Sitzenfrei, R. and Deletic, A. (2017) Modelling transitions in urban water systems. *Water Research* 126, 501–514, https://doi.org/10.1016/j.watres.2017.09.039.

Rittel, H.W.J. and Webber, M.M. (1973) Dilemmas in a general theory of planning. *Policy Sciences* 4(2), 155–169, https://doi.org/10.1007/BF01405730.

Roefs, I., Meulman, B., Vreeburg, J.H.G. and Spiller, M. (2017) Centralised, decentralised or hybrid sanitation systems? Economic evaluation under urban development uncertainty and phased expansion. *Water Research* 109, 274–286, https://doi.org/10.1016/j.watres.2016.11.051.

Särkilahti, M., Kinnunen, V., Kettunen, R., Jokinen, A. and Rintala, J. (2017) Replacing centralised waste and sanitation infrastructure with local treatment and nutrient recycling: Expert opinions in the context of urban planning. *Technological Forecasting and Social Change* 118, 195–204, https://doi.org/10.1016/j.techfore.2017.02.020.

Sasse, L. (1998) *DEWATS Decentralised Wastewater Treatment in Developing Countries*, Bremen Overseas Research and Development Association (BORDA), Germany.

Tchobanoglous, G. and Leverenz, H. (2013) *Source Separation and Decentralization for Wastewater Management.* Larsen, T.A., Udert, K., Lienert, J. (ed), pp. 101–115, IWA Publishing, London, UK.

Trapp, J.H., Kerber, H. and Schramm, E. (2017) Implementation and diffusion of innovative water infrastructures: obstacles, stakeholder networks and strategic opportunities for utilities. *Environmental Earth Sciences* 76(4), 154, https://doi.org/10.1007/s12665-017-6461-8.

Urich, C., Bach, P.M., Sitzenfrei, R., Kleidorfer, M., McCarthy, D.T., Deletic, A. and Rauch, W. (2013) Modelling cities and water infrastructure dynamics. *Proceedings of the Institution of Civil Engineers: Engineering Sustainability* 166(5), 301–308, https://doi.org/10.1680/ensu.12.00037.

US-EPA (2005) *Decentralized Wastewater Treatment Systems: A Program Strategy*, US-EPA, Washington, D.C..

Wang, S. (2014) Values of decentralized systems that avoid investments in idle capacity within the wastewater sector: A theoretical justification. *Journal of Environmental Management* 136, 68–75, https://doi.org/10.1016/j.jenvman.2014.01.038.

Weitzman, M.L. (2007) A review of "the stern review on the economics of climate change". *Journal of Economic Literature* 45(3), 703–724.

Wolf, M. and Störmer, E. (2010) Decentralisation of wastewater infrastructure in Eastern-Germany. *Network Industries Quarterly* 12(1), 7–10.

Zheng, J. and Lienert, J. (2018) Stakeholder interviews with two MAVT preference elicitation philosophies in a Swiss water infrastructure decision: Aggregation using SWING-weighting and disaggregation using UTAGMS. *European Journal of Operational Research* 267(1), 273–287, https://doi.org/10.1016/j.ejor.2017.11.018.

Zheng, J., Egger, C. and Lienert, J. (2016) A scenario-based MCDA framework for wastewater infrastructure planning under uncertainty. *Journal of Environmental Management* 183, 895–908, https://doi.org/10.1016/j.jenvman.2016.09.027.

9

MANAGEMENT OF URBAN DRAINAGE INFRASTRUCTURE

Nelson Carriço, Maria do Céu Almeida, and João Paulo Leitão

9.1 Introduction

Urban drainage systems provide essential public services for sewerage disposal ensuring public health, safety, and environmental and social goals (Grigg, 2003; Porse, 2013). These systems are composed of a large number of components (or assets) of different types (e.g., sewers, manholes, pumps). The assets deteriorate due to current deterioration processes (e.g., degradation of materials over time) and non-controlled processes (e.g., sewer's poor-quality production and external actions, such as construction works). These processes contribute to component failures and increased sewer systems' maintenance and operational costs as well as all consequences of failure, including urban flooding, pollution, sewer collapses, and blockages (Carriço et al., 2012). Thus, water utility engineers and managers face the challenge of keeping their systems operational, efficient, and reliable to provide a good quality of service. Therefore, water utilities should adopt infrastructure asset management (IAM) processes to plan their actions on the assets (e.g., repair or rehabilitation) in order to provide an adequate level of service and ensure their sustainability (Alegre & Coelho, 2012). The levels of services can be defined considering European standards such as EN 752:2017 and regulations including the Water Framework Directive (Directive n. ° 2000/60/EC) or the Urban Wastewater Treatment Directive (Directive n. ° 91/271/EEC).

Several concepts and methodologies of IAM may be found in the literature (Alegre & Coelho, 2012; Almeida & Cardoso, 2010; Beuken et al., 2020; BSI, 2008; IAM, 2015; IPWEA & NAMS, 2006, 2015; Matthews et al., 2012; Osman, 2012; Ugarelli et al., 2010). The International Organization for Standardization (ISO) has developed the first family of international standards for asset management, which include ISO 55000, ISO 55001, and ISO 55002 (ISO, 2014a, b, 2018b). According to these standards, asset management is the coordinated activity carried out by an organisation to realise value from its assets and involving a balancing of costs, risks, opportunities, and performance benefits (ISO, 2014a). In the context of urban drainage infrastructures, IAM is the activity of managing the infrastructure assets providing a service to users, achieving required levels of service, while minimising the total cost of ownership, operating and maintaining these assets, protecting the environment, and ensuring safety (Alegre & Coelho, 2012; Cardoso et al., 2017; Fjeldhus & Ugarelli, 2014). Planning asset rehabilitation is of utmost importance to improve the level of service in terms of structural performance (e.g., increase of structural integrity), hydraulic performance (e.g., reduction of roughness or infiltration),

DOI: 10.4324/9781003057574-12

and environmental performance (e.g., reduction of exfiltration of sewage to adjacent ground) (Almeida et al., 2015). Rehabilitation is the process of upgrading the asset condition to its "as-new" condition, if practicable (IPWEA & NAMS, 2006, 2015). The term "condition" may refer to different meanings, such as structural condition, water quality, hydraulic capacity, serviceability, location, and economics (Thomson et al., 2013). Condition assessment is an essential part of any IAM since it provides the critical information needed to assess the physical condition and functionality of an urban drainage system, allowing estimation of the remaining service life and asset value (Feeney et al., 2009). An asset rehabilitation plan is a proposal of scheduled works designed to upgrade assets to a good condition. Rehabilitation techniques may be classified into three different types: replacement, renovation or refurbishment, and repair according to standards EN 752 (CEN, 2017) and EN 15885 (CEN, 2018). A renovation consists of an asset intervention that incorporates, either totally or partially, the existing asset material. A replacement consists in substituting the existing asset with another new material, which adopts the function of the disabled one. A repair consists of the correction of the asset's anomalies.

There are approaches of urban water IAM focused on asset rehabilitation that meet the requirements of the standards, in which asset rehabilitation must be done in an integrated and proactive way and consider the three decision levels of the water utilities (i.e., strategic, tactical, and operational, see Figure 9.1). According to ISO 55000 (ISO, 2014a), the strategic asset management plan is derived from the organisational plan and specifies how organisational objectives are to be converted into asset management objectives, the approach for developing asset management plans, and the role of the asset management system in supporting the achievement of the asset management objectives. Consequently, asset management planning has a greater focus on the tactical and operational levels.

Thus, IAM is an organisation-wide approach that should ensure a balance among the dimensions of performance, risk, and cost in a long-term perspective, requiring a coordinated intervention between the different levels of planning (strategic, tactical, and operational). Hence, it is a multidisciplinary approach, the main skills involved being the management (including economics and sociology of organisations), the engineering (e.g., civil, environmental, mechanical), and the information (e.g., information management, communication, computing). Regarding information, it is crucial for any IAM methodology to have reliable data about the existing assets since condition assessments rely on a deep knowledge of them (Carriço & Ferreira, 2021; Dawood et al., 2020; Fitchett et al., 2020).

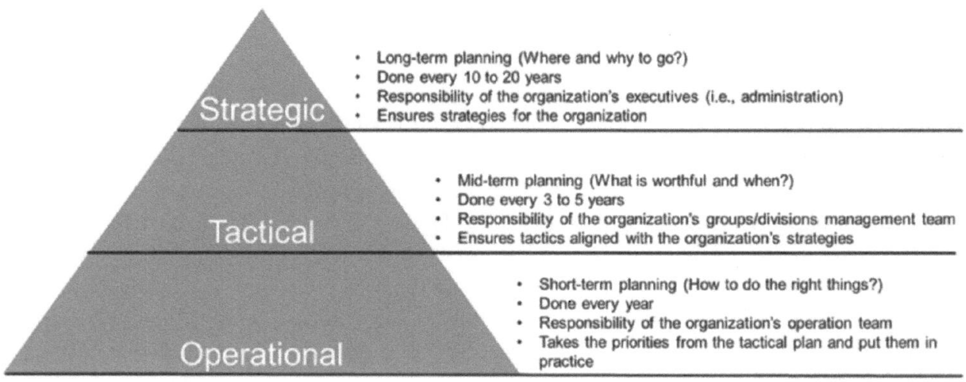

Figure 9.1 Infrastructure asset management planning levels.

After this introduction, the data requirements to assess the condition of an urban drainage system are presented. This data may be collected in different ways and their analysis facilitates determining structural and functional issues, and also the performance of the system. The causes and consequences of failure events are also needed for the condition assessment, being the subject of the risk section. Any intervention to the urban drainage system has a cost that must be assessed. Finally, decisions about urban drainage system assets rehabilitation need to be addressed in considering the different points of view or aspects that are often contradictory. This may be addressed by using multicriteria decision analysis methodologies.

9.2 Data requirements

Data about urban drainage system assets are of utmost importance. However, many assets lie buried and their condition assessment must be based on accurate information. Condition assessment is based on a diagnosis that aims to identify the anomalies present or signs of the existence of problems. These problems may involve four primary aspects of performance: hydraulic, environmental, structural, and operational (WRc, 2001). The identification of these problems can be done through analysis of the records of past events and other relevant information. Examples of relevant information may include records or studies of flood events, clearing procedures, sewer collapses, pumping station failures, employee accidents, sewer damages, inspections, user complaints, hydraulic behaviour analysis, and monitoring (Almeida & Cardoso, 2010). Additional information on the diagnosis scope may be necessary and includes relevant legal requirements and licences, operation and maintenance plans, mathematical modelling and computational simulation studies, inspection reports, urban drainage flowrates, rainfall, quality monitoring reports, and urban development plans.

The basic data needed to make a diagnosis may be categorised as follows (Almeida & Cardoso, 2010):

- **Cadastral** – refers to static data (i.e., that does not vary greatly in time) related to the design and building phases of the different assets, such as identification, type, location, dimensions, shape, material, depths, connections, etc. This information may exist in several formats, including paper or digital support, and it may or may not be complete and updated.
- **Functional and physical condition** – refers to dynamic data (i.e., that vary in time) of the different assets more related to their functionality and structural condition, which can be obtained from inspection information (reports, videos, photos), studies about the hydraulic behaviour, such as mathematical modelling and hydraulic and hydrological measurements, surveys of illicit industrial effluents discharge situations and locations, illegal discharges on the receiving environment, performance assessment reports (assessment period, performance indicators used, and targets considered), and structural condition assessment reports.
- **Operational and maintenance** – refers also to dynamic data, but most especially related to the daily operations and maintenance activities of the different assets, such as cleaning and maintenance activities of sewers (e.g., dates, locations, types of sediment removed, depth of the sediment layers, the volume of removed sediment), sewer blockage and collapse reports, incident reports (e.g., civil protection response due to flooding, polluted discharge in receiving medium), data complaints (e.g., regarding clogging, flooding or odours).
- **Inflow (quantity and quality)** – refers to dynamic data, obtained from the monitoring of different variables (e.g., level, velocity, flow rate, rainfall), such as domestic sewage flowrate, rainwater flowrate, industrial or commercial sewage flowrate, and groundwater flowrate.

These data are usually collected, stored, managed, and analysed using information systems that are often dispersed in different divisions of the utility. Over the last decades, water utilities have made significant investments in implementing information systems to address the increasing complexity of daily control, operation, management, and planning (Halfawy, 2008). Consequently, IAM is data intensive and uses a plethora of information systems (Haider, 2007), such as geographic information system (GIS), customer relationship management (CRM), enterprise resource planning (ERP), supervisory control and data acquisition (SCADA), and computerized maintenance management system (CMMS), among others.

The huge amount of assets and data involved makes data and information management a challenging task. New tools and processes are often needed to collect, gather, manage, analyse, and use asset data, creating and using these tools can stimulate and improve knowledge within the organisation and the decision-making process (Alegre et al., 2015). These developments in information systems are motivated by the growing need to make the utilities more resilient and flexible, with greater transparency and rationality in the decision-making process, as well as to respond to challenges more quickly and efficiently. However, data duplication should be avoided through good planning and coordinated development. Nevertheless, and regardless of these developments, engineers and managers often make their decisions based on incomplete, inaccurate, or out-of-date information (Carriço et al., 2020), with consequences for the decision-making process that eventually can compromise the quality, risk, and cost of the service provided.

Data collection is an activity consuming significant human, technological, and financial resources. Furthermore, frequently important data are neither collected nor recorded, or the available data do not respond to the needs or do not fulfil the requirements for the different stakeholders. For that reason, the water utility should previously identify the necessary information requirements to fulfil its objectives to rationalise the use of resources and maximise data collection effectiveness.

9.3 Structural condition assessment

A sewer is exposed to the action of several factors that may cause degradation of its physical condition until reaching the extreme situation of collapse. Therefore, a diagnosis is essential to identify critical situations (e.g., critical sewers identified as having a high risk of collapse) and is a core piece of a IAM methodology. According to the Institute of Public Works Engineering Australasia (IPWEA), condition assessment facilitates estimating the remaining service life, given the associated risks of failure and the future needs of the population, as well as work alongside risk management, maintenance planning, and data collection techniques to achieve robust and accurate results (IPWEA & NAMS, 2006).

In this context, physical condition assessment not only reflects asset structural integrity in a system, but also incorporates the consequences of a potential failure. This assessment consists of collecting the data and information through direct inspection, observation, and investigation and indirect monitoring and reporting and analysing the data and information to decide the structural status of the asset (USEPA, 2005). Thus, the adoption of a methodology that allows associating structural performance metrics to the system assets to create an information source of the whole system condition is necessary (Almeida & Cardoso, 2010).

Many condition assessment methodologies exist (Ana & Bauwens, 2010; Davies et al., 2001; Khan et al., 2010; Tran et al., 2007), but most of them converge in some steps, namely, setting objectives for the condition assessment, identification of the assets and available data, asset inspection, data analysis, and decision making (Feeney et al., 2009). The output result of any condition assessment method is generally given in a condition rating scale or condition index, which allows quantifying the asset condition (Amani et al., 2012; IPWEA & NAMS, 2015) and identifying priority assets. The condition rating scale may be carried out by using, or adapting, a known rating scale established by a regulation authority, a research institution, or a utility with recognised best practices.

Since it is impossible to inspect all linear metres of an urban drainage system due to prohibitive costs, many condition assessment methods focus on high consequence/high-risk approaches. In this case, the first decisions to be made are defining which sewers to inspect. There are several methods to define the priorities of sewers for inspection. Most methods evaluate both criticality (consequence of failure) and condition (likelihood of failure) (Feeney et al., 2009).

One of the most widespread methods used to assess structural conditions is the one recommended by the British Water Research Center – WRc (WRc, 2001). This was initially proposed in the 1980s and has been gradually improved. The method ranks the severity of sewers defects found during an inspection. Later, European authorities adopted the WRc method as their benchmark sewer defect coding standard. Thus, the approach indicates three criteria to select the assets for an inspection plan. They consider:

- The **consequences of failure**, which can be estimated without the need to carry out an inspection. Sewers are ranked depending on the economic impact resulting from the occurrence of their failure: A – sewers with high consequences in a failure case, the costs are at least two times higher than the pre-failure status; B – sewers where a failure has lesser consequences than Grade A, but in which it is desirable to avoid the collapse; C – all other situations. The first two grades (i.e., A and B) are critical and the last one (i.e., C) is non-critical.
- The available information on the determining factor for the **likelihood of failure**, in which the aim is to identify the assets that are most likely to have a failure (e.g., sewers at a small depth or low record of failures occurrence).
- Both consequences of failure and available information on the determining factor for the likelihood of failure, aiming to choose the sewers with a high **risk of failure**.

After the inspection, the observations are coded according to standard EN 13508-2 (CEN, 2003) and then the defects observed are ranked according to their severity using a numerical system ranging from 1 to 5, 1 being the best and 5 the worst, as shown in Table 9.1.

Functional and physical performance can then be determined from the level of structural condition.

9.4 Performance dimension

The main purpose of condition assessment is to identify defective assets and rehabilitate them. This rehabilitation will have an impact on the urban drainage system's performance, which will

Table 9.1 Structural condition ranking for inspected sewers (WRc, 2001)

Rank	*Condition*
5	Collapsed or collapse imminent
4	Expected collapse in the near future
3	Collapse not expected in the near future
2	Very low collapse expectation in the near future but potential for further deterioration
1	Acceptable structural condition

also impact the quality of service provided by a utility. So performance assessment is a pathway to quantify objectively the capabilities and defects of a system, providing support to rehabilitation actions. In addition, it promotes establishing independent and normalized comparisons (Cardoso, 2007). A performance assessment system (PAS) is the set of data, calculations, performance metrics, and contextual information that allow the evaluation and reporting of the performance of a single asset, a whole infrastructure, a service provided, or a utility (Alegre & Covas, 2010; Almeida & Cardoso, 2010).

In general, the performance metrics are the core component of the PAS, which can be grouped into three types (Alegre, 2007; Alegre & Covas, 2015; Almeida & Cardoso, 2010):

- **Performance indicators** – are quantitative measures of efficiency or effectiveness for the activity of a water utility. A performance indicator consists of a value resulting from an algebraic combination of different variables expressed in specific units. Performance indicators are typically expressed as ratios between variables: may be commensurate (e.g., in percentage) or non-commensurate (e.g., euros per cubic metre).
- **Performance indices** – are standardised and commensurable measures that may result from the combination of more disaggregated performance metrics (e.g., performance indicators or levels) or analysis tools (e.g., hydraulic simulation models, statistical tools, cost efficiency methods). Generally, a performance index aggregates into a single measure several aspects of an assessment.
- **Performance levels** – are qualitative measures typically built to express performance on an ordinal scale (e.g., excellent, good, fair, poor). Generally, they are adopted when not enough data or information exists to calculate performance quantitatively.

An urban drainage system's performance varies over time; therefore, the performance metrics (i.e., indicators, indices, or levels) should be calculated for the current condition and also for future scenarios of how its performance may vary over time. Moreover, performance metrics should always be associated with the quality of the data used to calculate them.

Performance indicators may be used to compare performance between the different areas of activity of a utility, to follow their evolution over time, or to check against predefined targets (Cardoso et al., 2004). This type of metric may also be used by an external entity, such as a regulator, to assess a utility's performance. Some countries have regulators (e.g., Portugal) that use a PAS based on performance indicators to evaluate the quality of service provided to users by the utilities (ERSAR, 2017).

According to Alegre et al. (2006), a PAS should comprise performance indicators, variables (i.e., data used in the calculation of the indicators), context information (i.e., information about aspects that can alter the indicator), and descriptive factors (i.e., indicators, variables, or other data that help to interpret the results of the indicators and to identify improvement measures). The standard ISO 24511:2007 recommends the use of a PAS in line with the IWA guidelines (Matos et al., 2003).

Each performance indicator requires:

- Precise definition, with concise meaning and unambiguous interpretation.
- The possibility of calculation by most of the utilities without significant additional effort.
- The possibility of verification by independent entities (e.g., regulator).
- Simplicity and ease of interpretation.
- Objective and unbiased measurement from a specific aspect to avoid subjective or distorted judgements.

Table 9.2 Example of a performance indicator definition sheet (adapted from Almeida & Cardoso, 2010)

wOp21 – Sewer renovation rate (%/year)
(Length of renovated sewers during the reference period × 365//reference period)/total length of existing sewer at reference date × 100 *wOp21 = (wD25 x 365 / wH1) / wC1 × 100* *wC1 – Total extension of sewer's network (km)* *wD25 – Length of renovated sewers (km)* *wH1 – Number of days in the reference period (day)*

The PAS should be defined considering the following requirements:

- Adequacy of representation of the main relevant aspects of the utility's performance.
- Absence of redundant meanings or objectives between the indicators.
- Refer to the same assessment period (one year is the most usual).
- Refer to the same geographical area.
- Applicability to utilities with different characteristics and degrees of maturities of management.

Each indicator and variable must be clearly defined as shown in Table 9.2.

Technical performance assessment, at the level required by engineering, is more suitable using performance indices instead of performance indicators. Some examples of performance assessment metrics, grouped by aspects, that may be used in an urban drainage system are as follows (Almeida & Cardoso, 2010):

- **Hydraulics –** water level, flow velocity, overflow volume, overflow peak and duration, the ratio between maximum wet weather flow and maximum dry weather flow.
- **Environmental –** concentration of pollutants, polluted overflow discharges, septicity.
- **Structural**—failure rate, leakage.
- **Economic**—maintenance costs, power costs.
- **Social –** urban drainage service disruption, complaints, and odours.

Performance indicators may be converted into performance indices through the application of a performance function or into performance levels when they are compared with reference levels. Consider the example of water level, which is one of the most important variables for the hydraulic performance assessment of sewers. This performance index is a direct measure of the sewer hydraulic behaviour and is also related to aeration efficiency and flooding potential. The selected performance function of the water level is shown in Figure 9.2. Considering that the average water level should not be above 0.5 of sewer diameter (D) for sewers sizes below or equal to 500 mm, and 0.75D otherwise. The best performance is achieved when the water level is below 0.5D for D ≤ 500 mm (or 0.75D for D > 500 mm). Water levels higher than these values are considered hydraulically less efficient. The limit of acceptance (a performance value of 2) will go up to D. Since most sewers can tolerate moderate pressure flow, despite its adverse effects, the function does not drop to zero at D. Instead, zero performance is assigned only when the water level reaches ground level (flooding), with a linear variation between. The performance function is applied to the calculated water level at each pipe.

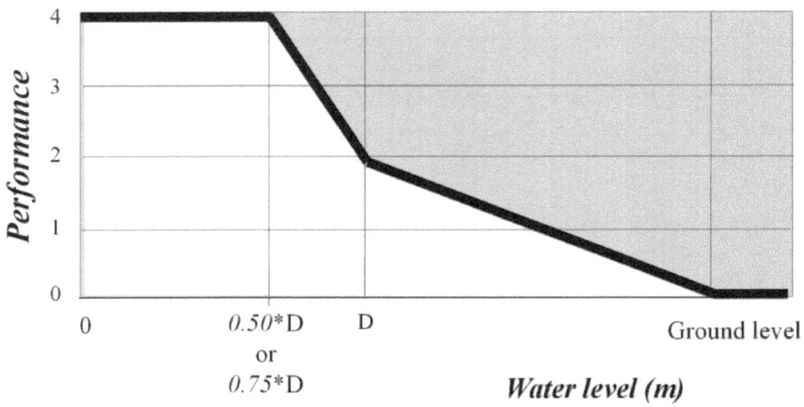

Figure 9.2 Water level performance function (adapted from Cardoso et al., 2005).

For the calculation of some of the hydraulic performance indices, it is necessary to resort to the use of mathematical modelling and/or computational simulation.

9.5 Risk dimension

Every activity of a utility involves risks that must be managed. The risk management process allows supporting the decision making considering the uncertainty and possible future events that are expected to have an impact on the utility's objectives. According to the standard ISO Guide 73:2009, the risk is understood as the effect of uncertainty on objectives (ISO, 2009). In engineering, the risk is expressed by the product of likelihood and consequence(s), being necessary to define a time frame for the analysis as well as outline associated consequences (Luís, 2014).

The rehabilitation activity is part of risk management since it aims to restore or improve the performance of the system or its assets, contributing to the fulfilment of the utility's objectives (Almeida & Cardoso, 2010).

Conceptually, risk management promotes preventive asset rehabilitation and involves several activities, such as:

- Identification of the main hazards.
- Establishment and monitoring of critical control points.
- Development of standardised operational procedures.
- Development and implementation of preventive maintenance plans.
- Guarantee that critical materials and equipment supplies are rapidly available.
- Development of emergency plans.

When defining rehabilitation actions, risk and reliability analysis methodologies are usually used. Risk and reliability are different but related concepts. The reliability of a system can be defined as the frequency or the likelihood that the system is in a satisfactory state. Sometimes reliability is considered the opposite of risk since risk or likelihood of failure may be simply one minus the reliability (Hashimoto et al., 1982). Risk and resilience may use the same analysis tools, such as Event Tree Analysis and Fault Tree Analysis. The most common risk management framework used under the reference standards AS/NZS 4360:2004 ISO 31000:2018 (AS/NZS, 2004; ISO, 2018a) is presented in Figure 9.3.

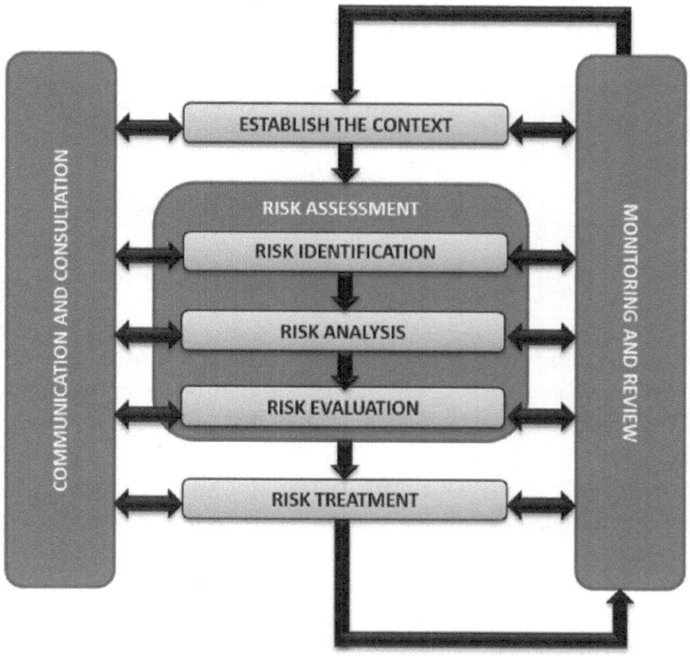

Figure 9.3 Risk management process (adapted from ISO 31000:2018).

The risk management process can be applied to support asset management planning as well as emergency response planning or specific plans to respond to relevant scenarios, including:

- Asset failure (e.g., sewer collapse, stoppage of a pumping station, WWTP process failure).;
- Incidents caused by external factors (e.g., fire or explosion in assets, facility intrusion, discharge of contaminants, or discharge of flammable substances into the sewerage).
- Incidents caused by natural hazards (e.g., earthquakes, floods).
- Context change scenarios (e.g., climate, social).

Risk assessment includes the processes of risk identification, risk analysis, and risk evaluation (ISO, 2018a).

Risk identification is the process of investigating, recognizing, and describing risks, and it includes identifying the risk sources, hazards, hazardous events, and factors affecting risk magnitude. Risk identification involves finding the sources of risk, events, situations, or circumstances that could have a significant impact on the objectives (of the system or organization) and the nature of that impact. This activity may involve the collection and treatment of historical data, theoretical analysis, and opinions of specialists. Risks can be identified using a wide variety of methods, such as checklists, scenario analysis, fault trees, and event trees (ISO, 2019). In addition to identifying risks, it is also important to identify the failure modes of critical assets that can lead to an aggravation of the consequences of an event. The use of reliability techniques, such as failure mode and effects analysis (FMEA) or failure mode, effects, and criticality analysis (FMECA), allows detailed analysis of asset failures with an emphasis on the identification of their causes.

Risk analysis is the process of estimating the risk for each event identified and consists in determining the likelihood and consequences of events combining them to obtain the level of risk.

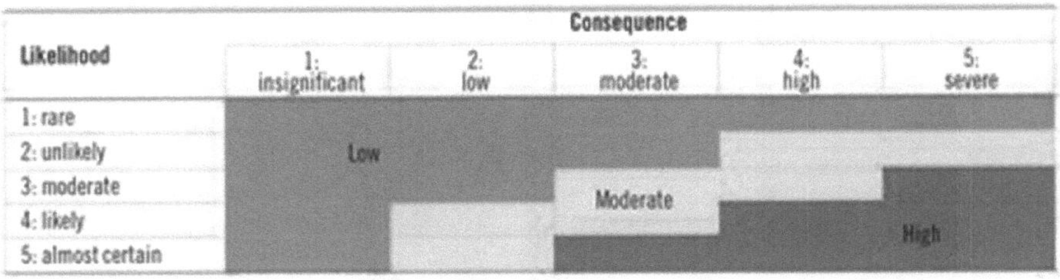

Figure 9.4 Example of a risk matrix.

The risk analysis is crucial for the subsequent stages of risk evaluation and decision making for risk treatment. Many available techniques can be applied for risk analysis, being grouped into qualitative (e.g., Delphi, fault trees, risk matrix) or quantitative (e.g., Monte Carlo simulation). One of the most used techniques in risk analysis is the risk matrix.

A risk matrix is a qualitative approach for estimating risk in situations where data are insufficient for detailed analysis or when consequences to be considered are non-numeric. This method is also widely used for ranking or screening risks when more than one risk is identified, providing a way to select those that require further analysis or to identify those that are broadly accepted. Application of the risk matrix requires a definition of appropriate scales for expressing both likelihood and consequence levels of potential events and a definition of the matrix to derive a risk level for each combination of likelihood and consequence, as in the example shown in Figure 9.4 (Leitão et al., 2013).

Risk evaluation requires an interpretation of the estimated risk in the previous stage of risk analysis aiming at a decision. This decision may be, for example, the need for risk control, the definition of risk prioritisations, or the definition of the alternative to be followed for risk treatment. The decision on the need for treatment and the type of treatment depends on the costs and benefits of taking risks and implementing controls.

According to the ISO 31000 standard (ISO, 2018a), the risk acceptance criterion should be defined in the phase of establishing the context (see Figure 9.3) and must take into account the following factors:

- The nature and types of causes and consequences that may occur and how will they be measured.
- How the likelihood is defined.
- The time horizon of the likelihood and consequence.
- How the level of risk is determined.
- The vision of the stakeholders.
- The level at which the risk is acceptable or tolerable.
- If combinations of multiple risks must be considered and, if so, how these combinations will be considered.

The simplest structure for defining the criteria of risk is a single level that divides the risks that need treatment from those that do not. If, on the one hand, this criterion provides attractively simple results, it may not, on the other hand, reflect the uncertainties involved in estimating the risk and defining the boundary between the risks that need treatment and those that do not need it.

The decision on the need for treatment and its type depends on the costs and benefits of taking risks and implementing controls. In this case, a common approach is to divide the risks into three levels:

i. A high level where the risk is considered intolerable; in this case, the risk treatment is essential regardless of its costs.
ii. An intermediate level where costs and benefits are taken into account.
iii. A low level where the risk is considered negligible or so small that no treatment is required.

Risk treatment is the process of selecting and implementing actions to modify risk and may involve:

- Elimination of the source or sources of risk.
- Changing the likelihood.
- Changing the consequence.
- Transferring, sharing, or retaining the risk.

Risk treatment options are not necessarily mutually exclusive or appropriate in all circumstances (ISO, 2019). The different options are integrated into decision methodologies together with other additional factors, namely, costs, social values, political and ethical principles, and legal issues. The treatment process includes two distinct moments:

i. Acceptance or non-acceptance of risk.
ii. The decision to select and rank risk intervention measures, determining the degree of control to prevent the aggravation of the current risk or mitigation to decrease the current risk.

9.6 Cost dimension

The cost dimension aims to assess:

- The deterioration of an infrastructure and quantification of the investment needs in rehabilitation.
- The total cost for comparison of the different alternatives of rehabilitation actions.

The infrastructure value index (IVI) may be used to assess the infrastructure deterioration and is given by the ratio between the current (fair) value of an infrastructure and the replacement cost on a modern equivalent asset basis (Alegre et al., 2014):

$$IVI(t) = \frac{\sum_{i=1}^{N} \left(rc_{i,t} \cdot \frac{rul_{i,t}}{eul_i} \right)}{\sum_{i=1}^{N} rc_{i,t}}$$

Where: t = reference time; IVI = infrastructure value index at time t; N = total number of assets; $rc_{i,t}$ = replacement cost of asset i at time t; $rul_{i,t}$ = residual useful life of asset i at time t; eul_i = expected useful life of asset i.

The IVI of mature, well-maintained infrastructures should be in the vicinity of 0.5 (0.4–0.6). IVI values from 0.5 to 1 correspond to "young" infrastructures with low rehabilitation needs. Values of IVI from 0.2 to 0.5 indicate that investments in rehabilitation should increase. IVI values less than 0.2 indicate that a considerable part of the infrastructure is not in an acceptable condition, requiring significant and urgent investment in rehabilitation.

An underlying assumption of this formulation is the availability of values for the following variables:

- Asset replacement costs (rc) are based on reference mean values. Time-specific market fluctuations shall be smoothed or eliminated; the use of unit cost functions for each type of asset is recommended (e.g., a unit cost function for small, medium, or large diameter pipes based on replaced length; for pumps, a function based on installed power; for sewers, a function based on the class of diameter and class of depth).
- Residual asset useful live (rul). In general, installation date and a linear depreciation are adopted.
- Expected asset useful live (eul) based on the current experience and information and not on the accounting / fiscal depreciation periods.

The assessment of the total cost of the different alternatives can be done through a typical cost-benefit analysis considering the costs and benefits over an extended period. The use of a common criterion for the calculation of the cost or benefit associated with each alternative in the analysis is necessary for comparison purposes. For that, the life cycle costing (LCC) and the whole-life costing (WLC) approaches may be used.

9.7 Mathematical modelling

Mathematical models are computational tools that allow simulating the behaviour of a system or process. The use of mathematical modelling of urban drainage systems requires specific knowledge and experiences, including:

- Systems functional performance requirements.
- Sewer and other assets hydraulics.
- Urban hydrology.
- Specific assumptions and conditions of use for the chosen computational application.
- Flow and other variables of interest measurement methods and their accuracy.
- Engineering solutions.

These models can be used at different levels, namely, at the planning level and the design and management of urban drainage systems level. Thus, they constitute essential tools in the implementation of an integrated methodology to the rehabilitation of these systems.

The models can be used for simulation of the behaviour of a real system or for forecasting using scenarios for the future, being the variables estimated based on past and present values of characteristic variables of the system.

Mathematical models are used to support the following activities, among others:

- **Planning** – to study strategic developments in the planning of urban drainage systems. In this case, models cannot be calibrated or validated.
- **Design** – the models facilitate evaluating, planning, and comparing design alternatives of new systems or expansion of existing systems. In this case, they are based on scenarios and projections and some data from other systems with similar characteristics can be used. These models cannot be calibrated or validated.
- **Operation and maintenance** – to study operation and maintenance scenarios either in normal or in emergency situations. These models should be based on statistically historical

information and should be calibrated and verified using for example time series of rainfall or flow.

- **Rehabilitation** – to diagnose existing flaws, namely, in the hydraulic and environmental aspects, in the study of corrective solutions, and in the assessment of interventions phasing scenarios. Like the previous models, these models should be based on statistically historical information and should be calibrated and verified using, for example, time series of rainfall or flow.

Typically, mathematical models aiming at rehabilitation interventions in urban water systems are used mainly to evaluate structural and hydraulic conditions. These models may be physically based or deterministic and statistically based (Malek Mohammadi et al., 2019). The first group are intended to understand the physical mechanisms that drive sewer condition, and the second use mathematical relationships to relate the history of sewer conditions with sewer deterioration factors (Caradot et al., 2017).

Models of urban water systems typically consist of (Coelho et al., 2006):

- A set of data describing the physical characteristics of the system, its demands (e.g., inflows), and its operational conditions.
- A set of mathematical equations that reproduce the (e.g., hydraulic) behaviour of individual components and the system, expressed in terms of main state variables (e.g., sewer flow) instantiated by the data referred to in (i).
- Numerical algorithms are suitable for solving mathematical equations.

The computational applications available, whether commercial or free of charge, make possible the development of models for specific systems as long as data referred in (i) are available.

Depending on the model function, it may be relevant to adopt different levels of system simplifications. These simplifications must be guided by specific criteria to maintain the useful response of the model in line with the real system (e.g., main sewers above a given diameter, simplified description of contributing drainage basins) while maintaining critical components (e.g., manholes where topological variations occur, measurement points, outfalls, structures with storage volumes, regulation, or control components).

Simplifications can have several advantages in reducing the time spent in data collection and verification and the volume of data to be manipulated. Also, they can reduce the errors associated with the manipulation of large data volumes and increase the ease of interpreting the system's functioning. For example, in an urban drainage system three types of simplification can be considered in hydraulic modelling:

- Omission of sewers from peripheral areas (i.e., smaller diameters) and consideration of inflow to the downstream sewer.
- Merging the consecutive sewers with the same characteristics into a single sewer.
- Replacement of a complex network by another simpler one but with a similar functional response.

The model development must be well documented to facilitate the identification of simplifications or specific conditions for later uses. Since mathematical modelling implies the use of large data volumes it is recommended to adopt appropriate procedures to minimize errors.

9.8 Multicriteria decision analysis

Traditionally, decisions on sewer system rehabilitation are based only on financial criteria, focused on the capital cost and the operation and maintenance (O&M) costs. The rehabilitation decision process of urban drainage systems based on a single point of view (e.g., cost analysis) is not adequate as it neglects the system's performance, the risks of failure, and the whole life costing of the system infrastructure. Lately, decision makers are acknowledging the importance of points of view (e.g., performance and risk) in considering the perceptions of the utility, other stakeholders, and the society at large; consequently, the complexity of the decision-making process increases. Generally, increasing performance and reducing risk and cost are conflicting objectives. Thus, the aggregation of criteria and metrics of different dimensions can significantly benefit from the use of multicriteria decision analysis (MCDA) tools. Some authors may use the term "multicriteria decision making" (MCDM) instead of MCDA and also use the term "multiple criteria" or "multicriteria" instead of multicriteria.

The main purpose of the MCDA methods is to help find solutions for real-life problems, often problems with conflicting points of view (Vincke, 1992). The multicriteria problem is related to the methods and procedures by which different criteria, points of view, and perspectives can be formally involved in the decision-making process. MCDA is generally divided into two different branches depending on whether the decision problems are discrete (multi-attribute decision making [MADM]) or continuous (multiobjective decision making [MODM]) (Bouyssou et al., 2006; Kazimieras Zavadskas et al., 2019; Malczewski, 2018).

An MCDA methodology is composed, typically, of three main phases: (1) identification of the decision problem; (2) problem structuring; and (3) problem evaluation, as shown in Figure 9.5. Problem structuring is a crucial stage for the outcomes. It includes identifying objectives, selecting scenarios within the analysis period, and defining problem options and the problematic type. Problem evaluation comprises the selection of assessment metrics, choice of an aggregation method, application of the method, and sensitivity and robustness analysis (Carriço et al., 2021).

The identification of the decision problem is based on the diagnosis carried out during the IAM planning. The objectives relevant to the decision problem under analysis should be identified. The external factors, such as political, normative-legal, economic-social, demographic,

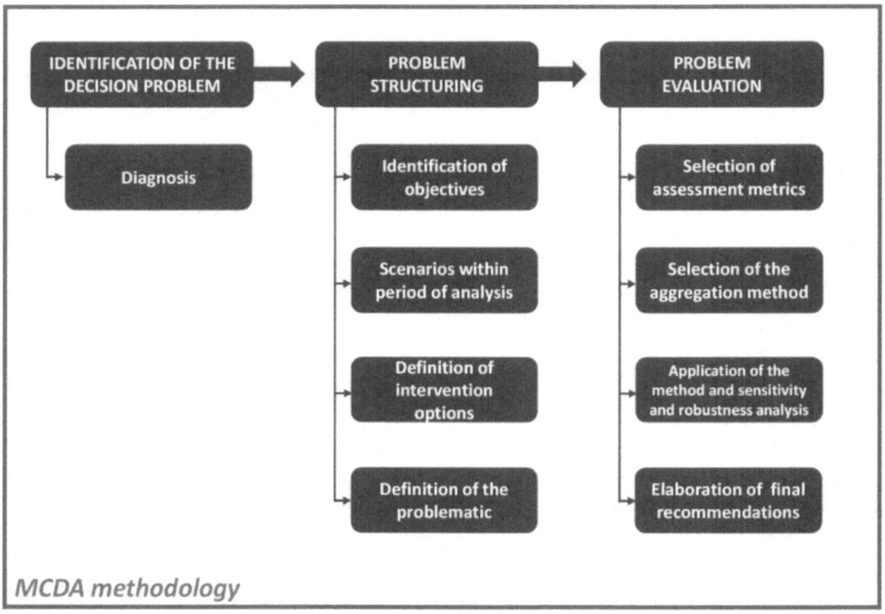

Figure 9.5 MCDA methodology (adapted from Carriço et al., 2021b).

technological, and environmental, should be identified and scenarios within the analysis period should be addressed. The decision intervention options may be of different types, such as infrastructural (e.g., rehabilitation of sewers), operational (e.g., modification of a pumping station operation mode), and/or maintenance (e.g., replacement of a pump in a pumping station). The definition of the problematic, which is the establishment of the type of problem (e.g., ranking, sorting, and selection), is a very important step in problem structuring.

The multicriteria methods require identification of key factors that form the basis of problem evaluation. These factors are referred to in different forms, such as values, objectives, criteria, and fundamental points of view. The different terminologies derive, mainly, from the coexistence of different schools of thought within MCDA. The key factors that arise from the problem structuring phase will be used to select the assessment metrics. This selection follows formal rules, and the resulting metrics are used to evaluate the consequences of each decision. Selected metrics can be aggregated in a later stage. The selection of the aggregation method understood as a mathematical algorithm to address this type of problem will essentially depend on the problematic, on the mode of how the problem was structured, and on the type of information to be modelled with the metrics.

Many MCDA methods have been developed, namely, weighted sum, multi-attribute utility theory (MAUT), multi-attribute value theory (MAVT), measuring attractiveness by a categorical-based evaluation (MACBETH), analytic network process (ANP), analytic hierarchy process (AHP), technique for order of preference by similarity to ideal solution (TOPSIS), and Élimination et choix traduisant la realité (ELECTRE). Most of these methods are suitable for ranking the problematic but others, like ELECTRE, are a family that can be used for different types of problematics.

9.9 Final remarks

This chapter briefly presents several concepts and methodologies to model urban drainage infrastructure rehabilitation. Nowadays, rehabilitation has become commonly grouped under infrastructure asset management. Several methods have been developed worldwide, most including a step of condition assessment of the assets of infrastructure. The condition assessment can use direct or indirect methods, the former requiring the inspection of the asset, and the latter requiring the collection and analysis of assets data and information. Direct methods are limited by costs. Indirect methods are used to predict asset conditions. After data and information analyses, the asset's remaining useful life and its long-term performance are carried out. Decision making about rehabilitation interventions or further inspections can be based on these assessments.

References

Alegre, H., Baptista, J.M., Cabrera Jr, E., Cubillo, F., Duarte, P., Hirner, W., Merkel, W., Parena, R. (2006). Performance indicators for water supply services, second edition, Manual of Best Practice Series, (p. 305). IWA Publishing, London, ISBN: 1843390515.

Alegre, H. (2007). *Gestão patrimonial de infra-estruturas de abastecimento de água e de drenagem e tratamento de águas residuais*. LNEC.

Alegre, H., & Coelho, S. T. S. T. (2012). Infrastructure asset management of urban water systems. In Avi Ostfeld (Ed.), *Water supply system analysis: Selected topics* (pp. 1–26). InTech. https://doi.org/10.5772/52377

Alegre, H., & Covas, D. (2010). Gestão patrimonial de infra-estruturas de abastecimento de água: Uma abordagem centrada na reabilitação. In ERSAR (Ed.), *Série Guias Técnicos* (Vol. 16). ERSAR.

Alegre, H., & Covas, D. (2015). *Integrated planning of urban water services: A global approach*. Deliverable D53.1a. Manual 1. TRUST

Alegre, H., Brito, R. S., Neves, C., & Almeida, R. (2015). Standarization in infrastructure asset management (in Portuguese). Conference paper. ENEG 2015, Porto. 01–04 December 2015.

Alegre, H., Vitorino, D., & Coelho, S. (2014). Infrastructure value index: A powerful modelling tool for combined long-term planning of linear and vertical assets. *Procedia Engineering*, 89, 1428–1436. https://doi.org/10.1016/j.proeng.2014.11.469

Almeida, M. C., & Cardoso, M. A. (2010). *Infrastructure asset management of urban drainage and stormwater services. An aproach based on rehabilitation* (in Portuguese). Technical Guide n.17. ERSAR, LNEC.

Almeida, M. C., Covas, D., & Beceiro, P. (2015). Rehabilitation of sewers and manholes: Technologies and operational practices. D53.1e. Manual 6. TRUST

Amani, N., Nasly, M. A., & Samat, R. A. (2012). Infrastructure component assessment using the condition index system: Literature review and discussion. *Journal of Construction Engineering and Project Management*, 2(1), 27–34. https://doi.org/10.6106/JCEPM.2012.2.1.027

Ana, E.V., & Bauwens, W. (2010). Modeling the structural deterioration of urban drainage pipes: The state-of-the-art in statistical methods. *Urban Water Journal*, 7(1), 47–59.

AS/NZS. (2004). *AS/NZS 4360:2004: Risk management*. Standards Australia.

Beuken, R., Eijkman, J., Savic, D., Hummelen, A., Blokker, M., & Evides, J. E. (2020). Twenty years of asset management research for Dutch drinking water utilities. *Water Supply*, 20(8), 2941–2950. https://doi.org/10.2166/ws.2020.179

Bouyssou, D., Marchant, T., Pirlot, M., Tsoukiàs, A., & Vincke, P. (2006). *Evaluation and decision models with multiple criteria: Stepping stones for the analyst*. Springer Science + Business Media, Inc.

BSI. (2008). *PAS 55–1:2008 Asset management. Specification for the optimized management of physical assets*. The Institute of Asset Management.

Caradot, N., Sonnenberg, H., Kropp, I., Ringe, A., Denhez, S., Hartmann, A., & Rouault, P. (2017). The relevance of sewer deterioration modelling to support asset management strategies. *Urban Water Journal*, 14(10), 1007–1015. https://doi.org/10.1080/1573062X.2017.1325497

Cardoso, M. A. (2007). *Avaliação do desempenho de sistemas de drenagem urbana: Vol. Doutoramen*. Instituto Superior Técnico (IST).

Cardoso, M. A., Brito, R., Ribeiro, R., & Alegre, H. (2017). *Infrastructure asset management: Maturity assessment of water utilities based on international standards*. ISO 55000.

Cardoso, M. A., Coelho, S., Matos, R., & Alegre, H. (2004). Performance assessment of water supply and urban drainage systems. *Urban Water Journal*, 1(1), 55–67. https://doi.org/doi:10.1080/15730620410001732053

Cardoso, M. A., Coelho, S. T., Praca, P., Brito, R. S., & Matos, J. (2005). Technical performance assessment of urban sewer systems. *Journal of Performance of Constructed Facilities*, 19(4), 339–346.

Carriço, N., Covas, D., & Almeida, M. do C. (2021a). Multi-criteria decision analysis in urban water asset management. *Urban Water Journal*. https://doi.org/10.1080/1573062X.2021.1913613

Carriço, N., Covas, D., & Almeida, M. do C. (2021b). Rehabilitation of an industrial water main using multicriteria decision analysis. *Water*, 13(22), 3180. https://doi.org/10.3390/w13223180

Carriço, N., & Ferreira, B. (2021). Data and information systems management for the urban water infrastructure condition assessment. *Frontiers in Water*, 3, 670550https://doi.org/10.3389/frwa.2021.670550

Carriço, N., Ferreira, B., Barreira, R., Antunes, A., Grueau, C., Mendes, A., Covas, D., Monteiro, L., Santos, J., & Brito, I. S. (2020). Data integration for infrastructure asset management in small to medium-sized water utilities. *Water Science and Technology*, 82(12), 2737–2744. https://doi.org/10.2166/wst.2020.377

Carriço, N. J. G., Covas, D. I. C., Almeida, M. C., Leitão, J. P., & Alegre, H. (2012). Prioritization of rehabilitation interventions for urban water assets using multiple criteria decision-aid. *Water Science and Technology*. 66(5), 1007–1014.

CEN. (2003). EN 13508-2: Conditions of drain and sewer systems outside buildings: Part 2: Visual inspection coding system. *Engineering*, 360, 226.

CEN. (2017). *EN 752:2017: Drain and sewer systems outside buildings: Sewer system management*. CEN.

CEN. (2018). *EN 15885:2018: Classification and characteristics of techniques for renovation, repair and replacement of drains and sewers*. CEN.

Coelho, S. T., Loureiro, D., & Alegre, H. (2006). Modelling and analysis of water supply systems (in Portuguese). IRAR, Lisboa.

Davies, J. P., Clarke, B. A., Whiter, J. T., & Cunningham, R. J. (2001). Factors influencing the structural deterioration and collapse of rigid sewer pipes. *Urban Water*, 3(1–2), 73–89. https://doi.org/10.1016/S1462-0758(01)00017-6

Dawood, T., Elwakil, E., Novoa, H. M., & Delgado, J. F. G. (2020). Artificial intelligence for the modeling of water pipes deterioration mechanisms. *Automation in Construction*, 120, 103398. Scopus. https://doi.org/10.1016/j.autcon.2020.103398

ERSAR. (2017). *Water and waste services quality assessment guide: 2nd generation of the assessment system*. ERSAR.

Feeney, C., Thayer, S., Bonomo, M., & Martel, K. (2009). White paper on condition assessment of urban drainage collection systems. www.epa.gov

Fitchett, J. C., Karadimitriou, K., West, Z., & Hughes, D. M. (2020). Machine learning for pipe condition assessments. *Journal: American Water Works Association*, 112(5), 50–55. Scopus. https://doi.org/10.1002/awwa.1501

Fjeldhus, K., & Ugarelli, R. (2014). Infrastructure asset management: A proposed framework for Norwegian water utilities. *VANN*, 04, 473–481

Grigg, N. S. (2003). Water, urban drainage, and stormwater infrastructure management. In *Water, urban drainage, and stormwater infrastructure management*. Lewis Publishers. https://doi.org/10.1201/9781420032338

Haider, A. (2007). *Information systems for engineering and infrastructure asset management*. Springer Gabler. https://doi.org/10.1007/978-3-8349-4234-0

Halfawy, M. R. (2008). Integration of municipal infrastructure asset management processes: Challenges and solutions. *Journal of Computing in Civil Engineering*, 22(3), 216–229. https://doi.org/10.1061/(asce)0887-3801(2008)22:3(216)

Hashimoto, T., Stedinger, J. R., & Loucks, D. P. (1982). Reliability, resiliency, and vulnerability criteria for water resource system performance evaluation. *Water Resources Research*, 18(1), 14–20. https://doi.org/10.1029/WR018i001p00014

IAM. (2015). *Asset management: An anatomy*. IAM.

IPWEA, & NAMS. (2006). *International infrastructure management manual* (2nd ed.). INGENIUM; IPWEA.

IPWEA, & NAMS. (2015). *International infrastructure management manual* (5th ed.). IPWEA; NAMS Group.

ISO. (2009). *ISO Guide 73:2009: Risk management: Vocabulary*. International Organization for Standardization.

ISO. (2014a). *ISO 55000:2014: Asset management: Overview, principles and terminology*. International Organization for Standardization.

ISO. (2014b). *ISO 55001:2014: Asset management: Management systems: Requirements*. International Organization for Standardization.

ISO. (2018a). *ISO 31000:2018: Risk management: Principles and guidelines*. International Organization for Standardization.

ISO. (2018b). *ISO 55002:2018: Asset management: Management systems: Guidelines for the application of ISO 55001*. International Organization for Standardization.

ISO. (2019). *ISO/IEC 3010:2019: Risk management: Risk assessment techniques*. International Electrotechnical Commission.

Kazimieras Zavadskas, E., Antucheviciene, J., & Kar, S. (2019). Multi-objective and multi-attribute optimization for sustainable development decision aiding. *Sustainability*, 11(11), 3069. https://doi.org/10.3390/su11113069

Khan, Z., Zayed, T., & Moselhi, O. (2010). Structural condition assessment of sewer pipelines. *Journal of Performance of Constructed Facilities*, 24(2), 170–179. https://doi.org/10.1061/(ASCE)CF.1943-5509.0000081

Leitão, J. P., Almeida, M. D. C., Simões, N. E., & Martins, A. (2013). Methodology for qualitative urban flooding risk assessment. *Water Science and Technology*, 68(4), 829–838. https://doi.org/10.2166/wst.2013.310

Luís, A. (2014). *Strategic risk management in water utilities: Development of a holistic approach linking risks and futures*. Cranfeld University.

Malczewski, J. (2018). 1.15: Multicriteria analysis. In B. Huang (Ed.), *Comprehensive geographic information systems* (pp. 197–217). Elsevier. https://doi.org/10.1016/B978-0-12-409548-9.09698-6

Malek Mohammadi, M., Najafi, M., Kaushal, V., Serajiantehrani, R., Salehabadi, N., & Ashoori, T. (2019). Sewer pipes condition prediction models: A state-of-the-art review. *Infrastructures*, 4(4), 64. https://doi.org/10.3390/infrastructures4040064

Matthews, J. C., Selvakumar, A., Sterling, R., & Condit, W. (2012). Analysis of urban drainage and water system renewal decision-making tools and approaches. *Journal of Pipeline Systems Engineering and Practice*, 3(4), 99–105. https://doi.org/10.1061/(ASCE)PS.1949-1204.0000114

Matos, R., Cardoso, A., Duarte, P., Ashley, R., Molinari, A. and Schulz, A. (2003). Performance indicators for wastewater services, Manual of Best Practice Series, (p. 192). IWA Publishing, London, ISBN: 9781900222907

Osman, H. (2012). Agent-based simulation of urban infrastructure asset management activities. *Automation in Construction*, 28, 45–57. https://doi.org/10.1016/j.autcon.2012.06.004

Porse, E. (2013). Stormwater governance and future cities. *Water*, 5(1), 29–52. https://doi.org/10.3390/w5010029

Thomson, J., Flamberg, S., & Condit, W. (2013). Primer on condition curves for water mains (EPA/600/R-13/080). U.S. Environmental Protection Agency. https://cfpub.epa.gov/si/si_public_record_report.cfm?Lab=NRMRL&dirEntryId=266113

Tran, D. H., Perera, B. J. C., & Ng, A. W. M. (2007). Neural network based prediction models for structural deterioration of urban drainage pipes *Land, Water and Environmental Management: Integrated Systems for Sustainability, Proceedings, Christchurch,* 1, 2264–2270

Ugarelli, R., Venkatesh, G., Brattebo, H., Di Federico, V., & Saegrov, S. (2010). Asset management for urban urban drainage pipeline networks. *Journal of Infrastructure Systems*, 16(2), 112–121.

USEPA. (2005). *Guide for evaluating capacity, management, operation, and maintenance (CMOM) programs at sanitary sewer collection systems*. USEPA.

Vincke, P. (1992). *Multicriteria decision-aid*. John Wiley & Sons, Inc.

WRc. (2001). *Sewer rehabiltation manual: Vol. 4th Edition*. Water Research Council Publications.

10

HISTORY OF TECHNOLOGICAL CHANGE IN URBAN WASTEWATER MANAGEMENT, 1830–2010

Jonas Hallström and Martin V. Melosi

10.1 Introduction

The evolution of humans has been a co-evolution with technology. From the dawn of humanity, solutions to problems have resulted in a made, technological world, designed to sustain, control, and expand human societies (e.g. Hughes, 2004; Nye, 2006). One recurring challenge for these societies has been the provision of freshwater and disposal of wastewater, which initially was very rudimentary but became more and more complex with the emergence of towns and cities in a few thousand years BCE in the Middle and the Far East. These urban societies around the Nile, Euphrates, Tigris, Indus, and Yangtse Rivers were heavily dependent on river water for agriculture, and, therefore, they have sometimes been labelled "hydraulic societies" (cf. Buhagiar, 2016). During mid- to late antiquity, Greek and Roman cities came to embody the state of the art of urban water technology, unsurpassed for more than a thousand years. With the advent of industrialisation and a new much more substantial urbanisation in western Europe and the United States in the 19th century, water supply and sewerage were designed as city-wide technological systems for the modern city (Hallström, 2009; Melosi, 2008).

This chapter describes and analyse the most important stages of technological change in urban wastewater management during the period 1830–2010, with examples taken primarily from Sweden, northern Europe, and the United States.

10.2 Background: Wastewater management before the modern era

Management of wastewater was a fundamental activity in the first urban human societies because it was located near large rivers with substantial seasonal variations in flow. These populations thereby learned to divert storm- and wastewater from one place to another. More advanced urban sewer systems were introduced from ca. 2,500 BCE in, for example, Mohenjo-Daro in present-day India; there were baths, sewers, and rudimentary water closets. In Mesopotamia, there also were networks of brick sewers in certain places by this time. In Persia, the widespread and sustainable water provision system of *qanats* – dug shaft wells connected by subterranean tunnels through a common aquifer – also doubled as drainage from farmlands. The Greeks

DOI: 10.4324/9781003057574-13

were very likely inspired by the Persian *qanats* when constructing their underground aqueducts, and during the Hellenistic period Greek wastewater technology was developed with drainage pipes, sewers, and simple toilets. The Romans are well known as the engineers of the apotheosis of water technology in magnificent, above-ground aqueducts and integrated, city-wide water networks in the capital of Rome. However, the sewer networks were integral to their way of building cities, for example, the large main conduit of Rome, the *Cloaca maxima* (Buhagiar, 2007; Harandi & de Vries, 2014; Mays, Koutsoyiannis, & Angelakis, 2007).

Despite the fall of the western Roman Empire and its institutions in the 5th century CE, administrative continuity continued in southern Europe in the form of the Roman Catholic Church. Moreover, in terms of infrastructure, trade routes, roads, and water and wastewater networks continued to be used and developed by Visigoths and later Arabs in southern Europe during the Early Middle Ages. There were also old Greek and Persian systems in use in Spain, Malta, and southern Italy (Buhagiar, 2016; Harandi & de Vries, 2014; Squatriti, 1998). In northern Europe, Roman systems were also in use throughout the Middle Ages, for example, in Paris, London, and some German and Nordic towns (e.g. Westholm, 1995).

All in all, however, the inhabitants of Europe had to rely on off-grid systems of wastewater management during the Middle Ages and well into the early modern and modern eras. The same also can be said about colonial and post-colonial America. In cities and towns, there were gutters for managing stormwater and household wastewater, and privy vaults, cesspools, and outhouses for human excrement, the latter of which were also employed as the primary means of sanitation in the countryside (Katko & Juuti, 2014; Melosi, 2008; Tarr, 1996). In the 17th and 18th centuries, there were so-called protosystems – fairly rudimentary technological solutions or systems that were privately or publicly managed or operated – but these were virtually always for water provision, not sewage management. There were sewers before the mid-19th century. However, they were generally designed for drainage of marshy land or stormwater and built in a piecemeal fashion to the nearest waterway, so they could hardly be called systems in the modern sense. A few major cities, such as London, also had sewers with water closets by this time (Hallström, 2009; Melosi, 2000).

The premodern attitude about the disposal of wastewater was thus mostly to drain liquid waste from wherever it was deemed unnecessary, which meant manual discharge in a street, cesspool, ditch, or pipe to the nearest watercourse. Before the mid-19th century, the sewerage of ancient civilizations, notably Rome, remained unmatched. The sewers of the early modern period were street gutters or drains rather than real underground sewers and the discharge of waste in the sewers was forbidden, although a great deal ended up there anyway. Street cleaning was also a way of cleansing the cities of wastes that could end up in the drainage by means of scavengers (Buiter, 2008; Melosi, 2008).

10.3 Technological change in modern urban wastewater management practices and systems

10.3.1 From drainage to sewerage: The British public health movement and the evolution of sewerage systems, 1830–1900

Englishman Edwin Chadwick was one of the earliest proponents of sewerage as a technological *system* in a modern sense. Originally a barrister, Chadwick developed an interest in sanitary reform amid the dirty, unhealthy urban environments of Great Britain in the 1830s and 1840s. His was a technological solution: technological water and sewerage systems should be connected to improve and cleanse the environment because a clean environment was seen as a way of preventing disease and poverty. His system visions borrowed from nature; his water supply and sewerage systems were compared with the artery and vein systems of the human body, and his thinking about recycling of waste was like "bringing as it were the serpent's tail into the serpent's mouth," that is, designing a holistic system (quoted in Binnie, 1981, p. 12). Contemporary English engineers did not take to Chadwick's ideas immediately because pipe-sewer trials in several English cities were anything but promising and there was resistance from local plumbers. However, Chadwick had a major influence on views about water and sewerage technology in Britain, the rest of Europe, and the United States during the mid- and latter- half of the 19th century. Sewerage was increasingly seen as an integrated technological system in theory if not in practice (Hamlin, 1992; Hughes, 2012; Melosi, 2008).

Although sewage pipes had been built in connected networks as early as antiquity, with the most advanced of these being constructed by the Romans, the 19th-century innovations were different in several ways. Modern wastewater networks included pipes that should be self-cleaning, initially either large egg-shaped brick sewers or small, round, glazed clay pipes (pipe sewers) in order to be used in large, centrally controlled urban systems. These sewerage systems also should transport waste quickly away and be as maintenance free and automated as possible. Sewerage and other modern systems were characterised by large capacities, speed, and automation as well as a dependence on a centralised municipal organisation with the legal right to apply charges or taxes, incur debt, etc. (Hallström, 2009; Tarr & Dupuy, 1988). Before their introduction, however, waste technology systems, such as sewers and toilets, faced strong resistance from those who relied on the old off-grid systems for their livelihoods, often those who manually collected and sold the waste from the cities. The extension of sewer systems was also uneven during this initial phase; the infrastructure was sometimes mixed with earlier solutions, and affluent neighbourhoods were generally networked before poorer ones (Hallström, 2011; Melosi, 2000; Tarr, 2020).

The ideas of Edwin Chadwick and the British public health movement provided the specific historical context for the diffusion of modern sewerage technologies in Europe and America in the mid to late 19th century. Modern sewer systems were thus needed to clean up the growing cities. Not only should they improve sanitary conditions and fight disease (mainly cholera), but they should also boost morality and prevent a working-class revolution. Britain was in this sense not only the great industrial model, but also the primary public health and engineering model at the time, although Germany and France also exerted some influence (cf. Katko & Juuti, 2014). In Sweden and most other Nordic countries, continental Europe, and the United States, the models for wastewater systems were therefore mainly drawn from Britain during this initial period up to the turn of the 20th century (Drangert, Nelson, & Nilsson, 2002). In particular, the new London sewer system with interceptors installed in the Embankment and designed by Joseph Bazalgette after the "Great Stink" in 1858, impressed European and American engineers (Buiter, 2008; Porter, 1998).

The most prominent Swedish civil engineers of the late 19th century came from the Royal Corps of Engineers. They went on study trips primarily to Great Britain and Germany but were mainly inspired by British role models. The situation was similar in America where engineers also were inspired by their colleagues in Britain. Sweden followed suit but the dominance of combined sewerage using the same pipes for both stormwater and household wastewater endured. J. G. Richert, who was the primary Swedish sewerage engineer (Richert, 1869), designed combined sewerage networks constructed mainly with glazed clay pipe sewers. This had been impossible in Britain where pipe sewers were smaller (Hamlin, 1992). Richert adapted the pipes he designed to suit Swedish conditions, which meant building networks of pipe sewers with greater pipe dimensions than in Britain and with brick sewers at critical junctions (Hallström, 2009).

In contrast with strides made in US waterworks, the development of underground wastewater systems there was limited until about 1880. Few cities were in a position to finance two major technologies of sanitation simultaneously. Private companies had footed the bill for the early development of water-supply systems, sometimes with governmental support. A similar path for sewerage systems was unlikely because its revenue-generating potential was limited. Eventually, underground sewers came to be recognised as valuable in staving off epidemics, in preventing flooding, and in making available connections to water closets. Privy vaults and cesspools had been relatively effective and inexpensive disposal options until piped-in water was available. Open ditches sufficed as storm drains in communities not yet experiencing rapid growth. Expanding urban populations and piped-in water began to challenge the old methods of sewage disposal as early as the 1830s. As in England, water closets appeared in some middle-class homes soon after running water became available. As running water became more common, the cesspool-privy vault systems began to fail. The controversy came to a head in the United States in the 1880s.

Combined systems, using one large pipe for household waste and stormwater, came first. A newer technology, the separate system, often used two pipes — a small one for household waste and a larger one for stormwater. In some separate systems, stormwater was simply diverted into street gutters. The first planned systems (although not city-wide) appeared from the late 1850s through the 1870s. Only combined systems were successfully built in the United States in the 1860s and early 1870s. (In both Britain and America, combined sewerage was the dominating design in the 19th century, largely because of the lack of successful separate systems.) At this time, disposal of sewage by dilution in large bodies of running water was the prevailing practice. Sewage irrigation and chemical precipitation were well established, and intermittent sand filtration was under study as early as 1868. But none of these methods were widely practised in the United States before the 1880s (Melosi, 2008).

Towards the end of the 19th century, it dawned on engineers and users that the downside of going from drainage to sewerage systems was pipe blockage and river pollution, and efforts were made to avoid these by regulating what the users put in toilets and by experimenting with different pipe diameters and flow rates. Water pollution became an urgent problem in Great Britain as early as the early 19th century, due to the large number of toilets and industries releasing untreated waste directly into watercourses, affecting both people and nature. In Continental Europe and in America, water pollution was recognised first as a nuisance and later as a health problem in the late 19th century, as industrial pollution and discharge from the increasing number of water closets spread. The same polluted water often entered the water systems of other towns farther downstream, which exacerbated these issues (Hamlin, 1992; Melosi, 2008).

Attempts were made to resolve the pollution of waterways with rudimentary wastewater treatment. Chadwick proposed that wastewater should be used as an agricultural fertiliser, and this was the most popular treatment technique from 1850 until the turn of the 20th century; sewage farms operated in Europe, for instance, outside of Edinburgh, Berlin, and Paris, and in the United States in Illinois, Wyoming, and California. However, this technique was plagued by many problems: large areas of farmland were required, irrigating the land with sewage spread an unpleasant smell, the waste came to farmers even during those months of the year when no fertiliser was needed, and the schemes did not support themselves economically. A similar method, intermittent filtration, was refined in the United States in the late 19th century but was confined to the Atlantic coast. Chemical treatment methods, for instance, chemical precipitation, were also developed during the latter part of the 19th century without really succeeding (Denton, 1877; Goubert, 1988; Melosi, 2008; Reid, 1991). On the European periphery, for example, in

Sweden, the existence of large waterways in combination with restrictions on the use of water closets, few industries, and a largely rural population meant that the problem of river pollution was only beginning to be recognised (Hallström, 2004).

The varied response to the pollution of rivers and lakes in both Europe and America also had to do with the dominating perceptions of water quality in the late 19th century (cf. Gurian & Tarr, 2011). Examples from Sweden testify to the strong influence of variants of the miasmatic theory of disease (disease emanating from filth and foul smells and thus non-contagious) until the turn of the 20th century. Together with prevalent methodologies of chemical analysis, the miasmatic theory affected the way in which water quality was conceived and the general notion of what kind of water was good. It had to be free from organic matter, not too hard, and taste and smell good (cf. Hamlin, 1990; Richert, 1869). Professionals were aware that contaminated soil also could affect groundwater and watercourses, but the maxim that "running water purifies itself" was generally valued. It was putrefying organic material that was considered the cause of disease, whether poisonous odours that emanated from it filled the air, permeated the soil, or made water stagnant. From the point of view of miasmatic aetiology, flowing water was thus not considered problematic. This was why in the 1860s August Almén could condemn the river water in the Swedish city of Norrköping because of the large organic content and still consider it "rather usable" (Almén, 1868, p. 102); the river would purify itself of the organic contaminants anyway.

10.3.2 The bacteriological health paradigm and the dawn of wastewater treatment, 1900–1950

With the advent of bacteriology (disease transmitted by germs, thus contagious) in the 1890s and the early 1900s, the parameters important for evaluating water became increasingly biological. What became important in the bacteriological paradigm was the origin of matter. Well water that previously had been condemned on chemical grounds as being too hard or containing ammonia could instead be approved. Pathogens were believed to be killed off as the water penetrated through the soil down to the groundwater. Previously, ammonia had invariably been taken as a sign that the water had been contaminated by animal organic matter, but now it was recognised that it could also have a vegetable origin, which was not considered dangerous. The biological processes of sand filters for water treatment were also understood, which made them essential not only in reducing organic material and making the water clear, but also in killing infectious bacteria (Almquist & Richert, 1898; Gurian & Tarr, 2011; Hamlin, 1990; Melosi, 2008).

With bacteriology came a scientific basis for understanding microbiological pollution. On the one hand, it was now generally acknowledged that cholera and other epidemics could spread via water due to germs. On the other hand, bacteriology also reinforced the significance of quickly removing potentially pathogenic rotting matter. As physicians were primarily concerned with the health of city populations, water closets were increasingly promoted to flush out wastes from cities and into the adjacent environment. Those who worried about river pollution and therefore opposed water closets were most often not experts but represented a range of other interests; agriculturists were concerned about the loss of nutrients, some water engineers were alarmed about the enormous increase in water consumption when connecting water closets to sewers, etc. There was also criticism due to technical problems with malfunctioning water closets (Hallström, 2004).

Generally, most professionals embraced water closets and worried little about water pollution. J. Gust. Richert, son of J. G. Richert and also a prominent Swedish wastewater engineer, famously said that "dirt does less harm in the water than in the dwelling, and the health of fish is less important than that of human beings" (*Förhandlingarna vid Nordiska teknikermötet i Stockholm den 15–19 juni 1897*, 1898, p. 233). The notion that running water purifies itself was so strongly held that it outlived the miasmatic paradigm, certainly because there was some truth to it, particularly when small amounts of filth were discharged in large streams, but also because medical and industrial interests promoted it. By the early 1900s, the self-purifying phenomenon had been translated to fit the bacteriological paradigm (e.g., Carlson, 1912), and, subsequently, water closets spread at an unprecedented rate both in Europe and America in the first half of the 20th century.

Thus, water closets were increasingly accepted, although at the same time concern over water pollution became more prevalent in both America and Europe. Some cities consequently started filtering their freshwater or changed their water supply to groundwater. The pollution from water closets and industrial sources was sometimes battled by introducing intercepting sewers but rarely with treatment. Swedish physicians and engineers made study tours to Germany, Great Britain, and Denmark to gain knowledge of new sewage purification methods, the most preferred of which became biological filters in combination with a common septic tank. Treatment in large, common facilities was thus preferred to private septic tanks or cesspools, although few treatment techniques were implemented in practice on a larger scale before the mid-20th century (Hallström, 2002; Isgård, 1998; Lundgren, 1974).

Overall, the introduction of biological techniques was a major step forward in sewage treatment. British wastewater engineers pioneered these techniques in the 1890s, in particular the contact filter that facilitated microbiological processes that could "destroy" sewage naturally (Hamlin, 1988). The first half of the 20th century saw the evolution of other biological methods both in America and in Europe, primarily variations and improvements of the septic tank, the most popular and effective of which became the German Imhoff tank for preparing sewage for secondary treatment.

The most important technique that would revolutionize wastewater treatment was the activated sludge process. Although inspired by American aeration methods, it was invented by Briton Gilbert J. Fowler who published his first results in 1913. This process achieved bacteria reduction of around 90% and required only small filter areas. Although the debate about sewer design had largely receded, discussion continued between proponents of filtration of freshwater and sewage treatment, so the construction of treatment facilities was slow during the period in America as well as in Europe. Insight grew among professionals that dilution and the self-purification of rivers were insufficient but that some measures needed to be taken to protect water sources, and these included the new biological treatment methods (Melosi, 2008).

In Europe and America, the "end-of-pipe" solution with combined (one large pipe) sewerage and water closets thus became increasingly prevalent in wastewater management and sanitation, especially in cities close to large waterways. Some cities, for example, Stockholm, had forbidden water closets in the late 19th century, but by the 1910s the ban had been lifted in most cities and water closets were increasing rapidly. In Sweden, the new treatment methods for purifying sewage involving microbiological processes, for example, activated sludge, were slow in coming on a large scale. In fact, there was little or no wastewater treatment at all in most Swedish cities before 1950. In the 1930s, the first big wastewater plant was built in Stockholm, but the main purification technique was still mechanical separation and chemical precipitation. There was also some experimentation in smaller towns with activated sludge and other biological methods but they failed to spread widely (Drangert & Hallström, 2002; Isgård, 1998).

10.3.3 Wastewater innovation and renovation: Suburban sprawl and the age of environmental awakening, 1950–2010

The breakthrough around the globe for effective, large-scale urban wastewater treatment did not come about until after 1950 (e.g. Frost et al., 2016). In the period after 1950, the models for virtually all technical solutions in Swedish wastewater management came from the United States, and sewage engineers went on frequent study tours to pick up the latest innovations. The main American purification technique was activated sludge, which had been invented in the early 20th century but became widespread only after 1950 due to its great potential in the growing American megacities. The method was successfully transferred to the largest Swedish cities in the 1950s, 1960s, and 1970s, when wastewater plants were built at an unprecedented rate. In the smaller towns, other purification techniques were tested, for example, biological dams or stabilization ponds. Even these were taken from the United States. Swedish wastewater engineers studied them in smaller cities in Nebraska and South Dakota, with a similar "Nordic" climate, and it became the most common treatment technique if one counts the number of plants. Parallel to this development, a transition from combined sewers to separate (two pipes or one pipe and a drainage ditch) or duplicate sewers began in order for the pipe networks to be able to lead only wastewater, not stormwater, to the new treatment plants (Isgård, 1998; Melosi, 2008).

The last two decades of the 20th century saw an increased influence from the environmental movement. The rapid expansion of sewage treatment in the preceding decades had precluded any expensive attention to environmental issues, but now this changed. Environmentalists showed convincingly that the eutrophication (excessive enrichment with nutrients such as phosphorus) of the Baltic Sea was due, in part, to discharge from wastewater plants in countries around the sea. Consequently, this was a time when sewage plants in Sweden were increasingly equipped with additional technology for the reduction of nitrogen and phosphorus (Drangert & Hallström, 2002). The sewer networks, parts of which had not been replaced or repaired since the introduction of sewerage in the late 19th century, were also in need of renovation. Thus, renovation schemes began in many Swedish cities, and the old clay pipes were replaced with modern concrete or plastic pipes (Isgård, 1998). Eutrophication also became an important concern in other developed countries.

The late 20th century also saw a focus on reduction of waste, reuse of nutrients such as phosphorus and nitrogen, and recycling of sludge and biowaste from treated wastewater (Drangert, 2020). An example of the latter is the production of biogas as fuel from sewage sludge, something that proliferated in the early 21st century. Ageing wastewater infrastructures, however, continued to be a global problem even into the 21st century (Katko, Juuti, & Tempelhoff, 2010).

While water supply and wastewater treatment and disposal remained largely public ventures in the United States by 2000, privatisation made some inroads. However, in 2003 only 5% of the water systems in the United States were privately owned, and only about 15 percent of the population was served by corporate water. Trends in England were different. In the 1970s onward, sewage plants were largely privatised under the Margaret Thatcher government, but some restructuring took place by the turn of the 21st century. In terms of service, few municipal wastewater systems in the United States achieved regional status (as they had in Great Britain), although service areas expanded overall. The limited regionalisation of wastewater systems in America was due largely to limited federal and state support, especially after the 1970s. Local government generally took administrative and financial responsibility for wastewater services, and, in 2004, there were more than 16,500 sewage treatment plants in the United States. Overloading of treatment facilities became a serious problem after 1970, although the actual quality of treatment improved. In 1996 about 88% of all plant capacity included secondary treatment or greater, and more than three-quarters of all Americans were being served by centralised sewage treatment facilities (Melosi, 2008; Schneider, 2011).

Despite broadening the reach of sewage treatment, experts continued to seek technical improvements. These included mechanical removal of sludge from sediment tanks, more effective control of coagulation to permit shortened settling time and higher rates of filtration, and computerisation of plant functions. Land treatment made a comeback in an era when recycling was becoming more popular. Into the 21st century, biological treatment of sewage (begun in the 19th century) has endured because of its adaptability, but it is made more difficult because of waste pharmaceuticals and other micropollutants. Nonpoint source pollution resulting from many diverse causes has proved to be perplexing because its origins were varied, existing treatment systems were ill-suited to deal with it, and prevailing regulations had long ignored it. This complicated the development of new technology to confront wastewater problems moving forward (Melosi, 2008; Schneider, 2011), although regulation initiatives has proved successful in some countries, such as Australia, where discharge quality standards have promoted water recycling initiatives (Frost et al., 2016). Today, mechanical, biological, and chemical purification methods are central steps in wastewater treatment processes, but watercourse eutrophication remains a major "bottleneck" in the sewerage system as a whole, for example, in the Baltic Sea or the Gulf of Mexico.

10.4 Concluding discussion

Many critical problems relating to water pollution have been resolved during the last 150 years (see summary of techniques, technologies, and systems in Figure 10.1), but a more complete solution still lies partly outside the actual sewerage system. For example, a considerable amount of nitrogen and phosphorus, which leads to eutrophication, are discharged directly into watercourses from agricultural land around, for example, the Baltic Sea and the Gulf of Mexico. It is obvious that a number of complicated internal and external factors impact large technological wastewater systems, and they must be taken into account. As such, today's wastewater systems – essentially designed almost 200 years ago – provide opportunities and also present potential threats, not the least regarding the treatment of their end products.

Sewage treatment systems of some sort have been key service technologies in almost every city around the world, especially in large cities. Biological sewage treatment, in particular, such as

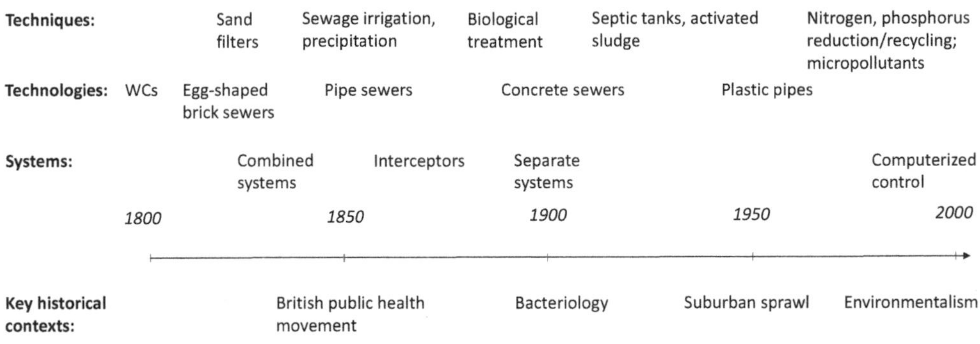

Figure 10.1 Visualization of techniques, technologies, systems, and key historical contexts in the history of urban wastewater management, 1830–2010.

the activated sludge approach, has had a wide impact in the 19th and 20th centuries and into the 21st. Ecologist Daniel Schneider has made the important observation that "Besides being a ubiquitous engineering component of the modern city, biological sewage treatment plants are also ecosystems. As such, they rely on the ability of microorganisms and other plants and animals to degrade sewage and produce a pure effluent" (Schneider, 2011, p. xv). On a larger scale, the adoption of biological sewage treatment points to a complex relationship between human society and the natural world. The process confronts an age-old problem of how to deal with waste products in a human society, often pitting promotion of technology against non-technical approaches. Biological sewage treatment is a hybrid of human engineering technology and natural systems. While not the singular solution to confronting wastewater and its impacts, such a melding of approaches may have implications for other environmental issues facing urban societies.

References

Almén, A. (1868). *Huru bör ett dricksvattens godhet bedömmas från sanitär synpunkt? Svar grundadt på jemförande undersökning af 80 i Skandinavien förekommande naturliga vatten*. Centraltryckeriet: Stockholm.

Almquist, E., & Richert, J. G. (1898). Om ozonbehandling af dricksvatten. In *Hälsovårdsföreningens förhandlingar, sammanträdet den 11 December 1897*. Stockholm.

Binnie, G. M. (1981). *Early Victorian Water Engineers*. London: Thomas Telford Ltd.

Buhagiar, K. (2007). Water management strategies and the cave-dwelling phenomenon in late-medieval Malta. *Medieval Archaeology*, *51*(1), 103–131.

Buhagiar, K. (2016). *Malta and Water (AD 900 to 1900): Irrigating a semi-arid landscape*. Oxford: British Archaeological Reports Oxford Ltd.

Buiter, H. (2008). Constructing Dutch streets: A melting pot of European technologies. In M. Hård & T. J. Misa (Eds.), *Urban Machinery: Inside Modern European Cities* (pp. 141–162). Cambridge, MA & London: The MIT Press.

Carlson, T. (1912). *Om syrets lösning i vatten och förhållande vid vattendragens självrening*. Centraltryckeriet: Stockholm.

Denton, J. B. (1877). *Sanitary Engineering: A Series of Lectures Given Before the School of Military Engineering at Chatham, 1876*. E. & F. N. Spon: London & New York.

Drangert, J.-O. (2020). Urban water and food security in this century and beyond: Resource-smart cities and residents. *Ambio*, 50(3), 679–692. doi:10.1007/s13280-020-01373-1

Drangert, J.-O., & Hallström, J. (2002). Den urbana renhållningen i Stockholm och Norrköping: från svin till avfallskvarn? *Bebyggelsehistorisk tidskrift*, *44*, 7–24.

Drangert, J.-O., Nelson, M. C., & Nilsson, H. (2002). Why did they become pipe-bound cities? Early water and sewerage alternatives in Swedish Cities. *Public Works Management & Policy*, *6*(3), 172–185.

Frost, L., Gaynor, A., Gregory, J., Morgan, R., O'Hanlon, S., Spearritt, P., & Young, J. (2016). *Water, History and the Australian City: Urbanism, Suburbanism and Water in a Dry Continent, 1788–2015*. Melbourne: Cooperative Research Centre for Water Sensitive Cities, Monash University.

Goubert, J.-P. (1988). The development of water and sewerage systems in France, 1850–1950. In J. A. Tarr & G. Dupuy (Eds.), *Technology and the Rise of the Networked City in Europe and America*. Philadelphia: Temple University Press.

Gurian, P. L., & Tarr, J. A. (2011). The origin of federal drinking water quality standards. *Proceedings of the Institution of Civil Engineers-Engineering History and Heritage*, 164(1), 17–26. doi:10.1680/ehah.9.00009

Hallström, J. (2002). *Constructing a Pipe-Bound City: A History of Water Supply, Sewerage, and Excreta Removal in Norrköping and Linköping, Sweden, 1860–1910*. Linköping: Dept. of Water and Environmental Studies, Linköping University.

Hallström, J. (2004). Historical perspectives on water pollution in the Baltic Sea: The case of the River Motala Ström, 1860–1910. In P. Eliasson (Ed.), *Learning from Environmental History in the Baltic Countries. The Baltic Sea Project - Learners' Guide No. 6* (pp. 51–61). Stockholm: UNESCO & Myndigheten för skolutveckling.

Hallström, J. (2009). Systemteori och teknik. En introduktion till stora tekniska system. In P. Gyberg & J. Hallström (Eds.), *Världens gång - teknikens utveckling. Om samspelet mellan teknik, människa och samhälle* (pp. 99–117). Lund: Studentlitteratur.

Hallström, J. (2011). Urban or suburban water? Working class suburbs, technological systems and environmental justice in Swedish cities in the late nineteenth century. In G. Massard-Guilbaud & R. Rodger (Eds.), *Environmental and Social Justice in the City: Historical Perspectives* (pp. 133–153). Cambridge: The White Horse Press.

Hamlin, C. (1988). William Dibdin and the idea of biological sewage treatment. *Technology and Culture, 29*(2), 189–218.

Hamlin, C. (1990). *A Science of Impurity: Water Analysis in Nineteenth Century Britain*. Berkeley: University of California Press.

Hamlin, C. (1992). Edwin Chadwick and the engineers, 1842–1854: Systems and antisystems in the pipe-and-brick sewers war. *Technology & Culture, 33*(4), 680–709.

Harandi, M. F., & de Vries, M. J. (2014). An appraisal of the qualifying role of hydraulic heritage systems: A case study of Qanats in central Iran. *Water Science and Technology: Water Supply, 14*(6), 1124–1132.

Hughes, T. P. (2004). *Human-built World: How to Think about Technology and Culture*. Chicago: University of Chicago Press.

Hughes, T. P. (2012). The evolution of large technological systems. In W. E. Bijker, T. P. Hughes, & T. J. Pinch (Eds.), *The Social Construction of Technological Systems: New Directions in the Sociology and History of Technology* (Anniversary Edition ed., pp. 45–76). Cambridge, MA & London: The MIT Press.

Isgård, E. (1998). *I vattumannens tecken. Svensk VA-teknik från trärör till kväverening*. Örebro: Ohlson & Winnfors.

Katko, T. S., & Juuti, P. S. (2014). History of water and sanitation services in the urban–rural context: The case of the city of Tampere, Finland. In T. Tvedt & T. Oestigaard (Eds.), *A History of Water: Series III, Vol. 1, Water and Urbanization*. New York: I.B. Tauris.

Katko, T. S., Juuti, P. S., & Tempelhoff, J. (2010). Water and the city. *Environment and History, 16*(2), 213–234.

Lundgren, L. J. (1974). *Vattenförorening. Debatten i Sverige 1890–1921*. Lund: Gleerups.

Mays, L. W., Koutsoyiannis, D., & Angelakis, A. N. (2007). A brief history of urban water supply in antiquity. *Water Science and Technology: Water Supply*, 7(1), 1–12.

Melosi, M. V. (2000). *The Sanitary City: Urban Infrastructure in America from Colonial Times to the Present*. Baltimore & London: The Johns Hopkins University Press.

Melosi, M. V. (2008). *The Sanitary City: Environmental Services in Urban America from Colonial Times to the Present*. Pittsburgh: University of Pittsburgh Press.

Nye, D. E. (2006). *Technology Matters: Questions to Live With*. Cambridge, MA: The MIT Press.

Porter, D. H. (1998). *The Thames Embankment: Environment, Technology, and Society in Victorian London*. Akron: University of Akron Press.

Reid, D. (1991). *Paris Sewers and Sewermen: Realities and Representations*. Cambridge, MA & London: Harvard University Press.

Richert, J. G. (1869). *Om vattenledningar och vattenaflopp*. Centraltryckeriet: Stockholm.

Richert, J. G., *Förhandlingarna vid Nordiska teknikermötet i Stockholm den 15–19 juni 1897*. (1898). Centraltryckeriet: Stockholm.

Schneider, D. (2011). *Hybrid nature: Sewage treatment and the contradictions of the industrial ecosystem*. Cambridge, MA: The MIT Press.

Squatriti, P. (1998). *Water and Society in Early Medieval Italy*, AD 400–1000. Cambridge: Cambridge University Press.

Tarr, J. A. (1996). *The Search for the Ultimate Sink: Urban Pollution in Historical Perspective*. Akron: University of Akron Press.

Tarr, J. A. (2020). Illuminating the streets, alleys, parks and suburbs of the American City: non-networked technologies, 1870–1920. *History and Technology, 36*(1), 105–128.

Tarr, J. A., & Dupuy, G. (Eds.). (1988). *Technology and the Rise of the Networked City in Europe and America*. Philadelphia: Temple University Press.

Westholm, G. (1995). Gaturenhållning, avfallshantering och stadsplanering. Medeltida teknik belyst av visbyfynd. *Nordisk arkitekturforskning, 8*(1), 7–18.

PART III

Regulation and economic perspectives

11

INSTITUTIONAL PERSPECTIVES ON WATER SERVICES

Sylvain Barone and Pierre-Louis Mayaux

11.1 Introduction

The institutionalist literature on water services is remarkably diverse.[1] In addition to defining institutions more or less broadly, it explores how institutions shape a wide range of outcomes, from service performance and service efficiency to patterns of interactions between key stakeholders, the social legitimacy of water arrangements, and the daily routines followed by water professionals.

To take the full measure of this diversity, it is useful to start with a broad definition of institutions. Douglas North famously defined them as "the rules of the game in a society, or more formally, the humanly devised constraints that shape human action" (North, 1990, p. 3). He thus made clear that institutions comprise not only "formal laws, rules, regulations and procedures" but also "informally-established patterns of behavior, norms, practices, and procedures" as the informal practices also become rules in their own right when they are accepted by a society (ibid., p. 4).

Two basic properties of institutions are common to all institutionalist work. First, institutions place clear limits on the range of feasible options available to an actor in a given circumstance. Institutions matter because they mandate or strongly encourage certain courses of action, while prohibiting or strongly discouraging others. Second, institutions are beyond the direct control of any individual actor who is subjected to them, whether a water utility, a city council, or a regulatory agency. In other words, institutions cannot be manipulated at will, including by powerful actors, otherwise they would lose exactly that which marks them as institutions.

Institutions are therefore distinct from organizations but the two are nevertheless intertwined. An organization is a collective actor that can have offices, personnel, financial resources, and a legal personality, such as a water association, a water provider, a ministry, or a bureaucracy. Institutions, by contrast, refer to the rules and the norms that organizations share or follow. But many institutions emerge and operate through the intervention of organizations and all organizations operate through institutions.

DOI: 10.4324/9781003057574-15

Box 1 Varieties of institutionalism: The challenge of different ontologies and research questions

Beyond the broad assumption that institutional environments shape what institutionally embedded actors can do, institutionalisms are rooted in diverse disciplines, and they involve both heterogeneous ontologies and research questions (Lowndes, 1996; Hall, Taylor, 1996). In terms of ontologies, institutionalisms span the whole spectrum from positivism to interpretivism (Hay, 2006). Thus, while many approaches conceive of the relationship between institutions and social behavior as a largely mechanical one, independent of the actors' own interpretation of the rules, others insist that this relationship is mediated by the sense-making of culturally complex actors. In terms of research questions, these diverse currents tend to focus on different dimensions of institutions, and hence on different mechanisms by which they shape actor' behaviors. Thus, economists are primarily interested in how institutions provide material incentives, rewards, and sanctions (Williamson, 2000). Political scientists tend to focus on how they entrench power relations and unequal forms of resource distribution (Mahoney, Thelen, 2010). Finally, sociologists and anthropologists pay considerable attention to how they provide identities, meanings, and daily routines to social actors (Douglas, 1986).

Importantly, these different forms of institutionalism do not focus on the same dependent variables. Thus, economists tend to be primarily interested in an institution's effects on service performance and service efficiency. Political scientists focus more on the patterns of governance and on the power configurations they shape. Sociologists and anthropologists are typically interested in how institutions derive their legitimacy, as well as in the everyday practices through which they are enacted, routinized, and continuously transformed.

Finally, these differences in research interest partly explain the use of different methodologies. Economists' variables are more easily captured by numerical indicators that allow statistical treatment. Although the design of indicators is also possible for other types of variables, it is often more difficult and reductive. While statistical treatments can be used, case studies and small-N comparisons are usually preferred by political scientists and sociologists, and they are typically based on archival research, interviews, and direct observations.

An institutionalist perspective of water services rests on the assumption that what hinders the development of water services, in the Global South and the Global North alike, is not merely the problematic attributes of actors or of actors' interactions – whether by these we mean limitations in financial resources and professional skills, political rivalries, or individual misconducts. Rather, institutionalists contend that these variables are themselves crucially shaped by the rules of the game that define the roles, responsibilities, sanctions, and rewards of the various stakeholders. This is one way to say that "the operation of water services largely depends on variables defined outside them" (Jaglin, Zerah, 2010, p. 13), even if there are feedback loops between actors and institutions as the former routinely seek to change the latter; and even though institutions should not be viewed as an ultimate source of explanation but, more productively, as one important factor, among others, in temporally dynamic and fundamentally multicausal service trajectories.

The chapter proceeds as follows. We begin by reviewing the structuring effects of the proximate institutional environment of water utilities. We then look at sector-specific institutions at the national level. In the third part, we further broaden the perspective by examining how larger political institutions shape the operation of water services at both urban and national levels. Finally, we briefly outline some avenues for future institutionalist research.

11.2 Contracts and indicators as core concerns of rational institutionalism

The great majority of water services in the world are publicly managed, with highly varying degrees of organizational autonomy granted to their public operator. Private companies account for only a small minority of water providers: urban populations served by private water operators reached 160 million in 2007, accounting for around 5% of the total urban population of the world (Marin, 2009). However, private management has garnered much research attention as it represented a major inflection point in the trajectory of many services. As such, it has served as a key leverage to theorize the broader role of institutions in water services.

Among institutionalist approaches to water services, rational institutionalism (RI) has largely dominated in terms of academic output, especially among economists. In a utilitarian perspective, it conceives of agents as being primarily motivated by their material interests, with clear and stable preferences in this regard, and the ability to design appropriate strategies to maximize these interests. RI has particularly guided research on the proximate institutional environment of water suppliers. Here, it should be emphasized that prior to any particular prescription, most RI scholars argue that water services should have *strong* institutions; that is, that one of the most fundamental problems plaguing water services are ill-defined mandates, ambiguous rules, and blurred responsibilities (Kessides, 1993; Guasch, 2004). This argument has led to widespread criticism of the unclear distribution of prerogatives between water providers and local authorities. Direct control of water services by local politicians, lacking both clear contractual provisions and operational autonomy, is widely seen as a major cause of inefficiency (Rouse, 2013). Sometimes, rule-based operation is even conceptualized, rather simplistically, as the polar opposite of "political interferences" (World Bank, 1994). Institutions, insofar as they provide predictability to stakeholders, lengthen their time horizons, and help them rank their heterogeneous preferences, are seen by institutionalists as intrinsically desirable.

Based on this shared assumption, a widespread prescription has been that the water provider should maintain an arm's length relationship with local political authorities. Local politicians are considered to have some inherently perverse incentives when it comes to water services; thus, if left to their own devices, they will indulge in problematic practices such as overstaffing the water company by making non-meritocratic appointments to reward friends and loyal supporters; underpricing services to avoid unpopular, albeit often necessary, price increases; and prioritizing infrastructure with highly visible effects (such as dams or the distribution network for drinking water) to the detriment of investments with less visibility and hence less politically profitable (e.g., routine maintenance of pipes, wastewater treatment). By contrast, strengthening the operator's autonomy from local politicians (with respect to senior appointments, mechanisms for career advancement, and the prohibition to use the company's revenues to finance non-water-related spending) is widely seen as a key to improving performance (Saleth, Dinar, 2004). In particular, independent managers have been found to be less reluctant to increase prices, thereby making it easier to raise the revenue required for investments and maintenance (Rouse, 2013). From a more sociological and anthropological perspective, some studies have highlighted the demoralizing effect of non-meritocratic appointments. It is one of the reasons why, for example, employment in a water utility in Indonesia has been symbolically associated with low social status. The resulting impact on morale has undermined the efficiency and productivity of water providers throughout the country (Martijn, 2005).

However, a company endowed with strong financial and operational autonomy might also be a company that is more difficult to control, especially if it is in a monopoly position, which tends to be the case for water services. Here, the threat of a vulnerable company, prone to be captured by opportunistic local politicians, can morph into its opposite, namely, that the water provider may be too powerful and prone to capture these very same politicians (Guasch, 2004). RI recommends two key approaches to mitigate this risk: to select the "right" firm by competitive tendering and to define specific and observable obligations for the provider in terms of investment targets and price-setting mechanisms (complete contracts).

Concerning the procurement process, RI tends to emphasize that in a natural monopoly such as water services, characterized by the high share of sunk costs related to operating costs, competition *for* the market, that is, regular and competitive tendering to select the management entity, should be well organized to offset the absence of possible competition at least partially *in* the market (Iossa, Rey, Waterson, 2019). Thus, to maximize contestability – the probability that several firms will submit quality bids, based on adequate information, with a non-trivial chance of replacing the incumbent provider – the procurement process should be as transparent as possible to mitigate the risk of corruption. Calls for bids should also be published well in advance of the submission deadline to allow all interested bidders to submit quality bids. Finally, they should also set out clearly measurable selection criteria to minimize local authorities' discretion in the selection process, such as the lowest possible prices or the highest number of new connections (Saussier, Tirole, 2015). These recommendations have been shown to have concrete effects. For example, in France, competitive bidding increased by more than 30% between 1998 and 2010 (Bolognesi, 2014).

Concerning the contractual provisions, most approaches stress the importance of defining performance indicators. This emphasis is consistent with the broader promises of performance management set forth by the doctrine of the new public management (NPM) (Moynihan, 2008). In turn, the level of target achievement should help determine the operator's remuneration through performance-based financing mechanisms. For example, in the United Kingdom, the Water Services Regulation Authority (Ofwat) sets limits on the prices charged for water and sewerage services while taking into account past performance, proposed capital investment schemes, and expected operational efficiency gains. Similar mechanisms have been developed in other countries, including Brazil and Mozambique. Some regulations rely on rewards but also on penalties to encourage good performance, as is the case of the ONEA in Burkina Faso and the SONES in Senegal.

The abundant literature on these mechanisms illustrates the more general fact that, compared to other forms of institutionalism, the pricing structure and price-related incentives are crucial to most rational institutionalists. For example, the rising block tariff structure that was initiated in the 1980s, for both equity and environmental reasons, was the focus of considerable attention and debate, as it was supposed to guarantee essential water consumption at low prices, while penalizing large consumers. However, it unexpectedly created a strong disincentive for water supply utilities to connect the poor to the supply network, as public fountains were charged higher volumetric tariffs than individual households, creating a counterintuitive cross-subsidy from poor to middle- and upper-class customers (Bakker, 2010). Likewise, indexation formulae linking local currency pricing to the US dollar have been strongly criticized as they triggered massive price increases during episodes of rapid currency devaluation, such as the one that occurred in Jakarta in 1998 or in Buenos Aires in 2001 (Iwanami, Nickson, 2008).

As a rule, RI has had the virtue of moving beyond general prescriptions (such as recommending public or private management or calling for more financing and training for mid-level managers), looking instead at the concrete, context-sensitive incentives faced by water providers or local authorities. However, it has come under regular criticism for its propensity to give an overly depoliticized account of water services (Pollitt, Bouckaert, 2011). Rational approaches to institutions often portray governance largely devoid of politics, i.e., "governance by numbers" (Supiot, 2017) where foundational disagreements, as well as the actors most directly in charge of mediating conflicts – the elected political representatives – are essentially perceived as a source of institutional disturbance. The first problem with this conception is that its parsimony comes at the cost of sociological realism, downplaying the fact that institutions are sites of everyday controversy and political skirmishes, not only over their perceived efficiency, but also over their distributive consequences, and the embedded notions of justice that they convey. The second problem lies in the neglect of the role of deliberative processes during the definition of priorities and the appropriate distribution of costs and benefits.

11.3 National regulatory institutions: Framing and enforcing the rules of water services

National water services institutions matter in two fundamental ways. First, they provide the general framework in which specific local rules are nested. Second, by defining both monitoring and judiciary mechanisms, they determine the degree of enforcement of all rules.

11.3.1 National frameworks for local rules

National rules take precedence over local rules to varying degrees and range from strict mandates to flexible guidelines. The stratification system in Colombia illustrates the case of an institutional system that strongly constrains local rules, even while leaving local actors substantial room for maneuver. Thus, pursuant to Law 142 of 1994, each municipality is responsible for classifying its dwellings in six categories ranging from "low-low" to "high" according to their physical properties, and each must follow the methods defined by the national government. The water rates for each stratum then vary widely, the three lowest strata being subsidized by the two highest strata. Water suppliers' business models and strategies are directly affected by stratification, which determines not only where it is more profitable to invest, but also the overall financial balance sheet of the services (Guevara, Shields, 2019).

Other types of national institutions influence water services in softer, more indirect ways. This is particularly the case of benchmarking, which has expanded considerably in the two last decades. Benchmarking is the process of measuring and comparing key performance indicators and practices to understand how and where water services need to change in order to improve their performance. So far, most benchmarking initiatives have focused on organizational efficiency, the use of resources, and quality and environmental performance and outcomes. They rest on the idea that an increase in transparency is important to change practices and improve performance. However, the scope, modalities, and actors in charge of national benchmarking systems vary significantly from state to state. For example, in the Czech Republic, the first standardized benchmarking project was carried out in 2015, when 1,437 operators were evaluated using 31 indicators under the supervision of the Ministry of Agriculture (Suchacek et al., 2018). Meanwhile, in the United Kingdom, Ofwat's service incentive mechanism (SIM) compares companies' performance based on customers' complaints and customer surveys.

However, the specific effects of benchmarking, as well as those of information and monitoring systems more generally, remain difficult to evaluate.

11.3.2 Enforcing the rules

Most institutional economists emphasize that contracts and indicators alone are not enough to align the disparate interests of key stakeholders. Rules have to be actually enforced. The economist Pablo Spiller, whose work has been influential at the World Bank, thus distinguished between institutions that generate incentives and institutions that comprise the governance framework of these incentives, and that allow them to be actually enforced. Two types of the latter institutions, in particular, have been the focus of attention: the existence of a more or less independent national regulator and the effectiveness of the judicial system in resolving disputes (Spiller, Savedoff, 1999).

Regarding national regulators, in the 1990s there was broad agreement that the creation of an agency as independent as possible from political authorities (whether from the executive or from the legislative branch) was eminently desirable. Such independence was supposed to ensure the stability of the rules of the game for the water industry and to reduce the uncertainties the industry faces. Independent regulatory agencies are governmental entities that possess and exercise some grant of specialized public authority, separate from that of other institutions, but are neither directly elected by the people nor directly managed by elected officials.

Thus, Sirtaine et al. (2005) included the degree of independence of the regulator as a positive criterion in their quality of regulation index, and they concluded that this index was positively correlated with a "proper" level of profitability for private water providers, that is, neither their capture of excessive rents nor profitability too low to allow adequate investment. In the same way, Guash, Laffont, and Straub (2008) estimated that the existence of a regulator significantly reduces the probability of contract renegotiation. However, Magetti (2007) cautioned against using formal, de jure independence as a proxy for de facto independence, as the former is in fact neither a necessary nor a sufficient condition for explaining variations in the de facto independence of agencies. Other factors, such as the life cycle of agencies, veto players, and international networks of agencies, have a more decisive impact.

On the other hand, many authors found mixed evidence for a relationship between the degree of independence of the regulator and the performance of infrastructure services in terms of accessibility, affordability, and quality (Stern, 2009; Trillas, 2010). Thus, while the central feature of some successful water reforms, especially in terms of accessibility, has been the establishment of an autonomous national regulatory agency (e.g., Albania, New South Wales in Australia, Mozambique, Portugal, and Zambia), other reform efforts that have arguably also been successful do not feature a national regulator and have much more dispersed and opaque regulatory arrangements (e.g., Brazil, Indonesia, the Philippines). In part, this is a question of the scale of the sector and the country's overall governance structure, which makes the independence of the regulator alone an insufficient condition to create enforceable rules where they are lacking. For example, in Colombia, the strongly decentralized structure of the water sector has rendered regulation costly and extremely demanding, requiring the two regulators to effectively regulate the 1,300 service providers over which they had oversight. It is also rare for a regulator, however independent and powerful, to obtain a monopoly on standard-setting and standard-monitoring. Much more frequent is the existence of multiple regulatory bodies establishing different expectations and restrictions, and the imposition of more than one method of regulation that eventually hinders performance improvements (Hood, Dickson, 2015).

More generally, the problem of independence should not obscure the broader issue of regulatory *capacity*. In this respect, Barone, Dedieu, and Guérin-Schneider (2016) highlighted the paradoxical effects of a series of reforms in France, inspired by New Public Management principles. For a long time, local services of the French state had provided technical support to local authorities for their drinking water and sewage infrastructure projects. The suppression of this engineering mandate in 2007 aimed to refocus the state on its core missions (especially monitoring and sanctions). While the introduction of performance indicators and commitments to results is often a strong principle of neo-managerial reforms, the dismantling of state engineering has led to a weakening of their use as a control tool and turned them into mere communication tools. In the end, the state's regulatory capacity has been weakened, in a context where France does not have an independent regulatory authority. Similarly, the loss of technical expertise has been a factor in weakening the state's ability to exercise control and enforce sanctions. Rather than benefiting incumbent service providers with a strong position vis-à-vis local authorities, the lack of regulations has made them more vulnerable to competition from low-cost bids that local authorities are sometimes no longer able to distinguish from good quality bids.

For its part, the importance of how the judiciary handles water services disputes has often been stressed in general but has not been the subject of much detailed study. Yet, the functioning of the judiciary can be crucial in channeling and mitigating conflicts, particularly between water suppliers and water users. For example, since 1990, Brazil has been implementing a quite robust consumer law, which today is supported by a dense network of public consumer offices (*procons*) headed by the judiciary of each state. Established in every major municipality, the *procons* play an important role in channeling user discontent, particularly with respect to water services (Mayaux, 2015). Likewise, Urueña (2012) showed that the Constitutional Court is the appropriate independent national agency to understand the emergence of a regulatory state in Colombia. It is a vehicle for the human rights discourse that balances the national efficiency-dominated landscape of water supply regulation represented by the independent agency, but it is also a forum where different views interact: "neo-liberal and neo-constitutional arguments collide in the Court's reasoning" (Urueña, 2012, p. 295).

11.4 How political institutions shape water services

The institutions that matter for water services are not just those that explicitly aim to shape the incentives, resources, and identities of water stakeholders. Rather, political institutions in the broadest sense – institutions that organize the representation of interests and worldviews across the political system, well beyond the water sector – also have far-reaching effects on water services.

11.4.1 Water services and urban governance

The relationship between the institutions of urban governance and water services has been studied from a variety of angles. First, a number of studies emphasized how water services are shaped by institutions governing other sectors of urban governance, such as planning use of land. Thus, in Peru, a strong correlation has been observed between the lack of access to water services and insecure land tenure. Conversely, land titling is associated with statistically significant increases in access to water (Meeks, 2018). This correlation can generate a vicious circle because, as Botton and Merlinsky (2006) point out, access to basic services is itself a vector of citizenship. Being deprived of a connection to the main network encourages a demoralizing self-perception as a second-class citizen, which undermines the motivation to engage in collective action, thereby further reducing the possibility of obtaining a connection in the near future. Likewise, Bakker et al. (2008) noted that the implementation of land use policies in Jakarta hindered the expansion of water services. Official development plans for the city encouraged an east-west pattern of urban development that attempted to avoid expansion into irrigated agricultural areas located north and south of the city. Yet, despite the failure of planning controls to stem urban sprawl, the expansion of water networks has been limited to the zones targeted by official planning, thereby excluding parts of the city where active development is actually taking place.

Beyond policies toward land use, many authors have stressed the importance of broader institutions, such as the quality of the local rule of law and the regulatory capacities of the municipality, including its ability to monitor the quality of the service delivered, to provide relevant contextual information to the water supplier, and to enforce national regulations (Franceys, Gerlach, 2008). Many case studies also point to the gap between lofty principles and objectives (adaptiveness, coordination, participation) and an urban water management that remains typically opaque and fragmented. Brown and Farrelly (2009) consider the barriers to be socio-institutional rather than technical, reflecting issues related to community, responsibility, knowledge, vision, commitment, and coordination. As a consequence, strategies of change should focus on "fostering social capital, inter-sectoral professional development, and inter-organizational coordination" (p. 839). In parallel, some authors have been exploring the synchronization of drinking water policies with the larger dynamics of urban development. For example, Rijke et al. (2013) argue that the most effective governance (whether more centralized or decentralized, formal or informal, etc.) depends to a great extent on at what stage along a transition process toward greater resilience a city finds itself.

However, some of these works have been criticized for still downplaying the political dimension of governance, implicitly conceiving governance as a "narrowly technical decision-making process" (Bakker, 2010, p. 8). Critics emphasize that, in addition to constraining actors, institutions may also enable them to act, organize, and mobilize, so long as they prove capable of seizing the opportunities embedded in the rules of the game. This is the case of participatory governance mechanisms in general. Thus, in Jakarta, the failure to expand water services has been exacerbated by the lack of participatory mechanisms that would allow the concerns of poor households to "filter up" to utility managers or political leaders, who may sometimes simply be unaware of the sheer number of unconnected residents or ignorant of the cost to the unconnected poor of obtaining water from alternative sources (Bakker et al., 2008). A similar conclusion was reached in the case of Dar-es-Salaam (Rugemalila, Gibbs, 2015).

The literature on water services points to a general lack of provider accountability – especially vis-à-vis users. It stresses how the lack of accountability and transparency leads to various types of dysfunctions and difficulties. Weak political accountability and weak feedback mechanisms from customers to policymakers are regularly invoked to explain the poor performance of water utilities (Aguilar-Benitez, Saphores, 2010). Conversely, many authors emphasize that strengthening accountability through extensive public participation allows water service providers to improve their performance in terms of pricing and investment (Lobina, Hall, 2007). Beyond narrowly conceived performance, the lack of transparency and accountability may generate non-adaptable systems, for example in the case of Los Angeles, where the water services system is obscure and users are accustomed to cheap, easily accessible water (Pincetl et al., 2016), or in the case of Switzerland, where this relationship between accountability and adaptability plays out in a more positive way (Lieberherr, Ingold, 2019). Likewise, institutional transparency has been found to be a key factor in successful management of water supplies, especially in reducing corruption (see for example Davis, 2004; Hutchings et al., 2015).

11.4.2 Beyond urban governance: Multilevel political institutions

Beyond urban governance, water services are shaped by complex, multilevel historical processes. Their analysis requires carefully considering a variety of institutional and political dynamics. For example, a significant change in the allocation of property rights or in the allocation of diverse resources (regulatory, financial, etc.) may alter the matrix of opportunities, constraints, incentives, etc. of different actors, and thus have a more or less direct impact on the organization and functioning of water services. Without claiming to be exhaustive, at least three dimensions are important to consider: the territorial organization of each state, state and public sector reforms, and bureaucratic cultures.

11.4.2.1 The territorial organization of the state

The territorial organization of each state (i.e., its form and degree of centralization, regionalization/devolution, or federalism) has far-reaching effects on water services. A comparison between European countries provides an excellent illustration (Barraqué, 1995). In France, until a reform in 2015 encouraged intercommunal groupings, water services were managed by municipalities and divided into thousands of services. The number of municipalities, or "communes," about 36,000, has no equivalent elsewhere in Europe, and their historical prerogatives, including over water, can be understood only in light of a long history going back to the Old Regime and the revolution of 1789. Throughout this period, institutional and political balances of forces were stabilized, making any attempt to weaken the communes politically very costly. The communes, especially rural ones, are major contributors to the college of elected officials who appoint senators. This municipal fragmentation has had many consequences, including the development of small water utilities with extremely limited technical and financial capacities, the necessary provision of technical support by the state, the appeal of rationalizing promises made by large private water companies, etc. Despite this clear historical correlation, in the literature on water services, the territorial organization of the French state is used more as a context or framework for analysis than as a direct explanatory variable.

11.4.2.2 State and public sector reforms

The general positioning of the state, which is reflected in the above-mentioned territorial organization, may also change as a result of broad administrative reforms. In many countries, the interpretation of recent changes in this regard has been the focus of considerable debate, with some authors speaking of a hollowing out of the state (Rhodes, 1994) while others conceptualized the development of a regulatory state (Levi-Faur, 2013). Whatever the specific interpretation, this far-reaching restructuring can have direct effects on water services. An emblematic case is the privatization, in 1989, of the 20 regional authorities in charge of drinking water management in England and Wales and the establishment of a national regulatory agency. In the Global South, calls by international funders to "clean up" public finances, especially through structural adjustment plans in Africa and South America in the 1980s and 1990s, resulted in a massive reduction in the investment capacity of public authorities in water services (Weber, 2004). It is thus clear that state and public sector reforms have an impact on the governance of water services, and they deserve to be studied as such.

11.4.2.3 Bureaucratic cultures

Bureaucratic cultures strongly influence water professional norms of behavior that constitute genuine informal rules, i.e., institutions. Many authors have focused on professional standards and institutional transfers in the water sector, particularly concerning integrated water resources management. They have studied them not only from a normative point of view, as a set of desirable objectives and institutional forms to target, but also from a more critical standpoint, for example as a catch-all category or "ideal" concept ill-suited to many contexts (see, for example,

Molle, 2008). These reflections have their more specifically urban derivations, such as the paradigm of the "urban resource nexus" (William et al., 2019). Moreover, the professional representations and norms that structure water policies are permeated by broader professional cultures and traditions of intervention. For example, in the field of water policy, France is emblematic of the many states that have inherited an institutional culture that is predominantly technical, hygienist, and productivist. This cultural dimension deserves to be explored in greater detail to go beyond simplistic reform prescriptions.

11.5 Conclusion: The contributions of institutionalism and avenues for future research

Institutionalist perspectives provide important insights into how water services operate in practice. Broadly speaking, they situate actors – their resources and their skills, their ways of doing things, their legal status as public or private entities – within the configuration of rules that provide them with obligations, incentives, and meaning. They highlight the way in which not only their strategies, but also their preferences themselves, are institutionally constructed. In so doing, they overcome the pitfalls not only of functionalism, but also of economism and technicism, which remain pervasive in the field of water governance, and mostly think in terms of technical inputs and outputs, types of infrastructures, and pricing.

Furthermore, institutional approaches rightly emphasize that institutions are path dependent for many reasons that include their propensity to produce specific types of learning and investments that are costly to reverse, their ability to strengthen their own support base by providing it with a reliable stream of benefits, and their capacity to limit the very possibility of imagining alternatives. Assuming the inherent difficulty of institutional reforms, they advance sophisticated ways of accounting for how change nonetheless happens. The contributions of Powell and DiMaggio (1991) on institutional isomorphism; of Hall (1993) on the different orders of change; of Pierson (2004) on timing and sequences; and of Streeck and Thelen (2005) on incremental but nonetheless transformative change, all attest to the sophistication of institutionalist theories of change.

Beyond these commonalities, institutionalist research on water services rests on different conceptualizations of institutions and how they structure behavior. It is therefore crucial that institutionalists carefully specify their own definition and the mechanisms linking institutions to behaviors, not only for its own sake but also to allow a better interdisciplinary dialogue.

Moreover, institutionalist research still overly focuses on particular institutional dimensions, and it does not pay sufficient attention to institutional constellations in all their complexity. To observe the interactions between different institutions, more multilevel case studies are needed, as well as small-N comparisons to generate new causal hypotheses on complex, joint institutional effects. In doing so, the overall consistency of institutional configurations should not be postulated by analysts. Rather, they should be attentive to mismatches and asynchronous development and to the potentially contradictory effects of different institutions.

More fundamentally, it is crucial not to elevate institutions to the status of an ultimate explanation for water services trajectories. Institutions shape actors' fields of action, but they themselves are shaped by other social processes, such as the changes in social structure, the dynamics of political conflicts and political settlements, or changes in dominant paradigms and assumptions (Kashwan et al., 2019). As Punjabi and Johnson (2019, p. 182) point out, institutions are not merely "rules, norms and artefacts of cooperative bargaining," but also "carriers of historical legacies, conveyers of state force, and a venue for struggles over state power." Analysis of social change should thus be dialectical, taking into account not only the role of institutions in political and social dynamics but also the way in which these dynamics, in turn, propel institutional transformations. This form of political institutionalism should not confine itself to purely institutional analyses of water services; rather, it should dialogue with other approaches or disciplines (especially those that study the economic, social, political, technical, and ecological *effects* of institutions), and/or build on the institutional analysis of water services to tackle broader political questions, such as the contemporary transformations of the state and modes of rule.

Note

1 For an overview of this diversity, see Box 1 below.

References

Aguilar-Benitez, I., Saphores, J.D. 2010. "Public accountability and performance of two border water utilities." *Water Policy* 12(2): 203–219.

Bakker, K. 2010. *Privatizing Water: Governance Failure and the World's Urban Water Crisis.* Ithaca, Cornell University Press.

Bakker, K. et al. 2008. "Governance failure: Rethinking the institutional dimensions of urban water supply to poor households." *World Development* 36(10): 1891–1915.

Barone S., Dedieu C., Guérin-Schneider L. 2016. "The removing of state public engineering in France: The paradoxical effects of a new public management reform in the water sector." *Politiques et management public* 33(1): 49–67.

Barraqué B. (ed.) 1995. *Les politiques de l'eau en Europe.* Paris, La Découverte.

Bolognesi, T. 2014. "The paradox of the modernisation of urban water systems in Europe: Intrinsic institutional limits for sustainability." *Natural Resources Forum* 38: 270–281.

Botton, S., Merlinsky, G. 2006. "Urban Water conflicts in Buenos Aires, Argentina: voices questioning the economical social and environmental sustainability of the water and sewerage concession." In: Barraqué B. (ed.) *Urban Water Conflicts*, pp.53–70. Paris, UNESCO.

Brown R.R., Farrelly M.A. 2009. "Delivering sustainable urban water management: A review of the hurdles we face." *Water Science and Technology* 59(5): 839–846.

Davis, J. 2004. "Corruption in public service delivery: Experience from South Asia's water and sanitation sector." *World Development* 32(1): 53–71.

Douglas, M. 1986. *How Institutions Think*, 1st edition. Syracuse, Syracuse University Press.

Franceys, R., Gerlach, E. (eds.) 2008. *Regulating Water and Sanitation for the Poor: Economic Regulation for Public and Private Partnerships.* London, Earthscan.

Guasch, J.L. 2004. *Granting and Renegotiating Infrastructure Concessions. Do It Right.* Washington, DC: World Bank Institute Studies.

Guasch, J.L., Laffont J.J., Straub S. 2008. "Renegotiation of concession contracts in Latin America: Evidence from the water and transport sectors." *International Journal of Industrial Organization* 26(2): 421–442.

Guevara, J., David, S., Shields, R. 2019. "Spatializing stratification: Bogotá." *Ardeth* 4: 223–236.

Hall, P.A. 1993. "Policy paradigms, social learning and the State: The case of economic policy-making in Britain." *Comparative Politics* 25(3): 275–293.

Hall, P.A., Taylor, R.C.R. 1996. "Political science and the three new institutionalisms." *Political Studies* 44(5): 936–957.

Harsono, A. 2005. "When water and political power intersect: A journalist probes the story of water privatization in Jakarta, Indonesia." *Nieman Reports*, 59(1): 45–47.

Hay, C. 2006. "Constructivist institutionalism." In: Rhodes, R.A. Binder, S.A., Rockman, B.A. (eds.) *The Oxford Handbook of Political Institutions.* Oxford, Oxford University Press.

Hood, C., Dixon, R. 2015. *A Government that Worked Better and Cost Less?.* Oxford, Oxford University Press.

Hutchings P. et al. 2015. "A systematic review of success factors in the community management of rural water supplies over the past 30 years." *Water Policy* 17(5): 963–983.

Iossa, E., Rey, P., Waterson, M. 2019. "Organizing competition for the market." *Toulouse School of Economics Working Papers* TSE-984.

Iwanami, M., Nickson, A. 2008. "Assessing the regulatory model for water supply in Jakarta." *Public Administration and Development* 28(4): 291–300.

Jaglin, S., Zérah, M.-H. 2010. "Urban water in the South: Rethinking changing services. Introduction." *Revue Tiers Monde* 203(3): 7–22.

Kashwan, P., MacLean, L.M., García-López, G.A. 2019. "Rethinking power and institutions in the shadows of neoliberalism." *World Development* 120: 133–146.

Kessides, C. 1993. "Institutional options for the provision of infrastructure." *World Bank Discussion Papers* 212.

Levi-Faur, D. 2013. "The Odyssey of the regulatory state: From a "thin" monomorphic concept to a "thick" and polymorphic concept." *Law & Policy* 35(1–2): 29–50.

Lieberherr, E., Ingold, K. 2019. "Actors in water governance: Barriers and bridges for coordination." *Water* 11(2).

Lobina, E., Hall, D. 2007. "Experience with private sector participation in Grenoble, France, and lessons on strengthening public water operations." *Utilities Policy* 15(2): 93–109.

Lowndes, V. 1996. "Varieties of new institutionalism: A critical appraisal." *Public Administration* 74(2): 181–197.

Maggetti, M. 2007. "De facto independence after delegation: A fuzzy-set analysis." *Regulation & Governance* 1(4): 271–294.

Mahoney, J., Thelen, K. (eds.) 2010. *Explaining Institutional Change: Ambiguity, Agency and Power*. Cambridge, Cambridge University Press.

Martijn, E. J. (2005). Hydraulic Histories of Jakarta, Indonesia (1949–1997): Development of piped drinking water provision. Vancouver: Unpublished research report, Department of Geography, University of British Columbia

Marin, P. 2009. "Public–private partnerships for urban water utilities: A review of experiences in developing Countries." *The World Bank, PPIAF, Trends and Policy Options* 8.

Mayaux, P. 2015. "The production of social acceptability: Privatisation of water services and social norms around access in Latin America." *Revue française de science politique* 65: 237–259.

Meeks, R. 2018. "Property rights and water access: Evidence from land titling in Rural Peru." *World Development* 102: 345–357.

Molle, F. 2008. "Nirvana concepts, narratives and policy Models: Insights from the water sector." *Water Alternatives* 1(1): 131–156.

Moynihan, D. 2008. *The Dynamics of Performance Management: Constructing Information and Reform*. Washington, DC, Georgetown University Press.

North, D.C. 1990. *Institutions, Institutional Change and Economic Performance*. Cambridge, Cambridge University Press.

Pierson, P. 2004. *Politics in Time. History, Institutions, and Social Analysis*. Princeton, Princeton University Press.

Pincetl, S. et al. 2016. "Fragmented flows: Water supply in Los Angeles County." *Environmental Management* 58(2): 208–222.

Pollitt, C., Bouckaert, G. 2011. *Public Management Reform: A Comparative Analysis in the Age of Austerity*. Oxford, Oxford University Press.

Powell W.W., DiMaggio P.J. (eds.) 1991. *The New Institutionalism in Organizational Analysis*. Chicago, University of Chicago Press.

Punjabi, B., Johnson, C.A. 2019. "The politics of rural–urban water conflict in India: Untapping the power of institutional reform." *World Development* 120: 182–192.

Rhodes R.A.W. 1994. "The hollowing out of the state: The changing nature of the public services in Britain." *The Political Quaterly* 65(2): 138–151.

Rijke, J. et al. 2013. "Configuring transformative governance to enhance resilient urban water systems." *Environmental Science & Policy* 25: 62–72.

Rouse, M. 2013. *Institutional Governance and Regulation of Water Services*. London, IWA Publishing.

Rugemalila, R., Gibbs, L. 2015. "Urban water governance failure and local strategies for overcoming water shortages in Dar es Salaam, Tanzania." *Environment & Planning C: Governement and Policy* 33(2): 412–427.

Saleth, R.M., Dinar, A. 2004. *The Institutional Economics of Water: A Cross-Country Analysis of Institutions and Performance*. Washington, DC, World Bank.

Saussier, S., Tirole J. 2015. "Strengthening the efficiency of public procurement." *Notes du Conseil D'Analyse Économique* 3(22).

Spiller, P., Savedoff W. 1999. *Spilled Water: Institutional Commitment in the Provision of Water Services in Latin America*. Washington, DC, Interamerican Development Bank.

Sirtaine, S. et al. 2005. "How profitable are private infrastructure concessions in Latin America?: Empirical evidence and regulatory implications." *The Quarterly Review of Economics and Finance* 45(2–3): 380–402.

Stern, J. 2009. "The regulatory and institutional dimension of infrastructure services." *CCRP Working Paper* 15.

Streeck W., Thelen, K. (eds.) 2005. *Beyond Continuity: Institutional Change in Advanced Political Economies*. Oxford, Oxford University Press.

Suchacek, T. et al. 2018. "Comparative analysis and benchmarking of water supply systems and services in Central and Eastern Europe." *Proceedings* 2(11): 597–604.

Supiot, A. 2017. *Governance by Numbers: The Making of a Legal Model of Allegiance*. Oxford, Bloomsbury Publishing.

Trillas, F. 2010. "Independent regulators: Theory, evidence and reform proposals." *IESE Business School Working Paper* 860.

Urueña, R. 2012. "The rise of the constitutional regulatory state in Colombia: The case of water governance." *Regulation & Governance* 6(3): 282–299.

Weber, H. 2004. "Reconstituting the third world? Poverty reduction and territoriality in the global politics of development." *Third World Quarterly* 25(1): 187–206.

Williams, J.G., Bouzarivski, S., Swyngedouw, E. 2019. "The urban resource nexus: on the politics of relationality, water-energy infrastructure, and the fallacy of integration." *Environment and Planning C: Politics and Space* 37(4): 652–669.

Williamson, O.E. 2000. "The new institutional economics: Taking stock, looking ahead." *Journal of Economic Literature* 38(3): 595–613.

World Bank. 1994. *Governance and Development: Issues and Constraints*. Washington, DC, World Bank.

12

FRAGMENTATION IN URBAN WATER GOVERNANCE

Navigating Legal and Normative Modalities

Lee Godden

12.1 Introduction: Changing dimensions of urban water governance and law

Urban water governance has changed significantly over time with the models for water supply, water allocation, and wastewater disposal reflecting wider politico-economic parameters and shifts in the patterns of urbanisation itself. The paradigm of large statutory public agencies being accorded legislative mandates to manage all aspects of urban water governance from supply point to varied industrial, commercial, and residential consumers in a tightly defined urban area has been modified in many urban centres. Indeed, infrastructure to source waters beyond the immediate city can produce impacts, such as rural displacements, well beyond the urban area, (Shi et al. 2021, p. 102354). Varying degrees of economic liberalisation in water services and the privatisation of water infrastructure have been a significant driver of such changes (Bakker 2010, pp. 4–5).

The law underpinning such systems also has changed in focus and content, with monolithic, omnibus water legislation often now augmented, or displaced, by generic utility regulation, environmental protection laws, water pollution and health legislation, consumer protection and competition laws, and water market rule setting. Patterns of water governance in many nations are now more fragmented than in the previous century, with hybrid public/private systems comprising a diverse set of laws and organisational arrangements. Despite increasing emphasis on the adaptive capacity of urban water systems (Gupta et al. 2010), the older, comprehensive model remains significant, even as legacy infrastructure and institutional forms that current laws must accommodate. These institutional forms constitute path dependencies to be accommodated by recent governance models, such as collaborative or market-based commercial water services (see Table 12.1). In turn, as levels of public investment in urban water supply have declined over time, and as commercial provision can be more ad hoc in its coverage, it has exposed significant gaps in the adequacy of urban water management, revealing underlying structural inequities in urban water governance (Shi et al. 2021, p. 102355). This inequitable situation has seen international policy attention turn to human rights laws and sustainability norms to address urban water governance gaps with the Sustainable Development Goals (SDG) a high-profile platform for addressing these problems (UNEP SDG 6 Clean Water and Sanitation and SDG 11 Sustainable Cities and Communities).

DOI: 10.4324/9781003057574-16

Table 12.1 Representative Urban Water Governance and Finance Laws

	Organisation	*Method of charging*	*Body overseeing utility pricing*
FRANCE	Municipalities are responsible for drinking water supply and sanitation services since 1885. Choice in how to manage the utility; can be either by a public authority, directly by the municipality, or delegated to public or private operator (contract). Municipalities always retain ownership of facilities.	Municipalities set tariffs. If management is delegated, the contract will define operational payments included in water prices paid by users. Ultimately, pricing varies due to local factors.	There is no national regulatory agency to approve tariffs. Price is controlled via a contracting process.
GERMANY	Municipalities in Germany are responsible for public water supply and sanitation due to right of self-government (art 28(2) German Basic Law). Five common forms of organisation: total municipal control; mixed operation; total private operation.[1] Municipalities may also cooperate.	Municipal levies legislation and municipal codes of the *Länder* set out the framework for calculating fees [limited by principles of public financial conduct].[2] No special provisions govern water price calculation, though case law applies calculation by reference to municipal levies acts.[3] Charges are increased by special state levies, such as an abstraction charge or wastewater levy paid by companies to the *Länder*.[4]	Different price controls apply depending on the contractual basis of the urban water use. Decisions of companies incorporated under private law are made by a supervisory board where the public has considerable say.[5] *Länder* anti-trust authorities (or German Federal Cartel Office) assess pricing for anti-trust characteristics.[6] See fairness test (s315 German Civil Code), which consumers can assert in court.
SOUTH AFRICA	Local governments are responsible for service delivery.[7] Two-tiers of water delivery. First, government-owned water boards (e.g., Rand Water) are responsible for bulk water supply to	Municipalities vary their pricing levels but all have a volumetric pricing structure, i.e., higher per-unit rates pertain to high water consumption, e.g., tariffs in Johannesburg are USD0.87/m^3 per month	The Minister of Water and Sanitation has powers under s10 of the Water Services Act 1997 to prescribe norms and standards in respect of tariffs, see 2015 prescription by Minister Mokonyane.[13] Regulations 6 and 7 set out principles

(*Continued*)

Table 12.1 (Continued)

	Organisation	*Method of charging*	*Body overseeing utility pricing*
	municipalities. Then, municipalities make provision for retail water to end-users for water supply services and sanitation. Municipalities are among 169 water services authorities under the Water Services Act 1997 (mixed, district, and local municipalities). Service delivery controlled entirely by municipality or delegated.[8] Contracting out provision of water services to private companies is extremely rare. Services usually direct municipal provision.	after the first 15m^3, then USD2.46/m^3 between 50-100m^3, and then USD3.35/m^3 after 100m^3.[9] Each bracket contains a fixed charge. Cape Town utilises fixed charges,[10] while Durban does not.[11] Free Basic Water policy (2001) in which every household is entitled to 6m^3/month for free. Municipalities may decide eligibility. In 2015–16, 95% of the population benefited; 66% of municipalities where policy was "indigent only."[12]	of sustainable and affordable pricing.
USA	State legislation grants municipalities the right to control the provision of water (and other public utilities).[14] Municipalities have broad leeway on how the water service is organised, including the permission to contract with private water providers.	Municipalities have broad leeway to determine how they charge for water due to the broad wording of their enabling statutes (i.e., the relevant state code).	Economic regulation is done by the relevant state's regulatory utility commissioner.[15] See, e.g., the rules requiring rates increases to be approved by the New York State Public Service Commission,[16] limiting late payment charges,[17] and otherwise regulating when water services may be cut off.[18]
CHILE	Chile's water sector moved to a market model with the Water Code of 1981, which maintained water as "national property for public use" but also granted permanent, transferable water-use rights to individuals.	The SISS decides on tariffs for drinking water and sanitation services.[21] Under the General Water and Sanitation Law (1988) tariffs must satisfy four objectives and abide by four principles. Objectives:	The Superintendencia de Servicios Sanitarios (SISS) is the economic regulator of urban water.[23] It has command and control functions for urban water. It enforces norms related to tariffs according to the tariff framework.[24]

(*Continued*)

Table 12.1 (Continued)

	Organisation	*Method of charging*	*Body overseeing utility pricing*
	There are 53 urban water service providers. Most of them are private utilities. A concessional regime exists where the private companies receive concessions to supply water services and so are responsible for maintaining, renovating, and building distribution networks.[19] Since 2001, the state has assigned concessions for 30 years. The concessional regime is implemented by the SISS.[20]	1. Full recovery of operation and maintenance costs; 2. Fund necessary infrastructure reposition and development plan investment; 3. Reductions when operators increase efficiency; and 4. Operational margins consistent with the opportunity cost of capital. The four principles: 1. Economic efficiency 2. Water conservation 3. Equity 4. Affordability [22]	To prevent monopolisation, providers are classified into three categories (big, medium, small) and no person or corporation is allowed to own more than 49% of the companies in a category. Affordability is met by a subsidy of between 15% and 85% of a water bill provided by municipalities to vulnerable households. This covers consumption up to 20m^3.[25]
COSTA RICA	Under Drinking Water Law 1953, municipalities are responsible for providing water and sanitation. The Costa Rican Institute of Aqueducts and Sewers (AyA) is a centralised federal public body responsible for water systems serving almost 50% of the population.[26] Municipalities supply water to 17.1% of the population.[27]	Tariffs for water and sewer services provided by AyA and private providers are subject to approval by ARESEP. Tariffs for services by municipalities are set by the municipalities themselves.[28] Given most suppliers are public, and almost all fees require regulatory approval, it indicates a cost-recovery model.	Major service providers are economically regulated by the Regulatory Authority for Public Services (ARESEP). It sets tariffs (along with technical regulations and monitoring compliance).[29]

[1]https://www.eureau.org/resources/publications/members-reports/5579-profile-of-the-german-water-sector-2020/file (24–25).

[2]Principles of equivalence (proportionality), cost-recovery, prohibition of cost overruns, equal treatment, and implementation in accordance with economic principles: see ibid., p. 28.

[3]Ibid.

[4]https://www.bdew.de/media/documents/6000_Sector_profile_of_the_German_water_sector_published_in_English.pdf (31).

[5]Ibid., 28.

[6]See German Act Against Restraints of Competition.

[7]Municipal Systems Act 2000.

[8]See Johannesburg Water, a legally and financially independent company wholly owned by the city of Johannesburg.

Table 12.1 (Continued)

[9]https://tariffs.ib-net.org/sites/IBNET/ViewTariff?CountryId=0&tariffId=61
[10]https://tariffs.ib-net.org/sites/IBNET/ViewTariff?tariffId=23742&countryId=24
[11]https://tariffs.ib-net.org/sites/IBNET/ViewTariff?tariffId=23746&countryId=24
[12]http://ws.dwa.gov.za/downloads/WS/P_I/Adhoc/AWensley/Water%20&%20Sanitation%20Tariffs_20152016%20portrait.pdf (32)
[13]https://cer.org.za/wp-content/uploads/2011/04/WSA-Draft-revision-of-norms-and-standards.pdf
[14]See California Public Utilities Code Ch. 5 Art. 1 (ss10001-10014), or the Texas Local Government Code Title XIII Subtitle A Ch. 552 Subchapter A (ss 552.001-552.003).
[15]See Illinois Commerce Commission, Public Utility Commission of Texas, New York State Public Service Commission, or California Public Utilities Commission.
[16]Codes, Rules and Regulations of the State of New York (NYCRR) Title XVI s531.5
[17]NYCRR Title XVI s531.5
[18]See NYCRR Title XIII Pt 14 (here), and Chapter 5 (this volume).
[19]https://www.oecd-ilibrary.org/the-governance-of-water-infrastructure-in-chile_5jfj4n5457vd.pdf?itemId=%2Fcontent%2Fcomponent%2F9789264278875-7-en&mimeType=pdf (290)
[20]Ibid. (313–314).
[21]Ibid. (291)
[22]https://www.researchgate.net/publication/282158270_Water_Pricing_in_Chile (9-10).
[23]Above n xix.
[24]Ibid. (313).
[25]Above n xxii (11–12).
[26]https://undocs.org/A/HRC/12/24/Add.1 at [19] and [41].
[27]Ibid.
[28]Ibid at [20].
[29]Ibid.

12.2 What is water governance?

Water Governance is a dynamic but contested term (Baker 2011, p. xv). Debates reflect socio-political conflicts, such as that pertaining to the relative effectiveness of public control or privatisation of water services as well as reflecting the myriad interests that communities, authorities, and businesses have in urban water. This shift to governance terminology signals a move from conventional concepts that have typically informed legal analyses. In the urban water context, the use of a governance lens may mean, for example, that a government departmental water regulator may be only one component of a water governance *system* that enforces elements of the water governance *regime* (for example, water licensing or pollution laws). Accordingly, "governance" may refer both to a legislative process (e.g., decision making) and an outcome (determining who gets water, when, and how). In reality, in urban water contexts, it is often both. Reflecting this duality, Jimenez et al. (2020, p. 829) suggest water governance is "a combination of functions, performed with certain attributes, to achieve one or more desired outcomes, all shaped by the values and aspirations of individuals and organisations." The changed analytical terrain of urban water law is better placed to reveal the spectrum of formal legal measures and informal normative modes that constitute the rules, the institutions, the networks, and the actors that are part of governing water in its many dimensions in urban settings (Mishra et al. 2020).

Governance approaches also accord with sociolegal research that utilises knowledge in the social sciences and the humanities to explore the embedded social or political power of legal arrangements for urban water. Pahl-Wostl (2015, p. 25) explores urban water governance definitions from political and social science that highlight specific aspects of governance systems, such as the means of making policy (and law), the translation of the differing preferences of actors, the institutions and rules that exist to shape the actions of actors, or the policies (and laws) employed. Her analysis seeks to encapsulate the complexity of societal dynamics and political factors. Governance definitions also focus on decision making across sectors and actors, where decisions are inflected by power differentials. It offers a platform also to critically interrogate how governance as a practice has been rather narrowly aligned to technical water management. An early Global Water Partnership report (Rogers and Hall 2003, p. 7), by contrast, suggests that "[w]ater governance refers to the range of political, social, economic and administrative systems that are in place to regulate development and management of water resources and provisions of water services at different levels of society."

Legal analysis of urban water ideally should reference sociolegal, pluralist, and empirically grounded accounts of law to iterate the complex normative structures that operate across sociolegal and biophysical dimensions in urban settings. Accordingly, this chapter draws on social learning practice (Godden and Ison 2019, p. 48) in which water governance exists as a contextual, dynamic, and relational process, and where such forms co-evolve. In many countries, the social dimensions of urban water reform have received less attention relative to other parameters, such as economic liberalisation or urban utility reforms. As a whole, engagement with the systemic complexity of water governance remains inadequate (Rubenstein et al. 2016, p. 82).

At its broadest, ideally, legal analyses of "urban water governance" should reflect the array of legal, regulatory, and behavioural measures that governments and other actors may undertake to ensure secure and sustainable urban water and its safe disposal or reuse for urban populations and ecologies. Approaches may switch dynamically to comprise market-based or conventional regulatory philosophies; incorporate evolving approaches such as integrated urban water management or sustainable urban water management (both of which imperfectly channel integrated water resource management; van der Brandeler et al. 2019, 1068). Interactions may occur between various levels of government, ranging from international conventions to local bylaws, and between state and nonstate actors. Urban water laws may apply to recognise the specificity of different urban water environments, such as urban wetlands or high-density urban areas (Bakker 2011, p. 52).

Given the diversity of situations for actualising urban water governance, this chapter provides a selective analysis of representative governance systems and the legal and institutional arrangements that typically pertain to them. The chapter analyses public and private water governance, including merged or hybrid forms and associated legal structures. It engages with the expanding range of institutions and actors in the urban water space that may have either a formalised legal presence or more normative influence on urban water governance. Finally, it examines how urban water governance can exhibit transnational dimensions under the impetus of commercialisation and globalisation, and with respect to legal movements such as the "right to water." Finally, it looks to future social equity challenges, exploring how water governance will contend with urbanisation pressures in the rapidly expanding cities of the Global South.

12.3 Transformations in urban water governance

The paradigm for urban water provision was built around a "hard engineering" model that required extensive capital investment in water infrastructure (Franceys and Hutchings 2017, p. 120). Its parallel in water law and regulation is a statutory agency model, whereby authorities, acting as corporations, undertake water and sanitation "services" for a defined (and apparently passive) urban population. Historically and geographically, this legal and organisational model was never, and it is not now, the only legal form for water management in urban centres. Many nations have a blurred and shared model of public and private corporations, including community organisations, involved in water services (Bakker 2011, p. 110), with corresponding regulation, often on a relatively localised scale. Drawing on regulatory theory, it is possible to categorise legally relevant modalities for urban water governance that interlace with technologies as four types: the law, social norms, the market, and architecture (or codes) that exist in regulatory environments as outlined by Brownsword (2011, p. 576).

The hard engineering model and accompanying bureaucratic legal and institutional forms for urban water governance are widely utilised by urban municipalities and/or provincial governments. The package also was a signature techno-legal instrumentality deployed by European colonising power during the 19th and early 20th centuries (Cullet 2018, p. 328) and by post-colonial nations in the post-independence era, when seeking to realise improved socioeconomic conditions and sustainability for their populations (Mishra et al. 2020, p. 162). For many nations, such investments in urban water infrastructure (physical infrastructure and as a combinate legal/institutional modality) remain rudimentary due to prohibitive costs (Franceys and Hutchings 2017, p. 121) or operate alongside historic (customary) systems – many of which fell into desuetude under colonial regimes.

From a different vantage, comprehensive statutory agency models comprised legal modalities that distinguished economies denoted by substantial investment in costly water infrastructure. Nonetheless, despite such prohibitive costs, many development interventions in nations continue as a form of implicit globalisation that is tied to this water governance ideal, "from which cities of the South are seen to deviate, but to which they are expected, eventually, to conform" (Bakker 2011, p. xx). Western scientific and technological practices thus promote an integrated (conjoined infrastructure and law/policy) ideal for water infrastructure provision (Cullet 2018, 329). By contrast, post-colonial research in cities reveals highly differentiated forms of urban water services, but with a splintering or exclusion from services, often on a race and class basis. A telling example of urban water "service provision" along racial and class lines occurred in South Africa under the apartheid regime, where black and coloured settlements were often totally bypassed by urban water infrastructure designed for white, elite neighbourhoods. A sociopolitical outcome that saw urban water reforms as a key political platform for the incoming ANC government (Bond 2012).

Other water governance research in cities of the South emphasises "urban informality" as a critical factor (Ranganathan 2014). Informality underpins differential access to urban services, such as water, and thus it constitutes a material expression of fragmentation and intersectionality but also intercultural relationships with water that coexist with more formalised legal regimes. Accordingly, much of the world's rapidly increasing global urban population remains dependent on (diminishing) local sources of water, especially groundwater, while systems of customary or subnational water governance are common and may operate in the interstices of water authority or commercial provision (Ranganathan 2014, p. 90). Such fragmentation in cities, particularly in the Global South, should not be viewed as chaotic or operating in a legal and regulatory void, but rather as utilising a spectrum of formal and informal modalities.

Nonetheless, the very recognition of the existence of these alternative forms of urban water governance has contributed to the rise of a discourse in financial and development circles that public water systems suffer from corruption, underinvestment, and a lack of necessary expertise, which justified reasons to move from such systems. Constraints on public expenditure also are touted as a major disincentive for continued reliance on public systems under the impetus of "the governance of thirst efficient management" (Bakker 2011, pp. 4–5). The advent of liberalised water markets has been one response, common to the cities of both the Global North and the Global South. Even so, both public and "liberalised" commercialised approaches still typically pivot upon modern technological modes and integrated water governance networks. The following sections discuss the conventional water law and regulatory modalities that typically reference that construct, before turning attention to the modalities that act as disruptors to conventional legal models.

12.3.1 Governing water quality in urban areas

Classically, much water law in urban settings has a technological focus on preventing pollution and contamination of water. Widespread concerns around safe water supply and the prevention of cholera and other water-borne diseases provided an important momentum for the adoption of large statutory water agencies that had a public health and urban planning foundation. That trend of disease and pollution control has continued with legal, policy, and scientific attention given to developing the technical aspects of urban water governance, in terms of finely calibrated and highly detailed pollution control laws. Howarth (2017, pp. 81–2) defines water governance as ensuring water quality and preventing pollution. "Pollution" can include a substance, an activity, or even a state of the environment in laws seeking to regulate water pollution (Howarth 2017,p. 80). Much regulation reflects an anthropocentric focus given the potential of pollution to interfere or damage human uses of the environment (see, e.g., EU Water Framework Directive, art 2 (33)).

If the urban water governance problem is given an overarching "pollution" framing, then typically the legal and policy response is the development of water quality standards by centralised agencies that regulate water pollution to achieve a satisfactory human health and environmental state. This extremely common model, with origins in rapidly urbanising cities during the mid-19th century, can range from prohibitive responses to industrial pollution via the criminalisation of various activities to attempts to preempt pollution through land use controls and to regulate polluting activities through conditional permits and licences. Only more recently has it involved setting positive objectives for the quality of water and surrounding environments (Howarth 87–89). In many "post-industrial" cities, this central pollutant control model has expanded to include numerous, identified pollution sources emanating from the cumulative impacts on water of the daily lives of citizens, such as plastic bag bans to prevent accumulation in urban waterways. Public health–oriented regulation in tandem with water laws may address pollutants, such as endocrinal or similar human pharmaceutical pollutants, and now – COVID-19 viral contaminants in wastewater (WHO Guidance 2020). A pollution-oriented model for urban water governance typically will form a component of dedicated water, health, or environmental responsibilities of government agencies at cascading levels of institutional responsibility or as delegated duties under contact or legislation for private organisations charged with public water supply services.

Nonetheless, widespread contamination of urban water in many cities continues despite the existence of sophisticated regulatory frameworks, with, at times, not even basic quality standards being met. Pollution, such as heavy-metal contamination of groundwater and drinking supply, or residential pollutants, such as sewage, in urban rivers are common problems. The resources available to monitor these harms and ensure compliance may be minimal, and similarly water authorities may lack the requisite powers to comprehensively address and constrain the activities of polluting industries in urban rivers (Islam and O'Donnell 2020). The conventional legal and regulatory modality fails at these critical points.

12.3.2 Governing water supply and allocation

The other strand to the conventional modality for urban water governance is law and policy directed to water "quantity." Water quantity captures both water supply and allocation. It can reflect the biophysical/environmental aspects of water security as well as the technical, engineering, socioeconomic and cultural allocation and distribution systems for urban water (Adams et al. 2020). Nonetheless, the core of urban laws governing allocation still tend to be a "standard agency" urban utility model (see table below). This legislative model posits a neutral system whereby water is "seamlessly" derived from a water supply source, such as a storage dam, reticulated evenly across the urban area to consumers who are able to pay for such services, and where "excess" water is removed through systems such as stormwater disposal. The "water grid" for urban catchments perpetuates this ideal in water law regulating water supply, allocation and urban water use in cities.

Underpinning such legal modalities for urban water allocation systems are conceptions of individual entitlements or rights – primarily couched in terms of water "property." While the classic division is a distinction between private and public property rights to water, in most allocation systems there will be a wider spectrum of legally endorsed rights and uses, not all of which attract a formal property designation. Indeed, there are contested legal interpretations of whether the term, "property" should be accorded to the complex situations of rights to access and use water. In recognition of the critical role of water, some societies treat water as a communal or public good (UNDP 2006) rather than a private right, with a focus on basic household water provision and community service. The accentuation of private property rights to water, accompanied by the expansion of urban water markets as a prevailing form of water distribution in corporate water governance modalities, has attracted much concern around water privatisation in neoliberal governance approaches (Sultana and Loftus 2012, Introduction). Viewing water through the lens of private property can heighten power inequities. When property rights to water operate to limit access to water (even when it is nominally free), corruption (Ranganathan 2014) and gendered inequalities can worsen. Gray and Lee (2018, p. 120) argue that water property and market systems do not adequately factor in social equity considerations. Markets reallocate resources towards those who are most economically competent in the pursuit of "efficiency." Accordingly, the potential for social inequities (e.g., limited water access for disadvantaged communities) is embedded in markets. This potential was taken to extremes in Bolivia, under a system of water privatisation where water prices rose exponentially and services ceased to those no longer able to afford water. The situation provoked civil unrest and violence and, ultimately, to the recognition of water as a human right in 2010 (Winkler 2012). Water markets as a governance mode, on the other hand, may not necessarily have the monopoly control characteristics identified with corporate global modalities. Instead, these "markets" may operate in relatively informal, social entrepreneur settings to provide urban water supplies for communities and local businesses (Garrick et al. 2019).

Attention to the effects of market modalities in urban water governance can obscure the intra-urban complexities of water allocation systems in urban settlements that range from localised customary laws governing access to water to highly stratified and nested organisational structures that distinguish water allocations in megacities (van den Brandeler et al. 2019). Legal models for water allocation systems in multilevel legal institutions where the various elements of allocation and access, including pricing, are filtered through institutional and corporate law structures, typically will contain highly detailed provisions that tightly define the responsibilities and duties for urban water provision (see for example, drought triggers and pricing in UK water legislation addressing drought planning).

Over time, these legal and regulatory modalities have garnered specific hierarchical organisational forms. In such water laws, typically there will be a Governing Board structure at the apex of a public or private corporate organisational form that oversees the operational, technical (i.e., engineering) and commercial aspects of the system – often subject to overriding government discretions (see table below). Such safety net requirements in governance systems may derive from water commons models or where there are statutory or constitutional requirements, such as guaranteeing minimum water supply to households. Notwithstanding legal safeguards, such as human rights (Clark 2020), the complex, multilayered large urban water governance systems may fail to adequately provide for vulnerable groups in urban areas (or fail at times to meet statutory requirements) due to socioeconomic and political factors, including globalisation and rapid urbanisation (Adams et al. 2020) and the growing mobility of capital that may seek higher investment returns than can be provided by many municipal water supply systems (Franceys and Hutchings 2017).

12.4 Governing cities as socio-ecological systems

The classic integrated grid and statutory agency modality has been challenged by growing acknowledgement of the socioecological and political power factors at play in urbanisation (Bakker 2011) and the value of more participatory forms of governance (Godden and Ison 2019). Accordingly, water governance in this sense relates to the social functions involved in regulating and managing water resources and the provision of water services to different levels of society (Özerol et al. 2018, p. 43). Such a "water governance *system*" is denied as "the interconnected ensemble of political, social, economic and administrative elements that perform the function of water governance." (Pahl-Wostl 2015, p. 26). Further, as Pahl-Wostl (2015, p. 27) notes a "water governance *regime*" is "the interdependent set of institutions (formal laws, societal norms or professional practices) that is the main structural component feature of a water governance system." This complexity is reflected in Australian legislation such as the Water Act 1989 (Victoria), whose purposes (s 1) include, to "promote the orderly, equitable and efficient use of water resources" as well as maximising community involvement in decision making and respecting Aboriginal cultural uses. Moreover, the Water Act 1989 contains strict obligations (s 141A) for the Melbourne Water Corporation regarding ensuring water provision to statutory water retailers (and thus to consumers) in the Melbourne metropolitan centre in Australia.

Robust community participation and the co-production of governance are distinctive features of governing water as a socioecological system (Adams et al. 2020, p. 1467). Generally, community participation will be encouraged but rarely mandated, including with regard to NGO representation in water law and decision making. This complex interface among law, institutions, and "the community" forms a coupled socio-ecological system. Complexity is heightened due to the multifaceted nature of many institutions with responsibilities for urban water that community groups must navigate. Urban ecology approaches offer a comparable perspective, drawing attention to water as an ecological process that supports human and environmental functions. In turn, water law for urban areas has slowly and sometimes reluctantly accommodated sustainability objectives for urban centres (Clark 2020). Trade-offs between environmental needs and human consumptive water uses in most water laws still tend to prioritise decisions favouring consumptive water uses rather than environmental or cultural/recreational purposes. Even so, the conventional legal resources and agency model hinged on infrastructural responses is now subject to potentially disruptive factors. Thus, legal and regulatory options need to comprehend a growing diversity of water sources, including recycled water, aquifer storage, desalination, and the politically unpalatable prospect of government-imposed water restrictions. Attention to water as a "flow" through urban areas (including a flow of wastes that might be recycled) has seen corresponding developments in water law that recognise alternative sources, such as stormwater reuse and aquifer recharge, and which require their own differentiated legal structures, planning, and governance processes (Bakker 2011, p. 17).

As Strang (2016, p. 296) argues:

> [I]n considering any regime of water governance and control and its infrastructural expressions, there is a need to recognise that all participants – the people, the material culture, the water and the wider environment and its nonhuman inhabitants and material things – are involved in fluid and sometimes transformational processes.

12.5 Transnational norms in urban water laws

Thus, legal and regulatory modalities increasingly need to be nimble in the face of escalating and compounding urban water governance challenges. Many existing national laws and institutions, including intergovernmental arrangements, can constrain the flexibility or feasibility of policy options and legal measures to operationalise complex governance responses. In such situations, international human rights laws and transnational water laws (including environmental protections) may offer normative guidance to influence the dynamics of national and local urban water supply, pollution, water allocation, and environmental protection laws.

A growing body of sociolegal research (see, e.g., Johns et al. 2010; Johns 2015) has identified an increasing trajectory in the utilisation of transnational and international legal and normative principles to inform water law and governance within national and regional contexts. The relevant systems may involve formal legal/institutional arrangements as well as informal behavioural codes and cultural and technological practices.

12.5.1 Human right to water

Given the pressing implications of water availability for sustaining life and for adequate sanitation that are held to underpin global development and public health goals, the United Nations has invoked human rights models to address these crucial entitlements. Second-generation human rights are oriented to socioeconomic needs. Water rights had a diverse pathway within that system, until ultimately securing the status of an independent human right with the UN General Assembly's 2010 Resolution on the Human Right to Water and Sanitation (Mason Meier et al. 2018, p. 102). This declaration recognises the vital necessity of water to humanity, including in urban settings. As a transnational legal principle and normative regime, it can be seen, in part, as a response to other transnational normative regimes that favour neoliberal resource economics and, ironically, sustainability. In 1993 a World Bank water resources report emphasized that "[w]ater is an increasingly scarce resource requiring careful economic and environmental management" (World Bank 1993, p. 9). The report set an agenda for water development with a central focus on privatisation and cost recovery, "to ensure the financial sustainability of water supply systems and to ration the withdrawal of scarce water" (World Bank 1993, p. 126). Given these guiding assumptions entailed in what became a pervasive market-based modality, water governance, especially in cities, became sensitised to addressing scarcity. As Bakker notes (2011, p. 110), scarcity, rather than being an absolute, is socially constructed but nonetheless very real. Thus, inequitable access to water may be embedded in the legal and sociophysical modalities of urban water networks. Exclusion of some groups from water access thus characterises public and private systems under processes of modernisation and urbanisation (Davis 2016, p. 357; Bakker 2011 p. 110). It was this market liberalisation process, as noted above, that precipitated a response from the public international law and normative regime to reiterate that water is an essential right that supports the dignity of individuals (Winkler 2012).

12.5.2 Incorporation of transitional norms in national settings

Normative models in water governance combining formal ("hard") law and soft law, behavioural change, and incentive-based models are increasingly common phenomena nationally (Johns et al. 2010), reflecting transnational influences as well as a decentering of the state as the primary "actor" in urban water governance. As these legal and normative modalities of human rights and social equity take hold in national urban water settings (Davis 2016, p. 58), transnational monitoring mechanisms can ensure state accountability for progressively realising rights to water and sanitation. Some caution is needed though regarding replacement of the formal legal (and therefore enforceable) elements of national urban water laws, despite policy moves towards more participatory and social equity forms of governance (Davis 2016, p. 358). Most water laws operative in urban contexts will retain "hard-edge" aspects, such as liability for harms such as water pollution under statute or general law of negligence, together with monitoring, enforcement, and sanctioning powers for government authorities. There may be some (limited) delegation of such functions to third parties or where outcomes are directed under legislation to be achieved via "market-disciplining" and commercial imperatives. Such blended governance forms display the institutional interactions and codes now common to emergent hybrid modalities of water governance. Moreover, even when urban water governance manifests in private law forms, an expectation remains of community accountability. The actual degree of community access to governance will largely depend on the legal models that are entrenched. For example, community water trusts may adopt more explicit community models (Owens 2016, p. 344) when compared to commercial utility sector companies. In the latter example, the inaccessibility of decision making in terms of effective community participation can be substantially increased as an inherent element of the character of corporations in general law. Another significant barrier to greater accountability in urban water governance lies in the embedded modalities of financial powers typically found in many of the institutional arrangements in water laws (Franceys and Hutchings 2017).

12.6 Financial parameters and models of water governance

The financial and resource management powers of organisations with responsibilities for urban water governance have received relatively little detailed research attention. Despite considerable innovation in water governance, the reliance on a conventional utility model for urban water management remains high in many countries. A factor that may embed utility models is that typically they are supported by specific financial and resourcing provisions in laws (including constitutional legislative power divisions) that are aligned to government authority or statutory agency structures. Finance and resourcing laws assume their own legitimacy as a modality of governance (Johns 2011, p. 392).

The following table provides selected examples from Europe, the United States, and countries from the Global South in terms of the relevant laws for urban water organisations. It outlines powers of setting water prices and the existence of equitable safeguards and/or regulatory oversight.

The examples outlined in the table, while clearly not representative of all global urban water laws, offer snapshots illustrating key points in relation to the four types of modalities, i.e., the law, social norms, the market, and architecture (or code). First, is the persistence of legal models grounded in institutional water utility models and corporate forms – these remain the backbone of municipal water services (France, Germany, South Africa, United States, and, to a lesser extent, Chile and Costa Rica). Second, is the incorporation of neoliberal, contractual modalities and economic regulators/ design codes (Chile, Costa Rica, United States, with some elements apparent in South Africa and a mixed model in France and Germany). Third, is the strong influence of social and transnational norms designed to provide equity of access and social safeguards (South Africa - comprehensive example, Costa Rica and Chile and Germany. The United States – standardised consumer protections – and surprisingly few measures in France). The fourth and final point is that the reproduction of conventional agency models has translated from the nations of the Global North to those in the Global South. Nonetheless, it is in the latter group of nations (at least in this small sample) where we see a greater degree of hybridisation, soft law, and code type innovations. These innovations in urban water governance are accompanied in several nations of the Global South by explicit legal protections of varying strength derived from transnational equity principles and social norms as well as explicit as human rights protections, including specific codifications of such measures as in South Africa.

12.7 Future challenges

While more than 50% of people now live in urban areas, the main wave of future urban expansion will occur in the Global South (Ranganathan 2018). The increasing intensity of urbanisation clearly poses challenges in established and in rapidly expanding cities in terms of ensuring continuity of adequate water services and in meeting new demands, including for participatory and socially responsive forms of urban water governance. Pervasive challenges arise from the transformations required to enhance equitable access to water and sanitation. Other transnational challenges include the need for urban water systems to respond to biophysical threats, such as pandemics and the identified risks of anthropogenic climate change that severely affect water availability. Further, such transnational challenges are magnified for rapidly urbanising populations in the Global South where inflows of people to cities and the development drive of governments place great pressure on the resources available for water systems. These myriad factors represent a substantial challenge to better adapting urban water governance systems (McDonald et al. 2014).

These challenges are not simply "external" threats to self-contained, "hard engineered" urban water systems and accompanying laws. Existing path-dependencies related to specific legal and regulatory forms of urban water governance aligned to corporate forms, both public and privatised often are entrenched through legal and financial measures. Yet, such modalities exist within a heterogenous and emergent mix of governance forms. Adopting a socio-ecological lens, it is more pertinent to consider urban water governance as a multidimensional phenomenon comprising combinations of formal law, (constitutions, legislation, and judicial decisions), together with soft law, social norms, the market, and various codes or prescriptions that influence behaviour.

Culturally, urban populations have a highly differentiated composition in demographic, racial, ethnic, and material wealth terms as well as differential experience of urban water governance – often deriving from diverse spatial distribution and equity of access to water and services. These socio-ecological factors contribute to individuals experiencing different material relationships and affective meanings that govern relationships and interactions with water (Strang 2016). At times, some modalities and experiences may dominate over others (e.g., water access as determined by material wealth). Adapting to the challenges of current complex urban water governance requires recognition of the socially constructed nature and multiple framing choices in a given urban water situation.

For cities of the Global South, informality is a significant factor at play in how urban populations may experience urban water governance and its material expression, i.e., actual access to water (Adams et al. 2020). Informality is not reducible to physical conditions, such as informal housing where water hard infrastructure is absent. As Ranganathan (2018, p. 2 of 16) notes, "informality" broadly refers to practices that fall outside the ambit of formal planning and official orders. Urban water informality is part of a wider negotiated practice involving public authority and the power of the state. The state is almost always intimately involved in some way in the maintenance and reproduction of informality, including in water governance. Further, the concept of urban informality is well recognised and embedded into the "sphere of development intervention" (Ranganathan 2014), including in the SDGs. To date, it has not fully translated into ensuring adequate water and sanitation access for many people living "informally" in cities. Despite significant transnational movements, such as human rights agendas (Clark 2020), together with initiatives by national governments, such as South Africa, vast disparities exist in the quality and performative outcomes of urban water governance – both formal and informal in expression. Adams et al. (2020, p. 1468) note the failure of conventional modalities used to address urban water insecurity in the Global South. These comprise, "public policy systems where a centralized state or local government is responsible for water delivery, private (market-based) systems where water provision is managed by a foreign or local investor, or a hybrid of the two systems."

Among the many legal, social, and cultural constraints to improving urban water governance is the recursive and self-perpetuating relationship between traditional legal and regulatory institutions, interventionist discourse, and technical practice that hinder transformative change (Rubenstein et al. 2016). Resistance to legally plural forms of governance within nation-states is a well-documented legal phenomenon (Berman 2015, p. 15) that persists despite efforts to introduce more diverse legal forms. Moreover, while many formally constituted urban water systems nominally remain tied to governance and funding modalities that are internal to nations, the financial locus of control increasingly may be ordered more strongly by transnational equity flows and urban investment patterns. Such patterns reflect the impetus of global urbanisation. As a counter, some regulatory transformations are occurring that are initiated in the social sphere in association with entrepreneurs and businesses, often in transnational contexts that cross multiple levels of governance and that confound traditional distinctions between state and market modalities (Morgan 2014). Urban water governance under such impetus will require augmentation of the social sphere to better explore how social, business, and cultural norms as well as diverse practices, actors, codes/designs, and organisational planning can coexist with formal legal and regulatory models to provide a more finely calibrated, and responsive modality of urban water governance that is alive to the situatedness of urban water in its socio-ecological context.

References

Adams, E., Zulu, L., and Ouellette-Kray, Q. 2020. 'Community water governance for urban water security in the Global South: States, lesson, and prospects.' *WIREs Water* 7(5), p. e1466.

Bakker, K. 2010. *Privatizing Water: Governance Failure and the World's Urban Water Crisis*, 1st ed. Ithaca and London: Cornell University Press.

Bakker K., and Cook C. 2011. 'Water governance in Canada: Innovation and fragmentation.' *International Journal of Water Resources Development* 27(2), pp. 275–289. https://doi.org/10.1080/07900627.2011.564969

Berman, P. 2015. 'Non-state lawmaking through the lens of global legal pluralism.' In *Negotiating State and Non-State Law*, edited by M. Helfand. Cambridge: Cambridge University Press, pp. 15–40.

Bond, P. 2012. 'The right to the city and the eco-social commoning of water: Discursive and political lessons from South Africa.' In *The Right to Water: Politics, Governance and Social Struggles*, edited by F. Sultana and A. Loftus. London & New York: Taylor & Francis, pp. 190–205.

van den Brandeler, F., Gupta, J., and Hordijk, M. 2019. 'Megacities and rivers: Scalar mismatches between urban water management and river basin management' *Journal of Hydrology* 573, p. 1067. DOI: 10.1016/j.jhydrol.2018.01.001

Brownsword, R. 2011. 'Responsible regulation: Prudence, precaution and stewardship,' *Northern Ireland Legal Quarterly* 62(5), pp. 573–598.

Clark, C. 2020. 'Global goal setting and the human right to water.' In *Oxford Research Encyclopedia of Global Public Health*, edited by D. McQueen. Oxford: Oxford University Press.

Cullet, P. 2018. 'Innovation and trends in water law.' In *The Oxford Handbook of Water Politics and Policy*, edited by K. Conca and E. Weinthal. Oxford: Oxford University Press.

Davis, M. 2016. 'Let justice roll down: A case study of the legal infrastructure for water equality and affordability.' *The Georgetown Journal on Poverty Law & Policy* 23, pp. 355–393.

Franceys, R., and Hutchings P. 2017. 'Governance and regulation of water and sanitation services provision.' *Routledge Handbook of Water Law and Policy*, edited by A. Rieu-Clarke, A. Andrew Allan, and S. Henry. Oxford: Taylor & Francis Group, pp. 120–134.

Garrick, D., O'Donnell, E., Damania, R., Moore,S., Brozović, N., and Iseman, T. 2019. *Informal Water Markets in an Urbanising World: Some Unanswered Questions*. Washington, DC: World Bank.

Godden, L., and Ison R. 2019. 'Community participation: Exploring legitimacy in socio-ecological systems for environmental water governance.' *Australasian Journal of Water Resources* 231, 45–5 . DOI: 10.1080/13241583.2019.1608688

Gray J., and Lee L. 2018. 'Water entitlements as property: A work in progress or watertight now?.' In *Reforming Water Law and Governance* edited by Cameron Holley, Darren Sinclair. Singapore: Springer, pp. 101–122.

Gupta, J, Termeer, C, Klostermann, J, Meijerink, S, van den Brink, M, Jong, P, and Bergsma, E. 2010. 'The adaptive capacity wheel: A method to assess the inherent characteristics of institutions to enable the adaptive capacity of society.' *Environmental Science & Policy* 13(6), pp. 459–471 https://doi.org/10.1016/j.envsci.2010.05.006

Howarth, W. 2017. 'Water pollution and water quality.' In *Routledge Handbook of Water Law and Policy*, edited by A. Rieu-Clarke, A. Andrew Allan, and S. Henry. Oxford: Taylor & Francis Group, pp. 78–94.

Islam, M.S., and O'Donnell, E. 2020. 'Legal rights for the Turag: Rivers as living entities in Bangladesh.' *Asia Pacific Journal of Environmental Law* 23(2), pp. 160–177.

Jimenez, A. et al. 2020. 'Unpacking water governance: A framework for practitioners', *Water* 12, pp. 827–848.

Johns, F. 2011. 'Financing as governance.' *Oxford Journal of Legal Studies* 31, pp. 391–415.

Johns, F. 2015. 'On failing forward: Neoliberal legality in the Mekong River Basin.' *Cornell International Law Journal* 48, pp. 347–383.

Johns F., Saul B., Hirsch P., Stephens T., and Boer B. 2010. 'Law and the mekong River Basin: A socio-legal research agenda on the role of hard and soft law in regulating transboundary water resources.' *Melbourne Journal of International Law* 11, pp. 154–174.

Kariuki M and Schwartz J., 2005. Small-scale private service providers of water supply and electricity: A review of incidence, structure, pricing, and operating characteristics. Policy Research Working Paper N. 3727, Washington DC: World Bank.

Mason Meier, B., Cronk, R., Luh J., Bartram, J., and de Albuquerque, C. 2018. 'Monitoring the progressive realization of the human rights to water and sanitation: Frontier analysis as a basis to enhance human rights accountability.' In *The Oxford Handbook of Water Politics and Policy*, edited by K. Conca and E. Weinthal. Oxford: Oxford University Press.

McDonald, R.I., Weber, K., Padowski, J., Flörke, M., Schneider, C., Green, P.A., and Montgomery, M. 2014. 'Water on an urban planet: Urbanization and the reach of urban water infrastructure.' *Global Environment Change* 27, pp. 96–105.

Mishra, B. et al. 2020. 'Urban water Governance concept and pathway sustainable solutions for urban water security.' *Water Science and Technology Library* 93, 161–174. ch 8 https://doi.org/10.1007/978-3-030-53110-2_10

Morgan, B. 2014. 'Water rights between social activism and social enterprise.' *Journal of Human Rights and the Environment* 5, pp. 25–48.

Owens, K. 2016. 'Reimagining water buybacks in Australia: Non-Governmental organisations, complementary initiatives and private capital.' *Environmental and Planning Law Journal* 33(4), pp. 342–355.

Özerol, Gül, et al. 2018. 'Comparative studies of water governance: A systematic review.' *Ecology and Society* 23(4), p. 43.

Pahl-Wostl, C. 2015. *Water Governance in the Face of Global Change: From Understanding to Transformation*. Switzerland: Springer International Publishing.

Ranganathan, M. 2014. '"Mafias" in the waterscape: Urban informality and everyday public authority in Bangalore.' *Water Alternatives* 7(1), pp. 89–105.

Ranganathan, M. 2018. 'Rethinking urban water (in)formality.' *The Oxford Handbook of Water Politics and Policy*, edited by K. Conca and E. Weinthal. Oxford: Oxford University Press. DOI:10.1093/oxfordhb/9780199335084.013.33

Rubenstein, N., Wallis, P.J., Ison, R.L., and Godden, L.. 2016. 'Critical reflections on building a community of conversation about water governance in Australia.' *Water Alternatives* 9(1), pp. 81–98.

Rogers, P., & Hall, A. W. (2003). Effective water governance. TEC Background Papers n. 7. Stockholm: Global water partnership.

Shi, L. et al. 2021. 'Shared injustice, splintered solidarity: Water governance across urban-rural divides.' *Global Environmental Change* 70, p. 102354.

Strang, V., 2016. 'Infrastructural relations: Water, political power and the rise of a new "despotic regime".' *Water Alternatives* 9(2), pp. 292–318.

Sultana, F., & Loftus, A. (2012). The right to water: Politics, governance and social struggles. Abingdon, Routledge.

UNDP. 2006. *Human Development Report: Beyond Scarcity: Power, Poverty and the Global Water Crisis*. New York: United Nations.

Winkler, I. 2012. *The Human Right to Water: Significance, Legal Status and Implications for Water Allocation*. Oxford: Hart Publishing.

World Bank. 1993. *Water Resources Management: A World Bank Policy Paper*. Washington, DC: World Bank.

World Health Organization, and United Nations Children's Fund. 2020. 'Water, sanitation, hygiene, and waste management for SARS-CoV-2, the virus that causes COVID-19.' *Interim Guidance*. 29 July 2020.

13

REVISITING THE THEORY ON THE REGULATION OF WATER UTILITIES

Evolution, challenges, and trends

Rui Cunha Marques

13.1 Introduction

Today, we are living in the age of water sector regulation. Regulatory agencies are being mobilized across the globe. Since the creation of the emblematic regulatory agency in England and Wales, the Office of Water Services (OFWAT), in 1988 (December 2021), more than 300 regulatory agencies have been created. In 2010, this figure was approximately 200 (Marques, 2010). For example, of the 20 Latin American countries, 17 have regulatory agencies (Marques, 2017a). In Brazil alone, there are approximately 80 regulatory agencies, including federal, state, intermunicipal, and municipal regulators (Narzetti and Marques, 2021). In Central and Eastern Europe and the Middle East, most countries have regulatory agencies. In recent years, several African countries (e.g., Angola and Cape Verde) have created regulatory agencies, or they are currently in the process of creating them (e.g., Ethiopia, Democratic Republic of the Congo, and South Africa).

The agencification that has taken place in Europe, particularly in the 1980s and 1990s, along with the regulation of the markets (by agencies) as a result of the liberalization and privatization of network industries reached the water sector slightly later than other sectors, mostly due to the following two reasons.

On the one hand, in the water sector, private sector participation was always less intense than public sector participation, and that involvement occurred mostly through public-private partnership (PPP) arrangements (e.g., concessions or leasing contracts). Therefore, regulation was carried out by contract, and it was believed that a contract where the rights and duties of the parties were settled would be enough to regulate the water utilities. However, contracts are imperfect and incomplete and liable to trigger ex post opportunism leading to renegotiations and disputes (Williamson, 1985; Hart, 1995). Thus, several countries, despite recognizing the importance of regulation by contract, have decided to complement it with regulation by agency, looking for the best of two worlds, as is the case in Portugal, Brazil, and Colombia (Stern, 2012; Marques, 2017b). Regardless, some countries, such as France, Spain, and Senegal, remain regulated exclusively by contract.

 DOI: 10.4324/9781003057574-17

On the other hand, as private sector participation is minor and even nonexistent in several countries, and water services are mostly provided by public utilities or directly by the local government, the creation of specific regulatory entities is not justified, since governments regulate them directly, giving direct instructions or orders to their directors. However, public companies are usually politically biased and inefficient and do not have incentives to avoid "quiet life" and X-inefficiency (Erbetta and Cave, 2007; De Witte and Marques, 2010). Therefore, several countries, such as Australia, Peru, and Argentina, have started to regulate public water utilities, addressing them as though they are privately owned. Regardless, other countries in Europe, such as the Netherlands, Sweden, and Germany, where the water sector is mostly publicly owned, do not have regulatory agencies.

The regulation of water utilities has been justified by the need to correct market failures (Berg and Tschirhart, 1989; Marques, 2005). Thus, the regulators intervene and aim to turn themselves into the "invisible hand" of the market to align the incentives of water utilities to pursue certain objectives or to behave in a certain way supposedly according to the public interest. The primary objectives of regulation, particularly cost recovery and economic sustainability through price regulation, have been mostly of an economic nature. Thus, most regulators have focused on providing incentives for being efficient while obtaining a fair and reasonable rate of return on investments. These objectives have been relatively successful in some jurisdictions, particularly in several high-income countries, but in other areas, economic regulation is usually a secondary issue or a theoretical objective, as the context and the level of development of the sector and therefore the priorities are different. Moreover, in recent years, economic regulation has been contested and redesigned since maximizing the efficiency of water utilities might be incompatible with their resilience and constrain innovation and/or investments in a circular economy. For example, climate change and the recent COVID-19 pandemic provide evidence that efficiency cannot be addressed in isolation and that incentives need to be aligned and fitted with contextual issues.

This chapter discusses the current state of water utility regulation across the globe, its evolution, the current problems, and the main challenges. Regulation is not a panacea; thus, regulatory governance needs to be strengthened, and its positive impact should be made clear. This chapter is structured as follows: the purposes and objectives of regulation are provided in Section 13.2, the dimensions and requisites of a regulator are given in Section 13.3, and Section 13.4 outlines the current major challenges in water utility regulation. Finally, in Section 13.5, conclusions and policy implications are discussed.

13.2 Theory of Regulation: Source, dimensions, and objectives

13.2.1 Source of regulation

Classic economic regulation theory postulates that public interest protection regulation is implemented to correct market failures (Joskow and Noll, 1981). Water utilities are prone to market failures, including natural monopoly features (e.g., subadditive costs), massive upfront and sunk investments, important externalities (positive and negative), and asymmetric information (moral hazard and adverse selection). Moreover, they provide a quasi-public good since their social value is much greater than their financial value (Marques, 2010). Water services are also endowed with significant economies of scale, scope, and density, and their performance (operational and financial) depends considerably on the local characteristics, particularly on the availability of adequate water resources (Carvalho et al., 2012). Furthermore, in contrast to other network industries, water services are minimally integrated due to the high cost of conveyance, and the structure of consumption (and revenues) is more homogenous between the different types of customers (e.g., residential and industrial customers), limiting cross-subsidization. Therefore, the market structure in the water sector is generally composed of several small-sized players with financial sustainability problems. Finally, this sector is highly politically influenced due to the close proximity between customers and elected officials who are responsible for water services. Consequently, low and nonsustainable tariffs (political prices) are frequently adopted.

Out of public interest, regulators intervene and try to mimic a market without failures (as an invisible hand) by allowing water utilities to recover costs and obtain a fair return on investments and by providing adequate incentives for the water utilities to be efficient, eliminating possible monopoly rents and avoiding a "quiet life" (Marques, 2010). In the water sector, the importance of regulation is exacerbated since it suffers from the possible detrimental influence of politicians, who have a short-run view that might jeopardize the sustainability of water services in the long run.

A different theory of regulation argues that regulation does not result from the attempt to correct market failures but rather that it is sought by some interest groups so that redistribution exists and favors them (Stigler, 1971). These interest groups can include regulated operators, customers, or even civil servants or bureaucrats who have a stake in external entities (regulatory agencies) making decisions that are good for them. Particularly, in politically biased sectors, such as the water sector, where tariffs are unbalanced and usually set according to political criteria, the water utilities, irrespective of their ownership, demand regulation. This is very common when the water sector is heavily subsidized or when it is very technically strong and engineers demand high-quality service and "gold plate" practices or solutions (Helm, 1993).

13.2.2 Dimensions of regulation

A regulatory system of a particular public service includes institutions, laws, and processes that allow a government to control it (Ogus, 2002). The regulatory system has two main dimensions: regulatory governance and regulatory substance (Brown et al., 2006). Regulatory governance, the "how" of regulation, is related to the institutional and legal design of the regulatory system and the framework within which decisions are made. It is comprised of the laws, processes, and procedures that outline the regulation undertaken. Regulatory governance "principles" have been discussed and proposed by different institutions, such as the OECD and the World Bank. For example, one OECD study proposed seven foundations for good governance: role clarity, preventing undue influence and maintaining trust, decision making and governing body structure for independent regulators, accountability and transparency, engagement, funding, and performance evaluation (OECD, 2015). Similarly, Marques and Pinto proposed a scorecard to measure regulatory governance based on ten criteria: transparency, predictability, consistency and proportionality, integrity, clarity of rules, regulatory coordination, authority, autonomy (financial, managerial, and operational), public participation, and accountability (Marques and Pinto, 2018).

Regulatory substance, the "what" of regulation, is the content of regulation (Levy and Spiller, 1994). It refers to the actual decisions, whether explicit or implicit, made by the regulatory agency corresponding to the rationale for the decisions. Regulatory substance encompasses decisions related to tariffs (levels and structures), the quality of service standards and benchmarking, the supervision of work and contracts, investment plans, the handling of customer complaints, accounting systems, periodic reporting systems, and social obligations, among other issues.

The evaluation of the regulatory system or the corresponding impact on regulation always needs to consider these two dimensions together. Initially, the focus of regulation was improving the decisions and, particularly, the methodologies and practices adopted (regulatory substance); however, it quickly became clear that how the decisions are made is equally important (regulatory governance), not only to avoid misconduct but also, above all, to provide legal certainty and allow for their legitimacy and acceptability among all stakeholders (Berg, 2013).

13.2.3 Objectives and the role of regulation

The main objective of water utility regulation is to correct the existing market failures and thereby protect customers' interests, ensuring a fair and reasonable return on the investments deployed. Customer interests are protected if the tariffs are affordable for them, reflecting efficient costs for a standard (high) quality of service provided. According to a previous president of the Water Industry Commission for Scotland (WICS): "The best way of promoting customer interests in a public sector model is by improving the economic efficiency of the industry, and thereby the value of money generated" (WICS, 2001). In reality, the objective of regulation of water utilities should be the provision of sustainable and resilient water services that balance the interests of customers (affordable and inclusive tariffs and high-quality service), the water utility providers (cost recovery and financial sustainability), and society (the protection of the environment) in the long run. Thus, the objectives of regulation are (a) the promotion of efficiency; (b) the protection of customer interests; (c) the financial sustainability of operators; and (d) the predictability and steadiness of the defined public policies (Marques, 2010). In sum, the traditional objectives of regulation are primarily economic regulation, following the belief that market failures can be eliminated and that their balanced functioning can maximize welfare.

Despite some successful examples in high-income countries, such as the United Kingdom or Australia, experience has provided evidence that the prioritized objectives in low- and medium-income countries cannot be the (only) ones mentioned. Regulators also aim to foster inclusivity, particularly through the universalization and provision of sustainable water services, all of which is in line with SDG no. 6 (Franceys and Gerlach, 2012). Furthermore, as governmental institutions are often unstable, regulatory entities can lead the sector and become a kind of "champion" of the sector, given its important functions of collecting, disclosing, and providing information on the sector and supporting the government in the development and implementation of sector public policies (Mumssen et al., 2018). This latter responsibility, which is sometimes the most important, can collide with, or be difficult to separate from, policy and regulation functions and reduce or harm the autonomy and independence of regulatory agencies, although this can be preferable in certain jurisdictions. Finally, the protection of the environment and particularly the green and circular economy in the last decade has become more relevant, as evidenced by, for example, accepting investments to improve water resource quality and availability to promote the reuse or use of sludges as fertilizers or in energy production (IWA, 2016). Finally, in recent years, a new objective and scope of intervention of regulators has become relevant in low-, medium-, and high-income countries. Climate change, particularly droughts, floods, and other extreme events, and more recently the COVID-19 pandemic, emphasizes the need and fundamental objective of regulation to make water services and their infrastructure more resilient (Lawson et al., 2020). Resilience is currently a buzzword, as water utilities need to be prepared to provide adequate water services when extreme events take place. Resilience is not an easy matter to address, and, as will be discussed in the following section, it can be opposed to the principle of efficiency, the primary and classic objective of regulation. Regulatory agencies should allow for and recognize investments and their impact on tariffs that, in some situations, can never be recovered (used). Raising tariffs is always very politically and socially complex and must be well justified. Thus, it is common to limit or postpone predicted investments and often necessary to ensure universal access and adequate levels of service to avoid jeopardizing the affordability of the service. Including investments in the tariff calculation that increase the resilience of the infrastructure but whose consequences are not immediate and may never take place, at least in the short run, is very controversial and always a tough regulatory decision.

13.3 Challenges in water regulation

13.3.1 Overall

Several issues and water regulation challenges are relevant and (should) be prioritized by regulatory agencies. Regulation must have a positive impact on the performance of the sector. The cost of regulation is high, regulators have wide powers and responsibilities, and regulators are not democratically elected. Thus, their legitimacy and acceptability rely on positive outcomes linked directly to their actions. In the long run, their survival is precarious. In addition to regulatory governance issues, which are important but beyond the scope of this chapter, it is important to discuss some regulatory substance issues that are still not adequately addressed (if at all) and that correspond to the most important challenges of the sector and consequently of the regulators. Among others, the most determinant here are related to universal access to water services and the achievement of SDG no. 6, which is a matter of human rights and universal access to regulation, including in rural areas and in informal settlements where the most vulnerable people live and where regulation is most needed. If this latter matter is one of regulation equity and the nondiscrimination of the poor, the third and fourth challenges are related to the homogenization of the regulation irrespective of the management model of water services, particularly in terms of how incentives can be provided to state- or municipal-owned companies or utilities and how regulation by an agency can be made compatible with regulation by contract and provide added value, which is typical in situations in which there is private sector participation. Finally, resilience and circular economy challenges are discussed from a regulatory perspective.

13.3.2 Universal service provision and achieving SDG no. 6

Without a doubt, ensuring universal water services in the developing world is one of the biggest challenges we face today, and it is one in which the role of regulators can be decisive. Securing access to water services is complicated, and although it is not always the priority, it is a human right, and its importance for public health, the environment, and the social and economic cohesion of the population is undeniable (Neto and Camkin, 2020). Several factors have prevented these goals from being achieved, and some of them are not directly related to the water sector. We could separate sanitation and water supply, as they are quite different, but that discussion would be beyond the scope of this chapter (Garlach and Franceys, 2010). Regardless, the contribution of regulation is key to achieving this goal.

Access to water services can be seen from two different perspectives, one related to physical access and the other to economic access. The former is connected with the existence of infrastructure and making the required investments, and the latter is connected with the affordability of the service for the population. The first is a necessary but not sufficient condition, and it is economic access that, in fact, enables the population's access to water services. Even if we ignore the technical (e.g., lack of water resources or of their quality) and legal (e.g., land tenure) constraints and focus on economic issues, there is no unique or magic solution. To eventually overcome this gap, several regulatory actions should be combined and taken simultaneously. In low- and medium-income countries, one unique alternative frequently consists of a cocktail of measures since each one by itself would not be sufficient. However, together they can help make decisive contributions to resolving the problem.

Therefore, the role of the regulation is key not only to deploying the required investments to allow for physical access but also to ensuring economic access. Concerning the former, the regulating agency should approve the plan of investments in considering the universalization targets and trajectory to pursue them. Concerning the latter, the regulating agency is responsible for price regulation, which includes designing the tariff structure and setting tariff levels, including all the incentives that this can provide to the market to reach particular objectives, such as low tariffs for particular customers and water conservation (Pinto and Marques, 2016). In particular, the subsidization scheme that allows for affordable tariffs is important for the regulating agency to prioritize in line with the social policies of the government (Narzetti and Marques, 2020). It can include one or more combined decisions, including those related to cross-subsidization, social/lifeline tariffs, funds, and direct subsidies to customers (monetary or volumetrics) or to water utilities to deploy investments (e.g., in relation to connections and metering; Andres et al., 2019). The major concerns are the effectiveness of the eligibility criteria and the funding mechanism, which can become a deadlock in low-income countries.

Finally, it should be noted that to be more effective in expanding coverage and access to water services, particularly to the poor, one-size-fits-all regulatory solutions do not work, and streamlined and adjusted solutions, including tariff structures, subsidization schemes, contractual arrangements, resale water tariffs, adjusted quality of service standards, and nonconventional technical solutions need to be found on a case-by-case basis.

13.3.3 Expanding regulation to peri-urban and rural areas

Another main challenge of regulation is regulating peri-urban areas, mainly composed of informal settlements. Across the globe, these informal areas are normally beyond the reach of regulatory agencies. Indeed, regulatory agencies do not "like" to regulate these areas, and, frequently, they are beyond the agencies' jurisdiction. They are "last in the queue," truly the "ugly duckling" of the sector (Marques, 2017a). However, the population covered by these systems is the most vulnerable and the one most in need of water utility regulation. Usually, they pay more for water and obtain worse service than other populations, despite sometimes representing the largest portion of the population. The regulatory agency must always be assigned to a geographical jurisdiction area and not to a particular type of operator.

Indeed, this market segment has to be better understood. It is important to recognize the contribution of the informal operators of these areas. They exist because they are needed and fill in a market gap or compensate for some state inefficiency or the absence of a "formal" operator (water utility). The strategy should be integrated by these "informal" operators into the formal system by incorporating them into the jurisdiction of the water utility or acknowledging their specificities and the fact that regulation there will inevitably be done in a different way (Trémolet and Hunt, 2006). Regardless, the aims of such regulation are similar and include controlling prices and supervising the quality of service (e.g., access to, reliability of, and quality of drinking water). It is true that this can be complex because the licenses and authorizations required might not be easy to obtain (e.g., land tenure, social security, or vehicle licensing) and can lead to increased water prices for final customers. Therefore, depending on the features of each situation, the characteristics of the water services to be regulated should be established, along with the way regulation should occur. In particular, the type of informal operator is essential for the regulatory design. Informal operators can be categorized into piped networks, point sources (kiosks, standpipes, or even household connections or private bulk water supply, such as wells or boreholes), and mobile distributors (tankers or trucks that purchase water from bulk suppliers or from private sources) (Kariuki and Schwartz, 2005). They can depend on the main utility for bulk water or own an independent bulk water source.

As far as rural systems are concerned, the intervention of regulatory agencies in these water services has importance equal to that in urban systems (Trémolet, 2015). However, the reality differs a bit different between the two. Normally, rural systems are small and scattered across the territory; therefore, incentives should be provided for their amalgamation. They are frequently managed directly by the community, and a lack of capacity and skills are usually the main problems and challenges these systems face. These rural systems usually work well until the moment problems appear (e.g., pump failure or water contamination). Therefore, regulation also has an important role in such areas.

13.3.4 Incentives and regulation of state- or municipal-owned utilities

The regulation of state- or municipal-owned enterprises is not an easy task. Indeed, it is much more complex than the regulation of private companies. Providing incentives is the main tool of regulatory agencies. However, incentives work differently with private and public companies. Rewarding outperformance (carrots) is much more difficult within the public sector because of the legal frameworks, traditions, and interests in this sector, and penalties (sticks) are likewise difficult to apply, as they frequently result in a transference of costs from the pocket of the users to that of the taxpayers. For example, labor unions are always barriers since they do not like either negative or positive discrimination. With private companies, shareholders benefit (reap profits) or suffer damage (losses) relative to their performance, which is a very effective and important incentive. However, they can also go bankrupt. Conversely, public companies do not suffer from bad performance in the same way, and thus, these incentives are not as effective (Marques, 2017a).

An alternative is to transfer to the board of directors or the president of the public company (or to the staff in general) the incentives and offer better salaries and amenities if performance exceeds what is expected and to apply salary cuts or reduce other amenities if the results do not meet expectations (Marques, 2010). Despite the difficulties in implementing this kind of new government practice in some countries (e.g., rigidity of public administration and opposition of labor unions), this effect is not as persuasive as that to shareholders because it is necessarily more limited in terms of earnings and losses. However, whenever possible, these compensatory schemes should be implemented.

When regulatory agencies are not truly independent authorities, which is most common, the umbrella is the same (the ministry, the secretary the state, or the mayor) as that of public-owned utilities. Therefore, all the pressure that can and should be imposed by the regulatory agency to improve and obtain better results disappears or loses intensity. Furthermore, when there are conflicts or sanctions must be applied, this may require the revocation of or reductions in exemptions (Marques, 2005).

In sum, regulating state- or municipal-owned utilities demands requisites different from regulating other utilities. There are two levels of intervention for improving regulations and making them more effective, one related to the regulatory governance of the regulatory agency, which needs to take this setting into account, and the other one associated with the corporate governance of the state- or municipal-owned companies, in which the regulatory agency can have some say (see OFWAT, 2019).

See above and the principles suggested by international best practices regarding regulatory governance (Marques and Pinto, 2017). As far as corporate governance is concerned, it is also important to endow the state- or municipal-owned utilities with a regulatory environment that has the right tools to keep the company efficient, effective, and concerned about the quality of service provided. Hence, certain aspects have to be ensured, including the integrity, competence, and merit of the board of directors. For example, the way the board is appointed is essential both inside (e.g., staff) and outside the organization (e.g., for all stakeholders).

Transparency is also very important. The main regulatory and administrative documents should be publicly available. In particular, policies about probity (e.g., a register of interests) and against corruption should also be available (Da Cruz et al., 2016). Benchmarking with other similar companies should be compulsory, and the results should be public, as should satisfaction surveys of customers and the population. Accountability is another very relevant principle, and companies should periodically present and discuss their performance with stakeholders.

Although state- and municipal-owned companies are publicly owned, they should work according to private law, fostering a commercial and customer-oriented culture (Irwin and Yamamoto, 2004). There should be legislation or rules in their statutes that limit political influence in the company and the development of noncommercial activities. Furthermore, performance payments should be established for directors and other staff. These companies require additional public reporting of performance and policies, either financial or technical, environmental and social. Given the public interest nature of these companies (providing an essential service), it should always be reported how this public interest is being pursued and what goals have been established.

13.3.5 Compatibility between regulation by agency and regulation by contract

The worst elements of two systems might predominate in countries where concession (or other arrangements) contracts have been adopted to introduce private sector activity in water services provision and regulatory agencies were simultaneously created to regulate these services. The mix of these two regulatory models, when not appropriately coordinated, is not clearly a positive sum game and may spur the development and success of privatization (Marques, 2017b).

In fact, on the one hand, contracts establish the rights and duties of both partners (public and private), but they are imperfect and incomplete and prone to ex post opportunism, and it is impossible to predict all future contingencies (Williamson, 1981). Therefore, regulation by contract might not be enough, and some kind of external ex post regulation may be required (Marques, 2018). On the other hand, regulation by agency is costly and gives too much discretion to the regulatory agencies, substantially increasing regulatory risks and leading these markets to become unattractive to the private sector in certain countries. Thus, if some of the rules of the game are set in a contract, these regulatory risks can be mitigated. Consequently, a possible solution may be a hybrid regulatory model, where contracts work in harmony with agencies. If this is done in a coordinated and effective way, the worst element of both worlds can be replaced by the best elements of both worlds. This hybrid solution, if properly implemented, allows for simultaneously overcoming the problems of contract incompleteness and regulatory risk.

Contracts with state- and municipal-owned companies are also becoming increasingly popular, precisely to avoid the patronage and political influence of the government and regulatory agency discretion (Cruz and Marques, 2013). They assume particular importance for the creditworthiness and financing of public companies, which require the steadiness of entities in the financial market and the assurance that future political opportunism will be mitigated. In this "public" model, contracts can be categorized into concession contracts, program contracts, partnership agreements, and contracts according to performance/objectives or other factors, since they ensure some rights and duties of both (public) partners (signers), thus mitigating the regulatory and political risk associated with the provision of these essential services (Pinto et al., 2015).

Tasks that remain within the scope of the regulatory agency should include renegotiation and tariff revisions, contract changes, the monitoring and supervision of service quality, contract compliance inspection, benchmarking and information disclosure, and addressing of complaints and conflicts (Marques, 2018). Bearing in mind the risk matrix of the contracts, the rate of return and financial conditions at the time of contract signature should also be kept within the purview of regulatory agencies.

Note that regulatory agencies should participate from the beginning of the project, that is, in the preparation of the public tender and in the draft design of the contract, influencing its contents while taking into account future regulation needs. This participation, despite its rarity, is fundamental for regulatory agencies to be effective and to avoid conflicts in the future. Indeed, regulatory agencies must defend contract sustainability rather than necessarily the interests of the public sector, which frequently has a short-term view corresponding to election periods.

13.3.6 Resiliency, efficiency, and populism

In the last two decades, climate change has set off sirens, and extreme events have become more frequent and much more acute. In particular, droughts and floods have had a great impact on cities, and water services have not always provided an adequate response (Gusman et al., 2017). Additionally, other extreme weather events have been more common and impact the assets and performance of the sector. Furthermore, the recent COVID-19 pandemic has had a great impact on the livelihood of the population (Antwi et al., 2021). This event has had several consequences that jeopardize the provision of water services, especially in low- and medium-income countries, where the social and financial sustainability of water services is severely threatened.

Resilience is ensured by oversizing the infrastructure and resources of water companies so that they can be utilized to accommodate extreme events (Pamidimukkala et al., 2021). This extra capacity to respond to requests demands greater investments and costs that have returns only when and if these events occur. Therefore, they constitute extra costs (Juan-García et al., 2017), which may be useful only sporadically and overutilized when these events do take place. However, in practice, they are considered inefficient. Note that efficiency, by definition, means to do more with less, and what is being proposed is doing the same with more while avoiding greater harm when extreme events occur.

Therefore, in a sector that is politically biased, prone to populism, and with a short-run view, the rise of tariffs is very complicated and consequently chronically suffers from postponements and underinvestment (Leflaive and Hjort, 2020). It seems obvious that compared to other, more visible and urgent investments concerning, for example, access to water services or improved service quality (e.g., continuity or drinking water quality), these investments are very difficult to accept and prioritize. As affordability is generally low, the willingness to pay for these additional investments will surely be very low.

For these reasons, regulatory agency is determinant and can help improve resilience not only by raising awareness about these issues, but also mainly by imposing and accepting investments and consequently to set and raise tariffs that allow for cost recovery and a fair and reasonable rate of return, avoiding political patronage and populism (NAO, 2015). Regardless, this is not an easy task, and a solid and strong regulatory agency is required.

13.3.7 Circular economy and environmental sustainability

Related to climate change, environmental sustainability and the development of the circular economy are also important challenges for water utilities and, consequently, for their regulation (Savini and Giezen, 2020).

The consequences of climate change demand important investments that must be paid by customers through tariffs and again in a politically biased sector, frequently characterized by low affordability and willingness to pay; these investments are not easy to carry out or to secure social and politically acceptance for them (Marques and Miranda, 2020).

Moreover, if some of these investments are imposed by law and need to be deployed by water utilities, others might be related to their own initiative, such as the development of reuse, the production of energy, and the recovery of nutrients and the use of sludge in fertilizers (Guerrini and Manca, 2020). The investments required for these actions have a high risk since the demand market is hardly controlled by the water utility. Therefore, incentives are required by both public policies and regulators. Moreover, the latter needs to be ready to accept them even if they are not financially sustainable and lead to an increase in tariffs.

Regulatory agencies have to ensure that these investments, irrespective of their outcomes, are included in the regulatory asset base and are recovered by water utilities. A similar situation is found for investments in innovations and technological changes.

13.4 Conclusions

This chapter discussed the state water utility regulation, including the main challenges and trends in this arena. Water service regulation has changed substantially over time. Initially, and particularly in high-income countries, the priority was economic efficiency, and regulators usually provided incentives trying to catch up water utilities to the best practices frontier.

Considering water utilities as natural monopolies and deeply affected by other market failures, regulation had a predominantly economic nature (price regulation), which caused actors to forget the political economy of water. This past focus of regulation of water utilities quickly proved that was not well founded and most countries, including the most developed ones, reformed their regulations, as they were not able to provide sustainable water services for all.

This chapter discusses several challenges and issues that need to be addressed by regulators to circumvent the past and current status quo of regulation of water utilities. On the one hand, challenges related to the political economy need to be better addressed by regulatory agencies, including the universalization of regulations (both geographical and for all management models of water services) and matching with other regulatory tools (e.g., contracts). On the other hand, environmental and social issues, comprising the universal access of the population to water services, resiliency, and environmental sustainability, must be at the core of the regulation of water utilities. Note that this does mean that economic efficiency is not important and should not be a concern; however, it cannot be the main and central objective. Water supply and sanitation are essential public services whose social and environmental values are much greater than their financial values, and regulators always need to bear this in mind.

References

Andres, L.A.; Thibert, M.; Lombana Cordoba, C.; Danilenko, A.; Joseph, G.; Borja-Vega, C., (2019). *Doing More with Less: Smarter Subsidies for Water Supply and Sanitation*. World Bank, Washington, DC.

Antwi, S.; Gettya, D.; Linnanea, S.; Rolstonb, A. (2021). COVID-19 water sector responses in Europe: A scoping review of preliminary governmental interventions. *Science of the Total Environment*, 762(25), 143068.

Berg, D.; Tschirhart, J. (1989). *Natural Monopoly Regulation: Principles and Practice*. Cambridge Surveys of Economic Literature.
Berg, S. (2013). *Best Practices in Regulating State-owned and Municipal Water Utilities*. Economic Commission for Latin America and the Caribbean (ECLAC/ CEPAL), New York .
Brown, A. C.; Stern, J.; Tenenbaum, B.W.; Gencer, D. (2006). *Handbook for Evaluating Infrastructure Regulatory Systems*. World Bank Publications, Washington, DC.
Carvalho, P.; Marques, R.; Berg, S. (2012). A meta-regression analysis of benchmarking studies on water utilities market structure. *Utilities Policy*, 21, 40–49.
Cruz, C.; Marques, R. (2013). *Infrastructure Public-Private Partnerships: Decision, Management and Development*. Springer, New York.
Da Cruz, N.; Tavares, A.; Marques, R.; Jorge, S.; De Sousa, L. (2016). Measuring local government transparency. *Public Management Review*, 18(6), 866–893.
De Witte, K.; Marques, R. (2009). Designing performance incentives, an international benchmark study in the water sector. *Central European Journal of Operations Research*, 18(2), 189–220.
Erbetta, F.; Cave, M. (2007). Regulation and efficiency incentives: Evidence from the England and Wales water and sewerage industry. *Review of Network Economics*, 6(4), 28.
Franceys, R.; Gerlach, E. (2012). *Regulating Water and Sanitation for the Poor*. Taylor & Francis, Abingdon: Routledge.
Gerlach, E.; Franceys, R. (2010). Regulating water services for all in developing economies. *World Development*, 38(9), 1229–1240.
Guerrini, A.; Manca, J. (2020). Regulatory interventions to sustain circular economy in the water sector. *Insights from the Italian Regulatory Authority (ARERA). H2Open Journal*, 3(1). 499–518
Guzmám, D.; Mohor, G.; Taffarello, D.; Mendiondo, E. (2017). Economic impacts of drought risks for water utilities through Severity-Duration-Frequency framework under climate change scenarios. *Hydrology Earth System Science Discussion*. 1–39
Hart, O. (1995). *Firms, Contracts, and Financial Structure*. New York: Oxford University Press.
Helm, D. (1993). The assessment: Reforming environmental regulation in the UK. *Oxford Review of Economic Policy*, 9(4), 1–13.
Irwin, T.; Yamamoto, C. (2004). *Some Options for Improving the Governance of State-Owned Electricity Utilities. Energy and Mining Sector Board Discussion Paper*. Paper no. 11. The World Bank, Washington, DC.
IWA (2016). *Water Utility Pathways in a Circular Economy*. IWA, Amsterdam.
Joskow, P.; Noll, R. (1981). Regulation in theory and practice: An overview. In Fromm, Gary (ed.), *Studies in Public Regulation*, The MIT Press, Cambridge, MA, 1–66.
Juan-García, P.; Butler, D.; Comas, J.; Darch, G.; Sweetapple, C.; Thornton, A.; Corominas, L. (2017). Resilience theory incorporated into urban wastewater systems management. State of the art. *Water Research*, 115(15), 149–161.
Lawson, E.; Farmani, R.; Woodley, E.; Butler, D. (2020). A resilient and sustainable water sector: Barriers to the operationalisation of resilience. *Sustainability*, 12, 1797.
Leflaive, X.; Hjort, M. (2020). *Addressing the Social Consequences of Tariffs for Water Supply and Sanitation*. Environment Working Paper no. 166, Organisation for Economic Co-operation and Development (OECD), Paris.
Levy, B.; Spiller, PT (1994). The institutional foundations of regulatory commitment: A comparative analysis of telecommunications regulation. *The Journal of Law, Economics, and Organization*, 10(2), 201–246.
Marques, R. (2005). *Regulação de serviços públicos*. Edições Sílabo, Lisbon.
Marques, R. (2010). *Regulation of Water and Wastewater Services: An International Comparison*. IWA Publishing, London.
Marques, R. (2017a). *State of the Art of Regulation of Public Utilities: Lessons to Argentina and Latin America*. Unpublished. The World Bank, Washington, DC.
Marques, R. (2017b). Why not regulate PPPs? *Utilities Policy*, 48, 141–146.
Marques, R. (2018). Regulation by contract: Overseeing PPPs. *Utilities Policy*. 50, 211–214.
Marques, R.; Miranda, J. (2020). Sustainable tariffs for water and wastewater services. *Utilities Policy*, 64, 101054.
Marques, R.; Pinto, F. (2018). How to watch the watchmen? The role and measurement of regulatory governance. *Utilities Policy*, 51, 73–81.
Mumssen, Y.; Saltiel, G.; Sadik, N.; Marques, R. (2018). *Regulation of Water Supply and Sanitation in Bank Client Countries*. The World Bank, Washington, DC.

NAO (2015). *The Economic Regulation of the Water Sector. Report by the Comptroller and Auditor General.* National Audit Office, London.
Narzetti, D.; Marques, R. (2020). Models of subsidies for water and sanitation services for vulnerable people in South American countries: Lessons for Brazil. *Water*, 12(7), 1976.
Narzetti, D.; Marques, R. (2021). Access to water and sanitation services in Brazilian vulnerable areas: The role of regulation and recent institutional reform. *Water*, 13(6), 787.
Neto, S.; Camkin, J. (2020). What rights and whose responsibilities in water? Revisiting the purpose and reassessing the value of water services tariffs. *Utilities Policy*, 63, 101016.
OECD (2015). *Inventory of Water Governance Indicators and Measurement Frameworks*. Organisation for Economic Co-operation and Development, Paris.
OFWAT (2019). *Board Leadership, Transparency and Governance: Principles*. Water Services Regulation Authority, Birmingham.
Ogus, A. (2002). *Comparing Regulatory Systems: Institutions, Processes and Legal Forms in industrialised Countries*. Working Paper no. 35, Centre on Regulation and Competition, University of Manchester, Manchester.
Pamidimukkala, A.; Kermanshachi, S.; Adepu, N.; Safapour; E. (2021): Resilience in water infrastructures: A review of challenges and adoption strategies. *Sustainability*, 13(23), 12986.
Pinto, F.; Marques, R. (2016). Tariff suitability framework for water supply services: Establishing a regulatory tool linking multiple stakeholders' objectives. *Water Resources Management*, 30(6), 2037–2053.
Pinto, F.; Da Cruz, N.; Marques, R. (2015). Contracting water services with public and private partners: A case study approach. *Journal of Water Supply: Research and Technology: AQUA*. 64(2), 194–210.
Savini, F.; Giezen, M. (2020). Responsibility as a field: The circular economy of water, waste, and energy. *Environment and Planning C: Politics and Space*, 38(5), 866–884.
Stern, J. (2012). The relationship between regulation and contracts in infrastructure industries: Regulation as ordered renegotiation' *Regulation and Governance*, 6(4), 474–498.
Stigler, G. (1971). The theory of economic regulation. *Bell Journal of Economics and Management Science*, 2(1), 3–21.
Trémolet, S. (2015). Regulation in rural areas. *Briefing Note, Building Blocks for sustainability Series*. IRC, The Netherlands.
Trémolet, S.; Hunt, K. (2006). Taking into account the poor in water sector regulation. *Water Supply and Sanitation Working Notes, Note no. 11*, The World Bank Group, Washington, DC.
WICS (2001). *Strategic Review Charges 2002–2006*. Water Industry Commissioner for Scotland (WICS), Stirling, Scotland.
Williamson, O. (1981). The economics of organization: The transactions cost approach. *American Journal of Sociology*, 87(3), 548–577.
Williamson, O. (1985). *The Economic Institutions of Capitalism*. The Free Press, New York.

14

TRENDS AND COMPARISONS OF OUTCOMES BETWEEN PUBLIC AND PRIVATELY OWNED UTILITIES

Germà Bel

14.1 Introduction

Water provision is a complex service; probably the most complex one that governments must provide to citizens. It is most frequently provided at the local level, and it is a critical infrastructure for economic activities taking place in their jurisdiction. It is an essential good and has been of primary relevance for the viability of urban settlements, quotidian life conditions, and economic prospects for development. Because of this, water availability has triggered conflicts between individuals and between communities in the past, and it is a potential source of major conflicts in the future (Swain, 2001; Farinosi et al., 2018; Flörke, Schneider, and McDonald, 2018). Indeed, it is the most sensitive public service that governments provide. Issues concerning accessibility, affordability, and equality are particularly relevant in this service. Also, debates and dilemmas are ongoing over public sector and private sector roles in its provision and delivery. Therefore, analyzing trends and outcomes for the water service is useful for gaining an understanding, not only about this service itself, but also about broader economic and social development.

The influential economist Joseph Schumpeter published in 1939 his magnum opus *Business Cycles: A Theoretical, Historical, And Statistical Analysis of the Capitalist Process*, in which he combines history, theory, and quantitative tools as appropriate methods to study economic and social developments. This road was taken by Albert O. Hirschman (1982) to study the oscillation between public and private solutions for economic and social problems. In Hirschman's view, permanent optimal policies do not exist (Hirschman, 1958). Even successful policy innovations, while initially improving the most relevant aspects of a specific social problem, tend to cause other unintended or unexpected consequences to emerge. These failures gain prominence and citizen dissatisfaction grows, until policy emphasis shifts to deal with newly emerged priorities. Most public services, and among them particularly water delivery, are quasi-markets (Boyne, 1998; Lowery, 1998). Even if they do not possess the attributes of pure public goods (non-rivalry and non-excludability; Stiglitz, 1986), market failures are frequent and relevant. Because of this, public and private interventions alternate, depending on which type of imperfection (e.g., inequality, inefficiency, insufficient innovation, corruption) gains priority at each point in time.

 DOI: 10.4324/9781003057574-18

In this chapter, a combination of history, theory, and quantitative knowledge is used to analyze the motivations and outcomes of public and private intervention in the delivery of drinking water services. Studies on the history of water distribution and delivery have been useful to build economic theories on public versus private delivery of public services. Furthermore, later empirical analyses have shed light on the actual outcomes of choices of delivery mode. Thus, to understand the changes between increased private involvement (i.e., privatization) and the shift back to higher public involvement (remunicipalization), I first begin by reviewing the history of water service delivery. Then, I discuss the main economic theoretical streams on that topic. After that, the empirical existing evidence on outcomes – costs and other effects – is presented and discussed. Finally, the main conclusions from the analysis are drawn.

14.2 History of water distribution and delivery modes: From private to public

The first urban water supply networks in the United States were privately owned and managed. Services began by the end of the 18th century. By 1800, more than 90% of urban water systems were private (Gómez-Ibáñez, 2003, p. 169). This hegemony, though decreasing, was maintained until the middle of the century. However, since the 1830s, an increasing number of cities developed centralized water systems owned and managed by municipalities (Melosi, 2012). As a result of this increase in the intervention of the public sector in water service, the share of public ownership and delivery increased, and, by 1875, the share of the public sector was 54%, thus overtaking that of the private sector (Gómez-Ibáñez, 2003, p. 169).

In 1873 a large number of bankruptcies of municipal bonds led the federal government to impose financial restrictions on cities, and, because of these financial restrictions, the expansion of the public sector diminished relative to that of the private sector (Jacobson and Tarr, 1995; Cutler and Miller, 2006). Expansion of municipal water service boomed in the last quarter of the 19th century, and between 1875 and 1890 the number of municipally owned and managed networks increased by 155%; but, privately owned systems grew by 350%, thus taking back the lead in the sector. Once the financial restrictions on municipalities imposed by the federal government were loosened in the last decade of the century, public initiative strongly recovered: by the end of the century, its share was well over 50%, and, by 1915, it surpassed two-thirds. This trend did not stop in the following decades, and, by the middle of the 20th century, public ownership was so dominant that virtually all new distribution networks were publicly owned from their inception.

Two different stages can be identified in considering the hegemony of public intervention in the water sector. Through most of the 19th century, expansion of public ownership was mainly due to the continuous increase in the creation of new water networks, a trend that accelerated in the third quarter of the century. After the impasse of 1875–1890, coincidental with the Gilded Age (probably best described in Veblen, 1899), an important change occurred: it was not only the new networks that were owned and managed by municipalities, but also a sharp increase in the municipalization of privately owned networks occurred in the Progressive Era (Lough, 2016), the 1896–1915 period that followed the Gilded Age. Data in Gómez-Ibáñez (2003, p. 106) show that while the number of privately owned urban waterworks was increasing throughout the 19th century (although at a much slower pace than public networks), between 1896 and 1915 the number of private waterworks decreased by 6% (while the total number of urban water networks increased by 40%). By 1900, only one out of the 11 cities in the United States with populations above 300,000 inhabitants had water service privately delivered (Jacobson and Tarr, 1995).

Historical data on the expansion of urban water networks in Europe is not as abundant as it is for the United States. The two countries with the most information are, probably, the United Kingdom and France. Water networks in London in the 18th and 19th centuries were owned and managed by private firms, until the private companies were nationalized by compulsory purchase after the Metropolis Water Act was passed in 1902 and the Metropolitan Water Board was created in 1903. In England, generally, the initial expansion of waterworks in the 19th century was mainly carried out by private firms, although, similar to the United States, public ownership and management were increasing as the century progressed (Millward, 2007). Because of a severe outbreak of cholera in 1848, the Public Health Act of the same year was passed. The act created a Central Board of Health and mandated the creation of local boards where the rates of cholera had been particularly high. Furthermore, the act made it possible to issue loans (to be paid from the rates) to invest in public health infrastructure (UK Parliament, 2020). Because of these changes, public ownership quickly expanded and soon became hegemonic.

Developments in France were different. A private company obtained a concession for the water service in Paris as early as 1782, and in many other cities private companies were in charge of developing waterworks. As in the United States and the United Kingdom, private ownership and management of water networks raised sharp criticism throughout the second half of the 19th century. Nonetheless, in contrast to the United States and the United Kingdom, the government did not engage as often in municipalization; rather, it switched from a concessional system to lease and management contracts (Pezon, 2012). That is, public intervention in network ownership and investments increased, but delivery tended to remain with private companies, who were responsible for operating and maintaining the water system. Government intervention was more complete by the end of the 19th century in the newly created water networks, which tended to be publicly owned and managed (Pezon, 2012). Countries heavily influenced by the French legal system and administrative tradition, such as Belgium, and most notably Spain, followed a similar path. In both cases, private companies were particularly active in the early development of waterworks, by the middle of the 19th century, and municipalization began later and was less intense than in other European countries (Núñez Romero-Balmas, 1996).

Other countries in Europe saw a higher degree of public involvement from the beginning of their water networks expansion. In Germany, a private concession was awarded in Berlin in 1856 (Roth, 1987), but the public sector was largely hegemonic in ownership and management of networks elsewhere (Barraqué and Kraemer, 2014). At the beginning of the 20th century, all water networks were publicly owned in Scandinavian countries, including Sweden and Denmark, and the government share was also very large in Italy (90%).

14.3 Public and private delivery of public services, economic theory, and water distribution

The first theory that explains the changing choices between private and public ownership and management of water networks is the *public interest theory*. This approach advocates a role for public intervention in the economy to protect and to foster public interest and to prevent it from being damaged by market failures. Among the most relevant factors are monopolistic conditions in public services and relevant externalities (Stiglitz, 1986), that is, effects of the service delivery (whether positive or negative) that private firms cannot internalize (either via revenues or via costs). Water distribution meets these conditions very well. On the one hand, urban water systems involve huge sunk costs (see, for instance, Helm, 2009; Ofwat, 2016), with network characteristics that involve relevant returns to scale, particularly economies of density.

Competition is largely absent, as duplicating water networks do not make economic sense. But monopolistic conditions can yield lower quantities and higher prices than under a competitive regime. Furthermore, water distribution has been directly linked to sanitation and health, helping to prevent the extension of epidemics. Both the monopolistic nature of the service as well as the relevant positive externalities involved in its expansion may lead the private sector to provide the service below what is socially desirable.

Support for the public interest approach to water municipalization has been provided by studies that analyzed changes after municipalization of waterworks in the largest cities in the United States. Boston, San Francisco, and Seattle were pioneers of municipalization among US large cities toward the end of the 19th century. They experienced an important increase in investment in water systems after the government took over the service (Jacobson, 2000). Similar patterns were observed in Baltimore, Houston, Los Angeles, and New York after municipalization of water. The largest cities were the places with the earliest and strongest increases in demand for water, and urbanization and agglomeration provided a more favorable space for the expansion of epidemics. Jacobson's (2000) interpretation is that municipalities were unsuccessful in their efforts to promote higher investment by private firms providing water service, given financial and contractual conditions. Hence, municipalization was a useful policy to deal with deficient investment by private firms that lagged behind demand.

Based on approaches that emphasize government failures rather than market failures (e.g., Stigler, 1971; Peltzman, 1976), a second theoretical stream is that of *private interest theory*. Its key tenet is that politicians do not adopt policies intended to favor public interest or to promote social welfare. Instead, politicians behave as self-interested individuals, seeking material rewards and political support. In this regard, public ownership and management of public services, and particularly that of water systems, can provide multiple opportunities for political patronage; that is, to provide jobs to the party machinery, namely, political adherents and allies (Menes, 2006). Furthermore, public ownership and management of water services could also facilitate populist policies intended to increase electoral support by means of politically targeted redistribution.

Several recent empirical studies aim to show that municipalization was rarely a consequence of addressing market failures. Regarding "public interest" claim's that monopolistic conditions in the water service were a key factor for municipalization, Troesken (1997) and Troesken and Geddes (2003) argue that the energy local service constituted a monopolistic sector as well, and, nonetheless, the frequency of municipalization was much lower than in the water service. Therefore, market failure because of monopoly would hardly be a reason for municipalization. With respect to public interest arguments related to health externalities, Troesken (1999) provides data that show that investment in filters to prevent typhus and other diseases by public companies was not higher, but even smaller in some cases, than that by private firms. Troesken and Geddes (2003) argue that, consequently, market failures because of externalities would not explain municipalization.

Concerning the most likely factors for municipalization of water services, Troesken (2001, 2002) concludes that the main driver was the transfer of income from taxpayers to users, as water prices were lower in public companies than in private companies, and this transfer was targeted in favor of areas of the cities from which municipal governments sought to increase electoral support. Furthermore, Troesken and Geddes (2003) contend that the main reason why private firms reduced their investment in water networks was the inability of municipal governments to credibly commit not to expropriate value from private companies after investments were made. In turn, the resulting underinvestment was used by municipalities as a pretext to municipalize private water companies.

Incomplete contracts and the transactions costs that they impose on the contracting partners have been thoroughly studied by Williamson (1979, 1981). Contracts involving long-term commitments, such as high sunk investments, involve huge transaction costs because many unforeseen changes can occur within the contractual agreement length. Due to the high degree of incompleteness of those contracts, they come with more risks and are more likely to fail than contracts that are less incomplete because agreements on how to deal with unexpected changes can be difficult to reach when interests are opposed. Furthermore, parties engaged in long-term contracts might be subject to the hold-up problem. Because both parties may be locked into the contract in the long run (either because investors have incurred high sunk investments or because the sponsor becomes highly dependent on the investor), instead of mutually beneficial cooperation, negative incentives arise. Among them, the most important is the incentive for the investor to underinvest, an incentive that is likely to be reinforced by the reduction of the contract duration.

This framework of incomplete contracts and transaction costs has been adopted by Gómez-lbáñez (2003: 157–158) to propose an alternative interpretation of why municipalization of water networks was particularly strong. Being unaware of limits to competition in water services, in the first phase of the development of the service municipalities entered into long-term contracts where rules were ambiguous, and basically they focused on establishing obligations on quantity, quality, and price of the services provided by private companies. Unsatisfactory development of first-generation contracts led to more detailed and specific second-generation contracts. But sharp increases in demand and fast technological change caused these contracts to quickly become obsolete. In the second phase of the expansion of urban services, public ownership became predominant in services under direct municipal control, such as water, because municipal governments had lower technical capabilities for regulating private providers. This stood in contrast to higher-tier governments that had gained responsibility for services such as electricity and gas. Providing subsidies from the municipal budget to private companies for extending the service to areas where it was financially unprofitable constituted a stronger political challenge for municipal governments than for governments at higher tiers. Given these conditions and constraints, municipalization was the most institutionally and politically viable policy for reforming water services.

14.4 Privatization of water services: Walking the way back

While in the first half of the 20th century theoretical insights and public policies promoted government ownership and management of public services (expanding as well to commercial goods and services), criticism of public interest theory and disturbing evidence regarding the results of government interventions grew after the 1960s and more intensely during the 1970s. Theoretical and empirical analyses originated in the Chicago School [Stigler (1971), Posner (1975), Peltzman (1976), Becker (1985)] and triggered reforms intended to shift the balance between public and private solutions to public service provision. In addition, legal regulation, as a replacement for public ownership, gained influence in the last quarter of the 20th century (Bortolotti and Perotti, 2007), after precursory reforms in that regard of the banking system and other industries Germany in the 1930s (Bel, 2010). Seen in historical perspective, the resurgence of privatization of public services in the last quarter of the 20th century could be interpreted as a reversal of the Progressive movement prominent toward the end of the 19th century, which paved the way for a long period of wide consensus on the public interest as a rationale for government intervention. This interpretation is much in line with Hirschman's (1982) views on oscillation between public and private solutions for dealing with economic and social problems.

An enormous quantity of theoretical and empirical analysis is available in the literature on the factors that have promoted the privatization of public services, and meta-analyses – literary (Bel and Fageda, 2007, 2017) as well as statistical ones (Bel and Fageda, 2009) – have reached conclusions worth reviewing here. Budgetary constraints and financial restrictions on local governments appear to be the most influential factors leading to privatization. Second, seeking higher efficiency and costs savings has been influential for privatization. In this case, however, a distinction must be made between scale economies (using private firms to join up volumes of production from nearby municipalities), so that returns to scale can be realized from which small municipalities would particularly benefit. On the other hand, transaction costs involved in privatization (Brown and Potoski, 2003) are relatively higher for small municipalities, which works against influence of scale economies. In recent literature, the search for efficiency has been weighted with the level of market competition, which would increase the likelihood of improving efficiency and obtaining cost savings (e.g., Hefetz, Warner, and Vigoda-Gadot, 2012; Hefetz and Warner, 2012). Furthermore, regarding transaction costs and their influence on privatization, it is worth noting that warnings have been issued that, when monitoring and supervision problems are high, reducing quality is a clear incentive for private providers to reduce costs (Hart, Shleifer, and Vishny, 1997; Levin and Tadelis, 2010).

Finally, political interests and ideological factors have been hypothesized as drivers of privatization. Efficiency factors related to politics have also been studied in the literature. Often political interests (i.e., seeking support to stay in office by paying attention to the closest interest groups' preferences) have been found to significantly influence privatization. For instance, most studies find a negative influence on the part of trade unions on privatization, even when addressing the potential endogeneity problems (that is, more privatization diminishes the relevance of trade unions), as in Bhatti et al. (2009), who find a negative association between public employees and privatization. However, in an interesting new development, Warner and Hefetz (2020) find that new contracts are higher among unionized municipalities, while contract reversals are higher among non-unionized municipalities.

Concerning ideology, consensus is wide on the scarce significance of ideological preferences on privatization of local public services (Bel and Fageda, 2007, 2009). Nonetheless, more recent literature somehow challenges this view following different public services are analyzed in a more detailed way. Several papers published in Scandinavia have distinguished between technical and social services, and they have found that while ideology does not significantly influence privatization of technical services, it seems to have relevance when social services are privatized (Petersen, Houlberg, and Christensen, 2015; Guo and Willner, 2017) because ideological preferences carry higher weight when deciding the delivery mode for this last type of service. Regarding technical services, within which water distribution lies, consensus on the lack of ideological influence is not as wide as in other technical services (such as waste management or road maintenance), and relevant studies, such as Picazo-Tadeo et al. (2012), have found ideology to be a relevant factor for privatization. It could well happen that the strong natural monopoly characteristics of water distribution, which make competition largely absent, require some ideological push in order to privatize.

14.5 Outcomes of privatization of water services

While few studies of drivers of privatization have been conducted for the water service, this service (together with waste management) has been the object of many empirical studies that analyze the effects of privatization. Studies focus most often on productivity and prices and, to a lesser extent, on water quality and affordability. Empirical literature on the relationship between urban water distribution and efficiency/productivity originated in the middle of the 1970s, first with studies focused on the United States. By the mid-1990s, after massive water privatization in England and Wales, the first econometric studies appeared for the United Kingdom. Later, empirical studies for countries beyond the United States and the United Kingdom appeared. Meta-analyses in Bel and Warner (2008) and Bel, Fageda, and Warner (2010) provide extensive details on the studies conducted until 2010. A few studies reviewed found public management was more efficient and others found private management to be more efficient. However, most studies did not find systematic differences in efficiency and productivity between public and private management.

Studies comparing efficiency have diminished in the last decade for the developed countries, and the view prevails that no significant differences exist because of ownership and management, as clearly stated in a very recent assessment by Helm (2020, p. 83) for the United Kingdom after 30 years of full privatization in England and Wales:

> The performance of the water industry since privatization has not, then, been unambiguously better than that which the public sector might have delivered. A nationalized industry would not necessarily have been much more inefficient; it would not have facilitated such widespread financial engineering; and the executive salary game would not have been permitted. It might have accessed private capital markets for debt.

Unlike research on developed countries, interest in comparing efficiency and productivity in developing countries has grown in the last years, as more data have become available. The resulting picture seems somewhat different; recent studies suggest that performance of public management in developing countries is lower than that in developed countries, so that comparison between delivery forms in developed countries is more favorable to private management (Carvalho, Pedro, and Marques, 2015; Cetrulo, Marques, and Malheiros, 2019; Zhao, 2020).

More recent empirical studies for developed countries have placed more emphasis on pricing, as can be seen in recent reviews by González-Gómez and García-Rubio (2018) and Porcher and Saussier (2018), which also address other dimensions of the service. Many studies have been conducted for France and Spain, the European continental countries (including the Czech Republic) where private management is more extensive, and they usually have found that water prices are higher with private management (e.g., Chong et al., 2006 for France; Martínez-Espiñeira, García-Valiñas, and González-Gómez, 2009, for Spain). However, other studies for France have found that higher prices with private management are only significant in smaller municipalities, but not in the largest ones (Chong, Saussier, and Silverman, 2015; Porcher and Saussier, 2018). Furthermore, Valero (2015) found that because private management is chosen for more complex service conditions, when controlling prices for delivery choice, the differential between private and public management disappears. Other studies have found higher prices with private management in European countries (Ruester and Zschille (2010) for Germany, and Silvestre and Gomes (2017) for Portugal). However, Romano, Masserini, and Guerrini (2015), who also control for choice of production form, did not find a significant association between management ownership and prices.

While research on prices in developing countries has been scarcer, more attention has been devoted to the impact of privatization on accessibility and affordability. The evidence available so far does not establish any robust conclusion on these issues. Positive influence of private management on coverage was found in Bolivia and Argentina by McKenzie and Mookherjee (2003) and Galiani, Gertler, and Schargrodsky (2005), whereas Clarke, Kosec, and Wallsten (2004) did not find any significant impact of private management in these countries. In the case of Malaysia, a negative effect of privatization on accessibility and affordability was found by Lee (2011).

14.6 Conclusion

After initial undertakings in a few cities in the late 18th and early 19th centuries, urban water service began to expand rapidly in the second half of the 19th century, parallel to heightened technological advances. Private initiative constituted the key actor in the initial phase of that expansionary process in most countries. However, pressures from intense urbanization and subsequent accessibility and affordability demands triggered dissatisfaction with private production. This, together with concerns over corruption, fueled the involvement of the public sector in water service delivery toward the end of the 19th and the beginning of the 20th centuries. By the end of the 20th century, however, concerns over the poor efficiency of publicly delivered services and, especially in developing countries, lack of financial resources for the extension and modernization of water service paved the way toward more privatization. This wave led to more privatization operations.

As happened in previous phases of organizational change in public service delivery, the complexity of urban water service led to challenges. Sunk physical networks are required for an effective service delivery, thus implying long-term investments. Water service carries attributes of high asset specificity and high contract management difficulty (Brown and Potoski, 2005; Hefetz and Warner, 2012), which results in significant transaction costs for governments that rely on outsourcing. The potential for competition for contracts is poor, and citizens' sensitivity about this service is high (Hefetz and Warner, 2012).

Water privatization has spawned a large number of empirical studies on the effects of privatization and comparisons of outcomes between public and private delivery. Available evidence suggests that while efficiency does not systematically differ between public and private production, especially in most developed countries, users pay higher prices with private delivery, which may be associated with higher subsidies from the budget under public delivery (sometimes underwriting debt from investments by water public companies). Regarding accessibility and affordability, robust evidence on the comparison between public and private production is scarcer; results are mixed, and no clear differences exist.

Privatization of public services has declined as a tool for policy reform in the 21st century, as other alternatives have emerged (Warner and Hebdon, 2001). Dissatisfaction with outcomes of privatization has been particularly intense in the water sector. On the one hand, this is because of higher prices paid under private management. On the other hand, it is because corruption and political favoritism have intensified under private delivery of water service. Furthermore, and beyond material concerns, the perception that water is a crucial good that must remain under public control has increased. Because of these reasons, pressures for remunicipalization (Hall, Lobina, and Terhorst, 2013) or reverse privatization (Hefetz and Warner, 2004) have been particularly acute in the water sector, as happened in the remunicipalization process in the Progressive Era at the turn of the 20th century.

It is beyond the scope of this chapter to assess what have been the effects of recent remunicipalization experiences. While these have generated a large number of cases, some of them in important cities like Paris and Berlin, a trend toward remunicipalization does not seem to be increasing (Clifton et al., 2021). Nonetheless, the tension between public and private delivery of water services has reemerged and will likely remain high in the future. A robust assessment of the outcomes of recent remunicipalization experiences in the water sector is called for in research efforts going forward when enough data become available.

References

Barraqué, Bernard & Kraemer, R. Andreas. 2014. Les services publics d'eau en Grande-Bretagne et en Allemagne: origine commune, trajectoires différentes. *Flux* 97–98(3): 16–29.

Becker, Gary S. 1985. Public policies, pressure groups, and dead weight costs. *Journal of Public Economics* 28(3), 329–347.

Bel, Germà. 2010. Against the mainstream: Nazi privatization in 1930's Germany. *Economic History Review* 63(1), 34–55.

Bel, Germà. 2020. Public versus private water delivery, remunicipalization and water tariffs. *Utilities Policy* 62(100982), 1–8.

Bel, Germà, Fageda, Xavier, 2007. Why do local governments privatize local services? A survey of empirical studies. *Local Government Studies* 33(4): 517–534.

Bel, Germà, Fageda, Xavier, 2009. Factors explaining local privatization: a meta-regression analysis. *Public Choice* 139(1): 105–119.

Bel, Germà, Fageda, Xavier, 2017. What have we learned from the last three decades of empirical studies on factors driving local privatisation? *Local Government Studies* 43(4): 503–511.

Bel, Germà, Fageda, Xavier, Warner, Mildred. 2010. Is private production of public services cheaper than public production? A meta-regression analysis of solid waste and water services. *Journal of Policy Analysis and Management* 29(3), 553–577.

Bel, Germà, Warner, Mildred, 2008. Does privatization of solid waste and water services reduce costs? A review of empirical studies. *Resources, Conservation & Recycling* 52(12): 1337–1348.

Bhatti, Yosef, Olsen, Asmus L, Pedersen, Lene H. 2009. The effects of administrative professionals on contracting out. *Governance* 22(1): 121–137.

Bortolotti, Bernardo, Perotti, Enrico. 2007. From government to regulatory governance: Privatization and the residual role of the state. *The World Bank Research Observer*, 22(1), 53–66.

Boyne, George A. 1998. Competitive tendering in local government: A review of theory and evidence. *Public Administration* 76(4): 695–712.

Brown, Trevor L., Potoski, Matthew. 2003. Managing contract performance: A transaction costs approach. *Journal of Policy Analysis and Management* 22(2): 275–297.

Brown, Trevor L., Potoski, Matthew. 2005. Transaction costs and contracting: The practitioner perspective. *Public Performance & Management Review* 28(3): 326–351.

Carvalho, Pedro, Pedro, Isabel, Marques, Rui C., 2015. The most efficient clusters of Brazilian water companies. *Water Policy* 17(5): 902–917.

Cetrulo, Tiago B., Marques, Rui C., Malheiros, Tadeu F., 2019. An analytical review of the efficiency of water and sanitation utilities in developing countries. *Water Research* 162: 372–380.

Chong, Eshin, Freddy, Huet, Saussier, Stéphane, Steiner, Faye. 2006. Public private partnerships and prices: Evidence from water distribution in France. *Review of Industrial Organization* 29(12): 149–169.

Chong, Eshien, Saussier, Stéphane, Silverman, Brian. 2015. Water under the bridge: Determinants of franchise renewal in water provision. *Journal of Law, Economics, and Organization* 31(suppl 1): i3–i39.

Clarke, George R.G., Kosec, Katrina, Wallsten, Scott, 2004. Has private participation in water and sewerage improved coverage? Empirical evidence from Latin America. *Journal of International Development* 21(3): 327–361.

Clifton, Judith, Warner, Mildred, Gradus, Raymond, Bel, Germà. 2021. Re-municipalization of public services: trend or hype? *Journal of Economic Policy Reform*, forthcoming. DOI: 10.1080/17487870.2019.1691344

Cutler, David M., Miller, Grant. 2006. Water, water everywhere. Municipal finance and water supply in American cities. NBER Chapters, in: *Corruption and Reform: Lessons from America's Economic History*. National Bureau of Economic Research, Cambridge (MA), pp. 153–184.

Farinosi, F., Giupponi, C., Reynaud, A., Ceccherini, G., Carmona-Moreno, C. De Roo, A., Gonzalez-Sanchez, D., Bidoglio, G. 2018. An innovative approach to the assessment of hydro-political risk: A spatially explicit, data driven indicator of hydro-political issues. *Global Environmental Change* 52: 286–313.

Flörke, Martina, Schneider, Christof, McDonald, Robert I. 2018. Water competition between cities and agriculture driven by climate change and urban growth. *Nature Sustainability* 1(1): 51–58.

Galiani, Sebastian, Gertler, Paul, Schargrodsky, Ernesto. 2005. Water for life: The impact of the privatization of water services on child mortality. *Journal of Political Economy* 113(1): 83–120.

Gómez-lbáñez, José A. 2003. *Regulating Infrastructure. Monopoly, Contracts and Discretion*. Harvard University Press, Cambridge, MA.

González-Gómez, Francisco, García-Rubio, Miguel A., 2018. Prices and ownership in the water urban supply: a critical review. *Urban Water Journal* 15(3): 259–268.
Guo, Ming, Willner, Sam. 2017. Swedish politicians' preferences regarding the privatisation of elderly care. *Local Government Studies* 43(1): 1–21.
Hall, David, Lobina, Emanuele, Terhorst, Philipp. 2013. Re-municipalisation in the early twenty-first century: water in France and energy in Germany. *International Review of Applied Economics* 27(2): 193–214.
Hart, Oliver, Shleifer, Andrei, Vishny, Robert W., 1997. The Proper scope of government: Theory and an application to prisons. *Quarterly Journal of Economics* 112(4): 1127–1161.
Hefetz, Amir, Warner, Mildred. 2004. Privatization and Its Reverse: Explaining the Dynamics of the Government Contracting Process. *Journal of Public Administration Research and Theory* 14(2): 171–190.
Hefetz, Amir, Warner, Mildred. 2012. Contracting or public delivery? The importance of service, market and management characteristics. *Journal of Public Administration Research and Theory* 22(2): 289–317.
Hefetz, Amir, Warner, Mildred, Vigoda-Gadot, Eran. 2012. Privatization and intermunicipal contracting: the US local government experience 1992–2007. *Environment and Planning C: Government and Policy* 30(4): 675 – 692.
Helm, Dieter. 2009. Infrastructure investment, the cost of capital, and regulation: an assessment. *Oxford Review of Economic Policy* 25(3): 307–326.
Helm, Dieter. 2020. Thirty years after water privatization: Is the English model the envy of the world? *Oxford Review of Economic Policy* 36(1): 69–85.
Hirschman, Albert O. 1958. *The Strategy of Economic Development.* New Haven, CT: Yale University Press.
Hirschman, Albert O. 1982. *Shifting Involvements: Private Interest and Public Action.* Princeton, NJ: Princeton University Press.
Jacobson, Charles D., 2000. *Ties that Bind. Economic and Political Dilemmas of Urban Utility Networks 1800–1990.* Pittsburgh, PA: University of Pittsburgh Press.
Jacobson, Charles D., Tarr, Joel A., 1995. *Ownership and Financing of Infrastructure: Historical Perspectives. Policy Research Working Paper Series, 1466.* Washington, DC: The World Bank.
Lee, Cassey. 2011. Privatization, water access and affordability: Evidence from Malaysian household expenditure data. *Economic Modelling* 28(5): 2121–2128
Levin, Jonathan, Tadelis, Steven. 2010. Contracting for Government Services: Theory and Evidence from U.S. Cities." *Journal of Industrial Economics* 58(3): 507–541.
Lowery, David. 1998. Consumer sovereignty and quasi-market failure. *Journal of Public Administration Research and Theory* 8(2): 137–172.
Lough, Alexandra W. 2016. The politics of urban reform in the gilded age and progressive era, 1870–1920. *American Journal of Economics and Sociology* 75(1): 8–22.
Martínez-Espiñeira, Roberto, García-Valiñas, María A., González-Gómez, Francisco. 2009. Does private management of water supply services really increase prices? An empirical analysis in Spain. *Urban Studies* 46(4): 923–945.
McKenzie, David, Mookherjee, Dilip. 2003. The distributive impact of privatization in Latin America: Evidence from Four Countries. *Economia. Journal of the Latin American and Caribbean Economic Association* 3(2): 161–234.
Melosi, Martin V. 2012. Full circle? Public responsibility *versus* privatization of water supplies in the United States. In: Barraqué, Bernard (ed.). *Urban Water Conflicts.* London: UNESCO-IHP. CRC Press. Taylor & Francis, pp. 39–56.
Menes, Rebecca. 2006. Limiting the reach of the grabbing hand: Graft and growth in American Cities, 1880 to 1930. In Glaeser, Edward and Goldin, Claudia, eds., *Corruption and Reform: Lessons from America's Economic History*, pp. 63–93. Chicago: University of Chicago Press.
Millward, Robert. 2007. La distribution de l'eau dans les villes en Grande Bretagne au XIXe et XXe siècles : le gouvernement municipal et le dilemme des compagnies privées. *Histoire, économie & société* 26(2): 111–128.
Núñez Romero-Balmas, Gregorio. 1996. Servicios urbanos colectivos en España durante la segunda industrialización: entre la empresa privada y la gestión pública. In Comín, Francisco, Martín Aceña, Pablo. eds., *La empresa en la Historia de España.* Madrid: Civitas, pp. 399–419.
Ofwat. 2016. *Water 2020: Our Regulatory Approach for Water and Wastewater Services in England and Wales.* London: Ofwat. https://www.ofwat.gov.uk/publication/water-2020-regulatory-approach-water-wastewater-services-england-wales/ (accessed 18 August 2021).
Peltzman, Sam. 1976. Toward a more general theory of regulation. *Journal of Law and Economics* 19(2): 211-240.

Petersen, Ole H., Houlberg, Kurt, Christenssen, Lasse R. 2015. Contracting out local services: A tale of technical and social services. *Public Administration Review* 75(4): 560–570.

Pezon, Christelle. 2012. Public-private partnerships in courts: the rise and fall of concessions to supply drinking water in France, 1875–1928. In: Barraqué, Bernard (ed.). *Urban Water Conflicts*. London: UNESCO-IHP. CRC Press. Taylor & Francis, pp. 57–68.

Picazo-Tadeo, Andrés J., González-Gómez, Francisco, Guardiola Wanden-Berghe, Jorge, Ruiz-Villaverde, Alberto. 2012. Do ideological and political motives really matter in the public choice of local services management? Evidence from urban water services in Spain. *Public Choice* 151: 215–228.

Porcher, Simon, Saussier, Stéphane. 2018. *Public Versus Private Management in Water Public Services: Taking Stock, Looking Ahead*. WP EUI RSCAS; 2018/64; Florence School of Regulation, Florence.

Posner, Richard A., 1975. The social costs of monopoly and regulation. *Journal of Political Economy* 83(4): 807–827.

Romano, Giulia, Masserini, Lucio, Guerrini, Andrea. 2015. Does water utilities' ownership matter in water pricing policy? An analysis of endogenous and environmental determinants of water tariffs in Italy. *Water Policy* 17(5): 918–931.

Roth, Gabriel J., 1987. *The Private Provision of Public Services in Developing Countries*. Oxford: Oxford University Press.

Ruester, Sophia, Zschille, Michael. 2010. The impact of governance structure on firm performance: An application to the German water distribution sector. *Utilities Policy* 18(3): 154–162.

Schumpeter, Joseph A. 1939. *Business Cycles: A Theoretical, Historical, And Statistical Analysis of the Capitalist Process*. New York: McGraw Hill.

Silvestre, Hugo C., Gomes, Ricardo C. 2017. A resource-based view of utilities: The key-determinant factors for customer prices and organizational costs in the Portuguese water industry. *Water Resources and Economics* 19: 41–50.

Stigler, George J., 1971. The theory of economic regulation. *Bell Journal of Economics and Management Science* 2(1): 3-21.

Stiglitz, Joseph E. 1986. *Economics of the Public Sector*. New York: W.W. Norton & Co.

Swain, Ashok. 2001. Water wars: fact or fiction? *Futures* 33(8–9): 769–781.

Troesken, Werner. 1997. The sources of public ownership: Historical evidence from the gas industry. *Journal of Law, Economics & Organization*, 13(1): 1–25.

Troesken, Werner. 1999. Typhoid rates and the public acquisition of private waterworks, 1880–1920. *Journal of Economic History* 59(4): 927–948.

Troesken, Werner. 2001. Race, disease, and the provision of water in American Cities, 1889–1921. *Journal of Economic History* 61(3): 750–776.

Troesken, Werner. 2002. The limits of Jim Crow: Race and the provision of water and sewerage services in American cities, 1880–1925. *Journal of Economic History* 62(3): 734–772.

Troesken, Werner, Geddes, Rick. 2003. Municipalizing American waterworks, 1897–1915. *Journal of Law, Economics & Organization* 19(2): 373–400.

UK Parliament. 2020. The 1848 Public Health Act. London: UK Parliament (https://www.parliament.uk/about/living-heritage/transformingsociety/towncountry/towns/tyne-and-wear-case-study/about-the-group/public-administration/the-1848-public-health-act/ downloaded 19 September 2020).

Valero, Vanessa. 2015/6. Les écarts de prix de l'eau en France entre les secteurs privé et public. *Revue économique* 66: 1045–1066.

Veblen, Thorstein. 1899. *The Theory of the Leisure Class: An Economic Study of Institutions*. MacMillan.

Warner, Mildred, Hefetz, Amir. 2001. Local government restructuring: Privatization and its alternatives. *Journal of Policy Analysis and Management* 20(2): 315–336.

Warner, Mildred, Hefetz, Amir. 2020. Contracting dynamics and unionization: Managing labor, political interests and markets. *Local Government Studies* 46(2): 228–252.

Williamson, Oliver E (1979). Transaction-cost economics: The governance of contractual relations. *Journal of Law and Economics* 22(2): 233–261.

Williamson, Oliver E (1981). The Economics of Organization: The Transaction Cost Approach. *The American Journal of Sociology* 87(3): 548–577.

Zhao, Jinjin. 2020. Productivity change in the privatized water sector in China (1999–2006). *Journal of Productivity Analysis* 53: 227–241.

15

INSTITUTIONAL, ECONOMIC, AND SPATIAL BARRIERS TO WATER SERVICES DELIVERY IN URBAN SLUMS AND INFORMAL SETTLEMENTS

Ellis A. Adams and William F. Vásquez

15.1 Introduction

In many low- and middle-income countries (LMICs), urban water insecurity – the lack of access to safe, affordable, readily available, and reliable supply of potable water in cities and urban areas – remains a critical public health and developmental challenge. Urban water insecurity is even more chronic in the rapidly expanding, unplanned, and low-income areas, including the peri-urban, informal, and slum settlements where most of the poor and vulnerable urban population lives. In many of these cities, wide gaps in water demand and supply have persisted for decades despite numerous global initiatives and frameworks for action, including Goal 7 of the Millennium Development Goals (MDGs), which sought to reduce by half the global population without access to basic water and sanitation services by 2015, the relatively more recent Sustainable Development Goals (SDGs) seeking universal basic safe water and sanitation for all by 2030, and the Human Right to Water Declaration by the United Nations in 2010.

Despite the popularity of pro-poor water supply initiatives and policies, public water utilities are often not fully committed to improving water services in low-income urban areas (Boakey-Ansah et al. 2019). Reluctance by public water utilities to improve services for the urban poor stems from a concern that customers may be unable to pay their bills regularly. In some cases, they deliberately marginalize the urban poor by pricing them out with high connection fees (Kundu and Chatterjee 2020). In informal urban settlements, residents may be paying, on average, twice or thrice the price of water from publicly shared standpipes. Outdated infrastructure causes supply to be intermittent, leading to long waiting times (Adams 2018b). The gaps in formal water service delivery in the disadvantaged urban areas are often filled by small-scale private water vendors of various forms, including tanker truck and cart operators, and door-to-door operators who sell water by the gallon. Water from informal vendors generally costs more because there is no price regulation, and the quality of water they sell can be poor and unsafe. Other water sources commonly used in slums are boreholes, protected wells, public taps, and water resale from private tap owners (Adams 2018b).

DOI: 10.4324/9781003057574-19

Population growth, rapid urbanization, and poverty, combined with other physical and social vulnerabilities, continue to exacerbate water insecurity in the urban areas of many developing countries (Adams et al. 2020). It is estimated that 23.5% of the global urban population lives in urban slums, many of them in LMICs in sub-Saharan Africa, South Asia, and Latin America (UN 2018). In this chapter, we adopt the UN Habitat's definition of a slum, which is a general term for any urban settlement that lacks one or more of the following services: improved water, improved sanitation, sufficient living space, durable housing, and secure tenure (UN Habitat 2003, UN Habitat 2018). Among the many reasons why slums have continued to proliferate and persist in these regions despite their squalid living conditions and repeated eviction threats is that they are influential in electoral politics, as the case in Mumbai demonstrates (Zhang 2018).

Currently, sub-Saharan Africa is the region with the highest percentage of urban population living in slums, followed by South Asia and the East Asia and Pacific region (see Figure 15.1). The regions with the most rapid urbanization and a disproportionately high number of urban residents in slums are also the most vulnerable to water insecurity. Between 1975 and 2010, the urban population in low-income regions nearly tripled, and although the urbanization rates in those areas have slowed down in the last decade, growth is still considerable in low-income countries.

Global and national statistics on water insecurity in slums are hardly available. In many LMICs, national statistics on water access tend to aggregate figures for urban and rural areas with no separate numbers for slums, leading to policy initiatives that favor wealthier urban areas. Generally, statistics from global monitoring outlets indicate near-universal urban coverage for improved water services in all regions except sub-Saharan Africa, where 84% of the urban population has access to basic water services (see Figure 15.2). Even so, many have observed that these numbers may be exaggerated considering that they overlook issues of intermittency, limited availability, and safe sources that may be contaminated by household handling (Adams

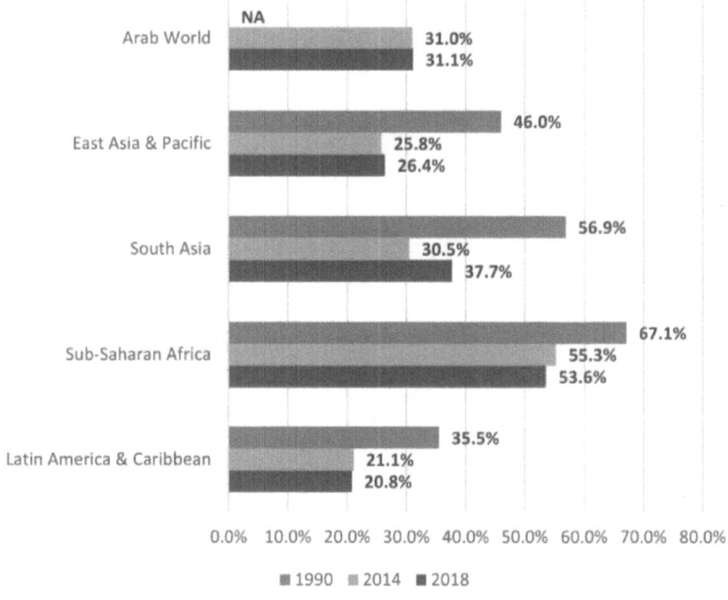

Figure 15.1 Percentage of urban population living in slums in selected regions (Source: (Data extracted from World Bank World Development Indicators).

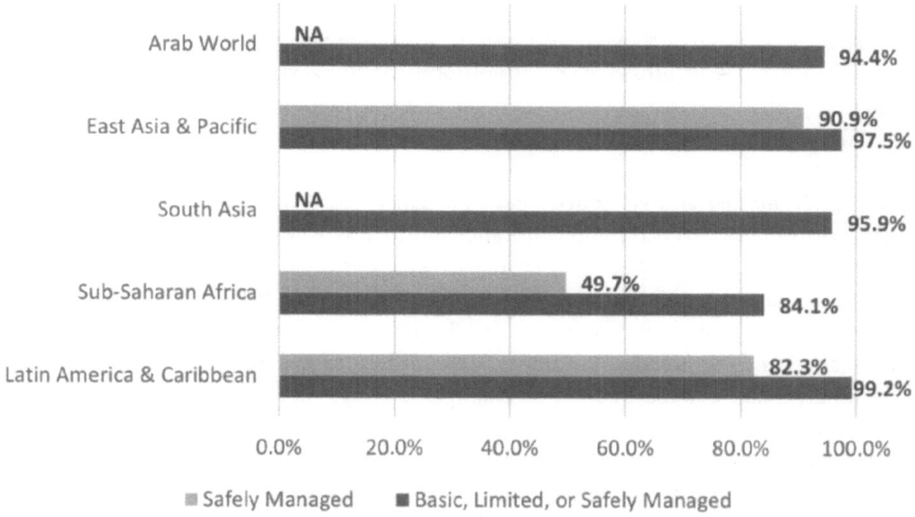

Figure 15.2 Percentage of urban population with access to improved water services as of 2017. (Source: World Bank World Development Indicators).

2018a; Smiley 2016). Thus, near-universal provision cannot imply universal coverage of safely managed services (i.e., access to water at home). Undoubtedly, the challenge is most severe in sub-Saharan Africa, where, even by conservative estimates, more than 15% of the urban population is without access to improved water services.

To advance the targets under the Sustainable Development Goal 6, we need to better understand the myriad obstacles to urban water service delivery, particularly for those in low-income areas where a complex set of spatial, political, economic, social, institutional, and informational factors impede state, municipal, and national efforts to address the water demand–supply gap. These obstacles are often highlighted in individual studies rather than synthesized in detail (e.g., see Sinharoy et al. 2019). The aim of this chapter is to highlight some of the major barriers to the improvement of urban water services in vulnerable and low-income areas. Using evidence from the literature and from our own research, we examine how geographical, spatial, and economic barriers impede efforts to both improve and extend new water services to underserved urban areas. The chapter highlights how these barriers disincentivize public utilities from extending new network connections to poor urban households as well as the institutional weaknesses that have historically undermined the achievements of major water sector reforms, including public, neoliberal, market-based, public–private partnerships, and community-based approaches. Finally, we offer policy lessons for overcoming barriers and advancing SDG 6.

15.2 Geographical and spatial constraints to urban water services improvements

In urban slums and informal settlements, settlement patterns, as well as density and spatial configuration of housing, make water service extension challenging even if there is state support to improve services. The size, location, and nature of slums may vary depending on when they were established. Older slums tend to have more people and to be located at the center of urban areas, although there are exceptions. Conversely, new slums tend to be smaller and more peripheral (Roy et al. 2014). Extending water infrastructure to slums may be relatively expensive when they are too distant from the main water infrastructure. Some old slum settlements lie in historical neighborhoods or districts that have been neglected by city planning officials. As those settlements have grown extensively, it has become even more difficult to extend water services because of the weak or nonexistent infrastructure. During the colonial era, water infrastructure planning in many cities of colonized countries was highly differentiated and was purposefully meant to serve the needs of the European colonizers, not the colonized (Hungerford and Smiley 2016). In South Africa, decades of apartheid policies have resulted in modern-day inequalities in water access in the country's slums and informal settlements (Robins 2019). As a result of decades of uneven planning that prioritized white neighborhoods over black and rural communities, slums, informal settlements, and other vulnerable communities continue to struggle with unequal water access (Millington and Scheba 2021). For example, during the apartheid era, hundreds of thousands of minorities and black people were relocated to communities with poor housing, some of them in flood-prone areas with limited access to water and sanitation services; these challenges have persisted (Enqvist and Ziervogel 2019).

The development of slums is usually unplanned, resulting in settlements with high population density and narrow streets (see Figure 15.3). In attempting to provide water services for slums, water utilities are confronted with the spontaneous emergence of new settlements that may encroach on old service areas (Boakye-Ansah et al. 2019). The construction work required to extend water infrastructure may be difficult in those settings due to limited access and the fragility of the built environment from continuous flooding. Slums are often located in marginal areas that are unsuitable for urban development and prone to natural disasters, e.g., riverbanks, steep slopes, rail tracks, under bridges, wetlands, and dumping grounds (Mahabir et al. 2016;

Figure 15.3 Slum in Brazil.

Rahaman 2018; Roy et al. 2014). Resilient infrastructure, which could be more expensive, could be required in areas prone to natural disasters. Hence, extending water infrastructure to slums may require substantial investments due to their topographical characteristics and physical conditions.

The ecology of urban slums imposes additional spatial constraints on water service improvements. It is typical for slums to be overcrowded and have pit latrines, leaking in the open, shared by multiple households. In slums where open defecation is common and waste dumps or landfills are not properly managed, poor sanitary conditions can become barriers to water service expansion. Greywater from bathrooms and overflowing wastewater from other sources can create swampy, unsightly, and challenging environmental conditions that are simply unsuitable for constructing pipes. Although nearby surface water sources could be earmarked by utilities for piped supply to slums, increased contamination could discourage utilities from extending new pipes due to potentially high operational costs, coupled with uncertainties in cost recovery largely due to low expectations that locals are willing to pay for improved water services.

15.3 Economic challenges of improving water services in slums and informal settlements

Local and central governments may lack the financial resources to invest in water infrastructure, and they are often concerned about the chances of cost recovery following investments into maintenance and new services in slums because residents are presumably unwilling or unable to pay for water services (Dos Santos et al. 2017). While it is true that income poverty is prevalent in slums, those households tend to spend more on water from alternative sources than the water bill paid by households connected to water systems. For instance, in Dhaka (Bangladesh) and Kumasi (Ghana), slum households' expenditures on water are up to 14 times higher than water tariffs charged by the water utility (Rahaman and Ahmed 2016; Adusei et al. 2018). These observations suggest that unwillingness to pay for new services is not necessarily contingent on income and can be influenced by many other factors.

Contingent valuation[1] studies suggest that the notion that slum households are not willing to pay official water tariffs cannot be generalized because willingness to pay depends on the local context. In Kano, Nigeria, Ahmad (2017) observed that slum household's reluctance to pay the monthly flat fixed tariff set by the Kano State Water Board was partially due to previously failed water projects. In contrast, Adusei et al. (2018) found that 90% of slum households in Kumasi, Ghana, are willing to pay official tariffs for improved water services because informal suppliers charged 14 times the official tariff. In Dhaka, Bangladesh, Rahaman and Ahmed (2016) found that slum households are willing to pay approximately 3.4 times more than the water price charged by the water utility. This suggests that there is a latent demand for private connections to improved water systems or "safely managed" water services, as denoted in the SDG water ladder (UN 2020).

In situations where it is not possible to provide safely managed water services, which are sources located on premises, available when needed, and free from contamination, households may be more willing to pay for alternative service levels. A major economic constraint to a willingness to pay may be that interventions focus mainly on the best possible services, i.e., household taps, without experimenting with alternative service levels that may be cheaper and more feasible in the short term. In Accra, Ghana, given that current regulations do not allow slum households without formal tenure to apply for private water connections, Vásquez and Adams (2019) investigated households' preferences for improved water services delivered by community standpipes in the slum of Nima-Maamobi. They found that households would pay up to \$1.25 for 20L of safe drinking water, and \$1.07 for 20L of water needing to be treated at home before drinking it, regardless of how distant those standpipes would be located. Those estimates are consistent with local prices of sachet water. They also found that households would prefer a community-based committee or a nongovernmental organization to manage the standpipes over the water utility and the municipal assembly. Their estimates can be compared against supply costs to assess the economic feasibility of alternative business models for improved water provision. In areas where slum households are not able to pay for improved water services, cross-subsidies can be explored as an alternative to cover supply costs. Pro-poor cross-subsidies programs have been successfully implemented in Zambia and Burkina Faso. In Kenya, Acey et al. (2019) found that water users are willing to pay an 8% increase in their water bill (approximately \$2.90) to subsidize sanitation in low-income settlements.

Water losses, illegal connections, and other economic constraints make slums and low-income urban areas less attractive for public water utilities. In slums with access to piped water, water companies may be unable to recover supply costs, which puts their financial autonomy and sustainability at risk. Leakages, illegal connections, and other forms of non-revenue water (NRW) are significant in urban slums. For instance, NRW accounts for an average of about 34% of the total water production in major metro cities of South Africa (Makaudze and Gelles 2015), and up to 50% in Dhaka, Bangladesh (Hossain and Ahmed 2015). In Dar es Salaam, nearly half of the water supplied to the city is lost or unaccounted for due to leakages and illegal connections (Rugemalila and Gibbs 2015). Improvements in access will require huge investments in the maintenance of old and leaky pipes, which may discourage public water utilities from extending new service lines. Economic losses from NRW can increase the likelihood of utilities sidelining slum areas where recovering costs is even more difficult.

Nonpayment of water bills by household tap owners is another economic challenge in urban slums that water providers deal with. Mukherjee et al. (2015) report that, in a slum in Bangalore, India, approximately 37% of households with tap connections did not pay for water. In some slums, there may be a high percentage of water users with outstanding payments bills. In a slum in Nicaragua, Vásquez and Alicea-Planas (2017) found that approximately 88% of water users were late in paying their water bills, with an average of 37 outstanding monthly payments. They found that nonpayment was due to income and that the penalty of service disconnection did not deter families from late payment or nonpayment. In the same study, some households acknowledged that even if their water services were disconnected, they would illegally reconnect to the system. Makaudze and Gelles (2015) report similar issues of non-affordability and a culture of nonpayment for water services in South Africa.

When coverage is not universal, some households connected to the piped water system resell to unconnected households at high prices. In Accra, Ghana, the water company has a program in which connected households can become water vendors. They are required to pay commercial, higher water prices to run their own water business. This helps the water utility in recovering supply costs while providing unconnected households with access to water. However, several of those water vendors do not register as commercial vendors, thus paying domestic rates to the utility even as they operate commercially and charge high rates to buyers (Allen et al. 2006; Tutu and Stoler 2016). In addition, complexities in tariff structures may also deter utilities from extending piped water to slums. Many tariff systems are such that poor households pay subsidized domestic rates to utilities, often resulting in inadequate funds to finance new connections.

15.4 Institutional, governance, and regulatory challenges to urban water services delivery

It is widely recognized among water scholars that governance and institutional failures are largely responsible for the urban water crisis in developing areas (Bakker 2010; Bakker et al. 2008). Some of the major sources of governance failure that undermine water service delivery are the factors that prevent households from connecting to urban water supply systems, including connection fee policies, costs, and housing and tenure policies (Bakker et al. 2008). The concept of governance failure underscores the institutional and historical failures that undergird poor water service delivery in urban areas, particularly in settings where allocation and distribution rather than physical water scarcity is the major issue. Governance failure can explain both why utilities are reluctant to connect new households and why poor urban households choose not to connect even when services are available. In addition, governance failure also provides a lens through which to discuss why many national and international water governance reforms have fallen short of expectations. Below we discuss some of the institutional impediments to water service improvements, focusing on both historical and contemporary factors.

Complex tenancy arrangements and lack of property rights in urban slums are major institutional impediments to extending water services (Boakye-Ansah et al. 2019). Tenancy in urban slums is very transient in nature due to insecure arrangements. For a household to apply for and be granted a tap, they usually must demonstrate proof of homeownership, something many slum dwellers simply do not have because they are either renting or squatting. Households with rental status or those that occupy structures on illegal land are unable to apply for new services, yet in many national jurisdictions, housing and tenure issues and utility services are managed by different institutions. Water utilities may be reluctant to increase connections to slums due to rapid turnover in rental and ownership status. By virtue of their location, some slums are viewed as illegal and are not legitimate beneficiaries of new water, sanitation, and hygiene (WASH) improvement programs (Sinharoy et al. 2019). Indeed, sometimes the state may not sanction water improvements because it wants to discourage the formation of new slums and an influx of new migrants into old slum settlements.

Decades of public and private water sector reforms and institutional changes have been unable to address the everyday water supply challenges in urban slums. Historically, three main formal institutional approaches have been used to supply water to the urban areas of LMICs: public or state-centered delivery, private and market-based privatization of water supply, and a hybrid of the two in what is often referred to as public–private partnerships. In public water systems, the state, a state-supported entity, or a local municipal operator is usually in charge of urban water service delivery. The relationship between the state and such operators differs in different contexts, and, in some cases, the exact partnership arrangement may not be clear. In Malawi, water boards are considered parastatal, which, in practice, implies both a separation from the state and service to the state. The water boards are expected to self-finance their operations using revenue from their water distribution. In most of Latin America, public water and sewerage systems are overseen by municipal authorities, although the specific institutional arrangements and structures differ by country. Public water systems in LMICs are often assumed to be more pro-poor compared to private systems as they generally tend to have more affordable water-pricing models. Despite these advantages, public water systems in many developing and underdeveloped countries face significant challenges in trying to improve water access for low-income urban residents. They struggle with weak infrastructure, disinvestment, low technical and personnel capacity, aging and weak infrastructure, poor management, and corruption, and, in many cases, they have been unable to cope with growing water demand, especially in low-income urban areas where cost recovery is a challenge (Dos Santos et al. 2017).

Privatization of water, predicated on the failure of the state to effectively manage public water systems, has largely proven ineffective for urban slums. In the early decades of the 20th century, private sector involvement in water services delivery was promoted widely in LMICs to increase capital investment in the water sector, improve access, and address the limitations of public and municipal systems. Multinational organizations, including the World Bank and the International Monetary Fund (IMF), actively promoted private water models as they cited poor management on the part of public utilities. The World Bank and the IMF conditioned loans and grants on the privatization of water (and sanitation) services. Driven by structural adjustment programs, a global neoliberal agenda, and cost recovery and profit to sustain operations, large-scale private sector involvement to date has been largely predicated on the quest to address inefficiencies in urban public water systems. While their overall objective was to improve water delivery in urban areas, private companies strategically cherry-picked wealthier urban areas to increase the chances of cost recovery, leaving out underserved urban settlements and slums with the greatest need for water service improvements. The fact that such settlements were dominated by low-income residents with limited ability to pay for services discouraged investments in those areas. Altogether, little evidence has been found that private sector involvement is more efficient than publicly managed urban water systems in the low-income countries (Prasad 2006), let alone for underserved communities. Experiences from many developing countries, including Bolivia, Ghana, Colombia, Philippines, and Tanzania, where private water contracts were terminated prematurely, have cast further doubt over the potential of private sector involvement to improve water delivery in urban areas.

Obsession with a modern infrastructural ideal that privileges public, universal water supply over heterogeneous small-scale infrastructures disregards abundant evidence that more slum dwellers utilize the latter (Allen 2010). In urban slums and informal settlements, households patch from different water sources to meet their daily water needs. Relatively affluent homes may purchase large poly tanks and fill them with water purchased from tanker services, while poor households may depend more on small-scale vendors. But vendors may purchase their water from formal or public sources and, in turn, supply it through informal markets and mechanisms. In more affluent urban areas, some residents still depend on the same informal supply networks to cope with intermittent water supply. Described variously as archipelagos (Bakker 2003), gray zones (Truelove 2019), or heterogeneous water infrastructures (Alba, Kooy and Bruns 2020; Rusca and Schwartz 2018; Kundu and Chatterjee 2020), water supply and access dynamics in cities and slums are widely diverse and combine multiple formal and informal networks with unclear boundaries. No one form of supply modality is universally able to satisfy water demand in slums, yet water sector reforms continue to focus on public and private, market-based models of water service delivery.

In the heterogenous arrangements, numerous non-state actors, especially NGOs, water vendors, community associations, municipal governments, and faith-based groups play critical roles in helping poor urban households secure their daily water needs. Such complex and overlapping water delivery modalities underscore that the formal and informal dualisms and the public–private dualisms are of little use in describing water delivery in many cities of LMICs (Truelove 2019). As Bakker et al. (2008) and others have argued, the rigid use of formal and informal to describe water delivery in the cities of the South may only serve to obscure the broader weaknesses in poor water governance. Most residents in urban areas and slums in LMICs do not own household taps and often live outside the main piped network.

Despite their ubiquity and importance, informal water providers in many cities of LMICs are neither regulated nor their activities monitored by formal agencies. Indeed, in many cases, public water utilities or municipal water units have no clear mandate to oversee the activities of informal vendors. In many cases, informal vendors mobilize into mafias with deep connections to the state. For example, Ranganathan (2014) compellingly shows in a case study of Bangalore how water scarcity in slums has been deliberately created for state workers to participate in water mafia schemes. Powerful local politicians and state agency workers are active participants in these mafia schemes. The Bangalore case is only one example of how vendors exploit institutional and regulatory lapses, participate in corruption schemes, and connive with state officials to extort the very neighborhoods they are expected to improve services for. The lack of regulation of these corrupt practices may be deliberate and a function of colonial and postcolonial planning (Bakker 2010). In Cape Town, South Africa, poor governance and institutional failures have been responsible for decades of poor water supply in informal settlements (Enqvist and Ziervogel 2019).

If given the needed institutional support, improvements in community-based water governance (CWG) models could help fill gaps in service delivery for urban slums and informal settlements. These models have become particularly critical considering the failures of public, private, and public–private partnership to service all urban areas, including slums and informal settlements. Recent work has demonstrated that CWG has the potential to augment service delivery, and, if scaled up effectively, could improve water quantity and supply reliability; reduce non-revenue water; lead to transparent and affordable pricing schemes; empower communities and create platforms for them to demand better services from the state; and help communities to create innovative partnerships with other actors for better service delivery (Adams et al. 2020b). Some of these benefits are in line with the benefits proponents of community-based natural resources anticipate, which include transparency, accountability, equitable benefit sharing, and increased social capital. However, without the right institutional support, community-based approaches, despite their potential, replicate the very challenges associated with public and private water delivery. For example, although communities may mobilize to meet the state halfway, schemes that inherit the same old and weak infrastructure fail to improve water delivery and these challenges have been observed in Kisumu, Kenya (Nzengya 2015), Lilongwe, Malawi (Adams and Boateng 2018), Manila, Philippines (Cheng 2013), and Mumbai, India (de Bercegol and Desfeux 2011). These challenges underscore that, although certain urban slum communities can mobilize, they can barely thrive in the same institutional weaknesses that gave rise to poor water supply in the first place.

15.5 Conclusion and policy lessons

Despite longstanding, global efforts to improve the living conditions of impoverished people in the framework of the Millennium Development Goals, and more recently the Sustainable Development Goals, water insecurity remains pervasive and precarious in slums and informal settlements worldwide. As demonstrated above, a complex set of geographical, physical, social, economic, and institutional barriers has impeded the extension of improved water services to those excluded areas until now. Rather than being generalizable, the prevalence of those impediments, and how they intertwine with each other, depends on the local context. Consequently, a variety of solutions will be required to achieve the universal provision of improved water services.

As we have outlined, the geographical and physical attributes of slums and informal settlements may represent a challenge to install water infrastructure to deliver drinking water at home. These geographical challenges call for better integration of mainstream urban planning with water services infrastructure and planning. This approach must be based on finding the right mix of physical infrastructure and social services delivery. Additionally, cost analysis should play a central role in how policymakers make decisions. For example, given that most underserved urban areas in LMICs have very old water infrastructure, there is the need for a careful cost analysis to decide whether it is more cost-effective to update old infrastructure or build new infrastructure altogether.

Insufficient cost recovery is another concern because slum households are often considered unable or unwilling to pay for water services. The SDG water ladder acknowledges that it will not be feasible to connect every home to piped water systems, and that alternative service levels will play an important role in securing water for those who currently have no access to improved water sources. In addition to technical analyses, local information on household preferences for alternative levels of improved water services will be required to design and implement water projects, tariff structures, and management arrangements responsive to households' preferences, willingness, and ability to pay. Vásquez and Adams (2019) present an example of how to gather that information. Interventions based on local preferences are more likely to gather public support and increase water revenues. When local ability to pay is insufficient to cover supply costs, different financing (e.g., cross-subsidies between water users) and cost reduction mechanisms (e.g., locals' willingness to work in community-based water organizations) can be tested. Official transfers to local water systems cannot be ruled out.

Alternative management approaches also seem warranted given that traditional forms of service provision (e.g., public and private water companies) have failed to extend water services in slums and informal settlements. While informal vendors have filled the gap in many urban areas, their water prices are often too high relative to official water tariffs and, in some cases, they promote corruption at different levels of the water sector. This calls for formalizing water vendors to deter rent-seeking behaviors. However, in practice, it may be difficult for local authorities to regulate and monitor water vendors due to the lack of financial resources and institutional weaknesses.

Water access for most urban residents happens in hybrid zones where diverse formal and informal delivery modalities coexist and interact with no clear separation (Allen 2010; Ahlers et al. 2014; Misra 2014; Schwartz et al. 2015). Formal and informal arrangements may be mutually constitutive, interrelated, and dependent on each other to thrive (Misra 2014). Residents of formal areas may equally depend on informal supply vendors (Cheng 2014; Alba et al. 2019). Formal utilities can partner with informal providers to extend their services to low-income areas as a means to increase revenue. Utilities may also sanction "informal" and "illegal" practices because they enable them to expand their operations and recover costs (Cheng 2014; Mitlin and Walnycki 2019). While informal water vendors are recognized for their contribution to water delivery in low-income urban settings, there are concerns over pricing and quality of the water they supply. It remains imperative that the right institutional environment is created not just to regulate informal water vendor activity, but also to properly integrate them into the water supply landscape and accord them better recognition. In many urban areas in LMICs, water supply without informal vendors will be extremely difficult given how underdeveloped the water infrastructure is at the moment.

Community-based approaches seem promising when governmental agencies and non-governmental organizations provide communities with technical and managerial support. Institutional support is critical for successful community engagements for water delivery in urban slums and informal settlements. Among the key pitfalls, community-based water governance fails to improve access in slums because of a weak capacity to address the same infrastructural challenges that undermine public and private water delivery. Institutional improvements aimed at water services delivery should therefore take seriously the needs, capacity issues, and weak infrastructure that all too often undermine well-intentioned community efforts. There is an urgent need for alternative forms of service provision based on principles of inclusiveness, sustainability, affordability, and equity. For innovative approaches to emerge and function, it is imperative to have the political will to strengthen the institutional framework required to facilitate coordination among different stakeholders.

Note

1 The contingent valuation method (CVM) consists of hypothetical scenarios designed to estimate the value of goods and services that are not traded in a market setting.

References

Acey, C., Kisiangani, J., Ronoh, P., Delaire, C., Makena, E., Norman, G., Levine, D., Khush, R., and Peletz, R. 2019. Cross-subsidies for improved sanitation in low-income settlements: Assessing the willingness to pay of water utility customers in Kenyan cities. *World Development*, 115, 160–177.

Adams, E. A., & Boateng, G. O. (2018). Are urban informal communities capable of co-production? The influence of community–public partnerships on water access in Lilongwe, Malawi. *Environment and Urbanization*, 30(2), 461–480.

Adams, E.A. 2018a. Intra-urban inequalities in water access among households in Malawi's informal settlements: Toward pro-poor urban water policies in Africa. *Environmental Development*, 26, 34–42.

Adams, E.A. 2018b. Thirsty slums in African cities: Household water insecurity in urban informal settlements of Lilongwe, Malawi. *International Journal of Water Resources Development*, 34, 869–887.

Adams, E.A., et al. 2020. Water insecurity and urban poverty in the Global South: Implications for health and human biology. *American Journal of Human Biology*, 32, e23368.

Adams, E.A., et al. 2020b. Community water governance for urban water security in the Global South: Status, lessons, and prospects. *WIREs Water*, 7(5), e1466.

Adusei, P.K., Oduro-Ofori, E., Amponsah, O., and Agyeman, K.O. 2018. Participatory incremental slum upgrading towards sustainability: An assessment of slum dwellers' willingness and ability to pay for utility services. *Environment, Development and Sustainability*, 20, 2501–2520.

Ahlers, R.; Cleaver, F.; Rusca, M. and Schwartz, K. 2014. Informal space in the urban waterscape: Disaggregation and co-production of water services. *Water Alternatives* 7(1): 1–14

Ahmad, M.T. 2017. The role of water vendors in water service delivery in developing countries: A case of Dala local government, Kano, Nigeria. *Applied Water Science*, 7, 1191–1201.

Alba, R, Bruns, A, Bartels, LE, et al. (2019) Water brokers: Exploring urban water governance through the practices of tanker water supply in Accra. *Water* 11(9): 1919.

Alba, R., Kooy. M., and Bruns, A. 2020. Conflicts, cooperation and experimentation: Analysing the politics of urban water through Accra's heterogeneous water supply infrastructure. *Environment and Planning E: Nature and Space*, 5(1): 250–271.

Allen, A. 2010. Neither rural nor urban: Service delivery options that work for the peri-urban poor. In *Peri-urban Water and Sanitation Services* (pp. 27–61). Springer.

Allen, A., Dávila, J.D., Hofmann P. 2006. The peri-urban water poor: Citizens or consumers? *Environment and Urbanization*, 18, 333–351.

Bakker, K. 2003. Archipelagos and networks: Urbanization and water privatization in the South. *Geographical Journal*, 169, 328–341.

Bakker, K. 2010. *Privatizing Water: Governance Failure and the World's Urban Water Crisis*. Cornell University Press.

Bakker, K., Kooy, M., Shofiani, N.E., and Martijn, E-J. 2008. Governance failure: Rethinking the institutional dimensions of urban water supply to poor households. *World Development*, 36, 1891–1915.

Boakye-Ansah, A.S., Schwartz, K., and Zwarteveen, M. 2019. Unravelling pro-poor water services: What does it mean and why is it so popular? *Water Sanitation & Hygiene for Development*, 9, 187–197.

Cheng, D. 2013. (In) visible urban water networks: The politics of non-payment in Manila's low-income communities. *Environment and Urbanization*, 25(1), 249–260.

Cheng, D. 2014. The persistence of informality: Small-scale water providers in Manila's post-privatisation era. *Water Alternatives* 7(1): 54–71.

de BERCEGOL, R., & Desfeux, A. 2011. *An Alternative to Conventional Public Water Service:" User Group Networks" in a Mumbai Slum.*

Dos Santos, S., Adams, E.A., Neville, G.,Wada, Y., de Sherbinin, A., Bernhardt, E.M., Adamo, S.B., 2017. Urban growth and water access in sub-Saharan Africa: Progress, challenges, and emerging research directions. *Science of the Total Environment*, 607, 497–508.

Enqvist, J. and Ziervogel, G. 2019. Water governance and justice in Cape Town: An overview. *WIREs Water*, 6, e1354.

Hossain, K.Z. and Ahmed, S.A. 2015 Non-conventional public-private partnerships for water supply to urban slums, *Urban Water Journal*, 12(7), 570–580.

Hungerford, H. and Smiley, G. 2016. Comparing colonial water provision in British and French Africa. *Journal of Historical Geography*, 52, 74–83.

Kundu, R. and Chatterjee, S. 2020. Pipe dreams? Practices of everyday governance of heterogeneous configurations of water supply in Baruipur, a small town in India. *Environment and Planning C: Politics and Space*, 39(2): 318–335.

Mahabir, R., Crooks, A., Croitoru, A., and Agouris, P. 2016. The study of slums as social and physical constructs: Challenges and emerging research opportunities. *Regional Studies, Regional Science*, 3(1), 399–419.

Makaudze, E. and Gelles, G. 2015. The challenges of providing water and sanitation to urban slum settlements in South Africa. In: Grafton, Q., Daniell, K.A., Nauges, C., Rinaudo, J.-D., Chan, N.W.W. (Eds.) *Understanding and Managing Urban Water in Transition* (pp. 121–133). Springer.

Misra, K. 2014. From formal-informal to emergent formalisation: Fluidities in the production of urban waterscapes. *Water Alternatives* 7(1): 15–34.

Millington, N., & Scheba, S. 2021. Day zero and the infrastructures of climate change: Water governance, inequality, and infrastructural politics in Cape Town's water crisis. *International Journal of Urban and Regional Research*, 45(1), 116–132.

Mitlin, D., & Walnycki, A. (2019). Informality as experimentation: Water utilities strategies for cost recovery and their consequences for universal access. *Journal of Development Studies*, 56(2), 259–277.

Mukherjee, M., Chindarkar, N., and Grönwall, J. 2015. Non-revenue water and cost recovery in urban India the case of Bangalore. *Water Policy*, 17, 484–501.

Nzengya, D.M. 2015. Exploring the challenges and opportunities for master operators and water kiosks under Delegated Management Model (DMM): A study in Lake Victoria region, Kenya. *Cities*, 46, 35–43.

Prasad, N. 2006. Privatisation results: Private sector participation in water services after 15 years. *Development Policy Review*, 24, 669–692.

Rahaman, M. 2018. Economic status and housing conditions of slums in Kalyani Town, West Bengal. *Indian Journal of Landscape and Ecological Studies*, 41(2), 112–121.

Rahaman, M.M., and Ahmed, T.S. 2016. Affordable water pricing for slums dwellers in Dhaka metropolitan area: The case of three slums. *Journal of Water Resource Engineering and Management*, 3(1), 15–33.

Ranganathan, M. 2014. 'Mafias' in the waterscape: Urban informality and everyday public authority in Bangalore. *Water Alternatives* 7(1): 89–105

Robins, S. 2019. 'Day Zero', hydraulic citizenship and the defence of the commons in Cape Town: A case study of the politics of water and its infrastructures (2017–2018). *Journal of Southern African Studies*, 45(1), 5–29.

Roy, D., Lees, M.H., Palavalli, B., Pfeffer, K., and Sloot, M.A.P. 2014. The emergence of slums: A contemporary view on simulation models. *Environmental Modelling & Software*, 59, 76–90.

Rugemalila, R. and Gibbs, L. 2015. Urban water governance failure and local strategies for overcoming water shortages in Dar es Salaam, Tanzania. *Environment and Planning C: Government and Policy*, 33, 412–427.

Rusca, M. and Schwartz, K. 2018. The paradox of cost recovery in heterogeneous municipal water supply systems: Ensuring inclusiveness or exacerbating inequalities? *Habitat International*, 73, 101–108.

Schwartz, K., Tutusaus Luque, M., Rusca, M. and Ahlers, R. (2015), (In)formality: the meshwork of water service provisioning. *WIREs Water*, 2: 31–36.

Sinharoy, S.S., Pittluck, R., and Clasen, T. 2019. Review of drivers and barriers of water and sanitation policies for urban informal settlements in low-income and middle-income countries. *Utility Policy*, 60, 100957.

Smiley, S.L., 2016. Water availability and reliability in Dar es Salaam, Tanzania. *The Journal of Development Studies*, 52(9), 1320–1334.

Truelove, Y. 2019. Gray zones: The everyday practices and governance of water beyond the network. *Annals of the American Association of Geographers*, 109, 1758–1774.

Tutu, R.A. and Stoler, J. 2016. Urban but off the grid: The struggle for water in two urban slums in greater Accra, Ghana. *African Geographical Review*, 35, 212–226.

United Nations. 2020. *The Sustainable Development Goals Report 2020*. https://unstats.un.org/sdgs/report/2020/The-Sustainable-Development-Goals-Report-2020.pdf

UN-Habitat. 2003. *The Challenge of Slums: Global Report on Human Settlements 2003*. Earthscan, London.

UN. 2018. *SDG 11. Sustainable Cities and Communities*. Available at https://unstats.un.org/sdgs/report/2021/goal-11/#:~:text=Between%202014%20and%202018%2C%20the,over%201%20billion%20slum%20dwellers

UN-Habitat. 2018. *SDG Indicator 11.1.1 Training Module: Adequate Housing and Slum Upgrading*. United Nations Human Settlement Programme (UN-Habitat).

Vásquez, W.F. and Adams, E. 2019. Climbing the water ladder in poor urban areas: Preferences for 'limited' and 'basic' water services in Accra, Ghana. *Science of the Total Environment*, 673, 605–612.

Vásquez, W.F. and Alicea-Planas, J. 2017. Factors associated with nonpayment behavior in the water sector of Nicaragua. *Utilities Policy*, 47, 50–57.

Zhang, Y. 2018. The credibility of slums: Informal housing and urban governance in India. *Land use policy*, 79, 876–890.

PART IV

Political processes

16

ACTOR NETWORKS IN URBAN WATER GOVERNANCE

Manuel Fischer, Karin Ingold, Mert Duygan, Liliane Manny, and Katrin Pakizer

16.1 Introduction

Whenever a variety of institutions, ideas, technical elements, and, most importantly, different societal actors are relevant to a policy sector – as is the case for urban water management – a network lens can be useful. A network lens additionally provides a popular way of operationalizing the complex and very general concept of *governance*. Many different definitions and forms of governance exist, but when adopting a structural perspective, governance can be defined as an arrangement or network of actors, be it organizations, individuals, or both. The network concept emphasizes, besides actors, stable relational patterns, thus the role of interrelations or interactions that represent the stable structural aspects of any type of problem-solving or management (Kenis and Schneider 1991).

For different reasons, this crucial role of relations, interdependencies, and interactions is particularly important in urban water governance. First, the urban water sector consists of different aspects and tasks, involves different policy sectors, and can, thus, be subject to cross-sectoral policy coordination problems that can typically be assessed or addressed by adopting a network perspective or analysis. Second, urban water management deals with infrastructure that in itself is interconnected. Ideally, the actor network governing the infrastructure would take those technical dependencies into account. Third, and strongly related to the former two, different types of private and public actors, resource and infrastructure owners and users, and managers and consumers are concerned with, affected by, or involved in urban water governance. Thus, the cross-sectoral and technically interconnected character of urban water governance also calls for networks at the actors' level. Ideally, the actor networks would "fit" with the underlying policy-related or technical issue. In sum, network approaches are useful to systematically describe and analyze the structure of interactions among this variety of actors and other elements that are connected to each other in complex and dynamic ways. Therefore, this chapter presents conceptual approaches, theoretical elements, and methodological tools related to actor networks in urban water governance (Fischer and Ingold 2020).

The focus on the urban context is particularly important, as the speed and intensity of global transformations, such as climate change, often manifest themselves at this level (Kaufmann and Sidney 2020). Urban areas also have considerable resources and leeway for developing and implementing environmental and infrastructure-related projects, and they can thus be areas of

DOI: 10.4324/9781003057574-21

transformative innovation (Brenner 2004; Friendly and Stiphany 2019; Kaufmann 2019; Hughes 2019). Urban governance strongly varies in terms of processes, institutions, and actor participation, and, thus, they can be fruitfully analyzed through a network lens (Ernstson et al. 2010; Blanco 2013; Van Meerkerk and Edelenbos 2014).

In the first part of this chapter, we discuss the main theoretical aspects and related research questions of network governance, including illustrative examples in urban water governance. Second, we shortly introduce the main concepts and measures in Social Network Analysis. Third, we discuss recent developments relevant to urban water governance: digitalization and decentralization as substantive innovations, and sociotechnical network approaches as a methodological innovation.

16.2 Conceptual thoughts linking networks to urban water governance

Governing complex issues such as urban water infrastructure requires interactions among a multitude of interested and specialized actors, representing different levels (e.g., state, city, neighborhood) or sectors (traffic, health, environment, etc.). This multitude of actors and their interactions can be represented as a network. Theoretically, the interactions of multiple actors and the resulting implications for governance have been dealt with under the labels of collaborative governance, network governance, governance networks, or polycentric governance, among others (Ansell and Gash 2008; Fischer and Leifeld 2015; Emerson, Nabatchi, and Balogh 2012; Berardo and Lubell 2016). The term of governance defines steering and decision making as action taken by a large number of public and private actors from different levels and functional areas of government and society (Jordan and Schubert 1992). Similarly, the related concept of network governance emphasizes the interactions of a diversity of public and private stakeholders who regularly interact together to engage in (goal-oriented) decision making (Ansell and Gash 2008). Translated to policy processes, this means that those actors are politically involved or interested and aim at influencing decisions in a specific field or sector (Marin and Mayntz 1991).

In urban water governance, besides the (policy) problem and related decisions that need to be taken at some point in time and engage actors during a limited period, there are many continuously ongoing management tasks. Thus, relevant actors involved in urban water governance are not only those attracted by salient (new) policy issues and related decisions, but also those involved in long-term responsibilities and activities, such as water use, distribution, or monitoring. Given that all those activities are very much linked to the geographical or spatial delimitation of the urban area under consideration, urban water governance networks might resemble environmental governance networks (Lubell 2003; Koppenjan, Koppenjan, and Klijn 2004). Environmental governance networks strongly depend on the area of the resource itself (forest pitch or river basin) where *management* happens within resource boundaries, but *governance* can also transcend them (for the difference between management and governance networks, see Ingold et al. [2018]).

Sometimes, discussions and research questions related to network governance are driven by slightly normative assumptions that the more interaction and collaboration among a large number, the better (Bodin 2017). However, actors do not only work together following claims of inclusive, collective, bottom-up problem-solving, but they interact because of a diversity of (individual) reasons. Network governance is shaped by factors such as institutions, power, resources, tasks, problem perceptions, or preferences of actors (Berardo and Lubell 2016; Berardo and Scholz 2010; Leifeld and Schneider 2012; Fischer and Sciarini 2016). One of the most prominent hypotheses in the study of networks is the homophily hypothesis claiming that actors with the same interests, goals, or worldviews tend to interact (McPherson, Smith-Lovin, and Cook 2001).

Network governance as deliberative, bottom-up ways of collective problem-solving are said to create trust and capacities for learning, reduce conflicts, and thereby improve the quality of outcomes (Ansell and Gash 2008; Emerson, Nabatchi, and Balogh 2012). Different structures of network governance can also influence policy change, governance innovation dynamics, policy diffusion and learning, or democratic legitimacy (Bodin et al. 2014; Victor, Montgomery, and Lubell 2017). With these claims in mind, approaches of network governance are also often prescribed by principles of sustainability (e.g., Sustainable Development Goals 17 on participatory governance), by rules and protocols (e.g., the European Union Water Framework Directive, see also Herzog et al. 2022), or fostered by forums and other collaborative institutions (Maag and Fischer 2018; Fischer and Leifeld 2015).

These main theoretical aspects of network governance and related concepts can apply to urban water governance, as illustrated with the following examples. For instance, Narayan et al. (2020) manage through the application of social network analysis to grasp decentralization aspects in India's urban water management sector. Lieberherr and Ingold (2019) study urban and rural water management and come to the conclusion that cooperation among different public and private, local and regional actors is enhanced when power structures align with coordination expectations. Also Fischer et al. (2019) include vertical (from local to international) and additionally horizontal (cross-sectoral) aspects and study the inclusion of a variety of actors in Vietnam's sanitation sector. Walters (2016) links (communication) network patterns to the quality of service provision in Nicaragua's sanitation sector. He further identifies most central actors in the network. Thus, in network governance studies, in general, and urban water governance, in particular, collaboration in networks, the horizontal and vertical inclusion of actors to address actors' conflicts, as well as the centrality of key players as a proxy for actors' power and influence are key.

16.3 Social network analysis

16.3.1 Main elements of networks

Concepts and measures from Social Network Analysis (SNA) (Wasserman and Faust 1994) have been frequently used for studying network governance, although not always: network governance is also a theoretical concept pointing at actor interactions without always describing them with tools of SNA (Berardo, Fischer, and Hamilton 2020). SNA is a method for describing structures of nodes and ties, and it has been increasingly used in a variety of fields from political science (Fischer and Sciarini 2016; Victor et al. 2016) to business marketing (Iacobucci 1996) to environmental governance (Bodin and Crona 2009; Sandström and Carlsson 2008) and, in particular, to (urban) water management (Ahmadi et al. 2019; Wutich et al. 2020). SNA is designed to deal with data on relations (ties) among entities (nodes), that is, data describing interconnected phenomena and consisting of interdependent observations on both nodes or ties.

The two defining elements of networks are nodes and ties, whereas the latter connect the first (Wasserman and Faust 1994). Nodes in network governance are social actors, such as individuals, organizations, or political-administrative entities, or a mix of all (Huxham et al. 2000). Organizational actors can be interest groups, political parties, administrative units, research centers, urban planners, (water) service providers or other actors involved in governance processes, and individuals can be representatives of these organizations. Finally, political-administrative entities, such as states or sub-states (Fischer and Jager 2020), can be nodes in networks of water governance, linked by water-related formal treaties. Besides social actors as nodes, the interdisciplinary and flexible nature of the network approach also facilitates the integration of other types of nodes besides social actors, for example, in order to take into account relations between social actors and issues (Brandenberger et al. 2020; Angst 2020; Metz, Angst, and Fischer 2020), institutions (Lubell 2013; Fischer, Angst, and Maag 2019), as well as ecological (Bodin and Tengo 2012; Bodin 2017; Ingold et al. 2018) or technical (Manny et al. 2021a) elements.

Network ties between social actors are usually ties of cooperation, coordination, information exchange, venue co-participation, conflict, or formal cooperation based on contracts. Some simplification is usually involved in the definition of a tie between network nodes in SNA. In reality, social relations can be very different. Ties can not only be present or absent, but they can also have different intensities and weights, such as different intensities of collaboration (Margerum 2008). Furthermore, cooperation and conflict, for example, can also exist in parallel, and they can be assessed in so-called multiplex networks, defined as networks composed of different types of relations (e.g., information, coordination) between the same set of actors (Hayes and Scott 2018). These multiplex ties can influence each other, as when, e.g., venue co-participation leads to more intense exchanges of information (Fischer and Sciarini 2016; Leifeld and Schneider 2012).

Nodes and ties can combine differently and give rise to different types of networks (Wasserman and Faust 1994). One-mode networks are the simplest form of networks, as they involve one type of node and (usually) one type of tie between these nodes. More complex types of networks are two-mode or bipartite networks where two different types of nodes are connected (without being connected among themselves). Finally, multilevel networks (Lazega and Snijders 2015) are networks with different types of nodes that are related to each other, and that include network relations within the different sets of nodes.

16.3.2 Descriptive tools of SNA

Networks can be analyzed on three different levels, the micro-level of individual actors in the network, the meso-level substructures within the network, that is, sets of nodes and ties in the network, or the macro-level of the entire network (Wasserman and Faust 1994; Borgatti et al. 2009; Borgatti, Everett, and Johnson 2018). The most common, important, and straightforward measure related to networks at the micro-level of network nodes is centrality. Most generally, centrality describes how central an actor is within the network, that is, the relational position of a given node in the overall network. Centrality, however, can take different forms (Freeman 1978). First, degree centrality is based on the simple number of incoming ties (in-degree) and outgoing ties (out-degree) of a node. Second, betweenness centrality takes into account the degree to which a node is located on the shortest path between any two other nodes in the network. Network nodes with high betweenness centralities are potentially important bridging actors or brokers in the network. Third, closeness centrality identifies the actors with the shortest paths (sequence of ties) to all other actors in the network, on average. Also in urban water governance studies, researchers are interested in more or less central actors, who might not only dominate the management processes, but also facilitate communication and finally the provision of (sanitation) services (Walters 2016).

At the meso-level of analysis, clustering is a relevant mechanism, as some actors will more strongly interact with others, resulting in the network being structured to some degree by denser clusters of nodes and less dense connections across these clusters. A first aspect of clustering is that it can be assessed in a bottom-up or in a top-down way. A bottom-up way would correspond to assessing clusters that are different (denser) from the rest of the network, and one can set different thresholds to identify them. By contrast, a top-down way of identifying clusters would correspond to predefining what the clusters should look like structurally. For example, one could define clusters as cliques, that is, a set of nodes in which every node is connected to every other node (Borgatti, Everett, and Johnson 2018; Wasserman and Faust 1994). These clusters could represent, e.g., groups of actors that deal with sectoral aspects of urban water governance but that lack cross-sectoral coordination ties.

Measures to describe the structural properties of a network at the macro-level of the entire network abound, and most of them correspond to some kind of aggregate or average measure of micro- or meso-level structures. First, network density corresponds to the proportion of observed ties as compared to all theoretically possible ties in the network. The measure thus assesses how many network ties there are based on the overall number of possible ties. Second, centralization of the network is an indicator to describe connectivity in the network. Based on potentially any centrality measure (see above), centralization assesses to what degree centrality measures in a network are equally distributed among nodes. The measure – if high – thus potentially also indicates whether there is a hierarchy in the network in terms of centralities, with few central actors and many less central actors. Third, reciprocity is actually a measure at the dyadic level (see below in relation to statistical network modeling), but it is often represented as an average degree across the entire network. Reciprocity indicates what proportion of ties among two actors are reciprocated, that is, if node *i* indicates a tie to node *j*, node *j* also indicates a tie to node *i*. This measure is valid only for directed networks where a tie points in one direction from A to B and/or from B to A. Information provision is a typical directed tie. When actor A provides actor B with information, this does not mean that actor B also has to share information with A. If that is the case, this would correspond to a reciprocated tie. By contrast, an undirected network would be co-participation in a meeting, where both actors at the end of a tie are supposed to equally be involved in this participation relation.

16.3.3 Statistical models for network data

The measures presented above all point to descriptive aspects of network analysis. An inferential approach to network data, that is, e.g., analyzing network ties as dependent variables influenced by a set of independent variables – is complicated by the fact that network ties cannot be assumed to be independent from each other. This somehow reflects the very nature of network approaches and data: that it represents situations where different network nodes are interrelated (Cranmer et al. 2017). In all statistical network models, "independent variables" (or covariates) usually used to explain the structure of networks in such models can be categorized into three types. First, node attributes can be important as they can detect the specific activity or popularity of given types of actors. Second, attributes of relations between two actors (actor dyads) can matter. An example for dyad-level substructures is the phenomenon of homophily. Homophily describes the effect that similar actors (from the same governance level, of the same organizational type, with similar specializations, etc.) tend to exchange information above average (McPherson, Smith-Lovin, and Cook 2001; Fischer and Sciarini 2016). Another example would be a tie of information exchange that (positively) influences a tie of support of some kind. Third, endogenous network structures independently of node or dyadic attributes matter, and they describe the core issue of statistically modeling networks: the potential dependencies of the network structure on the network structure itself. Reciprocity (the phenomenon of "tit-for-tat," that is, that an actor reciprocates a tie) or triadic closure (the phenomenon that the friend of my friend tends also to be my friend) are typical endogenous network level factors.

Different statistical approaches to network data have appeared in the recent past. There are differences to these approaches (Block et al. 2018; Cranmer et al. 2017). For example, dynamic models, such as Stochastic Actor-Oriented Models (SAOM) or Dynamic Network Actor Models (DyNAM), can take into account the evolution of networks based on network observations at different time points, and thus also model the rate of network change. Exponential Random Graph Models (ERGM) (Lusher, Koskinen, and Robins 2013) allow modeling network structures by comparing them to large sets of random networks. Temporal ERGMs (TERGM) can also model the evolution of networks over time (Leifeld, Cranmer, and Desmarais 2018). Statistical network approaches have been applied only rarely to urban water governance (Fischer, Nguyen, and Strande 2019).

16.4 Actor networks in urban water governance and new challenges

Urban water governance networks are increasingly challenged by ageing urban water infrastructures (and related treatment methods), demographic pressure on urban areas, hydro-climatic changes, and financial uncertainties. The required decisions about current and future infrastructure investments and management strongly relate to the involved actors and their interactions. We argue here that this actor network can be grasped and analyzed by SNA. We therefore study two challenges in urban water governance: decentralization and digitalization, and adopt a network perspective by studying them.

16.4.1 Decentralization

To establish more resilient and sustainable urban water management systems, decentralized water technologies could be integrated into existing centralized infrastructures, thereby creating hybrid water systems with beneficial features (Larsen et al. 2013; Hoffmann et al. 2020; (Larsen, Udert, and Lienert 2013; Hoffmann et al. 2020; Tomlinson et al. 2015). Through increased systemic variation, hybrid systems could provide more flexibility and adaptive capacities for future changes and challenges (Dunn et al. 2017) and prevent potential cascading failures of centralized systems (Buldyrev et al. 2010). Urban areas, most especially, can become catalysts for such innovative technological solutions (Wolfram 2018). Besides enabling tailored technical solutions to local contexts, decentralized technologies might create opportunities for new operational structures, engaging new public and private actors in the provision of water services that correspond to new nodes and ties in the urban water governance network, which can be analyzed through tools provided by SNA.

On the one hand, established actors (currently responsible for the management of the urban water infrastructures) might take on new tasks in regard to the decentralized water technologies, e.g., monitoring water quality standards or maintenance work. This way, the established actors remain responsible for the provision of water services, which would not significantly affect the existing governance network or their centrality within it. However, as decentralized water technologies could be located on private properties, which operators need to access, more direct relationships with the conventionally passive consumer of water services might become necessary, requiring more ties of cooperation, coordination, and information exchange to property owners, technology owners, as well as users. On the other hand, decentralized technologies might require specialized employees, potentially the development of new departments/agencies as well as the involvement of new public and private actors, which would affect the centrality of the present actors in the networks, increasing their overall complexity. For instance, we might see the development of private companies specialized in different operational services, as public authorities delegate certain services/tasks to private actors. Such new nodes in the network would likely be characterized by high betweenness centrality, as they interact with and bridge between different actors, such as property owners, technology developers and owners, users and public authorities. Furthermore, in this context, citizens might shift from consumers to co-producers of water services, as they take on a more active role in urban water governance, having more direct interactions with private and/or public actors, and thereby becoming more central in the network.

Decentralization of water infrastructure and its services might also lead to a partial decentralization of decision making, creating smaller administrative units, e.g., on neighborhood scale. Community associations and cooperatives might emerge, creating clusters on the meso-level of urban water governance networks. Such clusters could generate a platform for exchange, but they could also potentially reduce complexity for regulatory bodies (one representative instead of many individuals). A similar development can already be observed in the energy sector, where renewable energy cooperatives with extensive citizen participation have developed (Yildiz et al. 2015).

Despite the expected higher network density (more actual ties between the actors) and decreased network centralization (centrality distributed among more actors), it is very likely that public authorities (e.g., by a new or existing department or agency) will remain in some capacity responsible for the urban water infrastructures (and therefore stay central in the network), as infrastructures that provide essential services to society can be regarded as a common pool resource (Künneke and Finger 2009), e.g., by accrediting, supervising, or monitoring new actors in the field.

One important question related to decentralized systems is that of accountability: who is responsible for the decentralized water technologies in case of a malfunction, and how can we ensure the same level of access and quality of water services in an expanding actor network? Part of the solution to the issue of accountability could be intelligent, computer-based management using remote sensoring and ICT (Tomlinson et al. 2015), i.e., digitalization of urban water management.

16.4.2 Digitalization

Compared to other sectors, such as energy, digitalization in water management is still at its infancy (Knoblauch 2020). Yet, a growing hype surrounds digital technologies and the solutions they can bring to utilities in better managing the aging infrastructure, reducing leakages, improving supply, and monitoring water quality. Digital transformation is increasingly viewed as not optional but essential to manage complex sociotechnical systems, e.g., systems with interrelated components from society and technology or infrastructure. One example of such a sociotechnical system is the urban water sector and digitalization seen as one essential solution against the mounting pressure of climate change, urbanization, and population growth (Sarni et al. 2019).

The cornerstone of digital transformation is the use of sensors, big data analytics, Internet of Things (IoT), artificial intelligence, and machine learning to maximize the information needed to make better decisions and react faster in real time or even proactively to minimize the risks and social and environmental externalities. However, the use of advanced digital technologies requires enhanced data analytics and processing capacity to make sense of and exploit the large influx of data generated (Ingildsen and Olsson 2016). One of the challenging aspects thereof is to establish a certain degree of coordination among the entities producing, storing, sharing, and using the data (Manny et al. 2021b; Moy de Vitry et al. 2019). These can include private and public organizations from different sectors and levels of government. The interactions among these actors and the linkages across the horizontal and vertical dimensions of governance can be examined by SNA, providing insight into the factors leading actor networks to succeed or fail in delivering the desired outcomes concerning digital transformation.

Descriptive SNA and the assessment of the overall degree of coordination in a network already provide some insights into the degree of fragmentation of a network, and how the interaction between the utilities and other stakeholders works. In this regard, density is the simplest and at the same time one of the most prominent measures of interconnectedness among the actors in the network. It is defined as "the probability that a tie exists between any pair of randomly chosen nodes" (Borgatti et al. 2018, p. 150), thus representing the share of ties existent within a network compared to the amount of ties theoretically possible within this network. Whether higher density and interconnectedness lead to more efficiency of service delivery in the urban water system using digital tools is, however, an open empirical question.

Other measures, such as centralization or the core-periphery indicator, equally inform about how actors connect to each other when managing urban water systems, in general, and when managing them using digital tools, in particular. A core-periphery structure contains a core of highly interconnected actors, and a periphery of actors that are connected only to the center, but not among each other. In other words, the core of this network controls the majority of ties in this network. From an efficiency point of view, it might make sense to have a core of informed actors who lead the digital transformation in water governance. The periphery then typically intervenes at its convenience or to bring in innovation "from outside."

In addition to analyzing interactions among actors (i.e., human-to-human), the network approach can also be used to conceptualize and study the Internet ofThings (IoT), which refers to a set of smart things, such as sensors or devices, that collect, share, and communicate real-time data to control physical processes. IoT can have many applications fields in water sector, such as in smart irrigations systems (García et al. 2020) or for wastewater management (Verma 2020). Going a step further, studying networks of humans and things (i.e., human-to-things) can uncover how users interact with smart things in which they are no longer only consumers, but also producers of data and services (Ortiz et al. 2014). In the era of digitalization where the machines and devices are becoming increasingly capable and intelligent, network research can provide both theoretical and empirical means to design, model, and monitor interdependencies among human and nonhuman actors, e.g., through the approach of sociotechnical networks.

16.4.3 Sociotechnical networks

Both examples above illustrate mainly the potential of analyzing actor or governance networks, corresponding to the social rather than the technical aspect of urban water management. Decentralization and digitalization of urban water management, however, include both social and technical aspects. The latter can also be represented as interrelated units and thus as a network (Manny et al. 2021a). This idea of multilevel approaches to networks, including social actors but also technical elements, builds in part on the social-ecological networks approach (Bodin 2017; Bodin and Tengo 2012; Bodin et al. 2019). In this view, a more holistic network perspective relevant to urban water governance should add technical elements of urban water systems to the actor network. Similar to the concept of social-ecological systems (SES), the theory of sociotechnical systems (STS) (Geels 2004) suggests jointly examining social and technical elements and their complex sociotechnical interdependencies. For sociotechnical systems, we can analogously derive the definition of a sociotechnical fit (see Bodin and Tengo 2012; Ingold et al. 2018 for analyses of socioecological fit) that is present if the governance system aligns with the urban water system (Künneke, Ménard, and Groenewegen 2021). Such an alignment can be observed if, for example, actors are coordinating or exchanging information where they are responsible for connected or performance-dependent technical elements, e.g., pumping stations and a water treatment plant in a water system.

Here, a sociotechnical network (STN) (Elzen, Enserink, and Smit 1996; Guy, Marvin, and Medd 2011) refers to the operationalization of urban water governance as a sociotechnical system in its micro-level elements. As proposed by Manny et al. (2021a), an STN of an urban water system consists of social nodes that represent actors and technical nodes that define technical elements of urban water systems. Examples for technical elements in urban water systems are water reservoirs, pumping stations, storage tanks, or wastewater treatment plants (Loucks and Van Beek 2017). By contrast, social nodes refer to the diverse set of actors at the organizational level that are directly or indirectly linked to technical elements, i.e., involved in governing urban water systems. Such sociotechnical ties between social and technical nodes can represent various relations, for instance, responsibility or ownership of actors over technical elements. Besides these cross-level relations, a sociotechnical network further integrates relations at technical and social levels on their own. Ties between technical elements can be technical connections (e.g., water or wastewater pipes) while ties between actors create a "traditional" actor network describing the social relation of interest (e.g., collaboration, cooperation, or information exchange) (Manny et al. 2021a).

When it comes to analyzing empirical STN, adapting descriptive tools of SNA to the multilevel nature of STN may be useful. Relating to the concept of centrality, for example, it might be interesting to determine not only central or powerful actors at the social level, but also simultaneously identify central or critical technical elements to evaluate the overall sociotechnical fit (Eisenberg, Park, and Seager 2017). Figure 16.1 illustrates a simplified representation of a sociotechnical fit consisting of two related actors and two connected technical elements.

More complexity can easily be added to the description of an STN. With respect to current trends in "green infrastructures" (Lamond and Everett 2019), elements such as green roofs or retention ponds, which include both technical and ecological dimensions ("techno-ecological elements") can be incorporated. Further, relations focusing on ecological aspects, such as pol-

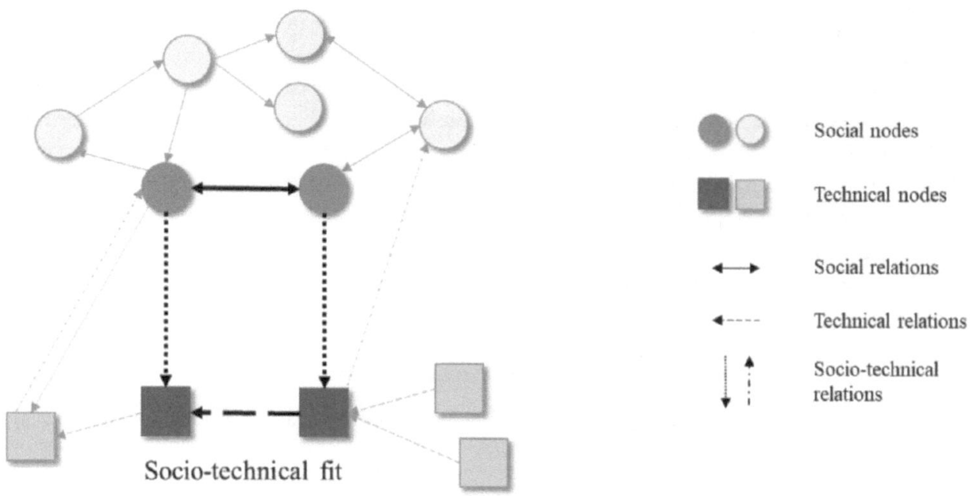

Figure 16.1 A simple representation of a sociotechnical fit (in color) consisting of two related actors and two connected technical elements within a sociotechnical network that includes social and technical nodes as well as social, technical, and sociotechnical relations.

lination or species movement, might be of interest when dealing with green infrastructures. Integrating ecological nodes and ties can consequently result in a socio-techno-ecological network. STN might also embody elements of cross-sector infrastructure systems, for example, electricity grids, gas, and telecom. Using a network perspective to analyze potential synergies between different infrastructures could be useful for innovation across sectors (Chopra and Khanna 2014). For example, studying system-overlapping technical elements, such as an electricity-producing wastewater treatment plant, can reveal how actors need to coordinate or whether entirely new (sociotechnically fitting) actor roles need to be created.

16.5 Conclusions

This chapter shows that complex urban water governance arrangements, and thus a variety of private and public actors, located at different decisional levels, can be seen as an actor network of more or less stable relations. The examples of decentralization and digitalization of urban water systems and the related challenges to and opportunities for urban water governance show that SNA can provide useful structural insights into these complex actor arrangements. SNA helps, for instance, to map the complex network of public and private actors, shows who is in the center of the network and might control information or resource flows, and who is at the periphery and might bring in innovation and novelty. The efficiency and quality of service delivery can depend on network characteristics (such as overall interconnectedness, the role of new actors, or the location of key actors), but it is ultimately an empirical question how "successful" a more or less connected network is in realizing digital or decentralized transformation.

SNA further facilitates linking both, the actor network and technical networks, in a so-called STN (sociotechnical network approach). Such an STN approach directly links to trends of decentralization and digitalization. For example, nodes representing digital technologies or ties reflecting data transfer between elements could provide insights into how digitalization affects urban water governance. STN may also prove useful in assessing sociotechnical interdependencies for urban water governance caused by an increasing number of decentralized technical elements.

Research gaps with respect to governance networks and (urban water) governance exist at the following front lines. First, whereas governance networks analyzed in the literature often tackle specific policy sectors, cross-sectoral and integrated governance across several issues or ecosystem services has been studied only more recently (Brandenberger et al. 2020; Metz, Angst, and Fischer 2020). Second, the comparative analysis of several network is still scarce, but it could be instructive in finding out what network structures lead to which kinds of outputs (Bodin et al. 2014; Victor, Montgomery, and Lubell 2017). Third, given the second point, it is also unclear how different types of instruments and institutions affect networks and, consequently, how networks can be steered in an innovative direction (Bache 2000).

References

Ahmadi, Arman, Reza Kerachian, Reyhane Rahimi, and Mohammad Javad Emami Skardi. 2019. 'Comparing and combining social network analysis and stakeholder analysis for natural resource governance', *Environmental Development*, 32: 100451.

Angst, Mario. 2020. 'Bottom-up identification of subsystems in complex governance systems', *Policy Studies Journal*, 48: 782–805.

Ansell, Chris, and Alison Gash. 2008. 'Collaborative governance in theory and practice', *Journal of Public Administration Research and Theory*, 18: 543–71.

Bache, Ian. 2000. 'Government within governance: Network steering in Yorkshire and the Humber', *Public Administration*, 78: 575–92.

Berardo, Ramiro, and Mark Lubell. 2016. 'Understanding what shapes a polycentric governance system', *Public Administration Review*, 76: 738–51.

Berardo, Ramiro, and John T Scholz. 2010. 'Self-organizing policy networks: Risk, partner selection, and cooperation in estuaries', *American Journal of Political Science*, 54: 632–49.

Berardo, Ramiro, Manuel Fischer, and Matthew Hamilton. 2020. 'Collaborative Governance and the Challenges of Network-Based Research', *The American Review of Public Administration*, 50: 898–913.

Blanco, Ismael. 2013. 'Analysing urban governance networks: Bringing regime theory back in', *Environment and Planning C: Government and Policy*, 31: 276–91.

Block, Per, Johan Koskinen, James Hollway, Christian Steglich, and Christoph Stadtfeld. 2018. 'Change we can believe in: Comparing longitudinal network models on consistency, interpretability and predictive power', *Social networks*, 52: 180–91.

Bodin, Örjan 2017. 'Collaborative environmental governance: Achieving collective action in social-ecological systems', *Science*, 357: 659.

Bodin, Örjan, and Beatrice I Crona. 2009. 'The role of social networks in natural resource governance: What relational patterns make a difference?', *Global Environmental Change*, 19: 366–74.

Bodin, Örjan, and Maria Tengo. 2012. 'Disentangling intangible social-ecological systems', *Global Environmental Change-Human and Policy Dimensions*, 22: 430–39.

Bodin, Örjan, Steven M. Alexander, Jacopo Baggio, Michele L. Barnes, Ramiro Berardo, Graeme S. Cumming, Laura E. Dee, A. P. Fischer, M. Fischer, and M. Mancilla Garcia. 2019. 'Improving network approaches to the study of complex social–ecological interdependencies', *Nature Sustainability*, 2: 551–59.

Bodin, Örjan, Beatrice Crona, Matilda Thyresson, Anna-Lea Golz, and Maria Tengö. 2014. 'Conservation success as a function of good alignment of social and ecological structures and processes', *Conservation Biology*, 28: 1371–79.

Borgatti, Stephen P., Martin G. Everett, and Jeffrey C. Johnson. 2018. *Analyzing Social Networks* (Sage).

Borgatti, Stephen P., Ajay Mehra, Daniel J. Brass, and Giuseppe Labianca. 2009. 'Network analysis in the social sciences', *Science*, 323: 892–95.

Borgatti, S. P., Everett, M. G., & Johnson, J. C. (2018). *Analyzing social networks*. Sage, Thousand Oaks.

Brandenberger, Laurence, Karin Ingold, Manuel Fischer, Isabelle Schläpfer, and Philip Leifeld. 2020. 'Boundary spanning through engagement of policy actors in multiple issues', *Policy Studies Journal*, 50, 35–64.

Brenner, Neil. 2004. *New State Spaces: Urban Governance and the Rescaling of Statehood* (Oxford University Press).

Buldyrev, Sergey V, Roni Parshani, Gerald Paul, H Eugene Stanley, and Shlomo Havlin. 2010. 'Catastrophic cascade of failures in interdependent networks', *Nature*, 464: 1025–28.

Chopra, Shauhrat S, and Vikas Khanna. 2014. 'Understanding resilience in industrial symbiosis networks: Insights from network analysis', *Journal of Environmental Management*, 141: 86–94.

Cranmer, Skyler J., Philip Leifeld, Scott D. McClurg, and Meredith Rolfe. 2017. 'Navigating the range of statistical tools for inferential network analysis', *American Journal of Political Science*, 61: 237–51.

Dolowitz, D. P., & Marsh, D. (2000). Learning from abroad: The role of policy transfer in contemporary policy-making. *Governance*, *13*(1): 5–23.

Dunn, G., R. R. Brown, J. J. Bos, and K. Bakker. 2017. 'Standing on the shoulders of giants: Understanding changes in urban water practice through the lens of complexity science', *Urban Water Journal*, 14: 758–67.

Eisenberg, Daniel A, Jeryang Park, and Thomas P Seager. 2017. 'Sociotechnical network analysis for power grid resilience in South Korea', *Complexity*, 15(4), 28.

Elzen, Boelie, Bert Enserink, and Wim A Smit. 1996. 'Socio-technical networks: How a technology studies approach may help to solve problems related to technical change', *Social Studies of Science*, 26: 95–141.

Emerson, Kirk, Tina Nabatchi, and Stephen Balogh. 2012. 'An integrative framework for collaborative governance', *Journal of Public Administration Research and Theory*, 22: 1–29.

Ernstson, Henrik, Stephan Barthel, Erik Andersson, and Sara T. Borgström. 2010. 'Scale-crossing brokers and network governance of urban ecosystem services: The case of Stockholm', *Ecology and Society*, 15.

Fischer, Manuel, and Karin Ingold. 2020. *Networks in Water Governance* (Springer).

Fischer, Manuel, and Nicolas W. Jager. 2020. 'How policy-specific factors influence horizontal cooperation among subnational governments: Evidence from the Swiss water sector', *Publius: The Journal of Federalism*, 50(4), 645–671.

Fischer, Manuel, and Philip Leifeld. 2015. 'Policy forums: Why do they exist and what are they used for?', *Policy Sciences*, 48: 363–82.

Fischer, Manuel, and Pascal Sciarini. 2016. 'Drivers of collaboration in political decision making: A cross-sector perspective', *The Journal of Politics*, 78: 63–74.

Fischer, Manuel, Mario Angst, and Simon Maag. 2019. 'Co-participation in the Swiss water forum network', *International Journal of Water Resources Development*, 35: 446–64.

Fischer, Manuel, Mi Nguyen, and Linda Strande. 2019. 'Context matters: Horizontal and hierarchical network governance structures in Vietnam's sanitation sector', *Ecology and Society*, 24(3), 17.

Freeman, Linton C. 1978. 'Centrality in social networks conceptual clarification', *Social Networks*, 1: 215–39.

Friendly, Abigail, and Kristine Stiphany. 2019. 'Paradigm or paradox? The 'cumbersome impasse'of the participatory turn in Brazilian urban planning', *Urban Studies*, 56: 271–87.

García, Laura, Lorena Parra, Jose M Jimenez, Jaime Lloret, and Pascal Lorenz. 2020. 'IoT-based smart irrigation systems: An overview on the recent trends on sensors and IoT systems for irrigation in precision agriculture', *Sensors*, 20: 1042.

Geels, Frank W. 2004. 'From sectoral systems of innovation to socio-technical systems: Insights about dynamics and change from sociology and institutional theory', *Research Policy*, 33: 897–920.

Guy, Simon, Simon Marvin, and Will Medd. 2011. *Shaping Urban Infrastructures: Intermediaries and the Governance of Socio-technical Networks* (Routledge).

Hayes, Adam L., and Tyler A. Scott. 2018. 'Multiplex network analysis for complex governance systems using surveys and online behavior', *Policy Studies Journal*, 46: 327–53.

Hoffmann, Sabine, Ulrike Feldmann, Peter M. Bach, Christian Binz, Megan Farrelly, Niki Frantzeskaki, Harald Hiessl, Jennifer Inauen, Tove A. Larsen, and Judit Lienert. 2020. 'A research agenda for the future of urban water management: Exploring the potential of nongrid, small-grid, and hybrid solutions', *Environmental Science & Technology*, 54: 5312–22.

Herzog, L., Ingold, K.& Schlager, E. (2022) Prescribed by law and therefore realized? Analyzing rules and their implied actor interactions as networks. *Policy Studies Journal*, 50, 366–386.

Hughes, Sara. 2019. *Repowering Cities: Governing Climate Change Mitigation in New York City, Los Angeles, and Toronto* (Cornell University Press).

Huxham, Chris, Siv Vangen, Christine Huxham, and Colin Eden. 2000. 'The challenge of collaborative governance', *Public Management an International Journal of Research and Theory*, 2: 337–58.

Iacobucci, D. (Ed.). (1996). *Networks in marketing*. Sage, Thousand Oaks.

Ingildsen, Pernille, and Gustaf Olsson. 2016. *Smart Water Utilities: Complexity Made Simple* (IWA Publishing).

Ingold, Karin, Andreas Moser, Florence Metz, Laura Herzog, Hans-Peter Bader, Ruth Scheidegger, and Christian Stamm. 2018. 'Misfit between physical affectedness and regulatory embeddedness: The case of drinking water supply along the Rhine River', *Global Environmental Change*, 48: 136–50.

Jordan, Grant, and Klaus Schubert. 1992. 'A preliminary ordering of policy network labels', *European Journal of Political Research*, 21: 7–27.

Kaufmann, David. 2019. 'Comparing urban citizenship, sanctuary cities, local bureaucratic membership, and regularizations', *Public Administration Review*, 79: 443–46.

Kaufmann, David, and Mara Sidney. 2020. 'Toward an Urban Policy Analysis: Incorporating Participation, Multilevel Governance, and "Seeing Like a City"', *PS: Political Science & Politics*, 53: 1–5.

Kenis, Patrick, and Volker Schneider. 1991. 'Policy networks and policy analysis: Scrutinizing a new analytical toolbox', in *Policy Networks: Empirical Evidence and Theoretical Considerations* (Campus Verlag).

Knoblauch, D., L. Felicetti and U. Stein 2020. *Policy Matrix: Screening Of Digital, Data And Water Policies* (Ecologic Institute).

Koppenjan, Johannes Franciscus Maria, Joop Koppenjan, and Erik-Hans Klijn. 2004. *Managing Uncertainties in Networks: A Network Approach to Problem Solving and Decision Making* (Psychology Press).

Künneke, Rolf, and Matthias Finger. 2009. 'The governance of infrastructures as common pool resources', in Workshop on the Workshop, 3–6.

Künneke, Rolf, Claude Ménard, and John Groenewegen. 2021. *Network Infrastructures: Technology Meets Institutions* (Cambridge University Press).

Lamond, Jessica, and Glyn Everett. 2019. 'Sustainable blue-green infrastructure: A social practice approach to understanding community preferences and stewardship', *Landscape and Urban Planning*, 191: 103639.

Larsen, Tove, Kai Udert, and Judit Lienert. 2013. *Source Separation and Decentralization for Wastewater Management* (Iwa Publishing).

Lazega, Emmanuel, and Tom A. B. Snijders. 2015. *Multilevel Network Analysis for the Social Sciences: Theory, Methods and Applications* (Springer).

Leifeld, Philip, and Volker Schneider. 2012. 'Information exchange in policy networks', *American Journal of Political Science*, 56: 731–44.
Leifeld, Philip, Skyler J. Cranmer, and Bruce A. Desmarais. 2018. 'Temporal exponential random graph models with btergm: Estimation and bootstrap confidence intervals', *Journal of Statistical Software*, 83(6), 1–36.
Lieberherr, Eva, and Karin Ingold. 2019. 'Actors in water governance: Barriers and bridges for coordination', *Water*, 11: 326.
Loucks, Daniel P, and Eelco Van Beek. 2017. 'Urban water systems', in *Water Resource Systems Planning and Management* (Springer).
Lubell, Mark. 2003. 'Collaborative institutions, belief-systems, and perceived policy effectiveness', *Political Research Quarterly*, 56: 309–23.
———. 2013. 'Governing institutional complexity: The ecology of games framework', *Policy Studies Journal*, 41: 537–59.
Lusher, Dean, Johan Koskinen, and Garry Robins. 2013. *Exponential random graph models for social networks: Theory, methods, and applications* (Cambridge University Press).
Maag, Simon, and Manuel Fischer. 2018. 'Why government, interest groups, and research coordinate: The different purposes of forums', *Society & Natural Resources*, 31: 1248–65.
Manny, Liliane, Mario Angst, Jörg Rieckermann, and Manuel Fischer. 2021a. 'Socio-technical networks of infrastructure management', *Journal of Environmental Management*, 54, 943–983. under review.
Manny, Liliane, Mert Duygan, Manuel Fischer, and Jörg Rieckermann. 2021b. 'Barriers to the digital transformation of infrastructure sectors', *Policy Sciences*, under review.
Margerum, Richard D. 2008. 'A typology of collaboration efforts in environmental management', *Environmental Management*, 41: 487–500.
Marin, Bernd, and Renate Mayntz. 1991. *Policy Networks: Empirical Evidence and Theoretical Considerations* (Campus Verlag).
McPherson, Miller, Lynn Smith-Lovin, and James M. Cook. 2001. 'Birds of a feather: Homophily in social networks', *Annual Review of Sociology*, 27: 415–44.
Metz, Florence, Mario Angst, and Manuel Fischer. 2020. 'Policy integration: Do laws or actors integrate issues relevant to flood risk management in Switzerland?', *Global Environmental Change*, 61: 101945.
Moy de Vitry, Matthew, Mariane Yvonne Schneider, Omar Farooq Wani, Liliane Manny, João P Leitão, and Sven Eggimann. 2019. 'Smart urban water systems: What could possibly go wrong?', *Environmental Research Letters*, 14: 081001.
Narayan, Abishek Sankara, Manuel Fischer, and Christoph Lüthi. 2020. 'Social network analysis for water, sanitation, and hygiene (WASH): Application in governance of decentralized wastewater treatment in India using a novel validation methodology', *Frontiers in Environmental Science*, 7: 198.
Ortiz, Antonio M., Dina Hussein, Soochang Park, Son N. Han, and Noel Crespi. 2014. 'The cluster between internet of things and social networks: Review and research challenges', *IEEE Internet of Things Journal*, 1: 206–15.
Sandström, Annica, and Lars Carlsson. 2008. 'The performance of policy networks: The relation between network structure and network performance', *Policy Studies Journal*, 36: 497–524.
Sarni, W., C. White, R. Webb, K. Cross, and R. Glotzbach. 2019. *Digital Water: Industry Leaders Chart the Transformation Journey* (International Water Association and Xylem Inc.).
Tomlinson, Bill, Bonnie Nardi, Donald J Patterson, Ankita Raturi, Debra Richardson, Jean-Daniel Saphores, and Dan Stokols. 2015. 'Toward alternative decentralized infrastructures', in *Proceedings of the 2015 Annual Symposium on Computing for Development*, 33–40.
Van Meerkerk, Ingmar, and Jurian Edelenbos. 2014. 'The effects of boundary spanners on trust and performance of urban governance networks: Findings from survey research on urban development projects in the Netherlands', *Policy Sciences*, 47: 3–24.
Verma, Sanjeev. 2020. 'How can the water industry benefit from IoT technology?', in *WaterOnline*.
Victor, Jennifer Nicoll, Alexander H. Montgomery, and Mark Lubell. 2016. *The Oxford Handbook of Political Networks* (Oxford University Press).
Walters, Jeffrey Paul. 2016. 'Exploring the use of social network analysis to inform exit strategies for rural water and sanitation NGOs', *Engineering Project Organization Journal*, 6: 92–103.
Wolfram, M. (2018). Urban planning and transition management: Rationalities, instruments and dialectics. In *Co-creating sustainable urban futures* (pp. 103–125). Springer, Cham.
Wasserman, Stanley, and Katherine Faust. 1994. *Social Network Analysis: Methods and Applications* (Cambridge University Press).

Wutich, Amber, Melissa Beresford, Julia C Bausch, Weston Eaton, Kathryn J Brasier, Clinton F Williams, and Sarah Porter. 2020. 'Identifying stakeholder groups in natural resource management: Comparing quantitative and qualitative social network approaches', *Society & Natural Resources*, 33: 941–48.

Yildiz, Özgür, Jens Rommel, Sarah Debor, Lars Holstenkamp, Franziska Mey, Jakob R Müller, Jörg Radtke, and Judith Rognli. 2015. 'Renewable energy cooperatives as gatekeepers or facilitators? Recent developments in Germany and a multidisciplinary research agenda', *Energy Research & Social Science*, 6: 59–73.

17

POLICY TRANSFER IN URBAN WATER MANAGEMENT

Evidence from ten BEGIN cities

Jannes Willems, Ellen Minkman, William Veerbeek, Richard Ashley, and Arwin van Buuren

17.1 Introduction

Cities across the world face the impacts of climate change and have no choice other than to adapt (Susskind, 2010). Therefore, "cities, rather than nation-states, may be the most appropriate arena in which to pursue policies to address global environmental challenges," including climate change (Betsill & Bulkeley, 2006: 143). One approach for creating climate-sensitive cities is the construction of multifunctional Blue and Green Infrastructure (BGI) that can store rainwater, but simultaneously support biodiversity, provide recreation opportunities, and stimulate public health (Ashley et al., 2018). BGI is a component of Nature-Based Solutions (NBS) that are being lauded as a more effective multifunctional infrastructure for the needs of today's and future citizens in urban areas, especially in light of the COVID-19 pandemic (NetworkNature, 2021). The delivery of BGI requires urban governments to develop new capacities and competencies, as it contrasts with the delivery of mono-functional grey infrastructure (Fletcher et al., 2015). To illustrate, the multifunctional nature of BGI means that the delivery of BGI requires multi-stakeholder involvement, in which, for example, private sector developers and community initiatives participate. To date, local governments face difficulties when designing and implementing BGI, possibly indicating that they lack capacities for developing an open, learning-based approach in which all the requisite stakeholders can participate (Frantzeskaki, 2019).

Urban governments are well aware that they are not the only ones struggling with the delivery of BGI. As a consequence, they are increasingly part of (trans)national networks in which those involved in urban water management and adjacent disciplines exchange experiences (Anguelovski & Carmin, 2011; Bulkeley, 2015). Transnational examples include the 100 Resilient Cities programme (Nielsen & Papin, 2021), the Cities for Climate Protection Programme (Zeppel, 2013), the EU Covenant of Mayors for Climate & Energy, and EU-funded networks such as Interreg (Hachmann, 2011). These networks are often voluntary and facilitate horizontal learning between local governments by exchanging knowledge concerning ideas, practices, and experiences. This learning can lead to policy transfer – defined as ideas, norms, programmes, or policy instruments that are developed based on policy examples from elsewhere

DOI: 10.4324/9781003057574-22

(Dolowitz & Marsh, 1996; Minkman et al., 2018). Policy transfer may help urban governments working on the delivery of BGI to build capacity by discovering and trying new practices.

A niche within policy transfer research concentrates on the study of city-to-city collaboration and learning, whereby cities engage in sharing knowledge, lesson drawing, and policy transfer (Haupt et al., 2020). Such transfers can take place bilaterally or in city networks. An example of bilateral policy transfer concerns the transfer of the "Water Square" concept from Rotterdam to Mexico City (Ilgen et al., 2019), while an example of policy transfer in a city network is the transfer of practical tools within the 100 Resilient Cities network (Haupt et al., 2020). These city-to-city learning networks exist across the globe and cover a wide range of policy fields.

Despite this increase in networks, the consequential policy transfer following from this has received limited attention in the field of urban water management so far. As such, we know relatively little about the effectiveness of these networks and *how* they facilitate city-to-city learning (Nielsen & Papin, 2021). Insights from the policy transfer and policy learning literature (e.g., Dunlop & Radaelli, 2013; Minkman et al., 2018) can help to overcome this knowledge gap. These bodies of literature discuss different forms of learning, their enablers and barriers, and how this can facilitate policy transfer. This chapter, therefore, aims to examine the relationship between climate adaptation networks, policy learning, and policy transfer in the field of urban water management. Our research question is: *How do networks for climate-sensitive cities facilitate policy learning, and how does this lead to policy transfer?* To answer this question, the European Interreg network BEGIN[1] (Blue-Green Infrastructure for Social Innovation) is examined. BEGIN comprises ten local governments[2] of various scales in mid-sized western European cities.

This chapter contributes to our understanding of the relationship between policy networks, policy learning, and policy transfer (van Herk et al., 2011; Bulkeley, 2015). Moreover, we provide new empirical insights from a European network. Since the participating municipalities involved in the network are spread across Europe, we are better able to take into account the context dependencies that influence policy learning and transfer. Ultimately, we are better able to understand the contribution of networks to local municipalities in helping to prepare to address the impacts of climate change.

17.2 Typology of city-to-city learning and transfer

This section outlines how learning in networks is the prime mechanism that underpins policy transfer. While policy transfer research traditionally focused on bilateral transfer between public actors, recent studies include private actors, think tanks, and transnational networks (Benson & Jordan, 2011; Stone, 2012). We understand policy transfer as "a process in which knowledge about institutions, policies or delivery systems at one sector or level of governance is used in the development of institutions, policies or delivery systems at another level of governance" (Evans, 2010: 65), and we are particularly interested in city-to-city networks, which consist of two or more cities that engage in a process of policy and knowledge exchange for a certain period of time (Ilgen et al., 2019; Shefer, 2019). City-to-city learning has been widely covered in environmental and urban water governance research (Haupt et al., 2020; Nielsen & Papin, 2021). According to van Herk et al. (2011), these networks have three functions in regard to learning: (1) the establishment of facts; (2) the creation of images; and (3) setting ambitions. Moreover, the set-up of the network can facilitate or obstruct policy learning. For example, the presence of a dense policy network, characterised by informal relations and face-to-face relations, ensures that resources can be mobilised (Vinke-De Kruijff et al., 2014). Perhaps, most importantly, in the light of city-to-city learning, networks may provide a learning platform by facilitating personal networking (Haupt & Coppola, 2019).

17.2.1 Policy learning in policy transfer

Although climate change manifests differently across places, similarities also exist and thus cities can learn from each other in ways to respond to climate change. Urban water managers may devise their own (policy) responses or they could look for ideas from elsewhere (Dolowitz & Marsh, 1996). Circulation of knowledge is key to policy transfer and key authors in the field have associated policy transfer with policy learning, policy-oriented learning, social learning, and lesson drawing (Hall, 1993; May, 1992; Rose, 1991). Moreover, exchange based on learning (instead of imitation, coercion, or promotion) is more likely to result in effective transfer (Minkman et al., 2018). In this chapter, therefore, we conceptualise policy transfer as a process that is primarily driven by learning (Stone, 2012), whereby we exclusively focus on horizontal voluntary transfer (Hertin & Berkhout, 2003). As Minkman et al. (2018) argue, the search for policies is usually bounded, as there is a specific idea of the kind of policies that one is looking for and/or the search is limited to locations with historical or trade ties, locations that are geographically close, or locations that brand their approach well. In addition, extensive policy evaluation, or piloting of transferred ideas, is an essential part of the process in order to ensure that transferred ideas are suitable for the policy problem at hand and fit in the receiving (institutional) context. This can be considered a learning process because actors learn about the characteristics and possible output of policy responses. A thorough evaluation results in well-considered decisions that receive broad domestic support, while less careful consideration may result in uninformed, inappropriate, or incomplete policy transfer (Dolowitz & Marsh, 2000). Consequently, this process takes time, causing learning to require considerable resources – including financial resources and an infrastructure to exchange knowledge – and is therefore not always attainable.

When looking at how learning takes place in networks, knowledge, ideas, and best practices are shared through various channels, including but not limited to meetings, newsletters, site visits, informal contacts, and conferences (Wolman, 2009). The networks have in common a multi-stakeholder and learning aspect (van Herk et al., 2011), but the issue at stake can have different interpretations and the types of stakeholders involved may differ between networks. Moreover, the learning process is dynamic and any assessment of which type of learning occurs is only a snapshot. Time is thus an important dimension of policy transfer and policy learning (Dool & Schaap, 2020). In addition, networks can facilitate learning, but the presence of a network alone will not lead to learning or transfer. Betsill and Bulkeley (2006) show, for instance, the difficulty of moving from the rhetorical commitment to effective new/improved policies and programmes. As a result, policy transfer theory alone cannot explain *how* networks contribute to learning.

Following Dunlop and Radaelli (2018), four forms of learning with their own merits and limitations can be distinguished (Table 17.1). The four types of learning are positioned in quadrants along two axes: problem tractability and certification of actors (Figure 17.1). Highly tractable problems are those for which "society (often public administration) has a repertoire of solutions, algorithms, or ways of doing things" (Dunlop, 2017: 261). In cases where there is a recognised knowledge authority, i.e., high certification of actors, learning will take place "in the shadow of hierarchy." Learning in hierarchy means that someone is being obeyed, most notably an authority that more or less imposes a set of standards on others, resulting in monitorable lessons to be transferred. This kind of learning is not likely to emerge in horizontal learning networks – as is the case for BEGIN – because participating cities have no authority over the decisions of the others. Similar to hierarchical learning, learning as a by-product of bargaining occurs in situations where problems are highly tractable. However, when no actor is qualified to make decisions, actors will start to negotiate with each other. This kind of learning results in

Table 17.1 Overview of characteristics of the types of learning based on Dunlop (2017), Dunlop and Radaelli (2018), and Minkman et al. (2018)

Type	*Reflexive*	*Epistemic*	*Bargaining*	*Hierarchy*
Characteristics/ facilitators	Individual open to change preferences; problem needs to be agreed on	Clear knowledge authority; policymaker open to suggestions	Dependency between state and non-state actors; no clear leader	Clear leadership
Barriers	Domination of certain arguments; lack of deliberative culture	Fragmentation of epistemic community; limited policy capacity	Lack of trust; presence of dogmas	Veto players
Learning by	All actors in the network	Policymakers	All actors in network (state + non-state)	Subordinates/ less powerful actors in the network
Type of lesson transferred	Lessons about social norms	Lessons about policy-relevant causal explanations	Lessons about preferences; cost of collaboration	Set of rules
Result	Agreement on problem (and solution)	Best practice shared by expert	Negotiation about way forward	Monitorable standards and norms

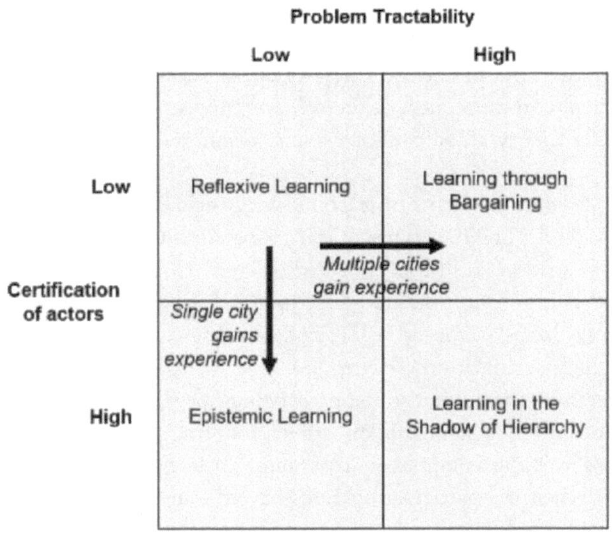

Figure 17.1 Typology of learning (adapted from Dunlop & Radaelli, 2018). Arrows indicate the anticipated changes over time of the network.

lessons about preferences and the costs of certain policy responses. However, both bargaining and learning in a hierarchy may take place within a city network (i.e., among actors involved within a city) rather than in a network of cities, for example, when a municipality adds BGI as a requirement for issuing building permits.

Contrary to bargaining and hierarchical learning, we argue that horizontal networks will form around issues with low tractability, i.e., where there is no established way of dealing with a certain issue yet. Problems with low tractability are characterised by high uncertainty and disagreement over goals and solutions (Leong, 2015). In the case of climate change adaptation in the BEGIN city-to-city learning network, we argue that the initial situation will be that of reflexive learning: all actors have the same level of experience and knowledge (namely limited) and they face problems for which no standard set of solutions is yet available. Actors start identifying the problem before exploring potential solutions. Reflexive learning thus sets requirements for the individuals involved and the institutional setting in which learning takes place. Individuals should join the process without predispositions and should be willing to change their preferences (Dunlop & Radaelli, 2018; Minkman et al., 2018). At the same time, the process design (i.e., roles in the network, way of interaction) should support this socialisation (Dunlop & Radaelli, 2018). Over time, some cities will develop experience with certain aspects of the problem. In some cases, this may result in epistemic learning, whereby experienced cities act as knowledge sources for cities with less experience. In the other cases, for example where multiple cities gain experience with a topic but in a highly different way, we expect learning to take the shape of a bargaining process. In these cases, the problem of tractability increases as multiple cities have developed their own way of working. However, the nature of the problem may vary between different contexts, which may result in diverging understandings of the problem and therewith diverging policy responses. In such a situation, no single authoritative source of knowledge emerges. Instead, multiple experts present their approach as "the" solution for all cities in the network. This results in negotiations about the way forward. See the arrows in Figure 17.1 for a visualisation of these dynamics.

17.2.2 Result of policy transfer and learning

Policy transfer can have several outcomes, including one-on-one adoption of policies, adaptation of ideas to fit the local context, and inspiration whereby elements of multiple policies are combined (Minkman et al., 2018; Rose, 1991). While the similarity of the original and new practices will decrease from adoption to inspiration, the degree of learning will increase. Where policy transfer analysis focuses on indicators that particular and tangible practices have been adopted, learning is an open and unstructured process and actors may engage in different learning processes at the same time (Shefer, 2019). When present, learning will result in the updating of beliefs based on experiences or social interaction (Dunlop & Radaelli, 2013). Nevertheless, learning should not be conceptualised as a binary variable that is either present or absent. Learning takes different shapes and several typologies exist. Following Minkman et al. (2018), the strength of learning depends on the process design, which requires attention throughout the exchange. Thus, process design and available resources are key for learning. A final cautionary note should be made that learning does not automatically equate to transfer. Learning may take place without policy transfer being observable. Learning may consist of negative lesson drawing, whereby the receiver has learned but decided not to adopt an external idea. Furthermore, other factors, such as dominant institutional players, may prevent learning from producing a significant impact on policy or governance structures in the receiving context (Shefer, 2019).

17.3 Methodology

17.3.1 Introduction to the case study

Local governments are increasingly operating in transnational networks to exchange experiences for creating climate-sensitive cities, of which policy transfer could be one outcome. An institutionalised form is the Interreg Community Initiative,[3] launched in 1990 by the European Union (Hachmann, 2011). As limited research exists on the actual policy transfer within networks, we expect that longer-running networks (such as Interreg) will provide more opportunities for transfer and, hence, that we are able to witness policy transfer in our case study of the North Sea Region Interreg project BEGIN. To illustrate, the BEGIN project builds further on the Interreg projects MARE, SKiNT, and FRC, which have aimed to develop learning and action alliances for professional stakeholders in flood risk management.

The Interreg project BEGIN (Blue-Green Infrastructure through Social Innovation; started in 2017 and operating in the North Sea region) consists of a consortium of ten local governments of mid-sized towns and boroughs, complemented by six knowledge institutes and consultancies. Each government participates with one specific case study. Some of the representations from the cities in BEGIN are rooted in flood management (Aberdeen, Bergen, Enfield, Gothenburg, Hamburg, and Kent), others in urban planning (Antwerp, Bradford, and Dordrecht) or environmental management (Ghent). City-to-city learning is a distinct element of the project, facilitated by a dedicated team (external to the governments). The premise is that knowledge should not occur merely accidentally, but become part of a more structured approach in which cities work together on common ambitions and responsibilities. Clustering the local governments into groups with common project characteristics and organising dedicated city-to-city learning meetings are part of the structure. An important assumption in BEGIN is that partner cities act as learning organisations that dedicate time and resources to discuss ideas and develop similar, yet tailored, approaches for tackling comparable challenges. The city-to-city learning approach is more targeted compared with other urban water management networks (e.g., the Rockefeller Foundation's 100 Resilient Cities or UNIDSDR's Resilient Cities Connect). Nevertheless, it is more loosely coupled than learning and action alliances, in which participating actors usually have already formulated a shared agenda (Haupt et al., 2020).

17.3.2 Data collection and analysis

To understand the policy learning taking place in the BEGIN network, we collected two streams of data. First, city partners filled in a questionnaire in 2016 and 2019 consisting of statements on the topics of participatory BGI development, landscape and urban design, BGI-driven business cases, cross-departmental cooperation and policy integration, and monitoring and maintenance. The statements were presented as a self-evaluation tool in which participants assessed to what extent they have developed new capacities and competencies required for delivering BGI (Fletcher et al., 2015; Frantzeskaki, 2019). The 10-criteria used to assess a city's maturation in incorporating BGI are adapted from the 10 essentials for making cities resilient (Haigh & Amaratunga, 2012; UNISDR, 2017) and the 17 principles for water-wise cities (IWA, 2016). While both lists cover a wide range of aspects, including urban design, financing, citizen participation, and governance, criteria for assessing the integration of operational monitoring and evaluation, which is common practice in the asset management of traditional grey water infrastructure, were missing. Subsequently, such criteria have been added to the survey used in the BEGIN project.

Each criterion is covered by two statements that reflect facilitation, provision of an enabling environment, and full integration into the practices. For instance, the use of economic valuation tools for monetising the co-benefits of BGI is considered the enabling factor for a complete assessment of the cost and benefits of BGI projects, while the ability to develop BGI-based viable business and financing cases would represent a further integration of a financial framework for developing BGI. The criteria have been assessed based on ordinal data derived from surveys across the BEGIN partner cities in 2016 and 2019, and they included participants from different departments involved in local BGI development. Values reflect the response per city based on a semi-quantitative scale between 0 (completely disagree) and 5 (completely agree).

Outcomes explain only partially learning processes in each city during BEGIN; although they also explain how cities perceive themselves compared with others. Yet, these outcomes do provide some insight into the evolution of learning during BEGIN. First, even after the first project meetings, cities were well aware of the state and progress each city had made in BGI-oriented projects. Second, the first questionnaire reviewed the range of city-to-city learning activities (such as workshops and field trips), in which at least three city partners joined. Some of the co-authors of this chapter hosted these events and we have used the field notes to complement the findings from the questionnaires.

The data analysis centred on three clusters of local governments,[4] based on their project scope. For each cluster, we defined the problem tractability and certification of actors at the initial stage, using the questionnaire and accompanying discussions from 2016. Subsequently, we used the field notes of the workshops to understand what type of learning took place, for which we distinguished between four types of learning (Table 17.1). As a final step, the changes in statements after the questionnaire from 2019 were used to identify alterations in the problem tractability and certification of actors, and what type of policy transfer took place (inspiration, adaptation, or adoption).

17.4 Results

The results section discusses three clusters of cities and traces back the learning process in each city (cf. Figure 17.1). Subsequently, the three clusters and learning processes are compared with each other.

17.4.1 Cluster 1 (Antwerp, Bergen, Bradford, Ghent, Dordrecht): Strategic projects of linear development along strips driven by urban redevelopment

Cluster 1 consists of the five municipalities that are planning major urban redevelopment. Municipalities are either transforming industrial into residential areas (Bergen) or creating or enhancing new green axes that run through their municipality (Antwerp, Bradford, Ghent, Dordrecht). The construction of blue and green infrastructure is a central element in the municipalities' plans: it is either the "glue" that connects different ambitions or it is an important subgoal of the project. To illustrate, Antwerp City Council aims to upgrade the left banks of the Scheldt River, in which BGI embodies nature development, flood protection, recreational opportunities, and economic development. In Bergen, climate adaptation is an important subgoal in the realisation of new neighbourhoods by creating space for BGI in public areas.

The average scores for Cluster 1 based on the questionnaire seem to show an almost stable perception of abilities (Figure 17.2). In all cases, municipalities lacked the experience of when and how to integrate BGI into their projects because of the complexity of the projects in which

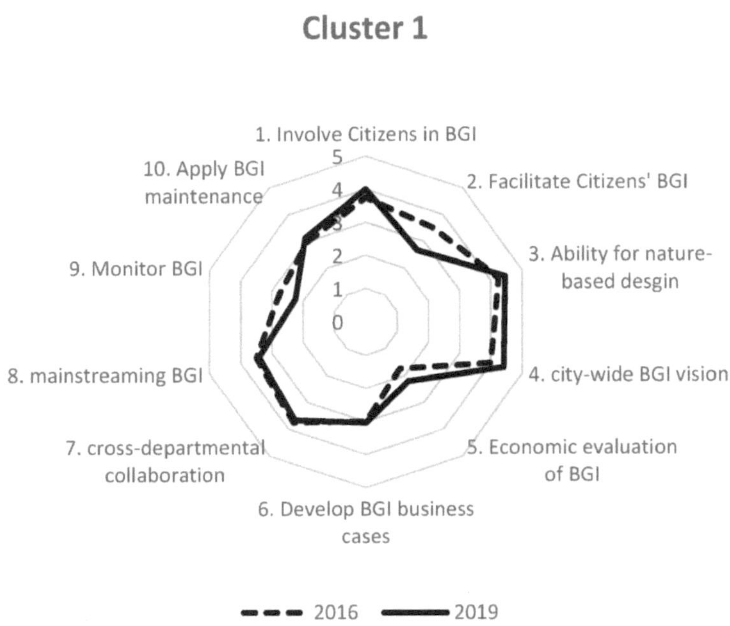

Figure 17.2 Average scores self-reflection 2016–2019 of Cluster 1.

each city is engaged. The primary challenge for each city is to embrace the multi-functionality of BGI (e.g., climate adaptation, liveability, social engagement, etc.), while still acknowledging that most of these projects are driven by housing or other developments.

As these projects were often in the early stages, the learning could be identified as reflexive learning, in which actors were making sense of integrative urban redevelopment projects, including BGI. The workshops organised for this cluster highlighted that the projects did not focus on urban drainage. Although the aim is to develop or enhance a blue-green corridor, with an emphasis on enhancing liveability and urban ecology, in three cities a substantial housing programme is realised along the green axes in Bradford, Bergen, and Antwerp that helps to finance the developments. When such a driver is absent, this could be an obstacle in developing BGI. The development of a blue-green axis-based narrative was initially proposed by Bradford. In this case, the Beck River creates a natural backbone along which different projects are located (highway, housing, park development, industrial restructuring). Other cities later adopted this approach. For example, Dordrecht developed a functional blue-green axis for sports, outdoor activities, and recreation. While the Bergen project was initially driven by a light rail development, the construction of a new urban river will actually become the lifeline of the project, which should act as a catalyst for broader urban development. Ghent also realised that a thematic development was needed, which would serve as a recognisable carrier along which projects could create a common denominator.

Overall, learning in this cluster concentrated on the need for a vision and integration with other objectives, which eventually resulted in policy adoption of the green-axis approach between cities. Because of significant problem tractability and better-certified actors, reflexive learning constituted a combination of epistemic learning and bargaining. The cluster of cities turned into a "community of practice" in which experiences were exchanged twice a year. In addition, the city partners in this cluster started to organise workshops dedicated to specific topics with which they were struggling, and they invited other city partners that were considered frontrunners. To illustrate, the city of Bradford has a multidisciplinary team led by a landscape architect. Other cities started to adopt such a team. For example, the city of Bergen invited the city architect to the planning of the Mindemyren project, even though the project was already three years underway.

17.4.2 Cluster 2 (Aberdeen, Enfield, Hamburg, Kent): Small-scale "applied" projects driven by drainage issues

Smaller cases of applied urban drainage projects are the focus of the second cluster of municipalities, in which nature-based solutions are preferred for flood risk management. The projects in Hamburg, Kent, and Aberdeen all focus on small-scale sustainable drainage systems (SuDS). The orientation is therefore engineering-driven and the city representatives in BEGIN share a similar professional culture and use similar valuation tools.

This cluster comprises teams of predominantly urban drainage experts with different levels of experience in the integration of nature-based solutions, ranging from extensive experience (Enfield) to limited (Hamburg). The municipalities focused their learning process on developing business cases for nature-based solutions, including the monetisation of co-benefits. This was largely driven by partners from the United Kingdom, which were looking for alternative funding sources to complement the limited public budgets. The cities involved started to invite knowledge partners within the BEGIN consortium to support them in monetising co-benefits, suggesting that the problem tractability was high but the certification of actors low. Technical partners, such as CIRIA, organised workshops dedicated to explaining valuation tools, such as B£ST and TEEB, and specialist consultants were appointed to carry out the valuations for Kent. One of the outcomes was that the municipalities realised that the magnitude of the co-benefits was important, but equally so were the associated categories of the co-benefits. To illustrate, increased property value as a co-benefit of BGI implies that existing property owners can be targeted to support BGI-based designs since they are the primary beneficiary.

Over time, the certification of actors increased, especially after exploratory applications of the tools, and learning took place in the shadow of hierarchy to exchange best practices. Those involved in this cluster may feel they gained the most from the project because of the hands-on knowledge and results they have obtained. This is reflected in the self-evaluation (Figure 17.3), where a clear progression can be observed. Policy transfer between the cities, therefore, resulted in inspiration, initially from technical partners.

The learning process was facilitated by the common language and perception because of the technical-financial orientation. Also, the involvement of the technical partners enhanced the learning process considerably. Yet, the orientation on drainage has also hindered the learning process: municipalities may perceive BGI from a limited perspective with a single focus. For example, especially the Hamburg case was perceived as somewhat trivial compared to the scale of other sustainability-driven urban redevelopment projects in the city. The case consisted of small flood retention basins to limit the flow of stormwater runoff from residential streets into adjacent housing. The limited scope allowed smooth progress for the project, but other city partners questioned whether the projects would lead to a broad and significant change in practices and policies. Although the projects may be ready for replication, scaling up will be a major challenge.

17.4.3 Cluster 3 (Gothenburg, Dordrecht, Ghent): City-wide approaches to public participation in BGI projects

The third cluster of cities brought together three municipalities that aimed to involve citizens as much as possible in the design, delivery, and maintenance of BGI through innovative

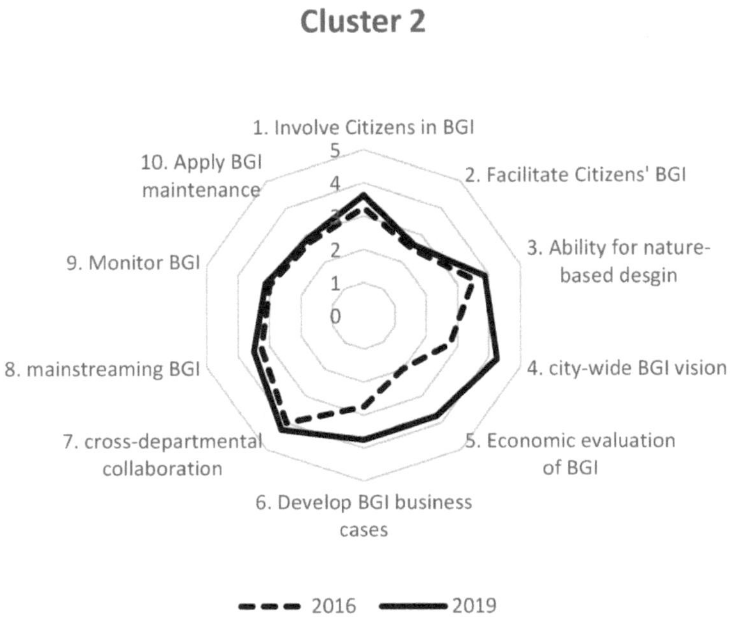

Figure 17.3 Average scores self-reflection 2016–2019 of Cluster 2.

participatory approaches. For instance, Gothenburg City Council handed over responsibilities to local residents in the development of the Frihamnen neighbourhood. Citizens could vote on which developments they preferred in this area (e.g., a sauna), and they were involved in the subsequent construction of it, facilitated by artists and a small project team. Some cities have limited experience yet with higher forms of participation (Dordrecht); others have ample experience yet gained in different policy domains (Ghent, Gothenburg). Therefore, the problem tractability is low and the certification of actors is mixed.

Given the experience in each city, the learning process was difficult to get off the ground. First, Ghent has a long tradition in social outreach, for example with so-called play streets, namely, streets that are closed to car traffic during holidays and are "handed over" to the neighbourhood to organise events, playgrounds for children, etc. Nevertheless, Ghent has not yet been able to adapt these types of initiatives to city greening projects that would coincide with a BGI agenda. Second, Gothenburg has ample experience with place-making approaches, which it successfully applied to create value and identity in the redevelopment of Frihamnen (a former port area transformed into a residential neighbourhood). While the area has become an example of small-scale initiatives that were co-developed with local residents (e.g., a sauna and outdoor classroom), the actual adoption of BGI-related principles is missing. Third, Dordrecht has been experimenting with citizen-driven bottom-up social initiatives to strengthen social cohesion in neighbourhoods with a lower socioeconomic status. Yet, instead of trying to bring citizens into BGI projects, Dordrecht attempted to add a BGI agenda to running social projects using enhanced liveability as the main co-benefit. This means that playgrounds and green zones are developed together with the previously mentioned projects. Because of the experience in each city, the self-assessment shows high scores on citizen involvement (Figure 17.4). The cities, therefore, seemed to rely on their own experiences rather than those of the other cities.

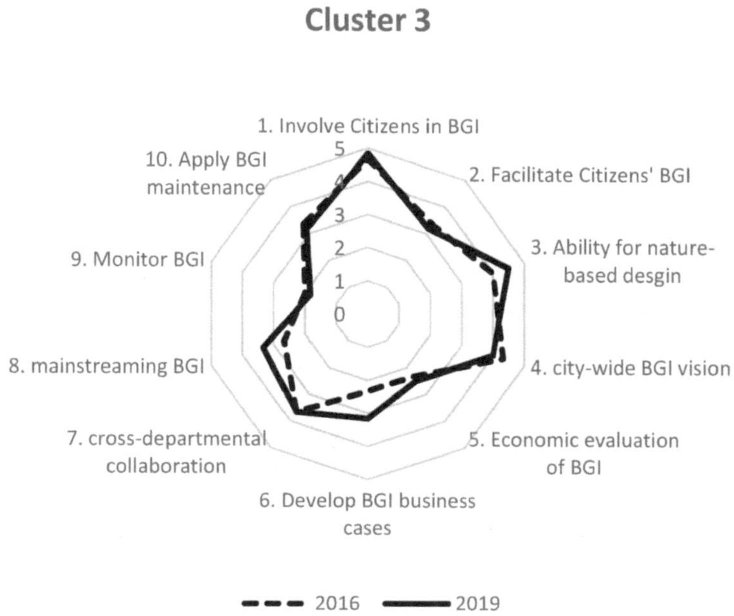

Figure 17.4 Average scores self-reflection 2016–2019 of Cluster 3.

The learning process was further complicated by the lack of urgency and undefined ambitions. City partners were more concerned with defining the scope of the project (see Cluster 1) before considering participation. Therefore, facilitating citizens in co-developing BGI has scored significantly lower in the self-assessment. Only Dordrecht and Gothenburg exchanged experiences, for example in regard to an open city lab. Based on the place-making experiences in Gothenburg, Dordrecht developed a studio space in the centre of the city where professionals and citizens could work together on urban climate adaptation and BGI, while fostering multi-stakeholder engagement in the development of requirements, design, and execution of BGI projects. In short, Dordrecht tried to actively incorporate features they observed in other city partners, and it could, therefore, be regarded as the most successful learner in the project. More than for the urban planning or engineering-driven projects above, social innovation seems a very localised, culturally dependent issue that is difficult to transfer. The learning process showed instances of reflexive learning, but problem tractability and certification of actors remained quite low. Altogether, limited policy transfer between municipalities occurred, or inspiration at best.

17.4.4 Comparison of clusters

Overall, the BEGIN network has facilitated learning seen in higher degrees of problem tractability and certification of actors. In Cluster 1 (BGI in urban redevelopment), reflexive learning shifted towards more epistemic learning ("a community of practice") and bargaining (specific city-to-city workshops). Policy transfer is seen in the adoption of approaches from one city to another, for example, concerning a shared narrative. Cluster 2 (urban drainage) started a learning process in which especially insights from technical partners were used by the municipalities, for example concerning the valuation of the co-benefits of BGI projects. Thus, policy transfer was predominantly inspiration-based marked by small-scale testing of the new tools (adoption). In

Cluster 3, reflexive learning never really took off since the local conditions for participation and social innovation were considered too context-sensitive. Despite the fact that some city partners identified themselves as frontrunners (e.g., Ghent and Gothenburg), no real policy transfer happened as, at most, only some elements of the other municipalities' approach were adopted.

As a result, our findings demonstrate instances of learning in regard to both problem tractability and certification of actors. The BEGIN network facilitated a process of joint sensemaking, in which common features of the projects were identified and experiences were exchanged. This led to not only a better definition of the issue at stake in each city, but also a diminishing sense of "uniqueness" of the project. Learning was hindered by both too narrow problem definitions (Cluster 2) and too broad (vague) problem definitions (Clusters 1 and 3). Operationalising the issue as primarily an urban drainage issue in Cluster 2 obstructed the breadth of learning, while the broad orientation of Clusters 1 and 3 (as requirements were not set in stone) made it challenging to move forward.

The learning process was heavily influenced by the resources available, of which we will discuss three: time, finances, and human capacity. Regarding time, the network demonstrated that it takes approximately two years to get to a shared understanding and a level of maturity. This makes it more difficult to get to real learning outcomes, taking into account the temporary (e.g., four-year) nature of such networks and given that these networks exist in parallel to the daily business of the participants (Dool & Schaap, 2020). The multidisciplinary nature of BGI makes learning more challenging: the issue is not confined to urban drainage solely but can relate to valuing benefits, participation, governance, and visioning. Stakeholders have to agree upon these matters, which require investments. Regarding financial resources, horizontal policy networks such as BEGIN provide additional finances to municipalities to foster learning between localities. However, many city partners are using these resources directly for their pilot projects, which suggests that learning is not considered a priority. For example, learning has not been defined as a deliverable for the local governments in obtaining European subsidies. This is further complicated by the fact that municipalities can switch between their case studies, which disturbs the learning process. The facilitators of the networks, meanwhile, lack the authority to demand that resources be invested in the learning process. This relates to the third and final point of human capacity. Most cities are not equipped to invest time in learning, so, for example, the frequency of workshops was scaled down and only a subset of the workshops was executed. Furthermore, setting up a cross-disciplinary team that participates in the learning process may benefit policy transfer. Some municipalities, such as Bradford and Bergen, are well set up for this and took much from the city-to-city learning network, but many others are not. Although networks would allow for deeper reflections, city partners often followed an instrumental rationale: learning had to directly benefit the implementation of the project, or it was otherwise discarded.

17.5 Conclusions

This chapter examined how networks for climate-sensitive cities facilitate policy learning and how this leads to policy transfer. To date, little is known about the effectiveness of these networks and *how* they facilitate city-to-city learning (Haupt et al., 2020; Nielsen & Papin, 2021). By using insights from the policy transfer and policy learning literature, we aimed to examine the relationship between climate adaptation networks, policy learning, and policy transfer in the field of urban water management. Our research question was: *How do networks for climate-sensitive cities facilitate policy learning, and how does this lead to policy transfer?*

Our findings demonstrate that the three clusters of city partners have moved towards a higher degree of problem tractability and certification of actors. Concerning problem tractability, the horizontal network enabled a process of joint sensemaking in which common themes regarding blue- green infrastructure could be defined. As such, the network facilitated capacity building in the cities, leading to a better understanding of issues and solutions (inspiration). Concerning actor certification, the joint sensemaking resulted in the identification of city partners as front-runners for specific themes. The subsequent city-to-city workshops with a few city partners as "problem owner" and a city partner or knowledge institute as "expert" led to tangible improvements in cities (adoption). Altogether, the BEGIN network has facilitated policy transfer in terms of both capacity building and tangible output.

The learning process was hampered by four main factors. First, resources available for learning are not necessarily invested in the actual learning process. Second, the multidisciplinary nature of BGI offers many opportunities for learning, yet it can also create an image among city partners that other parties cannot contribute to their unique project. Third, the learning network lacks authority or mandate, as it is supportive of the projects of the city partners. The investments in learning that city partners are required to make do not necessarily pay off directly, which can result in only partially committed partners. This complicates the initial joint sensemaking, as city partners will experience the benefits only at a later stage. Fourth, some of the cases that the cities brought in did not always lend themselves well to learn from. Ideally, the cases would appeal to all partners participating in the network.

Our findings confirm previous literature on policy transfer that underscores the need for abundant resources alongside prioritisation of learning. However, previous research suggests that city-to-city learning networks could make resources available that enhance policy learning and transfer, but our findings demonstrate the difficulty of accessing these resources. The institutionalised learning network of Interreg would allow for this, yet commitment and authority still have to be found and the absence of clear learning deliverables made learning not a priority. This implies that in four-year learning networks, almost half of the time has to be invested in getting to a shared sense of understanding. Our findings, however, underscore the need for a shared sense of understanding for BGI. The multidisciplinary nature of BGI has profound implications for urban water management, for example, in regard to urban development (Cluster 1), valuing of co-benefits (Cluster 2), or public participation (Cluster 3). City-to-city learning networks could reveal the shared challenges and approaches, for example, through self-assessment, which helps to enable policy transfer between localities.

This chapter has provided some insights into the effectiveness of networks for climate-sensitive cities and *how* they facilitate city-to-city learning. Future research could compare different forms of networks, for example, based on geographical scale (e.g., transnational versus local networks) or the set-up of networks (Haupt & Coppola, 2019). Also, the issue of time could be met with more longitudinal studies, in which networks are followed for longer periods of time (e.g., 5 to 10 years). For city-to-city learning networks, we offer two recommendations. First, learning could be placed closer to the heart of the network by defining, for example, better learning outcomes (deliverables) and by assigning more authority to learning facilitators. Participating in the network, then, will come with more obligations, which prevents city partners from shying away from challenging themselves (as seen in Cluster 2 in our case study). Second, by defining more upfront the learning outcomes, actors will be certified as knowledgeable, so the "problem owner" becomes more quickly identified (advantageous in particular for Cluster 1). Put differently, the learning process as discussed in the theoretical framework (Figure 17.1) will start at an earlier stage, i.e., already before the city-to-city network is formed, so a higher learning curve can be expected. It also forces cities to define their issue better beforehand, which hindered the learning process in Cluster 3. These suggestions can be taken up by future Interreg projects and related city-to-city learning networks.

Notes

1 See https://northsearegion.eu/begin/.
2 For reasons of readability we will refer to network members as "cities" and to BEGIN as a city-to-city network, even though there are various urban authorities involved. For example, there are suburbs (Enfield), counties (Kent), and metropolitan areas (Bradford) engaged in BEGIN.
3 https://www.interregeurope.eu
4 Governments can be part of multiple clusters.

References

Anguelovski, I., & Carmin, J. (2011). Something borrowed, everything new: Innovation and institutionalization in urban climate governance. *Current Opinion in Environmental Sustainability*, *3*(3), 169–175.

Ashley R. M., Digman C J., Horton B., Gersonius B., Smith B., Shaffer P., Baylis A. (2018). Evaluating the longer term benefits of sustainable drainage. *Proceedings of the Institution of Civil Engineers Water Management*, *171*(April 2018, WM2), 57–66.

Benson, D., & Jordan, A. (2011). What have we learned from policy transfer research? Dolowitz and Marsh revisited. *Political Studies Review*, *9*(3), 366–378. https://doi.org/10.1111/j.1478-9302.2011.00240.x

Betsill, M. M., & Bulkeley, H. (2006). Cities and the multilevel governance of global climate change. *Global Governance: A Review of Multilateralism and International Organizations*, *12*(2), 141–160.

Bulkeley, H. (2015). Can cities realise their climate potential? Reflections on COP21 Paris and beyond. *Local Environment*, *20*(11), 1405–1409.

Dolowitz, D., & Marsh, D. (1996). Who learns what from whom: A review of the policy transfer literature. *Political Studies*, *44*(2), 343–357.

Dool, L. van den, & Schaap, L. (2020). Urban network learning: Conclusions. In L. van den Dool (Ed.), *Strategies for Urban Network Learning International Practices and Theoretical Reflections*. Palgrave Macmillan, London.

Dunlop, C. A. (2017). Policy learning and policy failure: Definitions, dimensions and intersections. *Policy & Politics*, *45*(1), 3–18.

Dunlop, C. A., & Radaelli, C. M. (2013). Systematizing policy learning: From monoliths to dimensions. *Political Studies*, *61*(6), 599–619.

Dunlop, C. A., & Radaelli, C. M. (2018). The lessons of policy learning: Types, triggers, hindrances and pathologies. *Policy & Politics*, *46*(2), 255–272.

Evans, M. (2010). Public management in the postmodern era series. In W. Parsons (Ed.), *Public Management in the Postmodern Era Series* (pp. 64–96) Edward Elgar, Cheltenham.

Fletcher, T. D., Shuster, W., Hunt, W. F., Ashley, R., Butler, D., Arthur, S., ... & Mikkelsen, P. S. (2015). SUDS, LID, BMPs, WSUD and more–The evolution and application of terminology surrounding urban drainage. *Urban Water Journal*, *12*(7), 525–542.

Frantzeskaki, N. (2019). Seven lessons for planning nature-based solutions in cities. *Environmental Science & Policy*, *93*, 101–111.

Hachmann, V. (2011). From mutual learning to joint working: Europeanization processes in the INTERREG B programmes. *European Planning Studies*, *19*(8), 1537–1555.

Haigh, R., & Amaratunga, D. (2012). Making cities resilient. *International Journal of Disaster Resilience in the Built Environment*, *3*(2).

Hall, P. A. (1993). Policy paradigms, social learning, and the state: The case of economic policymaking in Britain. *Comparative Politics*, *25*(3), 275–296.

Haupt, W., & Coppola, A. (2019). Climate governance in transnational municipal networks: Advancing a potential agenda for analysis and typology. *International Journal of Urban Sustainable Development*, *11*(2), 123–140.

Haupt, W., Chelleri, L., van Herk, S., & Zevenbergen, C. (2020). City-to-city learning within climate city networks: Definition, significance, and challenges from a global perspective. *International Journal of Urban Sustainable Development*, *12*(2), 143–159.

Hertin, J., & Berkhout, F. (2003). Analysing institutional strategies for environmental policy integration: The case of EU enterprise policy. *Journal of Environmental Policy & Planning*, *5*(1), 39–56.

Ilgen, S., Sengers, F., & Wardekker, A. (2019). City-to-city learning for urban resilience: The case of water squares in Rotterdam and Mexico City. *Water*, *11*(5), 1–21.

IWA (2016). *The IWA Principles for Water Wise Cites*. International Water Association, London. Available from: https://iwa-network.org/publications/the-iwa-principles-for-water-wise-cities/

Leong, C. (2015). Persistently biased: The devil shift in water privatization in Jakarta. *Review of Policy Research, 32*(5), 600–621.

May, P. J. (1992). Policy learning and failure. *Journal of Public Policy, 12*(4), 331–354.

Minkman, E., Buuren, M. W. van, & Bekkers, V. J. J. M. (2018). Policy transfer routes: An evidence-based conceptual model to explain policy adoption. *Policy Studies, 39*(2), 222–250.

NetworkNature (2021). *Nature-based Solutions in Light of the Pandemic. Outcomes*. ICLEI, Freiburg. Available from: https://networknature.eu/sites/default/files/uploads/nn-s1-outcomes-website-final.pdf

Nielsen, A. B., & Papin, M. (2021). The hybrid governance of environmental transnational municipal networks: Lessons from 100 Resilient Cities. *Environment and Planning C: Politics and Space*, 39(4), 667–685.

Rose, R. (1991). What is lesson-drawing? *Journal of Public Policy, 11*(1), 3–30.

Shefer, I. (2019). Policy transfer in city-to-city cooperation: Implications for urban climate governance learning. *Journal of Environmental Policy and Planning, 21*(1), 61–75.

Stone, D. (2012). Transfer and translation of policy. *Policy Studies, 33*(6), 483–499.

Susskind, L. (2010). Responding to the risks posed by climate change: Cities have no choice but to adapt. *The Town Planning Review*, 81(3), 217–235.

UNISDR (2017). *How To Make Cities More Resilient: A Handbook For Local Government Leaders*. UNISDR, Geneva. Available from: https://www.unisdr.org/campaign/resilientcities/assets/toolkit/Handbook%20for%20local%20government%20leaders%20%5B2017%20Edition%5D_English_ed.pdf

Van Herk, S., Zevenbergen, C., Ashley, R., & Rijke, J. (2011). Learning and action alliances for the integration of flood risk management into urban planning: A new framework from empirical evidence from The Netherlands. *Environmental Science & Policy, 14*(5), 543–554.

Vinke-de Kruijf, J., Bressers, H., & Augustijn, D. C. (2014). How social learning influences further collaboration: Experiences from an international collaborative water project. *Ecology and Society, 19*(2).

Wolman, H. (2009). *Policy Transfer: What We Know About What Transfers, How It Happens, and How to Do It*. In The George Washington University Institute of Public Policy Papers (working paper 038).

Zeppel, H. (2013). The ICLEI cities for climate protection programme: Local government networks in urban climate governance. In *Climate Change and Global Policy Regimes* (pp. 217–231). Palgrave Macmillan, London.

18

RETHINKING URBAN WATER GOVERNANCE AND INFRASTRUCTURE IN EUROPE

Challenges and opportunities of regionalization and organizational autonomy

Eva Lieberherr, Frank Hüesker, and Katrin Pakizer

18.1 Introduction

Urban water governance structures the interdependencies of water and space (Schmidt, 2014). In dynamic urban contexts, the provision of water supply and sanitation entails the development of management processes and infrastructure systems that are resilient and can control hydrological variability at different spatial and temporal scales (Daniell & Barreteau, 2014). Historically, this has led to the development of large-scale infrastructures and centralized utilities, which are typically managed hierarchically by local public authorities (Romano & Akhmouch, 2019). While this sociotechnical system has physically and sociopolitically shaped urban space (Monstadt, 2007), the resulting spatial conditions, in turn, have influenced the development of infrastructural technologies and the institutions that govern them (Coutard et al., 2005).

Since the primary focus of urban water governance in contexts like Europe is no longer on initial and equal access to clean water services, contemporary urban water governance focuses on further advancing water infrastructures, including reregulation and calls for sustainability developments (Bolognesi, 2018; Schmidt, 2014). Similar to other network industries, the urban water sector is challenged by climate change, ageing infrastructure, technological innovations, and changing urban-suburban relations (Markard et al., 2012). Additionally, centralized water infrastructures, including wastewater, water supply, and (storm-) rainwater systems, are increasingly under pressure by political demands to introduce more sustainable practices (Sedlak, 2014).

From a governance perspective, the transformation toward more sustainable urban water governance requires the joint development of technological and institutional designs (Kiparsky et al., 2013). It raises questions about the "right" scale of water governance, including the degree of organizational consolidation and autonomy. In light of the current predominant local governance, the regional level is emerging as a focal area within which water governance (could) occur(s)[1] (Lieberherr, 2011; Mäding, 2012; Fürst, 2007; Bel & Warner, 2015). Arguments have grown for transitioning issue-bound governance forms into "regional governance" (Benz & Fürst, 2003), such as the creation of water management administrations at the basin or catchment

 DOI: 10.4324/9781003057574-23

level, rather than jurisdictional boundaries (e.g. Cumming et al., 2006; Hüesker & Moss, 2014). To address the increasing stress on water resources, engineers and water development agencies also often propose to widen the scale of water governance, which includes the centralization of management (Bolognesi, 2018; Feitelson & Fischhendler, 2009). Additionally, New Public Management (NPM) reforms, focusing on restructuring the public administration from within by increasing performance through efficiency gains (Pollitt, 2011),[2] have also generated mixed spatial responses in different parts of the water system (water supply, wastewater, stormwater). Some parts of the water system have been consolidated in an effort to increase economic efficiency, while others have been decentralized (Feitelson & Fischhendler, 2009; Klien & Michaud, 2019).

Regionalization and NPM reforms may stand at odds with each other, as the latter focuses on reforms within an organization that stipulates flat hierarchies rather than centralization with substantial horizontal coordination (Mäding, 2012). Although the two reforms represent different governance models (network and market or private governance), in practice, water governance is usually a hybrid of different models (Bolognesi, 2018; Romano & Akhmouch, 2019). For instance, with respect to operational modes of service delivery, regionalization typically includes elements from the NPM toolkit, such as organizational autonomy of service providers from municipal governments or sharing managerial and operational responsibilities between public and private partners (Thynne & Wettenhall, 2004; Silvestre et al., 2018). However, NPM reforms and measures such as organizational autonomy can also lead to more fragmented governance, if, for instance, single private providers take over small-scale infrastructure.

In light of these developments, we address the following question: What are the arguments for and against regionalization and organizational autonomy as two governance reforms? We tackle this question by reviewing the governance literature and looking at how these reforms work in practice, focusing on the German model of water governance. Compared to the other common models of urban water governance in Europe, the French and English models,[3] the heterogeneity in the German context regarding regionalization and organizational autonomy makes it particularly fitting for studying the arguments for and against these governance reforms (Bolognesi, 2014b, 2018; Schouten & Pieter van Dijk, 2007; Wackerbauer, 2007). Although some argue that the German case has experienced the least modernization reforms in the western European context, as service provision remains largely publicly owned with strong local governmental control, a "corporatization" trend has been evident since the mid-1990s (Bolognesi, 2014b), leading in some places to partial privatization in the form of shared ownership between public and private actors with private sector management and public supervision (Wackerbauer, 2007). However, the German model is more nuanced, as the federal government has not liberalized the water and sanitation sectors and the municipalities have the freedom to choose how they organize and manage these services, which can include regionalization. Before moving into the German model, we review the arguments about regionalization and organizational autonomy in the literature.

18.2 Reforming urban water governance

A large number of competing scales are found in water governance where trade-offs occur in an effort to reach decisions that are mutually beneficial for water and land management, and the people responsible for them (Daniell & Barreteau, 2014). There are scale mismatches, such as operational decisions being linked to short-term political agendas versus longer-term water planning and sustainable development needs (Daniell & Barreteau, 2014; Cumming et al., 2006). The spatial scale in governing water resources is critical, as it determines the area in which a governance regime exists and, thus, the issues and actors with which it has to contend. The scale of governance is socially constructed (Feitelson & Fischhendler, 2009); space is a social product, which embodies social relations of empowerment and disempowerment and the arena through and in which they operate (Swyngedouw, 1997; additionally see Chapter 22, "Political ecology").

In this context, regionalization has (re-)emerged as a paradigm for improving local public service provision, such as health care, waste collection, education, or water (Mäding, 2012; Fürst, 2007; Bel & Warner, 2015). As one form of regionalization, intermunicipal collaboration in urban regions might enable meeting the demands of sustainable water supply and wastewater disposal (Schmidt, 2014). Regionalization aims at establishing structures to enhance cooperation among municipalities and a variety of actors to co-manage service provision (Fürst, 2007; Frisken & Norris, 2001). As single municipalities are increasingly overloaded with new duties, conventional (fragmented) approaches are perceived as insufficient to accomplish tasks efficiently and effectively where regionalization is set forward as a solution for common problem-solving (Lieberherr, 2011). The regional level has increased in significance for urban water governance (Schmidt, 2014). In general terms, regional governance refers to policymaking becoming rescaled at the level of a region, where actors join together to tackle common tasks as a result of functional interdependencies rather than political boundaries (Fürst, 2007). An example of a region is a catchment where multiple municipal governments cooperate by joining together in a larger organizational form under joint management, typically a supra-municipal institution, to provide a service (Bel et al., 2014). Depending on the scope of regionalization there are different consequences for water utilities: providers might consolidate some or all activities (administrative cooperation or consolidation) and utility networks (physical structures) might be connected or left unchanged (Klien & Michaud, 2019).

Administrative consolidation per se does not have direct implications for public versus private principles with respect to the organizational type of regional service provision.[4] To understand organizational reforms, drawing on elements from the NPM toolkit becomes relevant (Pollitt, 2011). In this context, NPM addresses the increasing independence of service providers at the operational level and the decreasing direct influence of public authorities as well as democratic participation (Furlong & Bakker, 2010; Furlong, 2012). In other words, emphasis has been placed on the "clear separation of political and managerial roles" (Schedler, 2003: 332), resulting in increasingly autonomous organizations for public service provision (Thynne & Wettenhall, 2004). The operators no longer have to be embedded within the municipal administration, as municipal governments delegate their authority to a new organizational entity, which has its own legal personality and a statute (Bel et al., 2014). Administrative (also referred to as "corporate") consolidation of utilities (regardless of ownership) might involve utilities in close proximity; however, the utility networks do not necessarily have to be physically connected and the infrastructure can remain decentralized. This changes not only the number of customers served and the amount of water distributed, but also the number of towns that are served by a utility without changing the infrastructure (Klien & Michaud, 2019).

In the following, we differentiate between (1) the degree of *regionalization (intermunicipal cooperation)*; and (2) the degree of *organizational autonomy*, meaning the degree of decision-making freedom from government, as derived from the NPM literature (Schedler, 2003; Furlong, 2012). We review the literature explicitly addressing these two reforms (regionalization and organizational autonomy) in terms of arguments for and against more consolidated forms, which we summarize in Table 18.1.

Table 18.1 Positive and negative arguments for regionalization and organizational autonomy (own representation)

	Regionalization	*Organizational autonomy*
Positive	• Pooling of resources (capital, human) • *Economies of scale* • Internalizing externalities (coordinated decision making; potential synergies across sectors)	• Shorter decision-making processes • Better financial performance (access to external funds, longer-term financial planning)
Negative	• Coordination needs (transaction costs) • Less context-sensitive solutions (e.g., Less innovative regarding sustainable water technologies) • In case of physical consolidation: "Dis-economies of scale" (no clear benefits for already large utilities; high capital and energy costs)	• Absence of governmental oversight (potential need for reregulation) • Mismatch between corporate and public interests • Transaction costs

A key argument for regionalization is the *pooling of resources*, as multiple municipalities joining together is expected to lead to economies of scale, professionalization, and improved investment planning (Bel &Warner, 2015, Agranoff & McGuire, 2003).[5] The argument is that through a pooling of resources among multiple actors, the operator has more financial means and, hence, also human resources at its disposal than a single operator would; hence, the operator can increase investments to improve service quality (Agranoff & McGuire, 2003; Shih et al., 2004).

A recent overview of regionalization indicates that generally increased cooperation is significantly associated with lower financial costs (Bel & Warner, 2015; Bel et al., 2014). A study has also found that economies of scale strongly influences efficiency of the wastewater treatment process and positively impacts operational costs (Hernández-Chover et al., 2018). However, the profitability of regionalization is not a given, depending on the type and design of consolidation (Klien & Michaud, 2019). For instance, it depends on customer density and regional dispersion e.g., consolidating large high-density urban utilities with smaller rural utilities may not deliver the expected cost saving. Low-density consolidations and large geographical service areas reduce profitability and might even lead to "dis-economies of scale" (also unclear benefits if utilities are already large) (González-Gómez & García-Rubio, 2008). Moreover, physically connecting networks might negatively affect capital and energy costs (Klien & Michaud, 2019). Studies have also found that the positive effects of economies of scale depend on the size and ownership of utilities, where, e.g., dis-economies are more likely for publicly owned utilities than privately owned ones (Carvalho et al., 2012).

Another argument for regionalization is the *internalization of externalities* through regional organizations, which facilitate coordinated decision making and might also allow for exploiting further synergies between different sectors and issues, e.g., between flood protection and drinking water provision, spatial planning, and the installation of renewable energy plants (Furlong & Bakker, 2010; Bel & Warner, 2015; Agranoff & McGuire, 2003). Indeed, regionalization is seen as an opportunity to organize water supply and wastewater together, which are often organized in separate institutions, not utilizing potential intersections and synergies (Schmidt, 2014).

However, arguments can also be made against regionalization, as such cooperation may increase (transaction) costs due to *coordination needs*, such as negotiating contracts or organizational types, which often involves external support from legal advisers (Bel & Warner, 2015; Bolognesi, 2014b; 2018). Despite the existence of intermunicipal institutions, local authorities often remain in charge of the bulk of the tasks (Schmidt, 2014). Other impediments to regionalization are, e.g., political interests and power considerations, institutional path dependencies, or different conditions of the technical facilities, which potentially results in unwillingness to "inherit" a poorly maintained infrastructure system (Schmidt, 2014). Specifically, the more local governments join together and have delegates in the board of directors, the higher the transaction costs (Bel & Warner, 2015). Similarly, harmonizing programmes between local municipalities has been found to be a challenge in the context of consolidation (Furlong & Bakker, 2010). In sum, coordination can lead to increased costs, which may hinder or delay service delivery.

Another argument against regionalization is that when governance becomes more removed from a local context, then it is more likely to adopt less context-sensitive solutions, e.g., less sustainable, urban water management technologies (Green & Anton, 2012). Similarly, different localities might have a different hydrology, topography, and political economy (Bakker & Cook, 2011), which questions regional reforms. Finally, more local governance arrangements may encourage community participation (Feitelson & Fischhendler, 2009).

A central argument for increasing autonomy is that the separation of the operational management from the political system has been found to increase effectiveness of public service providers (Schedler, 2003). A key factor affecting this is the *shorter decision-making process* purportedly linked to higher autonomy. Due to the protocol of needing consent from public authorities or even citizen approval, when an organization lacks a legal personality, decision-making processes can become very lengthy and time-consuming (Bloomfield, 2006). When decision making is tied to the legislature period or election cycles this may lead to insufficient investments and to ineffectiveness (Furlong & Bakker, 2010; Pandey et al., 2007). Studies indicate that organizations that have their own legal personality under private law have a shorter decision-making process and are therefore able to implement more advanced technologies, which then positively affects the quality of public service delivery (Lieberherr, 2016). In terms of democratic control, research indicates that an organization where changes in bylaws are subordinate to a public vote is inferior in terms of goal achievement compared to one where decisions are made by the managing board of the organization (Lieberherr, 2016).

Increased autonomy from political authorities has also been found to positively affect financial performance of service providers (Araral & Wang, 2015). Factors include *access to external capital* and *longer-term financial planning* with increased financial autonomy and, freed from legislature period or election cycles, service providers also have available a broader range of financial instruments rather than being dependent on municipal funds (Furlong & Bakker, 2010).

However, other studies have found that devolving responsibility for expenditure and revenue can have negative implications for service provision (Kwon, 2013). A key factor is that *lacking governmental oversight* might negatively affect public goal achievement. Indeed, financial autonomy without governmental oversight can result in sudden insolvency of an organization, and it has been found to lead to high public costs (Bloomfield, 2006). While the government eventually holds the legal responsibility for service supply, it lacks the instruments for the early identification of financial imbalances (Pahl-Wostl, 2015). In addition, research has found evidence that, for example, public-private partnerships (PPPs) and private companies might pursue commercialized profit rather than a public service logic, which can adversely affect effectiveness, particularly if one of the public goals embraces natural resource protection (Lieberherr et al., 2012; Hüesker, 2011).

Thus, a potential also exists for *mismatch between corporate and public interests*, if PPPs and private companies operate against the citizens and public authorities' interests of affordability, accessibility, and quality (Lieberherr et al., 2012; Schiffler, 2015; Furlong & Bakker, 2010; Johnstone & Wood, 2001). Furlong and Bakker (2010) also found that when water management is delegated to a private actor, as can be the case in a PPP, then the municipality has to implement resource protection itself, as this is not safeguarded by the operator. With (partial) privatization an extra layer of regulation is required to ensure service quality, which increases *transaction costs* (Furlong & Bakker, 2010; Brown & Potoski, 2003). This issue is apparent with the high level of regulation in the context of privatization and NPM reforms (Menard, 2009). Lieberherr and Fuenfschilling (2016) came to the conclusion that reregulation can lead to ineffectiveness, while direct governmental oversight through the public administration has been found to be more efficient and, hence, effective. Following the "paradox of modernization" this could be due to the incoherence between the policy design and reregulation in the context of water reforms (Bolognesi, 2014a).

18.3 Considerations from the German model

Reliance on central infrastructures and equal access to water services are strong in Germany. So even if science and policymakers point out the many societal, technological, and other drivers to rethink water governance, measures like rescaling in water governance face resistance of the existing water governance structures. We will now illustrate the water governance challenges discussed above, providing examples from Germany according to the above arguments for and against regionalization and organizational autonomy (cf. Table 18.1).

Pooling of (capital/human) resources and internalizing externalities: River basin water boards in Germany's North-Rhine Westphalia state are broadly perceived as effective institutions for regionalized sustainable water management (Hüesker & Bernhardt, 2014; Moss et al., 2019). These powerful regional institutions are set in place, as the risk of polluted rivers in industrial regions or by floods was enormous and forced the federal state to disempower the municipalities. Especially the wastewater sector is regionalized and modern central infrastructure is constructed along the basin, e.g., resources for wastewater treatment or water loads below dams are pooled. The water boards, as integrated water governance, enable internalizing externalities: Water board members from enterprises and municipalities negotiate and decide on deals within the board. This is possible through trust – often developed over decades – among the members that decisions are equitable over the long term (Silveira et al., 2016).

Coordination needs: Usually the rescaling of institutions responsible for water governance and the empowerment of regional actors can be very difficult to implement in German practice. As it is a constitutional right of every municipality to organize public services independently on its territory, regionalization has to have strong drivers, such as powerful actors or water-related risks perceived as increasingly threatening (Hüesker & Moss, 2014). To act beyond the given administrative-territorial scales voluntarily is limited if members continue to individually influence the regional institutions. Many of the hundreds of German water boards therefore have only small and non-controversial tasks and are not powerful enough to enforce coordination. Urban-suburban cooperation in many German metropolitan areas means in these cases that massive and diverse coordination needs exist, often without a powerful organization to implement outcomes (Hüesker et al., 2011). Powerful actors, such as proactive municipalities like Munich, can replace this regional institution by taking leadership and leading to fragmentation. Relevant water governance issues can sometimes be solved by task-specific cooperation between urban and suburban actors, such as the shared use of wastewater infrastructures without regional institutionalization (Hüesker & Bernhardt, 2014).

Shorter decision-making processes: Large (traditional) municipal companies in the water sector are very often the opposite of dynamic, innovative, and competitive enterprises. Established decades ago to run water infrastructures safely and welfare-oriented, municipal companies are administering territorial monopolies without competitors, and the internal decision-making processes often depend too much on relations with political decision-makers (Hüesker et al., 2011). So, NPM reforms to strengthen organizational autonomy can refresh decision-making processes in urban water governance, as the former system often stagnated, at least in Germany.

Most management decisions in the water sector have impacts on public interests, and research in the German context shows that there is resistance to disconnect them from political influence completely (Hüesker, 2011; Hüesker et al., 2011). Indeed, the economic freedom for investors in urban water management is limited to a few decisions, as e.g., tariff calculation, including potential revenues that are heavily regulated by law. For instance, organizational autonomy can be used to run the wastewater treatment plant in an energy efficient manner or to improve rainwater management. Autonomy is organized in an effective way if water companies manage daily business and public authorities mainly monitor the outcomes.

Access to external funds and longer-term financial planning: Investment decisions connected to water governance are mostly long term and are quite heavily regulated in Germany (Hüesker et al., 2011). Investment in infrastructure can be refinanced by cost-covering tariffs in Germany. Hence, room for financial autonomy is limited. However, this remains interesting in the context of pushing water governance modernization. Due to, e.g., short-term political considerations, German municipalities often do not fully charge for watering in private gardens, a non-ecological exemption any independent company would probably not allow. Hence, privatization cases show that financial transparency increases with these reforms, as private actors try to get rid of "unnecessary" services like rainwater treatment, which should be paid by the citizens as a taxpayer and no longer as a water service consumer (Hüesker, 2011; Hüesker et al., 2011).

Absence of governmental oversight: NPM of urban water services and infrastructures can be implemented in many different ways in Germany, and small details of contractual arrangements might be decisive whether or not governmental oversight is finally guaranteed (Hüesker et al., 2011). Generally private investors negotiate contracts and corporate strategies supported by professional consultancies and law firms. But municipal governments might be constitutionally obliged to ensure a public veto right, at least regarding formal regulations, e.g., in the German wastewater sector. German cases also show that urban water governance was – before NPM reforms – hardly politicized or formalized. Privatization, commercialization, and liberalization of public service companies in the water sector can act like a wake-up call and lead to the repoliticization and formal regulation of public interest in water governance not existing in pre-NPM times. However, in the context of NPM reforms, it is difficult for municipal governments to act against the private interest (Hüesker, 2011; Hüesker et al., 2011). Here the role of the direct democratic influence – the citizens – can come centre stage, as the case of the partial privatization of Berlin showed (Lieberherr et al., 2012).

Mismatch between corporate and public interest: Investors in water management usually avoid acting like hedge funds as every company involved is aware that citizens would not accept treating water services as a usual commercial good (Hüesker et al., 2011). However, a potential mismatch exists especially if water services are commercialized in the context of NPM reforms (this means it is allowed to integrate a revenue in the calculation of water and wastewater tariffs). Again, the mismatch depends on the very details of how the revenue is calculated and what this implies for water management decisions. For example, in Berlin water consumption dropped rapidly from the 1990s onward, and it would have been in the public and ecological interest to minimize or reduce the water infrastructure to save money and water. However, as the revenue of partly privatized Berlin waterworks in Berlin in the 1990s was connected to the value of the water infrastructure, even larger wastewater treatment plants were constructed. Additionally, the public and the governmental interest are not necessarily the same' municipal governments and private companies can act in a common and nonsocietal interest together to increase revenues by investing in "golden" wastewater treatment plants or by promoting increased water consumption, as cases from Germany show (Hüesker, 2011; Hüesker et al., 2011).

18.4 Conclusion

In this chapter we have addressed the question regarding the arguments for and against governance reforms, focusing on the water governance reforms of regionalization and organizational autonomy. According to our literature review, key arguments for regionalization are pooling of resources and internalizing externalities. However, consolidation might also lead to transaction costs, particularly in the context of centralized infrastructures. Positive arguments for organizational autonomy are shorter processes, access to funds, and long-term planning. Negative arguments are found in terms of mismatches and absence of oversight.

We have discussed these arguments in the literature review in considering developments in Germany. We find that in the German context, regionalization on the river basin scale with state-founded water boards, institutionalized decades ago, supports the positive arguments underlying regionalization in the literature, such as pooling of human and financial resources. There are also negative arguments, like rising transaction costs, which, coupled with weak regional institutions, lead to often poor coordination between urban-suburban interaction. In terms of organizational autonomy, we find strong resistance to disconnecting water systems from political influence in Germany. Moreover, cases show mismatches between public and corporate interests, where, in the context of NPM reforms, local governments can become weak vis-à-vis private companies and where then direct democratic elements – the voice of the citizens – can play a key role in safeguarding the public interest. Conversely, cases show the positive impact of privatization on financial transparency and the potential politicization and formal regulation of public interest in water governance with NPM reforms. As the German illustrations show, central infrastructure paradigms remain powerful, such as the obligation to connect to municipal sewage systems or the water tariff calculation rules. In light of the pressing calls from science and politics to rethink centralized water infrastructures (Hoffmann et al., 2020; Sedlak, 2014; Domènech, 2011; Pahl-Wostl, 2015), we need to further explore the challenges posed to current and future urban water governance models and the potential resistance presented by incumbent actors.

Looking toward the future, a potentially increasing challenge for urban water management is the integration of decentralized technologies, which could enable using multiple water sources and feature greater path diversity, which otherwise would be hindered by the centralized design of conventional water infrastructures that are usually highly compartmentalized, lacking interconnectedness among infrastructures performing limited or specialized functions (Leigh & Lee, 2019). However, understanding is still limited of the type of water governance needed for regulating and operating decentralized water technologies (Kiparsky et al., 2013; Pakizer & Lieberherr, 2018; Pakizer et al., 2020). The established organizational structures, which are aligned with the centralized water infrastructure, might be overwhelmed with managing these new technologies, lacking the relevant knowledge and capabilities, e.g., skill development and training (Brown & Farrelly, 2009). For instance, in the particular case of decentralized water reuse technologies, coordination between multiple water bodies and other agencies might become necessary, such as building departments and environmental health agencies spanning different jurisdictions (Rupiper & Loge, 2019; Brown & Farrelly, 2009).

This chapter, dealing with the challenges and opportunities of regionalization and organizational autonomy reforms, provides building blocks for understanding how these governance reforms might contribute (or hinder) the integration of decentralized technologies in urban water systems. Regionalization has the potential for integrating and operating decentralized technologies, as municipalities could pool the necessary capital and human resources to develop the necessary organizational structures for operating such technologies. In addition, based on our discussion of governance reforms, organizational autonomy will likely also be important for integration of decentralized technologies, as it enables service providers to be more flexible, e.g., due to access to external capital and shorter decision-making processes. However, finding the "right" scale of water governance will most likely remain a highly contested topic, requiring continuous renegotiation and adaptation in light of existing paradigms alongside present and future challenges.

Notes

1 There are discrepancies between ideal scales and the scales implemented in practice (Feitelson & Fischhendler, 2009).

2 NPM reforms call for establishing organizations with increased autonomy from public authorities for service delivery, including privatization or contracting out to private firms (Furlong, 2016; Silvestre et al., 2018).

3 The English model is distinguished as an ideal type for regional organization, material privatization, and arms-length regulatory agencies. The French model is also demarcated by widespread private sector participation in the form of delegation (*affermage*), but the local government remains important and recent years have witnessed remunicipalization efforts (Lieberherr et al., 2016).

4 New regionalism, in contrast to previous reforms at the regional level, is characterized by an explicit focus on economic development, competitiveness, and governance arrangements that include cooperation between public and private actors and/or networks (Lackowska & Zimmermann, 2010).

5 By economies of scale we mean, e.g., pooling managerial tasks, namely, more efficient billing, improved purchasing (through higher negotiating power), bundled inventory, water testing operations, and financing (Shih et al., 2004).

References

Agranoff, R., & McGuire, M. (2003). *Collaborative Public Management: New Strategies for Local Governments*. Georgetown University Press.

Araral, E., & Wang, Y. (2015). Does water governance matter to water sector performance? Evidence from ten provinces in China. *Water Policy*, *17*(2), 268–282.

Bakker, Karen, and Christina Cook. 2011. "Water governance in Canada: Innovation and Fragmentation", *Water Resource Development*, 27(2): 275–289.

Bel, Germa, and Mildred E. Warner. 2015. "Inter-municipal Cooperation and Costs: Expectations and Evidence" *Public Administration*, 93(1): 52–67.

Benz, Arthur, and Dietrich Fürst. 2003. "Region – 'Regional governance' – Regionalentwicklung", in: Bernd Adamasche and Marga Pröhl (Eds.), *Regionen erfolgreich steuern*, Gütersloh: Bertelsmann-Stiftung, 11–66.

Bloomfield, Pamela. 2006. "The challenging business of long-term public–private partnerships: Reflections on local experience", *Public Administration Review* 66(3): 400–411.

Bolognesi, Thomas. 2014a. "The paradox of the modernisation of urban water systems in Europe: Intrinsic institutional limits for sustainability", *Natural resources forum*, 38(4): 270–281.

Bolognesi, Thomas. 2014b. "The results of modernizing network industries: The case of urban water services in Europe", *Competition and Regulation in Network Industries*, 15(4): 306–333.

Bolognesi, Thomas. 2018. *Modernization and Urban Water Governance: Organizational Change and Sustainability in Europe*. London: Palgrave Macmillan.

Brown, Rebekah R., and Megan A. Farrelly. 2009. "Delivering sustainable urban water management: A review of the hurdles we face", *Water Science & Technology*, 59(5): 839–846.

Brown, Trevor L., and Matthew Potoski. 2003. "Transaction costs and institutional explanations for government service production decisions", *Journal of Public Administration Research and Theory*, 13(4): 441–468.

Carvalho, Pedro, Rui Cunha Marques, and Sanford Berg. (2012). "A meta-regression analysis of benchmarking studies on water utilities market structure." *Utilities Policy* 21: 40–49.

Coutard, Olivier, Richard Hanley, and Rea Zimmerman (Eds.). 2005. *The Networked Cities Series. Sustaining Urban Networks: The Social Diffusion of Large Technical Systems*. New York/London: Routledge.

Cumming, Graeme S., David H.M. Cumming, and Charles L. Redman. 2006. "Scale mismatches in social-ecological systems: Causes, consequences, and solutions", *Ecology and Society*, 11(1): 14.

Daniell, Katherine A., and Olivier Barreteau. 2014. "Water governance across competing scales: Coupling land and water management", *Journal of Hydrology*, 519: 2367–2380.

Domènech, L. (2011) "Rethinking water management: From centralised to decentralized water supply and sanitation models", *Documents d'Anàlisi Geogràfica*, 57(2): 293–310.

Feitelson, Eran, and Itay Fischhendler. 2009."Spaces of water governance: The case of Israel and its neighbors", *Annals of the Association of American Geographers*, 99(4): 728–745.

Frisken, Frances, and Donald F. Norris. 2001. "Regionalism reconsidered", *Journal of Urban Affairs*, 23(5): 467–478.

Furlong, Kathryn. 2012. "Good water governance without good urban governance? Regulation, service delivery models, and local government", *Environment and Planning A*, 44(11): 2721–2741.

Furlong, Kathryn. 2016. *Leaky Governance: Alternative Service Delivery and the Myth of Water Utility Independence*. Vancouver/Toronto: UBC Press.

Furlong, Kathryn, and Karen Bakker. 2010. "The contradictions in 'alternative' service delivery: Governance, business models, and sustainability in municipal water supply", *Environment and Planning C: Government and Policy*, 28(2): 349–368.

Fürst, Dietrich. 2007. "Regional governance", in: Arthur Benz, A., Susanne Lütz, Uwe Schimank and Georg Simonis (Eds.), *Handbuch Governance*. Wiesbaden: VS, 353–365.

González-Gómez, Francisco, and Miguel A. García-Rubio. 2008. "Efficiency in the management of urban water services. what have we learned after four decades of research", *Hacienda Pública Española/Revista de Economía Pública*, 185(2): 39–67.

Green, Colin, and Barabara Anton. 2012. "Why is Germany 30 years ahead of England?", *International Journal of Water*, 6(3/4): 195–214.

Hernández-Chover, Vincent, Bellver-Domingo, Águada, & Hernández-Sancho, Francesco. (2018). Efficiency of wastewater treatment facilities: The influence of scale economies. *Journal of Environmental Management, 228*, 77–84.

Hoffmann, Sabine, Ulrike Feldmann, Peter M Bach, Christian Binz, Megan Farrelly, Niki Frantzeskaki, Harald Hiessl, Jennifer Inauen, Tove A Larsen, and Judit Lienert. 2020. "A research agenda for the future of urban water management: Exploring the potential of nongrid, small-grid, and hybrid solutions", *Environmental Science & Technology*, 54(9): 5312–5322.

Germà Bel, Xavier Fageda, Melania Mur. 2014. Does Cooperation Reduce Service Delivery Costs? Evidence from Residential Solid Waste Services, *Journal of Public Administration Research and Theory*, 24(1): 85–107.

Hüesker, Frank. 2011. *Kommunale Daseinsvorsorge in der Wasserwirtschaft -Auswirkungen der Privatisierung am Beispiel der Wasserbetriebe des Landes Berlin*. München: Oekom-Verlag (Hochschulschriften zur Nachhaltigkeit Bd. 51).

Hüesker, Frank, and Christoph Bernhardt. 2014. "State-founded water boards in industrialized Western Germany", in: Dave Huitema and Sander Meijerink (Eds.), *The Politics of River Basin Organizations. Institutional Design Choices, Coalitions and Consequences*. Cheltenham/Northampton: Edgar Elgar Publishing, 140–161.

Hüesker, Frank, and Timothy Moss. 2014. "The politics of multi-scalar action in River Basin management: Implementing the EU water framework directive (WFD)", *Land Use Policy*, 42: 38–47.

Hüesker, Frank, Timothy Moss, and Matthias Naumann. 2011. "'Managing water infrastructures in the Berlin-Brandenburg region between climate change', economic restructuring and commercialisation", *DIE ERDE*, 142(1–2): 187–208.

Johnstone, Nick, and Libby Wood. 2001. *Private Firms and Public Water: Realising Social and Environmental Objectives in Development Countries*. Cheltenham: Edward Elgar.

Kiparsky, Michael, David L. Sedlak, Barton H. Thompson, and Bernhard Truffer. 2013. "The innovation deficit in urban water: The need for an integrated perspective on institutions, organizations, and technology", *Environmental Engineering Science*, 30: 395–408.

Klien, Michael, and David Michaud. 2019. "Water utility consolidation: Are economies of scale realized?", *Utilities Policy*, 61: 1000972.

Kwon, Osung. 2013. "Fiscal decentralization: An effective tool for government reform?", *Public Administration*, 91(3): 544–560.

Lackowska, Marta, and Karsten Zimmermann. 2010. "New forms of territorial governance in metropolitan regions? A Polish–German comparison", *European Urban and Regional Studies*, 18(2): 156–169.

Leig, N. G., & Lee, H. (2019). Sustainable and resilient urban water systems: The role of decentralization and planning. *Sustainability*, *11*(3), 918.

Lieberherr, Eva. 2011. "Regionalization and water governance: A case study of a Swiss wastewater utility", *Procedia Social and Behavioral Sciences*, 14: 73–89.

Lieberherr, Eva, and Lea Fuenfschilling. 2016. "Neoliberalism and sustainable urban water sectors: Acritical reflection of sector characteristics and empirical evidence", *Environment and Planning C: Politics and Space*, 34: 1540–1555.

Lieberherr, Eva, Andreas Klinke, and Matthias Finger. 2012. "Towards legitimate water governance? The partially privatized Berlin waterworks", *Public Management Review*, 14: 923–946.

Lieberherr, Eva, Claudine Viard, and Carsten Herzberg. 2016. "Water provision in France, Germany and Switzerland: Convergence and divergence", in: Hellmut Wollmann, Ivan Koprić, and Gérard Marcou (Eds.), *Public and Social Services in Europe*. London: Palgrave Macmillan, 249–263.

Mäding, Heinrich. 2012. "Von der Zwischengemeindlichen Zusammenarbeit zur Strategischen Regionsbildung", in: Eckhardt Schröter, Patrick von Maravić, and Jörg Röber (Eds.), *Zukunftsfähige Verwalltung? Herausforderungen in Deutschland, Österreich und der Schweiz*. Opladen, Berlin & Toronto: Verlag Barbara Budrich, 253–280.

Markard, Jochen, Rob Raven, and Bernhard Truffer. 2012. "Sustainability transitions: An emerging field of research and its prospects", *Research Policy*, 41(6): 955–967.

Menard, Claude. 2009. "From technical integrity to institutional coherence: Regulatory challenges in the water sector", in: Claude Menard and Michel Ghertman (Eds.), *Regulation, Deregulation, Reregulation: Institutional Perspectives*, Northampton: Edward Elgar Publishing, Inc., 83–108.

Monstadt, Jochen. 2007. "Großtechnische Systeme der Infrastrukturversorgung: Übergreifende Merkmale und räumlicher Wandel", in: Dieter Gust (Ed.), *Wandel der Stromversorgung und räumliche Politik*. Hannover: Akademie für Raumentwicklung in der Leibniz-Gemeinschaft, 7–34.

Moss, Timothy, and Frank Hüesker. 2019. "Politicised nexus thinking in practice: Integrating urban wastewater utilities into regional energy markets", *Urban Studies*, 56(11): 2225–2241.

Pahl-Wostl, Claudia. 2015. *Water Governance in the Face of Global Change: From Understanding to Transformation*. Heidelberg: Springer International Publishing.

Pakizer, Katrin, and Eva Lieberherr. 2018. "Alternative governance arrangements for modular water infrastructure: An exploratory review", *Competition and Regulation in Network Industries*, 19(1–2): 53–68.

Pakizer, Katrin, Manuel Fischer, and Eva Lieberherr. 2020. "Policy instrument mixes for operating modular technology within hybrid water systems", *Environmental Science & Policy*, 105: 120–133.

Pandey, Sanjay K., David H. Coursey, and Donald P. Moynihan. 2007. "Organizational effectiveness and bureaucratic red tape: A multimethod study", *Public Performance & Management Review*, 30(3): 398–425.

Pollitt, Christopher. 2011. *Public Management Reform: A Comparative Analysis*. Oxford: Oxford University Press.

Romano, Oriana, and Aziza Akhmouch. 2019. "Water governance in cities: Current trends and future challenges", *Water*, 11(3): 500.

Rupiper, Amanda M., and Frank J. Loge. 2019. "Identifying and overcoming barriers to onsite non-potable water reuse in California from local stakeholder perspectives", *Resources, Conservation & Recycling: X*, 4: 100018.

Schedler, Kuno. 2003. "Local and regional public management reforms in Switzerland", *Public Administration*, 81(2): 325–344.

Schiffler, Manuel. 2015. "Berlin: Privatized to fill state coffers, remunicipalized at the State's expense", in: Manuel Schiffler (Ed.), *Water, Politics and Money*. London: Springer, 105–114.

Schmidt, Martin. 2014. "Regional governance vis-a-vis water supply and wastewater disposal: Research and applied science in two disconnected fields", *Water International*, 39(6): 826–841.

Schouten, Marco, and Meine Pieter van Dijk. 2007. *The European Water Supply and Sanitation Markets*. London: IWA.

Sedlak, David. 2014. *Water 4.0: The Past, Present, and Future of the World's Most Vital Resource*. New Haven: Yale University Press.

Shih, Jhih-Shyang, Winston Harrington, William A. Pizer, and Kenneth Gillingham. 2004. "Economies of scale and technical efficiency in community water systems", *Resource Future, Discussion Paper*, 04–15, 38p.

Silveira, Andre, Sandra Junier, Frank Hüesker, Fan Qunfang, and Andreas Rondorf. 2016. "Organizing cross-sectoral collaboration in river basin management: Case studies from the Rhine and the Zhujiang (Pearl River) basins", *International Journal of River Basin Management*, 14(3): 299–315.

Silvestre, Hugo C., Rui Cunha Marques, and Ricardo Corrêa Gomes. 2018. "Joined-up Government of utilities: A meta-review on apublic–public partnership and inter-municipal cooperation in the water and wastewater industries", *Public Management Review*, 20(4): 607–631.

Swyngedouw, Erik. 1997, "Excluding the other: The production of scale and scaled politics", in: Roger Lee, and Jane Wills. (Eds.), *Geography of Economies*. London: Arnold, 167–76.

Thynne, Ian, and Roger Wettenhall. 2004. "Public management and organizational autonomy: The continuing relevance of significant earlier knowledge", *International Review of the Administrative Sciences*, 70(4):609–621.

Wackerbauer, Johann. 2007. "Regulation and privatisation of the public water supply in England, France and Germany", *Competition and Regulation in Network Industries*, 8(2): 101–117.

19

SUSTAINABILITY TRANSITIONS IN URBAN WATER MANAGEMENT

Assessing the robustness of institutional arrangements

Aaron Deslatte, Margaret Garcia, Elizabeth A. Koebele, and John M. Anderies

19.1 Introduction

Urban water management regimes include the physical water storage and distribution infrastructure of a city as well as the governance arrangements that structure how resources and physical system components are utilized. As such, these regimes can be usefully characterized as coupled natural and human infrastructure systems. To maintain reliable outcomes within these systems under dynamic hydrologic conditions, cities and water utilities can make a variety of policy and management changes, such as raising water rates, implementing conservation measures, and investing in storage infrastructure. When these changes are significant and durable, and when they maintain or improve human standards of living while reducing pressures on the environment, they signal that the system has undergone a *transition toward sustainability* (Garcia et al. 2019; Malekpour, Brown, and de Haan 2015).

Sustainability transitions are long-term, fundamental transformational processes in which system performance is safeguarded or enhanced. Transitioning toward sustainability implies that actors can identify measures of desirable system performance and make decisions that move the system toward those measures. Performance is influenced by the complex interactions and feedback between the social components of the system (e.g., water users, regulators, and other policymakers) and the natural or built infrastructure (e.g., snowpack, reservoirs, etc.). Thus, transitions toward sustainability are triggered by some confluence of natural, built infrastructure and social, economic, and political factors. As many urban water management regimes across the globe increasingly face hydrologic stress, it is critical to systematically understand how these factors interact to hasten or inhibit transitions to more sustainable states.

Scholars have used many frameworks to study transitions. One of them, the Robustness of Coupled Infrastructure Systems Framework (Anderies, Janssen, and Ostrom 2004), has been used to organize the often non-linear and slow-moving forces underlying transitions, and, consequently, system sustainability. In this chapter, we apply this framework to demonstrate how interactions among natural, human, and built infrastructure have impacted the performance of urban water systems in three US metropolitan areas: Miami, Las Vegas, and Los Angeles.

 DOI: 10.4324/9781003057574-24

19.2 Robustness and resilience in urban water sustainability

Human and environmental systems are inherently complex. One may influence the other, often with emergent, non-linear, and irreversible processes (Ostrom 2009; Scheffer et al. 2009). The concepts of resilience and robustness are commonly used within theories that seek to explain the sustainability of such complex systems. This section highlights the value of these two concepts in relation to the capacity of urban water management systems to maintain adequate and reliable supply in the face of uncertainty and change.

We consider the *sustainability* of a system as the connection of favored measures of system performance to the decision-making processes likely to produce them (Anderies et al. 2013). Within urban water management, sustainability entails preserving the conditions for human well-being without damaging or depleting natural resources for future generations.[1] In other words, water managers make rate schedules, infrastructure investments, and (re)distributive choices in specific decision-making contexts to attempt to achieve collectively determined performance goals (Anderies et al. 2013). When some combination of climatic change, development pressure, or physical infrastructure decline threatens actual or future performance, users and managers may collectively change institutions – defined as their rules or shared strategies for behavior – in response (Muneepeerakul and Anderies 2017). Changes may be made to policies or actions assigned to specific actors as well as to constitutional or collective-action decision processes themselves (Garcia et al. 2019; Ostrom et al. 1994). The decision-making context is often critical to the scope of these alterations of rules, norms, or strategies, and many of the contextual parameters are often difficult to quantify (Anderies et al. 2007; Anderies, Ryan, and Walker 2006).

For a water management regime to transition toward greater sustainability, decision-makers must generate knowledge about these system dynamics and apply it to design the governance institutions and physical infrastructure necessary to achieve or preserve performance (Anderies et al. 2013; Pahl-Wostl 2007). Within this conceptualization, *resilience* refers to the system-level attributes that allow it to bounce back from, or transform in response to, shocks (Carpenter et al. 2001). Resilient systems are often self-organizing and can include attributes, such as a high capacity to learn and adapt. Systems which are resilient may be able to recover more quickly from droughts or flooding. Conversely, *robustness* refers to an intentional control process of reducing the sensitivity of a system's outputs to any perturbations from its inputs. Here, robustness could include the forecasted future availability of drinking water to meet the demands of some threshold of users. The robustness of a system depends on the feedback mechanisms or policies in place, and it may allow the system to appear unchanged, based on some standard of performance, even when shocks occur. Distinguishing between resilience and robustness allows us to link the fluidity of systems to the stability required for maintaining a desired level of performance. In other words, the concepts of resilience and robustness help to connect the transformative characteristics and dynamism of complex systems to the types of performance goals and decision-making premises embedded within the concept of sustainability.

Many frameworks used to analyze these types of complex systems attempt to organize and clarify how rules and norms shape human behavior to influence collective action (Anderies, Barreteau, and Brady 2019; McGinnis 2011; Ostrom 1990; Siddiki et al. 2011). The most well known is the Institutional Analysis and Development (IAD) framework developed by Elinor Ostrom and her colleagues (Ostrom 2011). The IAD focuses on "action situations" in which individuals and organizations collectively address shared-resource problems. Their outcomes are evaluated via "fast feedback" processes that may lead to reevaluations and adjustments intended to enhance system robustness. However, because the IAD treats the exogenous variables that

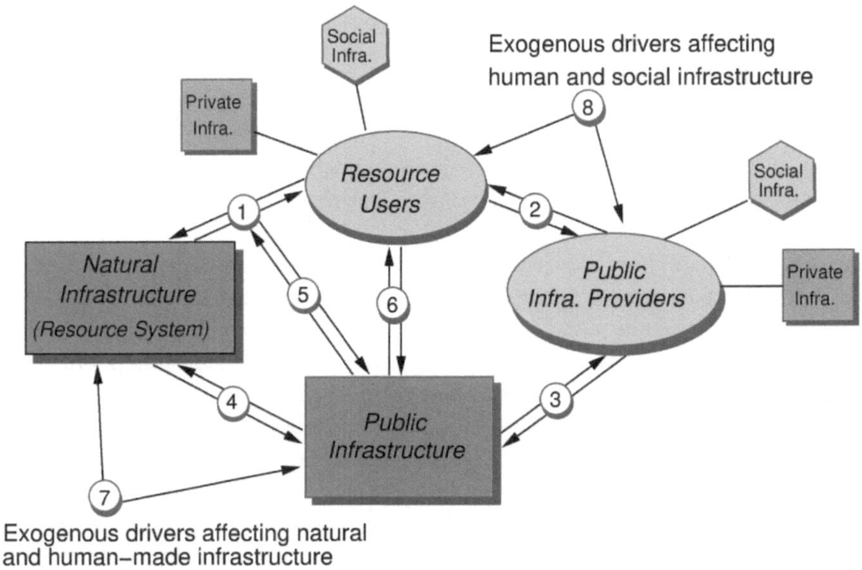

Figure 19.1 The Robustness of Coupled Infrastructure Systems Framework.

influence action situations (i.e., biophysical conditions, community attributes, rules in use) as static, it fails to account for slower system changes that affect system resilience over time and across multiple action situations (Anderies, Janssen, and Ostrom 2004; Anderies, Barreteau, and Brady 2019; Muneepeerakul and Anderies 2017).

Building on the IAD, the Robustness of Coupled Infrastructure Systems (hereafter "CIS") Framework (Figure 19.1) was developed to isolate slower-moving system dynamics that evolve over time in a coupled infrastructure system. Slow feedback, for example, may include changes to sources of water stress, shifts in public awareness of factors underlying droughts or other supply issues, and the effects of higher-level regulations placed on the operations of water utilities. To understand the processes through which these slow feedbacks affect system resilience, the CIS Framework explicitly attends to the interactions among the various system components.

In the CIS Framework, the IAD's "biophysical components" become the Natural Infrastructure (NI), such as waterways, and "hard" Public Infrastructure (PI), such as reservoirs. The IAD's "attributes of the community" are regrouped into Resource Users (RU), Public Infrastructure Providers (PIP), and some elements of the "hard" PI (e.g., water storage infrastructure or flood control systems). The IAD's "rules in use" are regrouped as "soft" PI (such as regulatory processes or bureaucratic culture). The numbered arrows between the components signify flows of information, resources, or authority in the system. Additionally, two types of disturbances are considered within the CIS Framework: external disturbances, such as floods, droughts, and climate change, which act upon the natural and human-made infrastructure (link 7); and socioeconomic (internal) changes, such as population growth, economic recessions/depressions, or political shifts that impact resource users and social infrastructure (link 8). Robustness within this framework links the feedback controls that water managers use to preserve system performance – and thus the hidden fragilities to novel stressors inherent in feedback systems – to system-level dynamics that characterize resilience. For example, designed feedback (strategies, policies, indicators) aimed at managing high-frequency (inter-annual or inter-decadal) variations in water supply can create fragilities to low-frequency variations (e.g., 100-year floods).

Further, such feedback may mask critical performance variation, in turn, suppressing learning processes and reducing resilience.

19.3 Analytic approach: Periods of accelerated change in three US cities

In the sections that follow, we draw on the CIS Framework to better understand the drivers of urban water management transition in three US cities: Las Vegas, Los Angeles, and Miami. External shocks and slow endogenous changes combine to challenge water supply in these cases. City level responses included both incremental changes to existing policies and infrastructures as well as the creation of new strategies. Applying the CIS Framework facilitates a better understanding of whether, and why, these responses reduce or increase resilience.

Each of the three selected cities experienced water supply threats and undertook significant water management changes during a similar period, making them useful in comparing similar systems (Garcia et al. 2019; Seawright and Gerring 2008). We constrain our analysis for each city to the period from 1991 to 2014 to hold a variety of factors constant (i.e., economic cycles, national-level regulation, and available technology). Guided by existing theories that elucidate drivers of transition in urban water systems (Hughes, Pincetl, and Boone 2013), we collected data on the hydrological, financial, social, and institutional conditions of each city and its water utility during the study period. We then analyzed longitudinal trends in these data using a variety of metrics, as described below.

To conceptually map urban water systems using the CIS Framework, we identify PIPs as the water boards, commissions, or councils that develop utility policies for water supply, transmission, distribution, and conservation, as well as the regulators or managers who implement and evaluate the policies. The NI consists of the water resources, including surface and groundwater. The primary RUs in each case are municipal water customers that use water in residential, commercial, and industrial applications. The PI comprises both "hard" infrastructure (wellfields, storage reservoirs, pump stations, and transmission and distribution mains) and "soft" infrastructure (rules, norms, and strategies), which create the circuitry for governing, managing, and enforcing resource constraints.

19.3.1 Characterizing robust controls

For each case, we identified one or more periods of accelerated change (PoAC), defined as spans of time when numerous, concentrated management or policy changes occurred (Garcia et al. 2019). Data used to detect these changes include interviews with water utility staff, utility comprehensive annual financial reports; water supply and demand data; regulatory documents, such as consumptive use permits; and media. PoACs may indicate transitions toward more sustainable water use because the system has moved beyond incremental adaptations to disturbances and toward new "pathways" as old ones become untenable (Anderies et al. 2013; Treuer et al. 2017). However, a PoAC does not mean these systems have become more resilient: while feedback processes have prompted clustered water policy changes aimed at meeting some performance goals, systems can become more vulnerable to novel shocks as they increase their robustness (Anderies 2003). As such, we characterize the events comprising the PoACs as "robust controls" that function at the operational level to generate and feed information back throughout the system.

19.3.2 Rules for governance in urban water regimes

Governance is the process by which rules, norms, and strategies for influencing behavior within a policy area are developed, applied, interpreted, and reformed (McGinnis 2011). Governance arrangements are shaped by the boundaries of the resource system, the cultural and political attributes of communities, how costs and benefits are divided among users, and the functionality of available monitoring, enforcement, and dispute resolution activities (Blomquist, Schlager, and Heikkila 2004; Schlager and Heikkila 2011). Within these arrangements, rules are both the outcomes of prior interactions (via feedback) and the working components of an action situation (Ostrom et al. 1994). Contained in institutional statements, rules constitute the formal basis of *positions*, the *choices* assigned to them, the *scope* of outcomes for which they are responsible, the *information* provided to them, the *aggregation* of decisions into outcomes, and the *payoffs* within action situations (McGinnis 2011).

To characterize the governance arrangements in our three cases, we compiled a subset of institutional statements from state constitutions, statutes, and city/county charters and ordinances (for detailed methodology, see Treuer et al. 2017). These statements represent the constitutional-choice level of water governance in each city, which legitimizes the collective-choice procedures for decision making by PIPs. From this corpus of statements, we identified three types of rules that could be considered as either authority, scope, or boundary conditions, described below, because of the central role they play in shaping collective actions.

First, *authority conditions* within water utility governing boards pertain to setting water rates, or how much users pay for water. This authority is typically captured in choice rules, which assign specific actions to positions. Variation in rate-making processes could ultimately influence important decisions about how RUs are informed of needs and how the PI is maintained, expanded, or depleted. Second, *scope conditions* pertain to the set of outcomes linked to choices and actors. These may include statements that prescribe desirable or avoidable performance outcomes for resource use or address the equity of resource allocation and decision-making processes. Third, we identified the *boundary conditions* for entering and exiting positions on the governing boards of the PIPs. These positions may be appointed by unelected officials, appointed by the elected representatives of the city (the mayor or council), appointed by elected officials outside the jurisdictions (regional or state officials), or directly elected by residents. Different boundary conditions for entering and exiting board positions can shape the incentives of actors and how they interpret the information flows between PIPs and RUs. For instance, RUs may directly appoint (via elections) those who occupy PIP positions, or resource managers within the PIP may be more insulated from RUs when they are appointed by other elected officials. PIP positions that are directly elected may be more responsive to RU's demands or preferences on shorter-term issues, while appointed regulators may be less responsive and more focused on longer-term issues. Thus, variation in the boundary conditions for resource managers/allocators within the PIP may materially affect the quality of information or material flow in the system.

19.3.3 Feedback through the natural system

To capture the stress placed on the natural resource system, we calculated the Water Supply Stress Index (WaSSI) for each city over time (McNulty et al. 2007). A composite measure of biophysical and regulatory-driven water stress calculated as total demand divided by total supplies, the WaSSI accounts for water sources available outside of the city's physical boundary as well as legal and infrastructure constraints (Garcia et al. 2019). Water supply is defined as the amount of water that may be legally withdrawn and is accessible to the water provider, making supply a function of water rights/allocations, restrictions, and infrastructure capacity for a respective year (link 4). The supply term captures variations in the NI (e.g., streamflow, link 7) as they are moderated by hard PI (e.g., storage and conveyance). Demand data (annual finished water supply) and supply data were gathered from utility reports and historic water permits. The demand term captures changes in the RU, including the number of resource users and the intensity of their use (link 1).

19.3.4 Feedback through the social system

Dynamics between the social components of the framework, the RU and PIP, were examined by compiling a metric of public attention to water issues, defined as "changes in the public's attitude toward [their] resources" (Hughes, Pincetl, and Boone 2013). We measured this via a proxy for issue attention: newspaper coverage of water issues. To gather this data, The major newspaper in each case study city (*Miami Herald*, *Los Angeles Times*, and *Las Vegàs Review-Journal*) was queried for the study period using an iteratively developed set of search terms (for details, see Garcia et al. 2019). Media salience can reflect information flows between the governing body and resource users (link 2), as well as actions taken to co-produce infrastructure (link 6).

We also examined feedback between the PI and PIP itself (link 3) by measuring the change in the financial net position of each utility over time. Net position is the sum of a utility's total assets (including capital assets, investments, and cash) minus total liabilities. Net position data were acquired from publicly available utility financial reports. These data reflect both financial and physical assets and investments, which may be either aggregated components or capacity produced via the "hard" public infrastructure. Table 19.1 describes the acronyms used throughout the chapter.

19.4 Empirical examples: Examining the robustness of water management regimes

Here, we use the CIS Framework to depict each case, understand and compare governance arrangements, and explore changes in robust controls in each system that may signal transition. A noteworthy observation is that, in all three cases, PoACs occurred without constitutional level governance reforms, suggesting that the adaptive capacity of each system was sufficient to maintain crucial system functions. Additionally, in all three cases, the number of utility water

Table 19.1 Acronyms of frameworks, concepts, and utilities

IAD	institutional analysis and development framework
CIS	The Robustness of Coupled Infrastructure Systems Framework
PI	public infrastructure, the hard (physical) and soft (rules) infrastructure of water systems
RU	resource users, households, farms, commercial and industrial users
PIP	public infrastructure providers, or governance decision-makers (i.e., utility boards, city councils)
PoAC	periods of accelerated change, indicators of transition
WaSSI	water supply stress index, equal to total water demand/total supplies
WASD	Miami-Dade Water & Sewer Department
SFWMD	South Florida Water Management District
MGD	million gallons per day
SNWA	Southern Nevada Water Authority
LVVWD	Las Vegas Valley Water District
LADWP	Los Angeles Department of Water and Power
LAA	Los Angeles Aqueduct
MWD	metropolitan water district

users (consumers) grew over the period while the water use per user decreased – a key metric of system robustness.

19.4.1 Miami

Miami's CIS relies on local groundwater, including the Biscayne Aquifer (primary source) and the Floridan Aquifer (backup supply). Surface and groundwater systems in Miami-Dade County are hydraulically connected, and surface water flows are carefully controlled to minimize both flooding and saltwater intrusion, major concerns for this low-lying coastal region (Hughes and White 2014). Human impacts, both local (draining the Everglades for land development) and global (sea level rise) have increased the risk of saltwater intrusion.

The Miami-Dade Water & Sewer Department (WASD) emerged from a gradual governance evolution that began in 1972 when Miami-Dade County voters approved the creation of a countywide water authority (link 6). The city of Miami transferred all city water and sewer properties to the authority in 1975. The authority was abolished in 1983 and absorbed by the county government (link 3), forming the current department. In Miami, WASD provides water to customers in Miami-Dade County. The population of resource users grew from 1.9 million in 1990 to 2.6 million in 2014.[2]

Miami's governance structure features a higher number of scope rules governing collective decision processes than Los Angeles and Las Vegas (Garcia et al. 2019). In this case, the majority of Miami's scope rules applied conservation goals to RUs seeking development permits (link 2), such as regulating landscape irrigation and use of high-efficiency plumbing fixtures for residential and commercial construction (link 1). Decision-making systems with greater discretion in how to achieve outcomes may enable quicker adjustments to changing environmental and political conditions (Novo and Garrido 2014). However, scope rules embedded in constitutional-choice institutional arrangements (the county charter) can also give these rules more enforcement power or legitimacy. Turning to the boundary conditions, the governance system also features a larger number of formal actors involved in collective-choice decisions. For instance, one actor, the South Florida Water Management District (SFWMD), is a regional body that oversees 16 counties, sets policy priorities, and permits and monitors WASD's use of their primary water source. While WASD is overseen by the elected county Board of Commissioners, SFWMD board members are appointed by the governor. The SFWMD also has distinct boundary conditions (gubernatorial appointment) likely shaping the incentives of its members.

Over the study period, the CIS experienced peak water stress (link 1) at two time periods (1993, 2006), and public attention to water issues (link 2) also peaked twice (2001, 2007). The one-year lag between water stress and public attention is indicative of the types of feedback that may trigger adaptation responses. Financial stress (link 3) peaked a year later in 2008, reflecting a decline in the net position (assets minus liabilities) of the utility. The alignment of water stress, public attention, and financial stress within the CIS coincided with a number of changes to robust controls, which we identify as a PoAC (2006-2011): Miami implemented durable irrigation restrictions, rate increases, and conservation surcharges; new leak detection and rebate programs were created; and a water use efficiency plan was adopted. Perhaps the most significant adjustment affecting the robustness of the CIS was the approval by SFWMD of the utility's first consolidated (systemwide) 20-year pumping permit, which extended the utility's planning horizon.

Robustness was also affected by investments in the hard PI (link 3). WASD pumps groundwater from 20 well fields in the Biscayne Aquifer, a shallow porous limestone aquifer. Additionally, four previously abandoned well fields in the Floridan Aquifer have been rehabilitated.[3] The utility operates three wastewater treatment plants, each of which has provided reclaimed wastewater since 2007. In 2013, WASD constructed an aquifer storage and recovery pilot project with the capacity to inject ten million gallons per day (MGD).

In summary, the CIS augmented robust controls (both hard and soft PI) and reduced per capita resource use. In the WASD service area, per capita water use decreased from 700 L/person/day in 1991 to 510 L/person/day in 2014 (Treuer et al. 2017).

19.4.2 Las Vegas

The Las Vegas CIS relies on the Colorado River as the primary water source. The Colorado River has been in drought since 2000. However, this drought is not just the result of hydrological cycles. Researchers estimate that from one-fourth to one-third of the water deficit is a result of above- average temperatures, a trend that is likely to continue (Udall and Overpeck 2017). Some now refer to this dry period as a period of aridification rather than drought as the drying trend is likely to persist as the climate warms (Overpeck and Udall 2020).

The governance arrangements have been largely stable since the Southern Nevada Water Authority (SNWA) was created by the Nevada legislature in 1991.[4] The SNWA is a regional wholesale utility that sells water, sets policy direction, and focuses on regional management; its board is composed of an elected official from each of the seven member utilities. One member utility is the Las Vegas Valley Water District (LVVWD), which was created by the state legislature in 1947 to replace a private utility and curb groundwater use.[5] The LVVWD operates at the city level and services unincorporated areas. Unlike Miami's regional water district, the SNWA cannot impose water restrictions on RUs. The composition of water use has not changed over the study period (Garcia and Islam 2018). However, the population of resource users has grown from 670,000 in 1990 to 1.6 million in 2015.

The institutional structure of the Las Vegas CIS features a higher proportion of choice (or authority) rules specifying the actions assigned to actors, in this case, the Nevada legislature, resource users, and the city council and utility. The choice statements assign actions primarily to the PIPs. The institutional statements themselves are less specific than in Miami, and they do not specify many outcomes or activities that could produce them. Pertaining to boundary conditions, Clark County Commissioners, who are elected, also serve as the LVVWD Board of Directors, although the Las Vegas City Council oversees rate-making decisions. As with Miami, the countywide election of the PIP reflects some effort to scale representation on the governing board to the broader scope of resource demands. Related to scope conditions, the utility is authorized (via link 3) to provide "reasonably adequate service and facilities" and set rates (link 2) it deems "just and reasonable." Constitutional-choice rules can be intended to legitimize the entities involved in collective or operational choice processes, and the Las Vegas CIS appears to do so via formalizing the process whereby residents (resource users) "may" attend any rate-setting hearing and provide evidence for or against a change (link 2).

Over the study period, the CIS experienced elevated water stress (link 1) from 2004 to 2009, peaking in 2005. Public attention to water issues (link 2) peaked three times (1991, 2003, and 2007) and lagged behind the increased water stress. Two periods of alignment occurred between water stress and public attention (2003–2004, 2007–2008). Meanwhile, financial stress (link 3) peaked from 2009 to 2012, reflecting hard PI investments and liabilities incurred. The alignment of water stress and public attention within the CIS coincided with a number of changes to robust controls during its PoAC (2005–2009): Las Vegas implemented durable watering restrictions, landscape codes, and golf course water budgets (impacting link 1); it also began incorporating climate change and demand uncertainty into planning processes (links 2 and 3).

The robustness of the CIS was also likely affected by investments in hard PI. The SNWA draws surface water from Lake Mead, a large reservoir on the Colorado River operated by the US Bureau of Reclamation. Las Vegas also benefits from flow regulation provided by upstream reservoirs, including Lake Powell. Las Vegas extracts water from Lake Mead via three intake structures, the third of which was completed in 2015 (link 3) to be used in the event the water levels dropped below the elevation of the existing intake structures (link 1). In addition, Las Vegas pumps groundwater from a series of wells in the Las Vegas Valley.[6] LVVWD also uses reclaimed wastewater, both directly and indirectly. LVVWD provides treated wastewater for golf courses, parks, and other large-turf facilities requiring irrigation. Additionally, LVVWD discharges treated wastewater back to Lake Mead, where they earn return flow credits allowing additional withdrawals (Garcia and Islam 2018). The risk of surface water reduction increased over this period with declining storage in Lake Mead, but all other water sources remained stable.

Similar to the other cases, the CIS has made both hard and soft PI changes and reduced per capita resource use. From 1990 to 2015, the per capita water use decreased from 1140 L/person/day to 682 L/person/day.

19.4.3 Los Angeles

Los Angeles relies on a portfolio of surface, groundwater, and recycled wastewater. Snowmelt from the eastern Sierra Nevada Mountains is a significant water source. The snowpack provides natural storage, allowing the system to balance out-of-phase water demands and precipitation. Snowpack in the eastern Sierras is projected to decline with warming projected under climate change (Bales, Rice, and Roy 2015). Snowpack also plays an important role in LA's other surface water sources, such as the State Water Project (Johannis et al. 2016). Intensifying wildfire is also changing the nature of CA's watersheds. Wildfire not only alters hydrology (temporarily lowering infiltration and evapotranspiration rates), but also impacts water quality by increasing nutrient and metal fluxes (Rust et al. 2018). Los Angeles also depends on water from the Colorado River and is therefore subject to the same drought periods and drying trends described for Las Vegas.

The governance structure of the CIS – centered around the Los Angeles Department of Water and Power (LADWP) – is the oldest among the cases, founded in 1902 to end private control of the city's water supply. Pertaining to boundary conditions, the Los Angeles City Council oversees the LADWP Board of Commissioners, who make water-rate decisions (link 2) and are appointed by the mayor. Thus, unelected managers play a larger role in robust control choices. The CIS also displays more complex, multilevel regulatory arrangements. Each of LADWP's water sources has an independent regulatory entity. For example, groundwater is overseen at the basin level, while surface water rights are overseen by the California Department of Water Resources and State Water Resources Control Board, which also set high-level policies and requirements. The LADWP provides water to customers in the city of Los Angeles. The population of resource users has grown from 3.5 million in 1990 to 4.0 million in 2015.

The institutional structure of the Los Angeles CIS features more actor types (14) than Miami (10) or Las Vegas (4), and it includes an almost equal number of choice and scope rules. The institutional statements governing collective-choice processes focus more on adjudication and enforcement of water rights (link 1), with procedural directions aimed at the state court system (link 3). They also spell out specific directives for the LADWP Board of Commissioners to grant water permits and fix rates (link 2), prescribe time and methods of payment (link 6), develop utility assets (link 3), and authorize decisions for "production and delivery" and water as well as conservation (links 1 and 4). Thus, more of the potential pathways for feedback are implicated at the constitutional-choice level, perhaps reflecting the complexity of water rights in the western United States and the diversity of the CIS water supply portfolio.

The LA CIS relies on several types of hard PI to draw from local groundwater (12%), recycled water (2%), imported water from the LA Aqueduct (LAA) (27%), and purchased water from the Metropolitan Water District (MWD) (59%).[7] LADWP has well fields and structures to facilitate recharge across five groundwater basins. Contamination in the San Fernando Basin in 2007 led to a reduction in groundwater availability. The city has used reclaimed water for irrigation since 1979, and its capacity for wastewater reuse increased 10-fold over the study period. Los Angeles has two sources of imported surface water: water the city controls from Owens Valley and water purchased from MWD. The city owns diversion licenses on streams in Mono Basin and Owens Valley and moves this water through two parallel aqueducts, collectively known as the LAA. There are no large storage reservoirs in this system, and, as a result, deliveries vary substantially with climate cycles (from 67 million cubic meters to 617 million cubic meters between 1980 and 2015, link 7). Purchased water from MWD originates from either the Colorado River or the State Water Project. Water from the Colorado River is conveyed via the Colorado River Aqueduct and benefits from storage on the main stem, such as Lakes Mead and Powell. The large scale of storage means deliveries are robust to short-term hydrological fluctuations; however, they have been impacted by both changing operational rules and long-term drought. The adoption of the 2006 Consolidated Decree, for example, confirmed that California is limited to 5,400 million cubic meters from the Colorado River, reducing MWD supplies available to Los Angeles (link 8). Water from the state water projects is conveyed from the Sierra Nevada Mountains through a network of canals, pumping stations, and pipelines, and it is stored in 34 reservoirs to balance intra- and inter-annual hydrological variability (LADWP 2015).

Over the study period, the Los Angeles CIS experienced three water stress peaks (link 1) in 1991, 2008, and 2013; two public attention peaks (link 2) in 1991 and 2013; and a financial stress peak (link 3) in 2003. High water stress and public attention aligned in 1991 and again in 2007–2009 during its PoAC (2007–2010). This PoAC featured numerous robust control measures throughout the CIS: durable rate increases (impacting link 1); rebates for turf grass and conservation education programs (production via link 6); piloted stormwater capture and recharge projects (links 3 and 4); and planning for reducing dependence on imported water supplies (links 3 and 4).

These intensified efforts allowed Los Angeles to meet state-mandated conservation targets during 2015–2017. From 1990 to 2015, the per capita water use decreased from 655 L/person/day to 432 L/person/day (Garcia et al. 2019).

19.5 Conclusion

In the three cases above, urban regions facing increased water stress made resource management adjustments to maintain system performance. All three cases experienced external changes, such as sea level rise, aridification, and groundwater contamination, impacting the NI and PI in ways that moved the system closer to the edge of its robust range (link 7). Further, external changes, such as federal rulings on interstate water allocations and state-level regulation, imposed new rules on the PIPs (link 8) and challenged the robustness of the systems.

In response, the PIPs invested in changes to PI and NI and altered the rules structuring the interactions between the four key components of these CISs represented in the CIS Framework. We identify PoACs when a cluster of these alterations is made. Identifying PoACs focuses the analysis on actions taken by the PIP to maintain robustness under changing conditions. The PIPs invested in hard PI (e.g., aquifer storage and recovery in Miami and a new intake structure in Las

Vegas), and in recovering lost NI functionality (e.g., groundwater remediation in LA). A range of rule changes – investments in soft PI – was also observed, though not at the constitutional level. In all three cases, the PIPs adopted combinations of voluntary and coercive measures to control water demand (link 5 or 6) and modified planning processes intended to increase strategic management of the resource (link 3).

To advance their sustainability, the three cities both reinforced existing feedback loops and adopted new robust controls intended to buffer outputs from changes to inputs. For example, Las Vegas constructed a new deeper Lake Mead intake structure, strengthening the existing feedback loop involving links 1, 4, and 6. Similarly, leak detection programs in Miami increase efficiency on the margins. In contrast, investments in all three cities to shift the perceptions and behavior of water users through link 6 and to alter planning processes to incorporate uncertainty and climate change through links 1, 2, 3, and 4, alter the structure of the system in ways that create new mechanisms to adapt the system to its environment. New approaches to the built environment can similarly create new mechanisms. For example, the strategy in Los Angeles of capturing local stormwater to serve as a supplemental water source, and Miami's piloting of aquifer storage and recovery with reclaimed wastewater, created new opportunities for water management operations. Exploring transitions across Miami, Las Vegas, and Los Angeles illustrates the potential of the CIS Framework to enable comparison across a diversity of cases and facilitate generalization on the redesign of robust controls as external conditions change.

However, the dynamics of coupled human and environmental systems are often unpredictable. The diversity of institutional arrangements is also a critical component of achieving sustainable system performance. It remains to be seen if differences in institutionally induced decision-making processes determine which types of robust controls are needed. However, by examining PoACs through the CIS Framework, we can better understand how different CISs respond to a set of projected exogenous changes. When their respective PoACs result in sustained system performance, we can identify design elements that may be generalizable to other urban regions facing similar social and climatic pressures. One limitation of our analysis is that the three PoACs observed are fairly recent and each city continues to experience both exogenous and endogenous change. Therefore, it remains unclear if the PoACs, in addition to increasing near-term robustness, have either increased or decreased system resilience.

Notes

1 https://www.epa.gov/sustainability/learn-about-sustainability#what
2 https://www.miamidade.gov/water/library/20-year-water-supply-facilities-work-plan.pdf
3 https://www.miamidade.gov/water/library/20-year-water-supply-facilities-work-plan.pdf
4 https://www.leg.state.nv.us/SpecialActs/67-SNevadaWaterAuthority.html
5 https://www.leg.state.nv.us/SpecialActs/62-LasVegasValleyWater.html
6 https://www.snwa.com/assets/pdf/water-resource-plan-printable-2019.pdf
7 https://s3-us-west-2.amazonaws.com/ladwp-jtti/wp-content/uploads/sites/3/2019/07/29154703/2018-Briefing-Book-Web-3.pdf

Bibliography

Anderies, John, Marco Janssen, and Elinor Ostrom. 2004. "A Framework to Analyze the Robustness of Social-Ecological Systems from an Institutional Perspective." *Ecology and Society* 9 (1). https://doi.org/10.5751/ES-00610-090118.

Anderies, John M. 2003. "Economic Development, Demographics, and Renewable Resources: A Dynamical Systems Approach." *Environment and Development Economics* 8 (2): 219–46.

Anderies, John M., Olivier Barreteau, and Ute Brady. 2019. "Refining the Robustness of Social-Ecological Systems Framework for Comparative Analysis of Coastal System Adaptation to Global Change." *Regional Environmental Change* 19 (7): 1891–1908.

Anderies, John M., Carl Folke, Brian Walker, and Elinor Ostrom. 2013. "Aligning Key Concepts for Global Change Policy: Robustness, Resilience, and Sustainability." *Ecology and Society* 18 (2). https://www.jstor.org/stable/26269292.

Anderies, John M., Armando A. Rodriguez, Marco A. Janssen, and Oguzhan Cifdaloz. 2007. "Panaceas, Uncertainty, and the Robust Control Framework in Sustainability Science." *Proceedings of the National Academy of Sciences of the United States of America* 104 (39): 15194–99.

Anderies, John M., Paul Ryan, and Brian H. Walker. 2006. "Loss of Resilience, Crisis, and Institutional Change: Lessons from an Intensive Agricultural System in Southeastern Australia." *Ecosystems* 9 (6): 865–78.

Bales, Roger C., Robert Rice, and Sujoy B. Roy. 2015. "Estimated Loss of Snowpack Storage in the Eastern Sierra Nevada with Climate Warming." *Journal of Water Resources Planning and Management* 141 (2): 04014055.

Blomquist, William Andrew, Edella Schlager, and Tanya Heikkila. 2004. Common Waters, Diverging Streams: Linking Institutions to Water Management in Arizona, California, and Colorado. Resources for the Future.

Carpenter, S., B. Walker, J. M. Anderies, and N. Abel. 2001. "From Metaphor to Measurement: Resilience of What to What?" *Ecosystems*. https://link.springer.com/article/10.1007/s10021-001-0045-9.

Garcia, Margaret, and Shafiqul Islam. 2018. "The Role of External and Emergent Drivers of Water Use Change in Las Vegas." *Urban Water Journal* 15 (9): 888–98.

Garcia, Margaret, Elizabeth Koebele, Aaron Deslatte, Kathleen Ernst, Kimberly F. Manago, and Galen Treuer. 2019. "Towards Urban Water Sustainability: Analyzing Management Transitions in Miami, Las Vegas, and Los Angeles." *Global Environmental Change: Human and Policy Dimensions* 58 (September): 101967.

Hughes, Joseph D., and Jeremy T. White. 2014. "Hydrologic Conditions in Urban Miami-Dade County, Florida, and the Effect of Groundwater Pumpage and Increased Sea Level on Canal Leakage and Regional Groundwater Flow." Reston, VA. https://pubs.usgs.gov/sir/2014/5162/.

Hughes, Sara, Stephanie Pincetl, and Christopher Boone. 2013. "Triple Exposure: Regulatory, Climatic, and Political Drivers of Water Management Changes in the City of Los Angeles." *Cities* 32 (June): 51–59.

Johannis, Mary, Lorraine E. Flint, Michael D. Dettinger, Alan L. Flint, and Regina Ochoa. 2016. "The Role of Snowpack, Rainfall, and Reservoirs in Buffering California against Drought Effects." *US Geological Survey*. https://pubs.er.usgs.gov/publication/fs20163062.

LADWP. 2015. *The 2015 Urban Water Management Plan*. Technical Report, Los Angeles Department of Water and Power, Los Angeles, CA.

Malekpour, Shirin, Rebekah R. Brown, and Fjalar J. de Haan. 2015. "Strategic Planning of Urban Infrastructure for Environmental Sustainability: Understanding the Past to Intervene for the Future." *Cities* 46 (August): 67–75.

McGinnis, Michael D. 2011. "An Introduction to IAD and the Language of the Ostrom Workshop: A Simple Guide to a Complex Framework." *Policy Studies Journal: The Journal of the Policy Studies Organization* 39 (1): 169–83.

McNulty, Steven G., Ge Sun, Erika C. Cohen, J. A. Moore-Myers, D. Wear, and W. Jin. 2007. "Change in the Southern US Water Demand and Supply over the next Forty Years." *Wetland and Water Resource Modeling and Assessment: A Watershed Perspective*, 43e56.

Muneepeerakul, Rachata, and John M. Anderies. 2017. "Strategic Behaviors and Governance Challenges in Social-ecological Systems." *Earth's Future* 5 (8): 865–76.

Novo, Paula, and Alberto Garrido. 2014. "From Policy Design to Implementation: An Institutional Analysis of the New Nicaraguan Water Law." *Water Policy* 16 (6): 1009–30.

Ostrom, Elinor. 1990. *Governing the Commons: The Evolution of Institutions for Collective Action*. Cambridge University Press.

———. 2009. "A General Framework for Analyzing Sustainability of Social-Ecological Systems." *Science* 325 (5939): 419–22.

———. 2011. "Background on the Institutional Analysis and Development Framework." *Policy Studies Journal: The Journal of the Policy Studies Organization* 39 (1): 7–27.

Ostrom, Elinor, Roy Gardner, James Walker, James M. Walker, and Jimmy Walker. 1994. *Rules, Games, and Common-Pool Resources*. University of Michigan Press.

Overpeck, Jonathan T., and Bradley Udall. 2020. "Climate Change and the Aridification of North America." *Proceedings of the National Academy of Sciences of the United States of America*. National Acad Sciences, 117(22), 11856–11858

Pahl-Wostl, Claudia. 2007. "Transitions towards Adaptive Management of Water Facing Climate and Global Change." *Water Resources Management* 21 (1): 49–62.

Rust, Ashley J., Terri S. Hogue, Samuel Saxe, and John McCray. 2018. "Post-fire water-quality response in the Western United States." *International Journal of Wildland Fire* 27 (3): 203–16.

Scheffer, Marten, Jordi Bascompte, William A. Brock, Victor Brovkin, Stephen R. Carpenter, Vasilis Dakos, Hermann Held, Egbert H. van Nes, Max Rietkerk, and George Sugihara. 2009. "Early-Warning Signals for Critical Transitions." *Nature* 461 (7260): 53–59.

Schlager, Edella, and Tanya Heikkila. 2011. "Left High and Dry? Climate Change, Common-Pool Resource Theory, and the Adaptability of Western Water Compacts." *Public Administration Review* 71 (3): 461–70.

Seawright, Jason, and John Gerring. 2008. "Case Selection Techniques in Case Study Research: A Menu of Qualitative and Quantitative Options." *Political Research Quarterly* 61 (2): 294–308.

Siddiki, Saba, Christopher M. Weible, Xavier Basurto, and John Calanni. 2011. "Dissecting Policy Designs: An Application of the Institutional Grammar Tool." *Policy Studies Journal: The Journal of the Policy Studies Organization* 39 (1): 79–103.

Treuer, G., Koebele, E., Deslatte, A., Ernst, K., Garcia, M., & Manago, K. (2017). A narrative method for analyzing transitions in urban water management: The case of the Miami-Dade Water and Sewer Department. *Water Resources Research, 53*(1), 891–908.

Udall, Bradley, and Jonathan Overpeck. 2017. "The Twenty-First Century Colorado River Hot Drought and Implications for the Future." *Water Resources Research* 53 (3): 2404–18.

PART V

Urban water governance and sustainability

20

URBAN METABOLISM AND WATER SENSITIVE CITIES GOVERNANCE

Designing and evaluating water-secure, resilient, sustainable, liveable cities

Steven J. Kenway, Marguerite Renouf, J. Allan, KMN Islam, N. Tarakemehzadeh, M. Moravej, B. Sochacka, and M. Surendran

20.1 Urban metabolism

20.1.1 What is urban metabolism?

The concept of "urban metabolism" has been around for some time, yet it has struggled to be applied in practice. In 1965, Abel Wolman used the concept to consider the then challenges of water and air pollution and public economic decisions. This was a visionary prescient of water security and energy challenges; however, over 55 years later, government and science struggle to assess "good urban water performance" considering "how do cities" (rather than utilities) manage water well.

Drawing on the definitions of others (Baccini and Brunner 1991, Kennedy et al. 2007, Wolman 1965a) we can define urban metabolism as "the process of resources flowing through and being transformed and consumed in urban settlements to sustain all the technical and socio-economic processes that occur within in it," as one might observe it in ecosystems or biological organisms. A distinguishing feature of urban metabolism is the emphasis on the "urban entity" as the functional unit (such as a precinct, an urban catchment, or a city). This contrasts with other city-related interests, which may consider subcomponent elements of the urban entity, such as water infrastructure systems that service the city.

A clear understanding and quantification of how "urban systems" perform with respect to resource utilisation are needed to inform policymaking and governance interventions. However, even basic concepts and definitions such as the meaning of "urban" can vary significantly and have an enormous impact on key metrics such as "urban population." For more complex aspects of cities, resource consumption data can be very patchy (Minx et al. 2011). Importantly, there is a lack of evaluation of resources consumption at the city scale (Minx et al. 2011). Evaluation frameworks can fill this gap by defining the problem, objectives, and methods and by communicating the results. They are crucial for creating knowledge about and quantifying the performance of

DOI: 10.4324/9781003057574-26

cities, which is essential for supporting city design and management and informing policy and governance interventions. In this chapter, we refer to a developed urban metabolism evaluation framework for water for understanding water impacts and the relationship between water flows in a city and other material and energy flows.

At the outset, we should define "urban," which can at the scale of a site up to entire regions. As urban water metabolism assessment relies on quantitative water mass balance analysis, it is essential to account for all water movement (physical and embodied) through or stored within a defined urban volume (Kenway et al. 2011a). For this, a clear three-dimensional "system boundary" is required comprising an area and its depth. For a city, we consider the urban entity to extend from the roofline and tree canopy to a depth of 1 m below ground (Renouf et al. 2017b). The system boundary of urban entities can have different scales – the city, the city-region (e.g., the metropolitan area comprising a given city with its satellites), an urban precinct (suburb, neighbourhood), or a specific site. Thus, a consistent definition of "system boundary" can be applied at the precinct (Farooqui et al. 2016) and site scales (Moravej et al. 2021), making it transferrable across scales. Bulk surface water storages located outside the urban entity and groundwater aquifers beneath it are excluded. This is to enable quantification of the exchanges of water between the "environment" and the "city" (Renouf et al. 2018). Similarly, inflows to and outflows from the urban entity via natural rivers and creeks (i.e., flows originating outside the urban area) are considered outside the "urban entity" system boundary.

These exclusions are essential for two reasons. First, water metabolism analysis aims to analyse how urban entities exchange water with "the environment." Consequently, the boundary between "urban" and "environment" must be clear. If major regional water storages are included within the "water balance," the effect would be that many "urban-environment" flows would be "internalised within the system," which would make tracking their impacts (or the influence of new design or management) much harder. Second, adopting a "tight" system boundary increases accuracy in the water mass balances because it helps detect smaller differences in the independently measurable total system inputs and outputs. Hence, it improves the accuracy of the assessment and improves the impact assessment of alternative urban designs and scenarios.

It should be noted that urban entities (described hereafter as "cities") are supported by regions, including surrounding peri-urban and rural areas. Cities depend on surrounding "regions" to provide freshwater (e.g., from water storages), energy (e.g., electricity and gas), food, and materials. Cities also depend on regions to accept wastes. The distinction between the city and its supporting region refers more to the functional relationship between them, rather than geographic proximity. The urban entity may be defined in a way that includes some areas outside the administrative boundary of a given city, e.g., smaller towns in a large agglomeration, while treating other areas, e.g., rural land, as located outside of the urban entity. A city-region is an example of this as it comprises a central city core surrounded by satellite cities within daily commuting distances; they are all considered the urban entity (Kennedy et al. 2014).

Urban metabolism is a useful concept for understanding how cities consume resources for several reasons. First, the "metabolism" metaphor brings the resemblance to an organism or ecosystem (Decker et al. 2000, Newman 1999, Wolman 1965a) and helps articulate a grand challenge of sustainability in the context of urban settlements. How do we reduce dependence on the environment to provide resources to cities and assimilate their wastes while simultaneously improving liveability and ecosystem well-being?

As an evaluation approach, an urban metabolism approach quantifies resource flows through urban entities, which could be water, energy, materials, or nutrients. Here, we are principally interested in water (Figure 20.1). When we refer to the water metabolism of the city, we mean

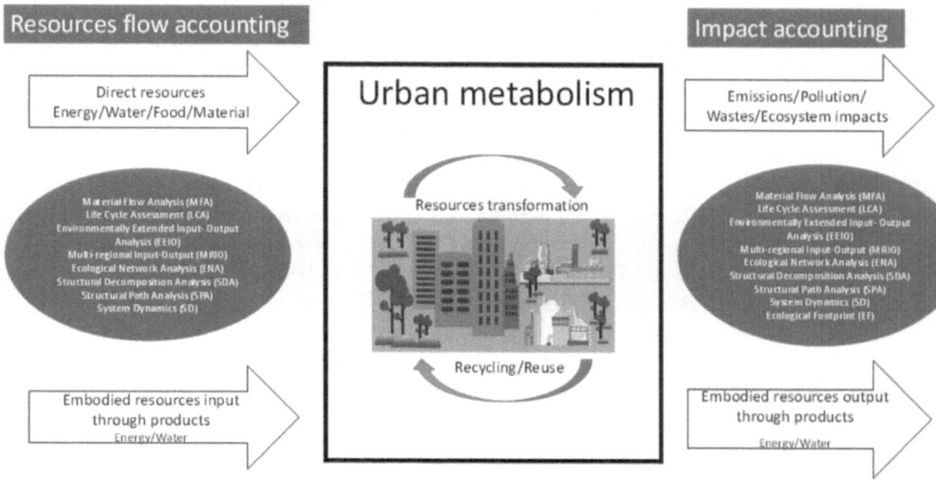

Figure 20.1 Resources flow and impact accounting in urban metabolism framework.

the process of water flowing through and being transformed and consumed by the urban entity to sustain all the technical and socio-economic processes that occur within it.

Urban water metabolism allows us to comprehensively and systematically account for the diverse flows (physical and embodied) of water into, within, and out of urban entities to understand the overall water efficiency. In addition, the impacts from water consumption, such as greenhouse gas (GHG) emissions and ecological footprint, can, by extension, also be quantified (Table 20.1).

As a robust analysis of water flows within a city, urban water metabolism can inform realisation of holistic city visions, such as Water Sensitive Cities, Water Wise Cities, or Sponge Cities. Improved urban design and planning options need to be considered within the broader context of citywide water efficiency, to optimise the function of water in cities, by bringing attention not only to citywide efficiency, but also to the function of water in cities. A metabolically efficient city in terms of water tries to emulate natural systems by making optimal use of the water supplies available to it, reducing water wastage, and recycling water and nutrients and energy carried in it. This chapter explores how urban metabolism evaluation offers a way of quantifying the progress cities are making towards this, to achieve resource-efficient, liveable, and productive Water Sensitive Cities of the future.

In 1965, the president of the American Water Association, Abel Wolman, used the concept of urban metabolism in a pioneering analysis to identify solutions to "shortages of water, pollution of water and air ... and [provide input to] public economic decisions" (Wolman 1965a, b). He defined urban metabolism as all the materials and commodities needed to sustain a city, including water, energy, and food. While he identified water as the dominant resource consumed in cities, he concluded that "there is plenty of water, but supplying it requires foresight." Over 50 years later, we are still grappling with long-term resource planning. But how has the concept of urban metabolism developed since then and how does it help address the challenges faced by cities today?

Since Wolman's foundation work, the concept has been applied as an evaluation approach, in a range of guises, to generate information and metrics about urban resource consumption (Table 20.1). For comprehensive reviews, see Kennedy et al. (2011), Baynes and Wiedmann (2012), and Renouf and Kenway (2017a, b). Newman et al. (1996) provided an early application of the

Table 20.1 Examples of urban water metabolism evaluation studies and methods used for resources flows and impact accounting

Criteria	*Methods*	*Selected Examples*
Direct and embodied water inflows/outflows to cities.	Material Flow Analysis (MFA) Life Cycle Assessment (LCA) Environmentally Extended Input-Output Analysis (EEIO)	(i) Direct water and energy flow in a hypothetical Australian city (Kenway et al. 2011b). (ii) Direct and embodied water flow in Beijing, China (Sun 2019).
Water hinterlands of cities.	Multi-regional Input-Output (MRIO)	(i) Water hinterlands of Sydney, Melbourne, and Brisbane, Australia (Islam et al. 2021).
Environmental and social impacts of water consumption.	Life Cycle Assessment (LCA) Environmentally Extended Input- Output Analysis (EEIO)	(i) Urban water system environmental and social impacts in Mexico City, Mexico (García-Sánchez and Güereca 2019).
Control-dependency relationship within the resources flow network.	Ecological Network Analysis (ENA)	(i) To what extent economic sectors are dependent on each other for energy in Beijing, China (Chen and Chen 2015).
Drivers and actors influencing the resources flow.	Structural Decomposition Analysis (SDA) Structural Path Analysis (SPA)	(i) Drivers of energy related emission changes in Singapore city (Su et al. 2017).
Impacts of varying socioeconomic parameters of resources consumption.	System Dynamics (SD)	(i) Water and energy-based interventions assessment in Penticton, British Columbia, Canada (Chhipi-Shrestha et al. 2017).
Impact of resources demand on natural capital and ecosystem.	Ecological Footprint (EF)	(i) Ecological Footprint (EF) and bio-capacity (BC) assessment of resources flow in Wroclaw, Poland (Świąder et al. 2020).

urban metabolism concept in framing indicators of urban performance in Australia's State of Environment reporting. Other applications have included highlighting the scale and trajectory of resource flows through cities, directly (Kennedy et al. 2007) and indirectly (Goldstein et al. 2013), understanding the urban impacts on regional resources (Billen et al. 2012), and identifying resource efficiency (Kenway et al. 2011a) and resilience (Pizzol et al. 2013) opportunities.

20.1.2 How is it different from other approaches?

Other urban resource analysis approaches adopt a different scope of analysis (Table 20.2). For instance, they focus on resource services within the city, resource consumption of the city-based consumers, or supply chains or economies centred on cities. As it is common for these different approaches to be presented under the banner of urban metabolism, to avoid confusion we suggest that a distinguishing feature of urban metabolism is the emphasis on the "urban entity" as the functional unit, which is defined by an urban boundary (Renouf et al. 2016, 2020a). Therefore, urban metabolism considers all resource-consuming activities within

Table 20.2 Different approaches to urban resource evaluation: approaches, perspectives, and methods (Renouf and Kenway 2017a)

Approach	*Perspective*	*Methods*
Resource system modelling	Resource services within the city (water, energy, waste)	**Material Flow Analysis (MFA) –system mass balance** Quantifies resource flows through urban infrastructure to manage supply against demand and plan infrastructure, etc.
Metabolism	City as producer and consumer	**Material Flow Analysis (MFA) – territorial, or city mass balance** Quantifies resource exchanges between the urban entity and its local / regional environment for sustaining urban functions
Consumption	City as consumer	**MFA extended with life cycle assessment (MFA + LCA) and environmental footprinting** Quantifies resource exchanges between the urban entity and the global environment, characterising impacts
	Urban consumers	**Environmental Input- Output Analysis (E-IO)** Quantifies the resource flows through economic supply chains at various scales using economic input-output tables
Complex systems	City as ecosystem	**Ecological Network Analysis (ENA)** Quantifies resource flow networks within the "urban system", to characterise the performance of the "system as a whole"
	Networks within the city	**System Dynamics (SD)** Quantifies trends in urban resource flows under varying socio-economic parameters

the urban boundary, and it focuses on the direct resource exchanges with the near hinterland. In comparison, other approaches may examine subcomponents of the urban entity or systems that cut across the urban boundary and consider indirect resource exchanges with distant/global hinterlands.

Among these other approaches, it is worth explaining how urban metabolism relates to footprinting techniques, such as ecological footprint (Wackernagel and Rees 1996) and water footprint (Hoekstra et al. 2011). Footprinting has its origins in the life cycle assessment (LCA) of products but has been used in the context of cities (Hoff et al. 2014). When applied to geographic entities, e.g., a city, a water footprint, for instance, is the total amount of water consumed, both direct physical flows and embodied flows associated with the goods and services consumed by city dwellers (Mekonnen and Hoekstra 2011). Thus, similarly to urban water metabolism, a water footprint has a strong consumption focus. However, water footprinting is more focused on water embodied in goods and services (influencing global water supplies) rather than water used directly in cities (influencing local/regional water supplies). Urban water metabolism evaluation can inform an indicator of the local/regional water footprint of a city. However, the use of the term "urban metabolism" in this context would not be in line with its purpose of enabling robust analysis of different water flows. In contrast, the footprint concept generates a single metric with little resolution to help characterise contributing flows and identify areas of concern (Sahely et al. 2003).

The most promising method for evaluating urban water metabolism is water balance material flow analysis (Renouf and Kenway 2017a). An urban water balance can be applied at different scales – to an urban water system for evaluating infrastructure (Bach et al. 2014), to urban catchments for understanding catchment hydrology (Haase 2009), or to the whole urban entity (Kenway et al. 2011a, Thériault and Laroche 2009). For urban metabolism applications, we are interested in the latter and refer to it as *city water balance* to distinguish it from the other approaches. A city water balance is a top-down evaluation of all inputs, outputs, and storage of both hydrological and managed water for the whole urban entity. It gives a complete picture of a city's hydrological performance.

In contrast, a water balance of an urban water system typically involves bottom-up modelling of flows managed by urban water infrastructure, often at catchment or neighbourhood scale (Bach et al. 2014), using models such as Aquacycle (Mitchell et al. 2001), Urban Developer (Hardy et al. 2005), and CRC for Water Sensitive City's Scenario Tool (formerly known as DAnCE4Water) (Rauch et al. 2017). They integrate potable water, wastewater, stormwater, and rainwater systems (thus also called *integrated urban water modelling*), and they are useful for assessing the influence of urban design on urban hydrology. However, it does not give a complete picture of the entire natural and human-influenced water cycle within the urban entity, as opposed to those based on an urban water metabolism framework, such as the Site-Scale Urban Water Mass Balance Assessment model (Moravej et al. 2021).

20.1.3 What do we know about the water metabolism of cities and its evaluation?

Currently, cities remain poorly understood in terms of the many flows of natural resources that occur within their boundaries. This is partially due to the lack of consistent definitions (e.g., of what is defined as "urban"). The fracturing or scattering of urban governance in many cities (some island nations are exceptions, e.g., Singapore) also means we often understand more about "states" and "countries" than we do about cities.

In general, there is a good understanding of the impacts of urbanisation on the water cycle, including increased stormwater run-off, increased risk of downstream flooding, reduced evapotranspiration, and reduced local water storage (e.g., reduced water stored in soil, vegetation, and wetlands). However, literature reviews (Renouf and Kenway in preparation, Renouf and Kenway 2017a) show that the array of evaluation approaches presented under the banner of urban metabolism gives a splatter-gun picture of urban water metabolism due to their highly varied scales and scopes. So, a first observation is that there is a need to harmonise the evaluation of urban water metabolism. We understand how natural hydrology is affected by urbanisation and small inroads have been to quantify these changes at the city scale (Bhaskar and Welty 2012). Models have been available for some time for evaluating flows managed by urban infrastructure (Bach et al. 2014). They are useful to evaluate managed urban water system components in an integrated way. Still, they are usually directed at urban components rather than the whole of the city, and they do not capture all flows through the urban landscape.

Emerging *urban water balance* models have started to address this gap by quantifying a broader spectrum of water flows (both natural and managed), from which metabolic indicators can be generated (Kennedy et al. 2014, Kenway et al. 2011a, Renouf et al. 2017a, van Leeuwen et al. 2012). While indicator development is still in its infancy, the work done shows that large flows of water pass through cities underutilised and that utilisation efficiency (e.g., water turnover rate) is a long way off from what is possible and it is not close to the optimal efficiencies of natural systems (Kenway et al. 2011a).

An *urban water balance* material flow analysis accounts for all flows across a defined three-dimensional urban boundary to achieve a mass balance (i.e., inflow = outflows + storage). It can be described by Equation 1 for a defined system, with boundary (*B*), as described above, (Kenway et al. 2011a).

$$\Delta S = (St1 - St2) = Qi(t1 - t2) - Qo(t1 - t2) \tag{1}$$

ΔS is the change in stored water volume (or mass) within the boundary, measured as the difference between water stored within (*B*) at the initial time (*St1*) and some following time (*St2*). ΔS also equals the difference in the total volume of inputs to *(Qi)* and outputs from *(Qo)* (*B*)—$Qi(t1 - t2)$ and $Qo\ (t1 - t2)$, respectively. For a system in which both the time interval ($t1 - t2$) and the boundary (*B*) have been defined, Equation 1 simplifies, and all units are expressed as fluxes: volumes or masses moving per unit time (Equation 2). This equation holds for any temporal and spatial scales.

$$\Delta S = Qi - Qo \tag{2}$$

20.2 Urban metabolism and Water Sensitive Cities

Cities move through transitional stages of water management, from an initial focus on water supply, sewerage, and drainage and then onto higher ambitions, such as waterway protection and resource conservation (Brown et al. 2009). A "Water Sensitive City" is defined as one that "integrates water supply, sewerage, stormwater and the built environment, and respects the value of urban waterways and whose citizens value water and the role it plays in sustaining the economy, environment and society" (Brown et al. 2009). A Water Sensitive City recognises diverse functions of water in the urban landscape. It encompasses multiple objectives of supply security, public health, flood control, environmental protection and repair, amenity, liveability, and sustainability (Brown et al. 2009, Rogers et al. 2020). This is consistent with the principles of liveable and sustainable cities (WSAA 2012) and the characteristics of a successful Australian water sector (National Water Commission 2011).

In the same way that the concept of a Water Sensitive City has been an evolution in the way we view and manage value water in the urban landscape, urban metabolism is an evolution in the way we evaluate and quantify water, energy, and other material flows. We identify three aspects that have changed, leading to a need for a metabolic approach to understanding and quantifying water management in cities:

(i) the changing face of supply security;
(ii) recognition of the multiple functions of water; and
(iii) the need to link cities to their regions.

Water supply security has traditionally meant the reliable availability of acceptable quality and quantity of water for health, livelihood, and production, though many definitions exist. Supply has traditionally focused on a centralised potable water system commonly fed by the capture in dams of rainfall from surrounding rural/forested catchments as well as groundwater. Hence, water security has conventionally been managed by planning future water availability from the defined supply catchment (usually outside the city) against future demands (from within the city). As centralised supplies have become prone to water shortages due to climate change and increasing water demand due to population growth. It is a necessity to expand the water supply sources from the centralised water sources to alternative (decentralised) water sources, such as stormwater, rainwater (roof run-off), and recycled wastewater, which are sourced within the system boundary (Wong and Brown 2009). This has made the picture of supply and demand more complicated, from one based on a single source of water with a uniform quality feeding all consumers to one that is based on multiple sources of water of varying qualities directed to different fit-for-purpose applications. Therefore, supply security needs to be evaluated in this context, as matching diverse supplies with the various functions of water in the urban landscape, to maximise the functionality that available water can deliver.

In addition to the local hinterland for water supply, a thorough knowledge of the dependence of cities on their global hinterland for embodied water through water-intensive goods, such as foods and clothes, is needed to understand the true water security of cities (Hoekstra et al. 2018). A thorough understanding of the embodied water supply chain of a city can help decision-makers cope with far-reaching water scarcity and prepare for any future disturbance in the product supply chain (Renouf and Kenway 2017a). This supply security aspect is essential because, in the absence of embodied water supplier regions, product supply security will be jeopardised in a city unless the same physical water is extracted within the city to produce the same quantity of products (Lenzen 2009). Urban metabolism offers a framework for this thorough understanding because it views urban water management holistically, including direct and embodied water flows. It could be operationalised to have a citywide water budget based on the city water balance described earlier.

The second change has been the recognition of the multiple functions of water in the urban landscape. Beyond its primary value for sustaining life, health, and productivity, we now recognise its value for liveability, amenity, recreation, environmental services, and cultural reasons. For example, the use of water for irrigation to maintain vegetation is central to mitigating urban heat island effects for improved amenity (Coutts et al. 2014, Rockström et al. 2009). Maintaining natural water flows in urban streams is needed for environmental services. Irrigation of sports fields can be important for community recreation and, hence, health and liveability. Urban metabolism offers a framework that allows us to quantify and account for these functions to evaluate the significance and feasibility of directing water to these functions.

The third change is the recognition of the connectivity between the city and its region. Cities are not isolated entities; they exist within and ultimately depend on their surrounding landscapes, which extend far beyond their built edge (Serrao-Neumann et al. 2017). The Water Sensitive City, therefore, needs to be understood in relation to its broader catchment and the landscape types that occur within that catchment (Serrao-Neumann et al. 2017) and their water-related needs. Greenspace is a feature that links the city with its region and is intimately linked with water. Understanding a city's metabolism of water, inclusive of its diverse functions, provides a mechanism for valuing greenspace and the connectivity it provides.

These above aspects are captured in a vision for a metabolically water and energy-efficient city (Kenway et al. 2013a), which can enhance our visions for Water Sensitive Cities:

> *Successful cities will be liveable, fun, and locally self-sustaining. They will meet regulated standards and optimise diverse energy and water supply portfolios affordably. The cities will achieve or strive towards net-zero energy (carbon) use and water waste. The city would be self-reliant and not draw on the surrounding environment for input resources or waste assimilation, noting that the degree of self-reliance is likely to depend on local conditions. What is critical is that the city operates within its local water and energy (carbon) budget. This efficiency paradigm is fully embraced at a "whole system" level. It fosters creativity, design innovation, business, and economic success.*

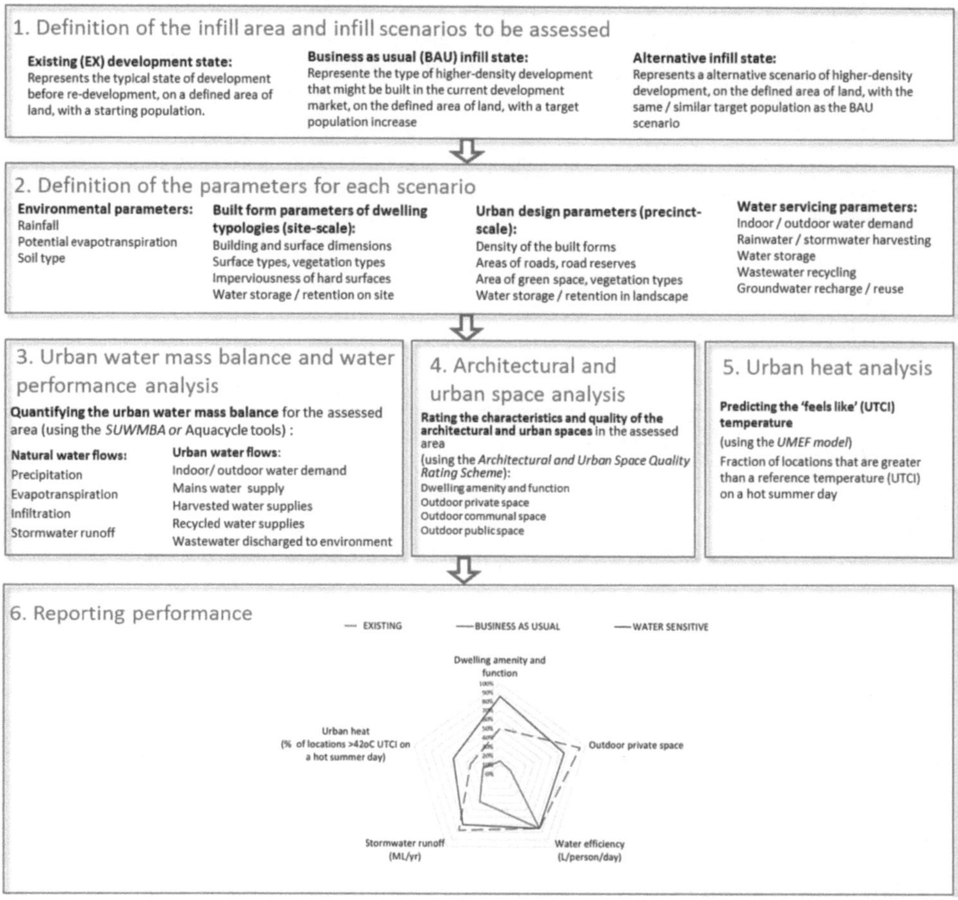

Figure 20.2 An example of evaluation framework based on urban water metabolism (Renouf et al. 2018) for designing water sensitive infill development adapted from Renouf et al. 2020a). See Renouf et al. (2020b) and London et al. (2020) for case study application of the framework.

In summary, urban metabolism offers a framework for articulating and quantifying progress towards achieving the attributes of a Water Sensitive City, by generating information to account for the diverse alternative water sources and functions of water in the urban landscape, linking cities to their supply catchments, and considering energy implications (Figure 20.2). It takes us from evaluating the water-sensitivities components of urban water to evaluating the city (or urban entity) as a whole (Renouf et al. 2020a).

20.3 An urban metabolism evaluation framework for water

This section introduces an *urban water metabolism evaluation framework* (Renouf et al. 2018). The framework articulates how quantifying the inflows and outflows across a defined urban boundary, based on an "urban water balance" approach (Kenway et al. 2011a) can support performance evaluation with quantitative indicators. This quantification allows for the generation of indicators representing metabolic characteristics (see examples in Table 20.3). The development of metabolic indicators is in its infancy. Still, it could include water use intensity for the urban

Table 20.3 Examples of metabolic indicators that can be generated from a "whole of city water balance" (adapted from Kenway et al. 2011a). These contain important measures, expressed as percentages, of performance such as "supply centralisation/self-sufficiency," which comments on how much water the city depends on water harvested from "within the boundary," e.g., from decentralised sources, as opposed to being provided from "outside the boundary." Similarly, "rainfall harvesting" gives an indication of what percentage of local rainfall is being captured and reused within the system. The conclusions that can be drawn from "metabolic" indicators include how well a city is being managed *collectively* for its water performance rather than "how good are its utilities for providing water supply."

%	*Supply centralisation* $C/(C+D_{R+G})$	*Reuse of anthrogenic input* $Re/(C+D_{R+G})$	*Rainfall harvesting* (D_R/P)	*Replaceability*	
				Stormwater run-off (Rs/C)	*Wastewater (W/C)*
Brisbane (SEQ)	100	2	0.1	104	48
Sydney	100	1	0.1	76	86
Melbourne	89	4	1	68	79
Perth	54	1	22	47	26

entity as a whole, water efficiency (water extraction per unit of functionality/productivity), degree of supply centralisation, the potential for self-sufficiency, water turnover rate, supply resilience, and degree of "naturalness" (Renouf et al. 2017a, 2019). Uniquely, the evaluation framework:

- accounts for flows of water into, out of, and stored within a clearly defined urban entity, to influence an "urban water balance";
- accounts for the diverse sources and uses of water within the urban landscape to generate an "urban water budget";
- connects water management in urban areas to the surrounding hinterland(s), including the peri-urban and rural areas;
- accounts for water-related energy implications; and
- generates repeatable and robust metabolic indicators.

An extension of this is also to consider links to surrounding peri-urban and rural landscapes. This recognises that cities are not isolated entities but exist within and rely on landscapes that extend beyond the urban boundary (Serrao-Neumann et al. 2017) and that urbanisation will expand into these landscapes as the population grows. Urban metabolism in this context highlights potentially competing uses of water in the city's hinterlands (food production, resource extraction, industrial production, energy generation, recreation, and natural habitat). This allows understanding whether and how the city's drain on water resources from surrounding landscapes compromises the functions it serves and what constraints may need to be placed on the future urban expansion to manage potential conflicts and trade-offs.

Like an ecosystem, water in the urban landscape is highly interconnected with natural and socioeconomic systems, internally across the various components, and externally with surrounding environments. Evaluating urban water metabolism at the city-entity or city-region scale allows us to see how individual components and changes to these components influence the metabolic efficiency of the entity as a whole and how to avoid a solution for one component becoming a problem for another component. An evaluation framework can inform a "city water balance," which quantifies flows of water across the urban boundary to affect a mass balance, and a "city water budget,' which accounts for the uses and users of water within this boundary.

Defining the boundary of the urban "entity" is crucial as it determines what is included in the evaluation, which is important for comparing, benchmarking, and monitoring cities. The boundary is both geographic and functional, and it is related to social and economic activities. For example, in a project that evaluated urban design scenarios of infill development at a precinct scale, the "boundary" was defined as "the physical boundary of urban entity consisting of (i) a horizontal boundary relates to the precinct/site boundary and (ii) vertical boundary ranges from 1m below the ground-level to the height of the tallest building or trees in the chosen location." Importantly, this required conceptualisation of the "system" as well, which was defined as "the combination of physical areas and technical systems associated with the urban area being assessed or evaluated. It includes built forms and landscapes within the physical urban area and the water services that draw from urban catchments, which may be outside the urban area being assessed" (Sochacka et al. 2021). The boundary must also consider what social and economic activities occurring within the geographic boundary are included. These features of scale- and boundary-making mark urban metabolism evaluation as different from existing modelling of urban water systems (Bach et al. 2014), which usually consider urban water systems at a precinct or catchment scale.

20.4 Governance and the urban metabolism framework

While there is a range of definitions and interpretations of governance, most simply, governance can be described as how decisions are made and how organisations operate to implement these decisions. The aim of good governance is to have transparent, participatory, inclusive, and responsive processes in place to meet the needs of the serviced community (ref). To take an integrated approach to decision making in an urban setting, it is imperative to understand the setting, including the relationships between resource flows. Considering governance through the lens of urban metabolism reveals insights into the benefits of one aspect of water (or related resources) in comparison to other influenced aspects. As such, it provides a pathway to optimise water management decision making in any urban envelope.

Sustainable urban metabolism requires robust governance, including coordinated institutional frameworks complemented by integrated policy packages (Davoudi and Sturzaker 2017, Dijst et al. 2018). On the one side, some consider that adopting urban metabolism as a theoretical basis is vital to effective strategic planning for cities and, ultimately, their sustainability (Davoudi and Sturzaker 2017, Zengerling 2019). On the other hand, others have gone further and suggested that metabolism and governance, when coupled with cultural considerations, form a basis for redefining the smart city paradigm (Allam and Newman 2018).

Traditionally, urban water management has been characterised by a siloed approach where different institutions manage different parts of the urban water cycle. While it has been recognised that there is a need for better integration and collaborative water governance, achieving this has been challenging (Serrao-Neumann et al. 2019). Fragmented responsibilities can make collaboration between institutional actors difficult to justify and achieve within the current business models. An urban metabolism framework could help build arguments for the role of collaborative and participatory governance and, in the process, improve the metabolic efficiency of urban entities. A vital strength of the urban metabolism framework is how it involves all sectors of society, recognises their different roles in impacting water quantity and quality, and demonstrates the flow-on effects to achieving desired social, economic, and environmental outcomes. Adopting and applying a metabolism lens to urban water management should enhance institutional collaboration by supporting the management of urban water on a holistic whole-of-water-cycle basis, and then extending this to enhance cross-sectoral integration.

A further advantage is the potential of the approach as a method for setting metabolic targets and benchmarks. An evaluation framework based on urban metabolism principles could provide a comprehensive and cohesive set of metrics, including hydrological performance and resource efficiency, to inform the design of incentive programmes targeting specific areas. One example is the "Infill Performance Evaluation Framework" used to evaluate water sensitivity of urban designs in an infill context (Renouf et al. 2020a). This approach to evaluation could, in turn, lead to specific targets for local developments and raise questions such as who is responsible for achieving and monitoring these targets.

What is clear is that the quantitative relationship among water, energy, and resource flows in cities is becoming better understood through the application of the urban metabolism framework. This enables decisions to be made from a solid evidence base, so risks and opportunities can be balanced and fairly traded – all key characteristics of good governance.

20.5 Next steps

There are many next steps for improving our knowledge and operationalising urban metabolism (AWE and ACEEE 2011, Kenway et al. 2011c, PMSEIC 2010) that relate to standards, funding and planning, education, training, technologies, and governance. Here, we focus specifically on the next steps needed to operationalise urban metabolism as one component contributing to the Water Sensitive evaluation, management, and governance of cities.

20.5.1 From frameworks to methods or methodologies

A greater agreement is needed on appropriate frameworks and the way they conceptualise and define "cities" and their resource flows. These frameworks shape the methods and tools we use to evaluate them. Without consistent and agreed upon frameworks it is difficult to generate meaningful and comparable information, because of "data chaos."

Examples from other fields are available of the directions in which urban water metabolism evaluation may head in its evolution from concept to method. For example, if we consider product life cycle assessment (LCA), the techniques for quantifying the environmental impacts of products started from the framework of "life cycle thinking" in the 1990s (UNEP 1999) and developed into a suite of recognised methods for product life cycle assessment (LCA) and "footprinting." These are now governed by international standards (Finkbeiner 2014), and they underpin environmental regulations and certification standards for products. There was an initial emphasis on GHGs (carbon footprinting), and now water footprinting methods are fast evolving (Hoekstra et al. 2011).

Similarly, methodologies and standards have recently been developed for quantifying greenhouse gas (GHG) emissions from cities (BSI 2013, World Resources Institute 2014), out of preexisting frameworks for GHG accounting. Following in the tracks of product LCA and footprinting, it is highly likely that urban water accounting will be influenced by urban metabolism frameworks for wide applications and benefit. Some progress has been made towards defining a standard set of urban water metabolism indicators (Kennedy et al. 2014). However, these indicators refer to various scales from site to city. To operationalise the urban metabolism framework into methods, what is required now is:

- engagement with international research efforts that are progressing harmonisation of frameworks and methods;
- learning from equivalent development in GHG accounting methods for cities; and
- leading the development of water-specific metabolism evaluation methods.

20.5.2 Data systems

Australia is a world leader in urban water data and performance management, thanks to the work of the National Water Commission, the Water Services Association of Australia, and its members (ref). However, data at the city scale in a form suited for use in an urban metabolism evaluation are still challenging to compile, as available data sets are commonly based on inconsistent frameworks.

Enabling data requirements for urban metabolism evaluation will require:

- consistent structures for urban water data reporting,
- interoperable data sets for urban water, energy, and other related flows, and
- public access to data.

In the long term, data acquisition could be facilitated by using the evolving "smart cities" and "big data" sources (Gemma and Mc Intosh 2014), such as the Australian Water Resources Information System (AWRIS,) the Australian Urban Resources Infrastructure Network (AURIN), and cloud-based economic flow models, such as the Industrial Ecology Virtual Laboratory (Lenzen et al. 2014).

20.5.3 Indicators for monitoring, benchmarking, and reporting

An important application of metabolism evaluation will be tracking the water performance of cities, comparing cities and design typologies, monitoring improvements over time, and monitoring the effectiveness of planning policies and water sensitive urban design initiatives. This seemingly straightforward task can be challenging and contentious due to the monitoring needs necessary for meaningful comparison and consideration of varying local conditions.

A starting point for this will be identifying a suitable set of indicators based on a specific framework and corresponding method (as discussed previously). Many indicators of different aspects of urban water performance have already been proposed, some specifically for water (e.g., City Blueprints [van Leeuwen et al. 2012]), but most are for a range of resource issues (examples are Asian Water Development Outlook (ADB 2013), Asian Green Cities Index (EIU 2011)). However, their utility for advancing urban water management goals is still to be demonstrated.

Progressing the development of metabolic indicators that can advance Water Sensitive Cities will involve:

- learning from indicators used in other resource management spheres;
- testing possible metabolic indicators across a range of urban design typologies and settings;
- consultation regarding the utility of indicators in the various spheres of influence, including urban planning and development and water infrastructure planning.

20.6 Summary and conclusions

Urban systems are a long way from the resource efficiencies of natural systems, and much change will be necessary to ensure their sustainability under future population growth and climate change pressures. In relation to water, it will be necessary to improve the metabolic efficiency of water use but also water-related energy. Many existing and new technologies can be deployed at household, precinct, and city scales for water and energy efficiency. To inform and optimise solutions, avoid unintended problem shifting, and monitor progress, we need to understand better and quantify the whole picture of water and water-related energy in cities. Urban metabolism frameworks could provide this understanding, which has implications for initiating collaborations between otherwise fragmented institutions and decision-makers.

Urban metabolism provides a rich conceptual and analytical framework against which the water performance of urban entities, including cities, could be much more thoroughly quantified, benchmarked, and managed. The urban water balance technique was identified as a good foundation for urban water metabolism evaluation. In this context, we refer to an *urban water balance* to distinguish it from other water balance applications such as supply-demand balance commonly used in engineering practices. The urban water metabolism evaluation framework gives a complete picture of urban water performance, including hydrological performance and resources efficiency, at the whole of city scale than existing urban water system modelling.

This chapter introduced an *urban water metabolism evaluation framework* based on an *urban water balance* that aims to account for the diverse sources and function of water in the urban landscape, connect urban water management to supporting hinterland, account for water-related energy, and generate metabolic performance indicators. It describes the potential links for this framework to guide urban water decision making and, hence, governance.

In the same way that the concept of a Water Sensitive City evolved in the way we view and value water in the urban landscape, urban metabolism is an evolution in the way we evaluate and quantify water (and related flows). For advancing Water Sensitive (or "Water Wise" or "Sponge") Cities, urban metabolism evaluation can (i) inform the strategic optimisation of the efficiency, supply security, and resilience of water and energy; (ii) include liveability and amenity as part of the functionality of water; and (iii) frame resource efficiency indicators for tracking progress towards Water Sensitive Cities.

Operationalising urban metabolism will require a significant effort to (i) further refine the proposed frameworks into methods alongside international efforts to harmonise the technique, (ii) inform consistent structures for urban resource data, including alignment with emerging "big data" system, and (iii) develop meaningful and repeatable indicators.

Acknowledgements

This chapter was possible due to support from the CRC Water Sensitive Cities in funding the research fellowship of Marguerite Renouf in the IRP4 project.

References

ADB (2013) *Asian Water Development Outlook 2013*, Asian Development Bank, Asia-Pacific Water Forum, Philippines.

Allam, Z. and Newman, P. (2018) Redefining the smart city: Culture, metabolism and governance. *Smart Cities* 1(1), 4–25.

AWE and ACEEE (2011) *Addressing the Energy-Water Nexus: A Blueprint for Action and Policy Agenda*, Alliance for Water Efficiency and American Council for an Energy Efficient Economy, Washington, DC.

Baccini, P. and Brunner, P.H. (1991) *Metabolism of the Anthrosphere*, Springer-Verlag, Dubendorf.
Bach, P.M., Rauch, W., Mikkelsen, P.S., McCarthy, D.T. and Deletic, A. (2014) A critical review of integrated urban water modelling: Urban drainage and beyond. *Environmental Modelling & Software* 54, 88–107.
Baynes, T.M. and Wiedmann, T. (2012) General approaches for assessing urban environmental sustainability. *Current Opinion in Environmental Sustainability* 4(4), 458–464.
Bhaskar, A.S. and Welty, C. (2012) Water balances along an urban-to-rural gradient of metropolitan baltimore, 2001–2009. *Environmental & Engineering Geoscience* 18(1), 37–50.
Billen, G., Garnier, J. and Barles, S. (2012) History of the urban environmental imprint: Introduction to a multidisciplinary approach to the long-term relationships between Western cities and their hinterland. *Regional Environmental Change* 12(2), 249–253.
Brown, R.R., Keath, N. and Wong, T.H.F. (2009) Urban water management in cities: Historical, current and future regimes. *Water science and technology,* 59(5), 847–855.
BSI (2013) *PAS 2070:2013. Specifications for the Assessment Greenhouse Gas Emissions of a City*, BSI Standards, London, UK.
Chen, S. and Chen, B. (2015) Urban energy consumption: Different insights from energy flow analysis, input–output analysis and ecological network analysis. *Applied Energy* 138, 99–107.
Chhipi-Shrestha, G., Hewage, K. and Sadiq, R. (2017) Water–energy–carbon nexus modeling for urban water systems: System dynamics approach. *Journal of Water Resources Planning Management* 143(6), 04017016.
Coutts, A., Demuzere, M., Tapper, N., Daly, E., Beringer, J., Nury, S., Broadbent, A., Harris, R., Gebert, L. and Nice, K. (2014) *The Impacts of Harvesting Solutions and WSUD on Evaporation and the Water Balance and Feedbacks to Urban Hydrology and Stream Ecology. Technical Report from CRC WSC Project B3.1*, Melbourne.
Davoudi, S. and Sturzaker, J. (2017) Urban form, policy packaging and sustainable urban metabolism. *Resources, Conservation and Recycling* 120, 55–64.
Decker, E.H., Elliott, S., Smith, F.A., Blake, D.R. and Rowland, F.S. (2000) Energy and material flow through the urban ecosystem. *Annual Review of Energy and the Environment* 25(1), 685–740.
Dijst, M., Worrell, E., Böcker, L., Brunner, P., Davoudi, S., Geertman, S., Harmsen, R., Helbich, M., Holtslag, A.A. and Kwan, M.-P. (2018) Exploring Urban Metabolism: Towards an Interdisciplinary Perspective. *Resources, Conservation and Recycling*, 132, 190–203.
EIU (2011) *Asian Green City Index. Assessing the Environmental Performance of Asian's Major Cities*, Siemens AG, Economic Intelligence Unit, Munchen, Germany.
Farooqui, T.A., Renouf, M.A. and Kenway, S.J. (2016) A metabolism perspective on alternative urban water servicing options using water mass balance. *Water Research* 106, 415–428.
Finkbeiner, M. (2014) *Background and Future Prospects in Life Cycle Assessment*, Klopffer, W. (ed), Springer Link, New York.
García-Sánchez, M. and Güereca, L.P. (2019) Environmental and social life cycle assessment of urban water systems: The case of Mexico City. *Science of the Total Environment* 693, 133464.
Gemma, P. and Mc Intosh, A. (2014) *Technical Report on Smart Water Management in Cities. SSC-0122-rev3*, International Telecommunications Union, Focus Group on Smart Sustainable Cities, Geneva.
Goldstein, B., Birkved, M., Quitzau, M.-B. and Hauschild, M. (2013) Quantification of urban metabolism through coupling with the life cycle assessment framework: Concept development and case study. *Environmental Reserach Letters* 8, 035024.
Haase, D. (2009) Effects of urbanisation on the water balance: A long-term trajectory. *Environmental Impact Assessment Review* 29(4), 211–219.
Hardy, M., Kuczera, G. and Coombes, P. (2005) Integrated urban water cycle management: The Urban Cycle model. *Water Science and Technology* 52(9), 1–9.
Hoekstra, A.Y., Buurman, J. and van Ginkel, K.C. (2018) Urban water security: A review. *Environmental Research Letters* 13(5), 053002.
Hoekstra, A.Y., Chapagain, A.K., Aldaya, M.M. and Mekonnen, M.M. (2011) *The Water Footprint Assessment Manual. Setting the Global Standard*, Earthscan, London.
Hoff, H., Döll, P., Fader, M., Gerten, D., Hauser, S. and Siebert, S. (2014) *Water Footprints of Cities: Indicators for Sustainable Consumption and Production 213–226.*
Islam, K.N., Kenway, S.J., Renouf, M.A., Wiedmann, T. and Lam, K.L. (2021) A multi-regional input-output analysis of direct and virtual urban water flows to reduce city water footprints in Australia. *Sustainable Cities and Society* 75, 103236.
Kennedy, C., Cuddihy, J. and Engel-Yan, J. (2007) The changing metabolism of cities. *Journal of Industrial Ecology* 11(2), 43–59.

Kennedy, C., Pincetl, S. and Bunje, P. (2011) The study of urban metabolism and its applications to urban planning and design. *Environmental Pollution* 159(8–9), 1965–1973.

Kennedy, C., Stewart, I.D., Ibrahim, N., Facchini, A. and Mele, R. (2014) Developing a multi-layered indicator set for urban metabolism studies in megacities. *Ecological Indicators* 47, 7–15.

Kenway, S., Gregory, A. and McMahon, J. (2011a) Urban water mass balance analysis. *Journal of Industrial Ecology* 15(5), 693–706.

Kenway, S., Lant, P. and Priestley, T. (2011b) Quantifying water-energy links and related carbon emissions in cities. *Journal of Water and Climate Change* 2(4), 247.

Kenway, S., McMahon, J., Elmer, V., Conrad, S. and Rosenblum, J. (2013a) Managing water-related energy in future cities: A research and policy roadmap. *Journal of Water and Climate Change* 4(3), 161–175.

Kenway, S.J., Lant, P., Priestley, A. and Daniels, P. (2011c) The connection between water and energy in cities: A review. *Water Science and Technology* 63(9), 1983–1990.

Lenzen, M. (2009) Understanding virtual water flows: A multiregion input-output case study of Victoria. *Water Resources Research* 45(9).

Lenzen, M., Geschke, A., Wiedmann, T., Lane, J., Neal, A., Baynes, T., Boland, J., Daniels, P., Dey, C., Fry, J., Hadjikakoub, M., Kenway, S., Malika, A., Morang, D., Murray, J., Nettleton, S., Poruschii, L., Reynolds, C., Rowley, H., Ugonj, J., Webb, D. and West, J. (2014) Compiling and using input-output frameworks through collaborative virtual laboratories. *Science of the Total Environment* 485–486, 241–251.

London, G., Bertram, N., Renouf, M., Kenway, S., Sainsbury, O., Todorovic, T., Byrne, J., Pype, M., Sochacka, B., Surendran, S. and Moravej, M. (2020) *Knutsford Case Study Final Report: Water Sensitive Outcomes for Infill Development*, Cooperative Research Centre for Water Sensitive Cities, Melbourne, Australia.

Mekonnen, M.M. and Hoekstra, A.Y. (2011) *National Water Footprint Accounts: The Green, Blue and Grey Water Footprint of Production and Consumption. Volume 1: Main Report*, UNESCO-IHE Institute for Water Education, Deft, The Netherlands.

Minx, J., Creutzig, F., Medinger, V., Ziegler, T., Owen, A. and Baoicchi, G. (2011) *Developing a Pragmatic Approach to Assess Urban Metabolism in Europe: A Report to the European Environment Agency prepared by* Technische Universität Berlin and Stockholm Environment Institute, Technische Universität Berlin.

Mitchell, V.G., Mein, R.G. and McMahon, T.A. (2001) Modelling the urban water cycle. *Environmental Modelling & Software* 16, 615–629.

Moravej, M., Renouf, M.A., Lam, K.L., Kenway, S.J. and Urich, C. (2021) Site-scale Urban Water Mass Balance Assessment (SUWMBA) to quantify water performance of urban design-technology-environment configurations. *Water Research* 188, 116477.

National Water Commission (2011) *Urban Water in Australia: Future Directions*, Australian Government National Water Comission, Canberra.

Newman, P., Birrell, R., Holmes, D., Mathers, C., Newton, P., Oakley, G., O'Connor, A., Walker, B., Spessa, A. and Trait, D. (1996) *State of the Environment Australia 1996*, State of the Environment Advisory Council, CSIRO Publishing, Melbourne.

Newman, P.W.G. (1999) Sustainability and cities: Extending the metabolism model. *Landscape and Urban Planning* 44(4), 219–226.

Pizzol, M., Scotti, M. and Thomsen, M. (2013) Network analysis as a tool for assessing environmental sustainability: Applying the ecosystem perspective to a Danish Water Management System. *Journal of Environmental Management* 118, 21–31.

PMSEIC (2010) *Challenges at Energy-Water-Carbon Intersections, Prime Minister's Science*, Engineering and Innovation Council, Canberra.

Rauch, W., Urich, C., Bach, P., Rogers, B., de Haan, F., Brown, R., Mair, M., McCarthy, D., Kleidorfer, M. and Sitzenfrei, R. (2017) Modelling transitions in urban water systems. *Water Research* 126, 501–514.

Renouf, M., Kenway, S., Bertram, N., London, G., Todorovic, T., Sainsbury, O., Nice, K., Moravej, M. and Sochacka, B. (2020a) *Water Sensitive Outcomes for Infill Development: Infill Performance Evaluation Framework p. 43*, Cooperative Research Centre for Water Sensitive Cities, Melbourne, Australia.

Renouf, M., Kenway, S., Serrao-Neumann, S. and Low Choy, D. (2016) *Urban Metabolism for Planning Water Sensitive Cities. Concept for an Urban Water Metabolism Evaluation Framework*, p. 42, Cooperative Research Centre for Water Sensitive Cities, Melbourne, Victoria, Australia.

Renouf, M., Moravej, M., Sainsbury, O., Lam, K.L., Bertram, N., Kenway, S. and London, G. (2019) *Quantifying the Hydrological Performance of Infill Development*, Australian Water Association, Melbourne, Australia.

Renouf, M., Serrao-Neumann, S., Kenway, S., Morgan, E. and Choy, D.L. (2017a) Urban water metabolism indicators derived from a water mass balance: Bridging the gap between visions and performance assessment of urban water resource management. *Water Research* 122, 669–677.

Renouf, M., Sochacka, B., Kenway, S., Lam, K.L., Serrao-Neumann, S., Morgan, E. and Low Choy, D. (2017b) *Urban Metabolism for Planning Water Sensitive City-regions. Proof of Concept for an Urban Water Metabolism Evaluation Framework Catchment Scale Landscape Planning for Water Sensitive City-regions*, p. 55, Cooperative Research Centre for Water Sensitive Cities, Melbourne, Australia.

Renouf, M.A. and Kenway, S. (2017a) Evaluation approaches for advancing urban water goals. *Journal of Industrial Ecology* 21(4), 995–1009.

Renouf, M.A. and Kenway, S.J. (2017b) Evaluation approaches for advancing urban water goals. *Journal of Industrial Ecology* 21(4), 995–1009.

Renouf, M.A., Kenway, S.J., Bertram, N., London, G., Sainsbury, O., Todorovic, T., Nice, K., Surendran, S. and Moravej, M. (2020b) *Salisbury Case Study Final Report: Water Sensitive Outcomes for Infill Development*, Cooperative Research Centre for Water Sensitive Cities, Melbourne.

Renouf, M.A., Kenway, S.J., Lam, K.L., Weber, T., Roux, E., Serrao-Neumann, S., Choy, D.L. and Morgan, E.A. (2018) Understanding urban water performance at the city-region scale using an urban water metabolism evaluation framework. *Water Research* 137, 395–406.

Rockström, J., Steffen, W., Noone, K., Persson, A., Chapin, F.S., Lambin, E.F., Lenton, T.M., Scheffer, M., Folke, C., Schellnhuber, H.J., Nykvist, B., de Wit, C.A., Hughes, T., van der Leeuw, S., Rodhe, H., Sorlin, S., Snyder, P.K., Costanza, R., Svedin, U., Falkenmark, M., Karlberg, L., Corell, R.W., Fabry, V.J., Hansen, J., Walker, B., Liverman, D., Richardson, K., Crutzen, P. and Foley, J.A. (2009) A safe operating space for humanity. *Nature* 461(7263), 472–475.

Rogers, B.C., Dunn, G., Hammer, K., Novalia, W., de Haan, F.J., Brown, L., Brown, R.R., Lloyd, S., Urich, C., Wong, T.H.F. and Chesterfield, C. (2020) Water sensitive cities index: A diagnostic tool to assess water sensitivity and guide management actions. *Water Research* 186, 116411. https://doi.org/10.1016/j.watres.2020.116411.

Sahely, H.R., Dudding, S. and Kennedy, C.A. (2003) Estimating the urban metabolism of Canadian cities: Greater Toronto Area case study. *Canadian Journal of Civil Engineering* 30(2), 468–483.

Serrao-Neumann, S., Renouf, M., Kenway, S.J. and Low Choy, D. (2017) Connecting land-use and water planning: Prospects for an urban water metabolism approach. *Cities* 60(2), 13–27.

Serrao-Neumann, S., Renouf, M.A., Morgan, E., Kenway, S.J. and Low Choy, D. (2019) Urban water metabolism information for planning water sensitive city-regions. *Land Use Policy* 88, 104144.

Sochacka, B., Kenway, S., Bertram, N., London, G., Renouf, M., Sainsbury, O., Surendran, S., Moravej, M., Nice, K., Todorovic, T., Tarakemehzadeh, N. and Martin, D. (2021) *Water Sensitive Outcomes for Infill Development: Final Report*, Cooperative Research Centre for Water Sensitive Cities, Melbourne, Australia.

Su, B., Ang, B. and Li, Y. (2017) Input-output and structural decomposition analysis of Singapore's carbon emissions. *Energy Policy* 105, 484–492.

Sun, S. (2019) Water footprints in Beijing, Tianjin and Hebei: A perspective from comparisons between urban and rural consumptions in different regions. *Science of the Total Environment* 647, 507–515.

Świąder, M., Lin, D., Szewrański, S., Kazak, J.K., Iha, K., van Hoof, J., Belčáková, I. and Altiok, S. (2020) The application of ecological footprint and biocapacity for environmental carrying capacity assessment: A new approach for European cities. *Environmental Science & Policy* 105, 56–74.

Thériault, J. and Laroche, A.-M. (2009) Evaluation of the urban hydrologic metabolism of the Greater Moncton Region, New Brunswick. *Canadian Water Resources Journal / Revue canadienne des ressources hydriques* 34(3), 255–268.

UNEP (1999) *Towards the Global Use of Life Cycle Assessment*, United Nations Environment Programme, Paris.

van Leeuwen, C.J., Frijns, J., van Wezel, A. and van de Ven, F.H.M. (2012) City blueprints: 24 Indicators to assess the sustainability of the urban water cycle. *Water Resources Management* 26(8), 2177–2197.

Wackernagel, M. and Rees, W. (1996) *Our Ecological Footprint: Reducing Human Impact on the Earth*, New Society Publishers, Gabriola Island.

Wolman, A. (1965a) The metabolism of cities. *Scientific American* 213(3), 179–190.

Wong, T.H.F. and Brown, R.R. (2009) The water sensitive city: Principles for practice. *Water Science and Technology* 60(3), 673–682.

World Resources Institute (2014) *Global Protocol for Communicty-scale Greenhouse Gas Emission Inventories. An Accounting and Reporting Standard for Cities, World Resources Institute*, C40 Cities Climate Leadership Group and ICLEI Local Governments for Sustainability, Bonn, Germany.

WSAA (2012) Cities of the future. *2011 Report and Recommendations*. Occasional paper 26, Water Services Assocation of Australia.

Zengerling, C. (2019) Governing the city of flows: How urban metabolism approaches may strengthen accountability in strategic planning. *Urban Planning* 4(1), 187.

21

LEVERAGING ARTIFICIAL INTELLIGENCE IN ADDRESSING WATER SAFETY CHALLENGES

Xu Wang

21.1 Introduction

Water is a vital resource for human survival and development. Sustainable water resources, water environment, and water ecology are crucial for human health and economic prosperity. However, over the past half-century, water security has become a globally important issue due to a series of factors, such as population growth, increased human activities, and climate change (Larsen et al., 2018). In September 2015, the United Nations (UN) Sustainable Development Summit formally adopted the UN Sustainable Development Goals (SDGs) for 2030 and called on countries and regions to cooperate and promote scientific and technological innovation for the effective implementation of the SDGs (United Nations, 2015). Among them, SDG 6 – "Clean Water and Sanitation" – aims to "provide and sustainably manage water and sanitation for all." In recent years, the rapid development of artificial intelligence (AI) technology (Perrault et al., 2019) has provided new ideas and tools to realize SDG 6. This chapter analyzes and summarizes the current situation and effects of AI applications in SDG 6-related fields in the context of the core connotations and process difficulties of SDG 6, explores the key issues that need to be addressed in using AI to advance SDG 6, and provides scientific suggestions for technological innovation and synergistic development in the fields of water, the environment, and AI.

21.2 The core essence and difficulties of SDG 6

21.2.1 The progressive nature of SDG 6 connotations

SDG 6 is a key component of the SDGs and the foundation for achieving the other 16 SDGs (Figure 21.1). SDG 6 contains 8 concrete targets and 11 monitoring indicators covering a wide range of topics, including water resources, water environment, water ecology, water facilities, and international cooperation related to water science and technology (United Nations, 2018) (Table 21.1).

Obviously, SDG 6 is a new, more comprehensive, systematic, and forward-looking development framework based on the historical and practical experience of the United Nations Millennium Development Goals and contains higher expectations for future water security.

DOI: 10.4324/9781003057574-27

Figure 21.1 United Nations Sustainable Development Goal 6 (SDG 6) and the other 16 SDGs.

Overall, the progressive nature of SDG 6 is reflected in the following four areas. **(i) Greater emphasis on fairness in its goal implementation.** The goal of SDG 6 shifts from "halving the global proportion of people without access to water and sanitation services" to "providing and sustainably managing water and sanitation for all," and proposes equal treatment in addressing the need for clean water and sanitation among vulnerable groups such as women and children and for capacity-building in the fields of water and the environment in developing countries. **(ii) Greater emphasis on the systematization and integrity of the water cycle management.** SDG 6 already explicitly calls for improving the efficiency of water resource usage, strengthening resources, energy recovery, and safe recycling in the process of water use, reducing the negative impact of sewage and wastewater discharge on the water environment and water ecology, and ensuring the safety of the water supply. **(iii) Paying more attention to the water security crisis in relation to human activities.** The challenges of water and environmental sustainability associated with population growth and increased human activities, such as frequent water pollution, water scarcity, and water ecological degradation, have become the focus of SDG 6. **(iv) Greater emphasis on the importance of transboundary, multisystem, and intersectoral collaboration for the integrated management of water resources and water systems and the realization of their equal benefits.** SDG 6 places greater demands on strengthening the protection of the ecological integrity of water in transboundary basins and promoting multisystem, and multisectoral synergistic management of water use and regeneration cycles.

21.2.2 Challenges in the implementation of SDG 6

SDG 6 provides a clear direction for sustainable global water and environment development in 2030, but it also faces new challenges. At present, the developmental stage of each country and region, as well as the scientific and technological level of different countries and regions in the

Table 21.1 Brief description on SDG 6

Goal		*Indicator*	
Code	*Description*	*Code*	*Description*
6.1	Achieve access to safe and affordable drinking water	6.1.1	Proportion of population using safely managed drinking water services
6.2	Achieve access to sanitation and hygiene and end open defecation	6.2.1a	Proportion of population using safely managed sanitation services
		6.2.1b	Proportion of population using a handwashing facility with soap and water available
6.3	Improve water quality, wastewater treatment, and safe reuse	6.3.1	Proportion of wastewater safely treated
		6.3.2	Proportion of bodies of water with good ambient water quality
6.4	Increase water-use efficiency and ensure freshwater supplies	6.4.1	Change in water-use efficiency over time
		6.4.2	Level of water stress: freshwater withdrawal as a proportion of available freshwater resources
6.5	Implement integrated water resources management	6.5.1	Degree of integrated water resources management implementation (0–100)
		6.5.2	Proportion of transboundary basin area with an operational arrangement for water cooperation
6.6	Protect and restore water-related ecosystems	6.6.1	Change in the extent of water-related ecosystems over time
6.a	Expand international cooperation and capacity-building	6.a.1	Amount of water- and sanitation-related official development assistance that is part of a government-coordinated spending plan
6.b	Support stakeholder participation	6.b.1	Proportion of local administrative units with established and operational policies and procedures for participation of local communities in water and sanitation management

field of water and environment, varies to different degrees, which makes the compound relationship between different targets within SDG 6 and with other SDGs mutually reinforcing or constraining (Kroll et al., 2019; Mulligan et al., 2020; Nerini et al., 2018). Therefore, a systematic understanding of the difficulties encountered in the implementation of SDG 6, and – taking this as a direction of focus – proposing solutions and programs according to local conditions, is an important basis for countries and regions around the world to jointly actualize SDG 6. Specifically, SDG 6 is facing three challenges in the process of its implementation.

(1) **Inadequate data monitoring and analysis and assessment tools seriously limit the ability of the United Nations organizations, countries, and regions to fully grasp the real status and effectiveness of water and sanitation development.** Effective measurement and monitoring of each SDG 6 target are important to guarantee the achievement of sustainable development of water and the environment. As of 2021, 4 of the 11 monitoring indicators of SDG 6 which include indicators 6.1.1, 6.2.1, 6.3.1, and 6.3.2 are still not available in most countries on a regular basis (AEG-SDGs n.d.). Currently, the metrics and monitoring of each SDG 6 target rely mainly on statistical or census data

(United Nations-Water, 2017). The former often lacks representativeness and spatial resolution due to factors such as unclear distinction between urban and rural areas (WHO-UNICEF, 2019), while the latter is limited by the labor and time costs of census work, resulting in poor real-time data, high uncertainty, and extremely limited representativeness of the data (Big Earth Data Program Chineses Academy of Sciences, 2019). Therefore, the scientific assessment of the SDG 6 progress urgently requires the development of long-term, multiscale, multidimensional, and high-resolution data monitoring and indicator modeling tools.

(2) **Difficulties in the construction, monitoring, simulation, evaluation, and overall optimal regulation of water cycle systems.** The complete water cycle system involves many natural and artificial water unit processes, such as surface water, groundwater, rainwater, and urban water supply and drainage systems, which is a hugely complex, diverse, dynamic, and interconnected system (Daigger et al., 2019). While traditional research thinking and management modes of water engineering is relatively closed and singular, there are differences in the scientific and technological capabilities of countries and regions over the world in related fields. This results in immense challenges during the construction, monitoring, simulation, evaluation, and overall optimal regulation of the water cycle system (刘俊新, 2015; Wang et al., 2018). Discussions on how to break through the traditional closed research ideas and modes, how to create an open water science and technology innovation ecology, and how to highlight the cross-fertilization of disciplines and international scientific and technological cooperation remain the core scientific issues in actualizing the integrated management of water resources and water systems.

(3) **Complexity of the water environment and water ecology risks.** With the rapid development of industrialization and the expansion and diversification of human consumption needs, water pollutants are characterized by increasingly complex types and spatial and temporal variability, and the risks to the water environment and water ecology have generally increased (Johnson et al., 2020; Guitton, 2017; Zhang et al., 2018b). How to rapidly identify, recognize, and efficiently address complex pollution and risks from extremely complex and variable water systems is a key technical challenge to be addressed in developing healthy, sustainable, and highly resilient water systems in the future.

21.2.3 Current status of research and application of AI in water and the environment

AI is a branch of computer science, which is a new technical science that studies and develops theories, methods, technologies, and application systems for simulating, extending, and expanding human intelligence (Stephen and Danny, 2020). In a sense, AI plays the central role in the whole "data collection-information derivation-information distribution" process. Thanks to the rapid development of Internet ofThings (IoT) technology and 5G network, massive raw data collected by all kinds of monitoring sensors could be transmitted to a cloud in real time; a computationally intensive AI system could integrate and treat those data through models into information rapidly using the compute resources provided by cloud computing; after that, information could be distributed to clients through channels to assist in decision making. In recent years, AI has been developing rapidly due to the large-scale development of computer computing power and the continuous breakthroughs of algorithms, which provides powerful tools for the research, development, and innovation of technologies such as water environment pollution prevention and control, water quality safety assurance, water-related facility optimization and reconfiguration, and watershed ecosystem management. Combing and summarizing the relevant literature at home and abroad over the past 10 years, it is found that AI technology currently plays an important role in research and application in four main areas.

21.2.3.1 Water environment pollution identification and risk response

Identifying and responding to water pollution events is an important prerequisite for efficient prevention and control of pollution in the water environment and a basic guarantee of water supply security.

(1) **Water quality indicator modeling and data fusion.** The application of AI in water quality indicator modeling and multidimensional, spatial-temporal data fusion has created new opportunities to improve the research, diagnosis ability, prevention, and control of water pollution (Figure 21.2). For example, the use of an artificial neural network adaptive selection method, characterized by water quality remote sensing and detection data, can realize the construction and application of non-linear water quality indicator models, providing the necessary basic data for water quality management and digital planning of water bodies (Chen et al. 2020a). The integration of AI algorithms, such as neural networks, support vector machine, and classification and regression trees, can provide integrated simulations of more complex water quality changes in water environments and their geobiochemical processes, providing important modeling tools for water quality protection and restoration in water bodies (Chou et al., 2018; García Nieto et al., 2019).

Figure 21.2 AI-oriented technology systems for water environment pollutant identification and risk response.

(2) **Risky substance detection and toxicity assessment.** Combining AI with spectral analysis technology is a current research hotspot. Near-infrared (NIR) spectroscopy can be used for rapid detection of water quality indicators, such as biochemical oxygen demand, while coupling with AI algorithms (represented by least squares support vector machine) to improve the accuracy of NIR spectroscopy in predicting water quality changes and to provide a quick solution for quantitative assessment of water pollution (Chen et al., 2020b); applying back propagation neural network and a *k*-means clustering algorithm to a laser-induced breakdown spectroscopy analysis provides new ideas and methods in obtaining an efficient, accurate, and low-cost estimation of heavy metals and other necessary indicators of surface water quality, a process that is traditionally expensive and time intensive (Zhang et al., 2018a). Meanwhile, the application of AI to environmental toxicology studies is also being explored at home and abroad, which provides new economical and efficient means for toxicity prediction and risk assessment of novelpollutants (Miller et al., 2018).

(3) **Water quality monitoring and construction of pollution emergency response plan.** With the rapid development of in-situ monitoring sensing technology and equipment, AI technology based on deep neural networks began to play an important role in spatial big data analysis, which provides powerful technical and decision-making support for optimizing water quality monitoring deployment schemes, improving pollution source resolution, and developing pollution emergency prevention and control systems (Hino et al., 2018; Jia et al., 2019; Ballesté et al., 2020; Bonansea et al., 2015).

21.2.3.2 Research and development of water quality safety and security technologies

With the continuous improvement of water treatment standards, the design and application of new functional materials for water purification, the analysis of pollutant removal mechanisms and the development of efficient technologies, targeted resource-energy conversion, and the regulation of pollutants have become research hotspots in the field of water treatment (Zodrow et al., 2017; Qu et al., 2017).

(1) **Design and application of new functional materials for water purification.** AI-based materials genomics technology has been developing rapidly to provide an efficient way for the design and development of new environmentally friendly, functional materials (De Luna et al., 2017). The computational simulation and optimization of the composition and properties of new materials through inverse learning of failed tests and historical data from the material development process (Smith et al., 2020), combined with target pollutant characteristics, is expected to abandon the traditional trial-and-error-focused material development paradigm, which will greatly facilitate the industrialization of new materials for water purification (Figure 21.3).

(2) **Analysis of pollutant removal mechanisms and development of high-efficiency technologies.** The migration and transformation mechanism of micro-pollutants, such as health-related drugs and personal care products, endocrine disruptors, and persistent organic compounds in municipal water treatment systems, is a key and challenging point in the development of efficient water treatment technologies (Qu et al., 2017; 许国栋 et al., 2016). The introduction of AI algorithms, such as random forest, least absolute shrinkage, selection operator regression, and feedforward neural networks, allows for the non-linear simulation and prediction of the behavior of micro-pollutants in the water treatment process, which provides new methods for enhanced water treatment technologies (Kulkarni et al., 2010; Raza et al., 2019). As the research on wastewater biological treatment mechanisms (based

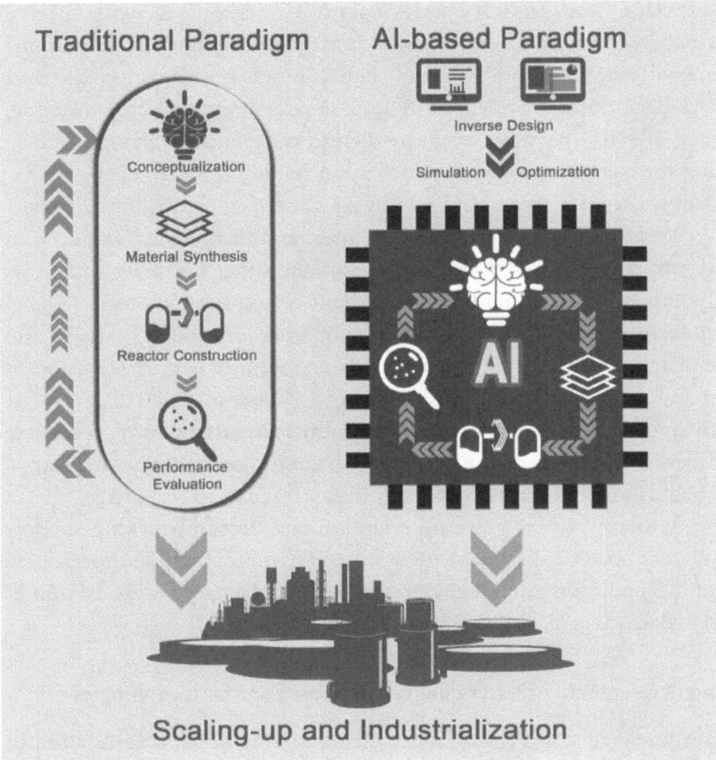

Figure 21.3 AI-assisted paradigm for R&D of new materials for environmental functions.

on molecular approaches such as macrogenomics and metabolomics) continues to advance (Han et al., 2020; Zhang et al., 2019), how to identify key functional microorganisms from the microbial big data of wastewater treatment systems has become a central difficulty in enhancing wastewater biological treatment. Combining AI technology with bioinformatics (Lesnik and Liu, 2017) provides important opportunities for information mining and microscopic parsing of water treatment systems and opens up new ways to elucidate wastewater biological treatment mechanisms (Figure 21.4). However, how to improve the accuracy and interpretability of information mining parsing is still a major challenge at present.

(3) **Pollutant-directed resource-energy transformation and regulation.** The core paradigm of water pollution control is gradually changing from pollutant removal to resource and energy transformation (Wang et al., 2015). Cutting-edge AI technologies of virtual and augmented reality, such as digital twin, will be expected to break through the technical difficulties of real-time simulation and synchronous control of the directional transfer and transformation of pollutants in water. However, there are still many technical problems to be solved (Zhao et al., 2020).

21.2.3.3 Optimal reconfiguration and integrated management of water-related facilities

With accelerated urbanization and socioeconomic development, urban water safety problems have become more prominent, mainly in terms of frequent water pollution, water shortage, and water ecological degradation (Rodell et al., 2018). The city is the center of human activity

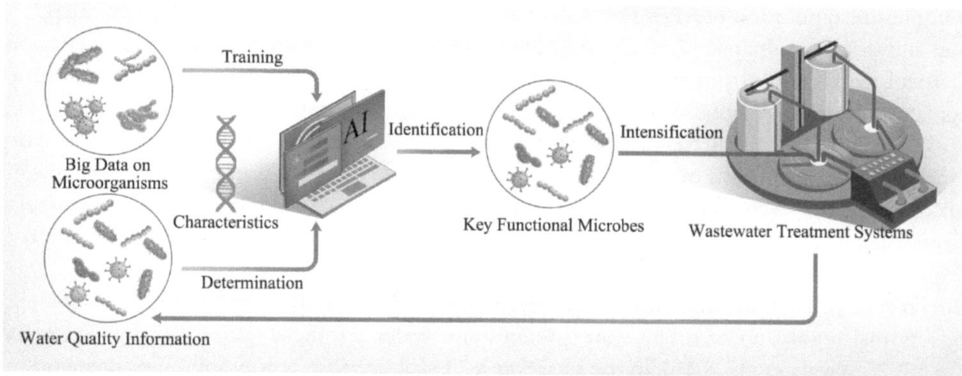

Figure 21.4 New ideas for research on biological wastewater treatment mechanisms and technologies based on AI.

and contains a complete water cycle system, with main characteristics of huge systems, complex processes, and water-related units that are closely interconnected and significantly affected by human activity (Daigger et al., 2017). However, traditional water system engineering takes water abstraction, water supply, and drainage as segmented objectives, and the paradigm for its research and management is both closed and singular – lacking a new paradigm for optimizing, managing, and reconstructing water-related facilities to meet the needs of cities' sustainable development from the perspective of system theory and holism. If traditional thinking continues, it will be difficult to achieve a substantial breakthrough in urban water security for a long time to come.

In the last two decades, the rise and iterations of information technology in the water industry, such as mechanistic models, sensors, and integrated analytics, especially the explosive development of AI in recent years, have provided key technologies to break through the bottleneck of optimal reconfiguration and integrated management of urban water systems (Figure 21.5). For

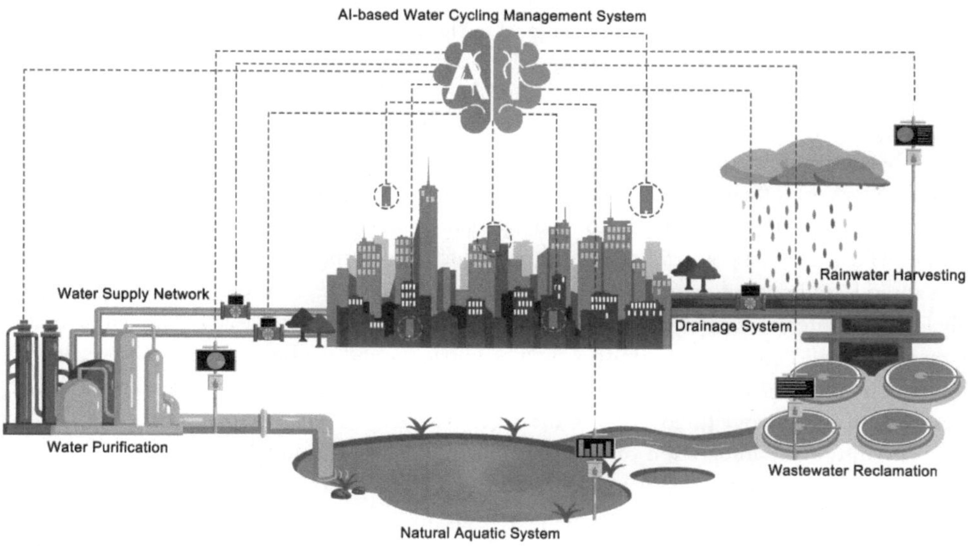

Figure 21.5 Data-driven model for integrated water cycle management with artificial intelligence.

example, the application of AI technologies, such as simulated annealing algorithms, to the planning and design of drainage systems and the management of stormwater resources can provide powerful technological support for the forward-looking layout, optimal design, and real-time regulation of drainage systems (Hassan et al., 2018; Ogidan and Giacomoni, 2016; Cunha et al., 2016). Further, by constructing a secondary optimization scheduling model based on a genetic algorithm and on the basis of achieving dynamic prediction of urban water consumption, it can also optimize the energy consumption of water supply, realize accurate control of water supply system operation costs, and effectively reduce its energy consumption and carbon emission (Abba et al., 2020; Bi et al., 2015; Zhang et al., 2019).

In recent years, AI technology has also been applied to research on integrated management and optimal regulation of urban water systems and water resources (Vamvakeridou-Lyroudia et al., 2020; Sayers et al., 2019); in the near future, it will be expected to build next-generation urban smart water systems with AI as the core to meet the changing needs of rapid urban development.

21.2.3.4 Simulation and integrated management of watershed ecosystem processes

Water and environmental processes are coupled processes involving the intersection of multiple layers, processes, scales, and elements of the earth. Their mechanisms are complex, involving a large amount of data, which is a major scientific problem for research frontiers in the field of earth and ecology environment (Reichstein et al., 2019; Xia et al., 2002). Among them, the watershed ecosystem, as a complex system formed by the interplay of water, soil, air, life, and human elements, is a microcosm of the coupled natural and social systems (Chen, et al., 2019), and an important scale for exploring the integrated management of water resources, water environment, and water ecology (刘昌明 et al., 2012). Ensuring the health of watershed ecosystems has important scientific value and practical significance for achieving the SDGs. In recent years, AI technology has been effectively integrated with earth observation technologies such as satellite communication, spatial positioning, remote sensing, and geographic information system (GIS) to realize the construction of a geoscience big data platform (Guo, 2018). which enables scientific simulation of large-scale hydrological cycle processes and their drivers such as natural precipitation (Pope and Gimblett, 2015), soil erosion (Shi et al., 2013), and glacial melt (Ji et al., 2013), thus providing a critical data base for process analysis and integrated assessment of watershed ecosystems.

Furthermore, how to conduct integrated simulation of natural-social-economic system inter-feed processes (Schluter et al., 2017; 傅伯杰, 2020) is the key to scientifically constructing integrated multiprocess and multifactor management of watershed ecosystems, and the leap forward in AI can provide powerful technical support for this purpose. For example, AI algorithms, such as random forest, gradient-boosted regression trees, and relevance vector machine can rapidly learn and predict the cascading responses of watershed ecosystems to dynamic factors, such as catchment land cover types, stressors such as nutrients, and seasonal evolution of vegetation (Sinha et al., 2019; Valerio et al., 2020) facilitating decision-makers in formulating watershed management objectives and governance measures.

In the future, the integration of AI algorithms with physical models of climate change and human activities (Giri et al., 2019; Liao et al., 2018; Romulo et al., 2018) and the integrated regulation of natural-social-economic systems at the watershed scale – based on the integration of geoscientific big data and socioeconomic indices – are expected to break through the technical system of green watershed construction and integrated management.

21.3 Critical issues to be addressed and countermeasure suggestions for AI applications in water and environment

The fourth industrial revolution is gaining momentum, and the rapid advancement of information technology with AI at the core has provided new opportunities for disruptive development in the field of water and environment – from traditional, empirical, and qualitative decision making to precise and quantitative intelligent decision making, creating the possibility for future-oriented reconstruction of healthy, sustainable, highly resilient, and intelligent water systems. The rapid advancement of AI technologies has injected new energy into the development and application of technologies from micro- to meso- and macro-scales, thus bringing a series of positive effects to accelerate the SDG 6 target process. Nevertheless, the development will also face many new challenges. Throughout the scientific exploration and practice described in the previous chapters, several key issues still need to be addressed for the future application of AI technologies in the water and environment sector.

(1) **Black-box effect and algorithm interpretability.** Although various AI techniques with machine learning as the breakthrough and deep learning as the implementation method have emerged with excellent predictive performance in the field of water and environment, their interpretability has been a shortcoming in the promotion of applications. For example, the high judgment capability currently possessed by deep neural networks is achieved by constructing multilayer non-linear mapping functions for layer-by-layer abstraction, and the black-box effect is its main feature (Jordan and Mitchell, 2015); in other words, although AI techniques with data-driven machine learning as the core can realize perception, learning, action, and even autonomous decision making, its technical effectiveness is mainly limited by the inability to explain to users the rationality behind its analysis and decision making, to assess the strengths and weaknesses of their models, to predict their generalizability to new tasks, and to even ensure their safety in future applications. In addition, researchers, engineers, and managers with backgrounds in water and environment-related disciplines often do not have relevant knowledge and technical experience in AI, which makes it more difficult for them to scientifically select, comprehensively evaluate, and cognitively understand AI technologies to solve water and environmental problems, thus leading to the practical value of AI technologies not being fully exploited. In the future, we should accelerate the breakthrough of new AI technologies with interpretability, as well as the development of theories and evaluation methods for system construction of AI technologies in the field of water and environment; thus, this is important to promote the comprehensive development of AI technologies in water and environment research, application, and education.

(2) **Large-scale computing power and negative environmental effects.** With the continuous development of monitoring, sensing, and modeling technologies, the operation mode of the water industry is gradually transforming to digital exploration (Will et al., 2019), which means that the volume of data related to water resources, water environment, and water ecology is showing a dramatic rise, and there is no shortage of data uncertainty and redundancy (Hoffmann et al., 2020). While AI has the ability to solve these challenges, as the volume of data in water systems continues to expand in the future, increasing data uncertainty and more complex connections among data, AI-based water, and environment solutions will consume a large amount of computational resources. In addition, the integrated management and collaborative regulation of water cycle systems must also rely on the application of computationally intensive AI technologies, such as deep neural networks, for which the development of large-scale computing power is a prerequisite. However, the

construction and operation of large-scale computing facilities can consume huge resources and energy and generate environmental issues such as carbon emissions. One study states that currently popular deep neural networks can emit more than 280 tons of carbon dioxide equivalent during massive training, which is five times the life-cycle carbon emissions of the average US car (Strubell et al., 2019). According to the US EPA, the world's built data centers currently consume 3% of the total global electricity consumption, and the consumption is rapidly growing at a rate of doubling every four years; meanwhile, the information and communications technology (ICT) industry account for about 2% of global greenhouse gas emissions, with a carbon footprint contribution equal to the combined carbon emissions of civil aviation (UN Environment Programme, 2008; Whitehead et al., 2014). The crude application of AI will likely exacerbate the global energy crisis and climate change, and even have unknown spillover effects on ecosystems, thus adversely affecting the implementation of energy and climate-related SDG goals. In fact, the human brain, when performing perception and cognition, not only processes current data, but also mobilizes relevant knowledge stored in the brain for understanding and reasoning, but the process consumes far fewer calories than the electrical consumption required to train AI models (Raissi et al., 2017). Introducing the water and environmental domain knowledge stored in the human brain for initial data screening and inference in the machine learning process can reduce unnecessary arithmetic-intensive processes, which helps to reduce problems, such as energy consumption and carbon emissions when AI is applied. In the future, the deep integration of AI technology with water and environmental domain knowledge should be strengthened, and water cycle management and risk prevention, and control systems and models based on AI arithmetic minimization, should be developed to reduce, or even avoid unnecessary spillover effects while solving water and environmental problems.

(3) **Data validity and standardization.** Another important foundation for AI, as a new data-driven technology, to be effective in the water and environment sector is to ensure the validity of data volume and quality. Currently, the global water industry is calling for and trying to promote the transformation of digital operation models (Blumensaat et al., 2019), especially emphasizing the monitoring and analysis of basic data, such as water quantity, water quality, and energy consumption in water treatment, wastewater treatment, and water supply and drainage processes, that creates favorable conditions for AI applications. However, for the majority of countries and regions, basic data such as water quantity and quality of urban water treatment systems generally rely on manual records, and the immediacy and validity of the data are poor; the migration and transformation process of pollutants in the water system is rapidly changing, and it is difficult to provide feedback on the immediate condition of the water system by only relying on manual record data. If the AI algorithm is trained in this way, the results will inevitably deviate greatly from the actual situation and lead to poor prediction performance. In recent years, the rapid development of online monitoring sensing devices and IoT technologies for water quantity, water quality, and energy consumption (Eggimann et al., 2017) has brought opportunities to solve the bottleneck of data immediacy and validity. Nonetheless, international standards in terms of data quality, interfaces, and protocols have not yet been unified, which is a key issue that needs to be addressed for the future integration of AI-based next-generation water systems with smart city systems. In the future, we should vigorously develop online data monitoring and sensing technology in the fields of water and the environment, and implement standardization of data quality, interfaces, and protocols; at the same time, on the premise that there are still differences in the level of development and human and material investments in different countries and regions, the expedient direction of devel-

opment at present should aim at breaking through the constraints imposed by the generally small data volume of current water systems and by developing new AI algorithms based on small data samples.

(4) **Restricted application and inequity issues.** Achieving equity is a common goal pursued by the SDGs, and the restricted application and diffusion of AI technologies may lead to or even exacerbate inequality problems in water and environmental governance in different countries and regions around the world (Perrault et al., 2018; Swathi et al., 2019). AI is considered one of the three cutting-edge technologies of the 21st century, but at present, AI-related knowledge education, technology development, and practical applications are all mostly found in developed countries. Especially in the case of transboundary watershed resources management, developed countries can use their AI advantages to enhance the capacity-building of water resource management and scheduling at the national level, so that they can take the lead and discourse power in transboundary watershed management, and thus have the ability to constrain the distribution of water resources in developing countries and regions, which is undoubtedly contrary to the goal of achieving equity in the UN SDGs (Rai et al., 2014). International cooperation in technology, education, and investment should be carried out to help developing countries and regions deploy capacity-building in emerging areas, such as AI, and to break through the inequalities caused by the restricted application of cutting-edge technologies, taking into account the common problems and major challenges of developing countries and regions in the field of water and the environment.

21.4 Future directions

With the purpose of implementing the UN SDGs by 2030, the use of AI and other emerging information technologies to realize the whole-system management of the water cycle of cities or urban clusters and the efficient prevention and control of environmental risks in a digital and intelligent mode is an important research direction and frontier for the environmental engineering discipline. Opportunities and challenges coexist, and the combination of data-driven algorithms and domain knowledge guidance should be strengthened in the process of deepening the application of AI technology in the domains of water and environmental pollution prevention and control, water quality safety assurance, water-related facilities optimization and reconstruction, and green watershed construction. It is important to enhance model interpretability on the basis of ensuring the accuracy of model prediction and to develop water system management, risk prevention, and control technology systems and operation modes oriented to AI arithmetic minimization. Development of standardized model algorithms, standardized construction theories, and effectiveness assessment methods for data quality and interface protocols are necessary to reduce or avoid unnecessary spillover effects while solving water and environmental problems. Further, a focus on the role of AI technology in building the capacity of water and environment governance in developing countries is crucial, especially in less developed countries and regions along the Belt and Road Initiative; at the same time, the expedient direction could aim at providing cloud computing resources for those developing countries to meet the resources required for AI deployment; thus, reducing the inequality of global governance caused by the application of AI in limited areas. Looking ahead, through the integrated innovation and synergistic development in the fields of water, the environment, and AI, it is expected that a healthy, sustainable, highly resilient, and intelligent next-generation water cycle system will be reconstructed on a global scale to meet the crucial need for enhancing the well-being of all human beings and to protect the water ecosystems.

References

Abba S, Pham Q B, Usman A G, et al. (2020). Emerging evolutionary algorithm integrated with kernel principal component analysis for modeling the performance of a water treatment plant. *Journal of Water Process Engineering*, 33: 101081.

Ballesté E, Belanche-Muñoz L A, Farnleitner A H, et al. (2020). Improving the identification of the source of faecal pollution in water using a modelling approach: From multi-source to aged and diluted samples. *Water Research*, 171: 115392.

Bi W, Dandy G C, Maier H R. (2015). Improved genetic algorithm optimization of water distribution system design by incorporating domain knowledge. *Environmental Modelling and Software*, 69: 370–381.

Big Earth Data Program Chineses Academy of Sciences. (2019). *Big Earth Data in Support of the Sustainable Development Goals*. Beijing: Big Earth Data Program Chinese Academy of Sciences.

Blumensaat F, Leitao J P, Ort C, et al. (2019). How urban storm- and wastewater management prepares for emerging opportunities and threats: Digital transformation, ubiquitous sensing, new data sources, and beyond - A horizon scan. *Environmental Science & Technology*, 53(15): 8488–8498.

Bonansea M, Ledesma C, Rodriguez C, et al. (2015). Water quality assessment using multivariate statistical techniques in Río Tercero Reservoir, Argentina. *Hydrology Research*, 46(3): 377–388.

Chen F H, Fu B J, Xia J, et al. (2019). Major advances in studies of the physical geography and living environment of China during the past 70 years and future prospects. *Science China Earth Sciences*, 49(11): 1659–1696.

Chen K Y, Chen H X, Zhou C L, et al. (2020a). Comparative analysis of surface water quality prediction performance and identification of key water parameters using different machine learning models based on big data. *Water Research*, 171: 115454.

Chen H Z, Xu L L, Ai W, et al. (2020b). Kernel functions embedded in support vector machine learning models for rapid water pollution assessment via near-infrared spectroscopy. *Science of the Total Environment*, 714: 136765.

Chou J S, Ho C C, Hoang H S. (2018). Determining quality of water in reservoir using machine learning. *Ecological Informatics*, 44: 57–75.

Cunha M C, Zeferino J A, Simões N E, et al. (2016). Optimal location and sizing of storage units in a drainage system. *Environmental Modelling & Software*, 83: 155–166.

Daigger G T, Murthy S, Love N G, et al. (2017). Transforming environmental engineering and science education, research, and practice. *Environmental Engineering Science*, 34(1): 42–50.

Daigger G T, Sharvelle S, Arabi M, et al. (2019). Progress and promise transitioning to the one water/resource recovery integrated urban water management systems. *Journal of Environmental Engineering*, 145(10): 04019061.

De Luna P, Wei J, Bengio Y, et al. (2017). Use machine learning to find energy materials. *Nature*, 552(7683): 23–27.

Eggimann S, Mutzner L, Wani O, et al. (2017). The potential of knowing more: A review of data-driven urban water management. *Environmental Science & Technology*, 51(5): 2538–2553.

García Nieto P J, García-Gonzalo E, Alonso Fernández J R, et al. (2019). Water eutrophication assessment relied on various machine learning techniques: A case study in the Englishmen Lake (Northern Spain). *Ecological Modelling*, 404: 91–102.

Giri S, Arbab N N, Lathrop R G. (2019). Assessing the potential impacts of climate and land use change on water fluxes and sediment transport in a loosely coupled system. *Journal of Hydrology*, 577: 123955.

Guitton M J. (2017). The water challenges: Alternative paths to trigger large-scale behavioural shifts. *The Lancet Planetary Health*, 1(2): e46–e47.

Guo H D. (2018). A Project on Big Earth Data Science Engineering. *Bulletin of the Chinese Academy of Sciences*, 33(8): 818–824.

Han Z, An W, Yang M, et al. (2020). Assessing the impact of source water on tap water bacterial communities in 46 drinking water supply systems in China. *Water Research*, 172: 115469.

Hassan W H, Jassem M H, Mohammed S S. (2018). A GA-HP model for the optimal design of sewer networks. *Water Resources Management*, 32(3): 865–879.

Hino M, Benami E, Brooks N. (2018). Machine learning for environmental monitoring. *Nature Sustainability*, 1(10): 583–588.

Hoffmann S, Feldmann U, Bach P M, et al. (2020). A research agenda for the future of urban water management: Exploring the potential of nongrid, small-grid, and hybrid solutions. *Environmental Science & Technology*, 54(9): 5312–5322.

IAEG-SDGs. *Tier Classification for Global SDG Indicators*. [2021-03-29]. https://unstats.un.org/sdgs/iaeg-sdgs/tier-classification/

Ji P, Guo H D, Zhang L. (2013). Remote sensing study of glacier dynamic change in West Kunlun Mountains in the past 20 years. *Remote Sensing for Natural Resources*, 25: 93–98.

Jia X L, Hu B F, Marchant B P, et al. (2019). A methodological framework for identifying potential sources of soil heavy metal pollution based on machine learning: A case study in the Yangtze Delta, China. *Environmental Pollution*, 250: 601–609.

Johnson A C, Jin X W, Nakada N, et al. (2020). Learning from the past and considering the future of chemicals in the environment. *Science*, 367: 384–387.

Jordan M I, Mitchell T M. (2015). Machine learning: Trends, perspectives, and prospects. *Science*, 349(6245): 255–260.

Kroll C, Warchold A, Pradhan P. (2019). Sustainable Development Goals (SDGs): Are we successful in turning trade-offs into synergies? *Palgrave Communications*, 5: 1–11.

Kulkarni P, Chellam S. (2010). Disinfection by-product formation following chlorination of drinking water: Artificial neural network models and changes in speciation with treatment. *Science of the Total Environment*, 408(19): 4202–4210.

Larsen T A, Hoffmann S, Luthi C, et al. (2018). Emerging solutions to the water challenges of an urbanizing world. *Science*, 352: 928–933.

Lesnik K L, Liu H. (2017). Predicting microbial fuel cell biofilm communities and bioreactor performance using artificial neural networks. *Environmental Science & Technology*, 51(18): 10881–10892.

Liao H, Sarver E, Krometis L A H. (2018). Interactive effects of water quality, physical habitat, and watershed anthropogenic activities on stream ecosystem health. *Water Research*, 130: 69–78.

Miller T H, Gallidabino M D, MacRae J I, et al. (2018). Machine learning for environmental toxicology: A call for integration and innovation. *Environmental Science & Technology*, 52(22): 12953–12955.

Mulligan M, van Soesbergen A, Hole D G, et al. (2020). Mapping nature's contribution to SDG 6 and implications for other SDGs at policy relevant scales. *Remote Sensing of Environment*, 239: 111671.

Nerini F F, Tomei J, To L S, et al. (2018). Mapping synergies and trade-offs between energy and the Sustainable Development Goals. *Nature Energy*, 3: 10–15.

Ogidan O, Giacomoni M. (2016).Multiobjective genetic optimization approach to identify pipe segment replacements and inline storages to reduce sanitary sewer overflows. *Water Resources Management*, 30(11): 3707–3722.

Perrault R, Shoham Y, Brynjolfsson E, et al. (2018). *AI Now Report 2018*. New York: AI Now Institute.

Perrault R, Shoham Y, Brynjolfsson E, et al. (2019). *The AI Index 2019 Report*. Palo Alto: Stanford University.

Pope A J, Gimblett R. (2015). Linking Bayesian and agent-based models to simulate complex social-ecological systems in semi-arid regions. *Frontiers in Environmental Science*,3: 55.

Qu J H, Zhao J C, Ren N Q, et al. (2017). Critical fundamental scientific problems in reclamation and reuse of municipal wastewater. *China Basic Science*, 1: 6–12.

Rai S P, Sharma N, Lohani A K. (2014). Risk assessment for transboundary rivers using fuzzy synthetic evaluation technique. *Journal of Hydrology*, 519: 1551–1559.

Raissi M, Perdikaris P, Karniadakis G E. (2017). Physics informed deep learning (part I): Data-driven solutions of nonlinear partial differential equations. *arXiv*, 1711, 10561.

Raza A, Bardhan S, Xu L, et al. (2019). A machine learning approach for predicting defluorination of Per- and Polyfluoroalkyl Substances (PFAS) for their efficient treatment and removal. *Environmental Science & Technology Letters*, 6(10): 624–629.

Reichstein M, Camps-Valls G, Stevens B, et al. (2019). Deep learning and process understanding for data-driven Earth system science. *Nature*, 566(7743): 195–204.

Rodell M, Famiglietti J S, Wiese D N, et al. (2018). Emerging trends in global freshwater availability. *Nature*, 557(7707): 651–659.

Romulo C L, Posner S, Cousins S, et al. (2018). Global state and potential scope of investments in watershed services for large cities. *Nature Communications*, 9(1): 4375.

Sayers W, Savic D, Kapelan Z. (2019).Performance of LEMMO with artificial neural networks for water systems optimisation. *Urban Water Journal*, 16(1): 21–32.

Schluter M, Baeza A, Dressler G, et al. (2017). A framework for mapping and comparing behavioural theories in models of social-ecological systems. *Ecological Economics*, 131: 21–35.

Shi Z H, Ai L, Li X, et al. (2013). Partial least-squares regression for linking land-cover patterns to soil erosion and sediment yield in watersheds. *Journal of Hydrology*, 498: 165–176.

Sinha J, Jha S, Goyal M K. (2019). Influences of watershed characteristics on long-term annual and intra-annual water balances over India. *Journal of Hydrology*, 577: 123970.

Smith A, Keane A, Dumesic J A, et al. (2020). A machine learning framework for the analysis and prediction of catalytic activity from experimental data. *Applied Catalysis B: Environmental*, 263: 118257.

Stephen L, Danny K, (2020). *Artificial Intelligence in the 21st Century*. 3rd ed. Virginia: Mercury Learning & Information, 750.

Strubell E, Ganesh A, McCallum A. (2019). Energy and Policy Considerations for Deep Learning in NLP//. In *Proceedings of the 57th Annual Meeting of the Association for Computational Linguistics*. Florence: Association for Computational Linguistics, 3645–3650.

Swathi B, Shoban S B, Monelli A. (2019). Artificial intelligence: Characteristics, subfields, techniques and future predictions. *Journal of Mechanics of Continua and Mathematical Sciences*, 14: 127–135.

UN Environment Programme. (2008). *Smart 2020: Enabling the low carbon economy in the information age*. Nairobi: UN Environment Programme.

United Nations. (2015). *Transforming Our World: The 2030 Agenda for Sustainable Development*. New York: United Nations.

United Nations. (2018). *Sustainable Development Goal 6: Synthesis Report on Water and Sanitation 2018*. New York: United Nations.

United Nations-Water. (2017). *Integrated Monitoring Guide for Sustainable Development Goal 6 on Water and Sanitation: Targets and Global Indicators*. New York: UN-Water.

Valerio C, De Stefano L, Martínez-Muñoz G, et al. (2020). A machine learning model to assess the ecosystem response to water policy measures in the Tagus River Basin (Spain). *Science of the Total Environment*, 141252.

Vamvakeridou-Lyroudia L S, Chen A S, Khoury M, et al. (2020). Assessing and visualising hazard impacts to enhance the resilience of critical infrastructures to urban flooding. *Science of the Total Environment*, 707: 136078.

Wang X, McCarty P L, Liu J, et al. (2015). Probabilistic evaluation of integrating resource recovery into wastewater treatment to improve environmental sustainability. *PNAS*, 112(5): 1630–1635.

Wang X, Daigger G, Lee D J, et al. (2018). Evolving wastewater infrastructure paradigm to enhance harmony with nature. *Science Advances*, 4(8): eaaq0210.

Whitehead B, Andrews D, Shah A, et al. (2014). Assessing the environmental impact of data centres part 1: Background, energy use and metrics. *Building and Environment*, 82: 151–159.

WHO-UNICEF. (2019). *Progress on Household Drinking Water, Sanitation and Hygiene 2000–2017: Special Focus on Inequalities*. Geneva: WHO-UNICEF.

Will S, Cassidy W, Randolf W, et al. (2019). *Digital Water: Industry Leaders Chart the Transformation Journey*. London: International Water Association and Xylem Inc..

Xia J, Tan G. (2002). Hydrological science towards global change: Progress and challenge. *Resources Science*, 1–7.

Zhang B, Wu L W, Wen X H. (2019). Potential Source Environments for Microbial Communities in Wastewater Treatment Plants (WWTPs) in *China. Environmental Science*, 40(8): 3699–3705.

Zhang X L, Zhang F, Kung H, et al. (2018a). Estimation of the Fe and Cu contents of the surface water in the Ebinur Lake Basin based on LIBS and a machine learning algorithm. *International Journal of Environmental Research and Public Health*, 15(11): 2390.

Zhang Y, Tang M, Tian Z, et al. (2018b). Research progress of removal technology of antibiotics from antibiotic production wastewater. *Chinese Journal of Environmental Engineering*, 12(1): 1–14.

Zhang Y Y, Gao X, Smith K, et al. (2019). Integrating water quality and operation into prediction of water production in drinking water treatment plants by genetic algorithm enhanced artificial neural network. *Water Research*, 164: 114888.

Zhao L, Dai T, Qiao Z, et al. (2020). Application of Artificial Intelligence to wastewater treatment: A bibliometric analysis and systematic review of technology, economy, management, and wastewater reuse. *Process Safety and Environmental Protection*, 133: 169–182.

Zodrow K R, Li Q, Buono R M, et al. (2017). Advanced materials, technologies, and complex systems analyses: Emerging opportunities to enhance urban water security. *Environmental Science & Technology*, 51(18): 10274–10281.

许国栋, 张婧怡, 陈珺, 等. (2016). 城市污水处理微污染物的挑战与对策. 给水排水, 52: 40–44.

刘俊新, 王旭. (2015). 城市污水处理的多目标管理. 给水排水, 41(9): 1–3.

刘昌明, 梁康. (2012). 作为水文科学基本理论的水循环研究若干探讨// 中国水文科技新发展——2012中国水文学术讨论会. 南京: 河海大学出版社, 4.

傅伯杰. (2020). 联合国可持续发展目标与地理科学的历史任务. 科技导报, 38: 19–24.

22
POLITICAL ECOLOGIES OF URBAN WATER GOVERNANCE

Julian S. Yates, Marc Tadaki, and Cristy Clark

22.1 Introduction

Political ecological analyses of urban water governance are produced by a broad community of scholars, practitioners, and policymakers with the aim of transforming knowledge into action (Robbins, 2020). Beyond a single theory, methodology, or approach, this community is held together by a series of common questions that orient studies relating to urban water. The aim of political ecological research and practice is usually – though not always – tied to a normative commitment to remaking socio-natural environments for the good of the (human and nonhuman) many, not the powerful few. Political ecologists frequently align with the interests of marginalized and vulnerable populations, including through theoretical commitments (e.g., Marxism), methodological rigour (e.g., critical chains of explanation), practical endeavours (e.g., community-engaged research), and creating impact beyond academic domains (e.g., action-oriented research; critical policy interventions). These commitments to pursuing socio-environmental justice mean that political ecology is not just concerned with understanding the world, but with changing it – often by affirming diverse knowledge and realities in relation to environmental change (Heynan & Van Sant, 2015; Loftus, 2015; Neimark et al., 2019).

Existing collections review these commitments, along with the topics, interests, and approaches common to political ecology in general (Bryant, 2015; Perreault et al., 2015; Robbins, 2020). In this chapter, we aim to demonstrate how political ecologies of urban water highlight the arenas, actors, and strategies that underpin struggle and contestation in pursuit of just urban water environments. Rather than attempt a comprehensive review of urban political ecologies of water, we identify several contemporary threads of academic debate that grapple with these political arenas in urban water governance, and we link those threads to some foundational avenues of theoretical orientation and enquiry. Exploring these evolving threads, we ask: In what forms does politics emerge, what opportunities exist for engaging with and changing it, and how have political ecologists approached these tasks?

Political ecological studies of urban water have incorporated the long-running influences of environmental history, hazards and vulnerability-focused studies, critical development studies, feminist studies, governmentality, post-colonial, and post-structural perspectives, and concerns regarding common property management practices and institutions. More recent attention has complemented these approaches by focusing on intersectional, embodied, and emotional

DOI: 10.4324/9781003057574-28

engagements with urban water, as well as by stressing the diverse relational engagements and ontological politics that underpin multiple urban water worlds.

These analyses are often concerned with exposing and challenging the explicit and implicit social structures, processes, and norms that perpetuate unjust environmental management. Political ecology often focuses on questions of power and politics in decision making at the human–environment intersection. In this chapter, we use the term "politics" to refer to contestation and struggle over knowledge, materials, and interests that make up water governance systems. Politics, in this sense, is much more than formal electoral politics or government decision making and policymaking. It includes everyday struggles over ideas about what water is, how it is managed, and who has access to it. Political ecological analyses often look beyond formal policies to explore alternative avenues in environmental management, such as by highlighting the needs of marginalized and vulnerable water user groups and communities.

While *urban* political ecology may be at a current crossroads in terms of its contemporary relevance and application (Gandy, 2021), we illustrate that political ecological analyses of water continue to be relevant for understanding *urbanization* as an ongoing process that unfolds across space. We highlight political ecological concerns as they apply to three contemporary threads of academic focus in urban water governance: extending political ecology insights beyond rural–urban divides; incorporating the materiality of water, infrastructure, and the roles of scientists in framing water governance issues; and understanding how urban water governance is increasingly acknowledging and attending to a plurality of water-related worlds and pieces of knowledge, including those of Indigenous peoples. Within each of these threads, we identify some of the real-world arenas where politics and contestation take shape, often among diverse actors with competing water interests and aspirations.

22.2 Metabolic relations and hydrosocial territories beyond a rural–urban divide

Urban political ecologists have studied urban spaces as flows within a circulatory system of energy, materials, labour, and capital accumulation. Rather than looking at the built and natural environments of urban space as static entities, political ecologists conceive urban environments as being continuously created through socio-spatial processes (e.g., capitalist production dynamics, property rights regimes and land use zoning, and international finance), which fuse objects, nature, and people together. This creates a range of geographically uneven outcomes, including for water access (Swyngedouw, 1996; Swyngedouw & Kaika, 2014). That is, urban water environments are continuously being made and remade, as flows of human and non-human influences and resources combine to produce unique – but continuously evolving – urbanizations.

This *process* of urbanization benefits some users at the expense of others because it hinges on historically produced conditions of access to and ownership of urban environmental amenities, including water and related infrastructure. At the heart of these approaches to understanding water within unequal urbanization processes lies a Marxian-influenced concept of urban metabolism. Marxian metabolism is not just a metaphor for the flow and processing of materials. It focuses attention on the complex and dynamic interchanges between human beings and nature as they take shape across space, paying particular attention to the central role of human labour in regulating these exchanges (Kennedy et al., 2011; Swyngedouw, 2006) (cf. Chapter 10 in this volume). Urbanization is therefore part of a process of *geographically arranged socio-natural metabolisms*, which redirect the flows of water, infrastructure, and social activities into uneven patterns across the city (Swyngedouw, 2006).

Matthew Gandy's research – on cities as diverse as Paris, London, New York, Lagos, Mumbai, and Los Angeles – illustrates metabolism at work, revealing the unequal outcomes that emerge at the intersection of capitalist development, state planning and regulation, and environmental concerns (Gandy, 1997, 1999, 2008, 2014). Focusing on the transformation of sewers in Paris, for example, he demonstrated how urbanization entailed planning processes, photographic representations, engineering advancements, household water consumption practices, and human waste collection. This combination transformed how wastewater flowed in the city, creating new opportunities for water management while foreclosing other water-related practices (such as bathing in the Seine River) (Gandy, 1999). The transformations were underpinned by metabolic relations of urbanization because they brought together historic layers of urban infrastructure and planning, capitalist production processes, class dynamics, and social norms concerning water and sanitation. This combination mediated the dynamic interchange among urban nature and human labour (such as through engineering expertise) to reshape the sewers of Paris.

Building on these political ecology approaches to urbanization as an uneven metabolic process, recent debates have begun to raise questions concerning what we consider to be "urban" about water governance. Recent contributions take us beyond the "urban" as a single site of relations to focus on "territorial urbanizations." For example, Millington and Scheba (2020) demonstrate how Cape Town's crisis of water scarcity is a material embodiment of combined pressures within and beyond the city. These pressures include infrastructure investment, climate change, a financial crisis, and shifting political commitments to the human right to water.

The intersection of these multiscalar pressures has been conceived recently in terms of hydrosocial territories. Hydrosocial relations are understood as those "through which symbolic formations are forged, social groups enrolled, and natural processes and 'things' entangled and maintained" (Swyngedouw, 2007, p. 10; 2009). The notion of hydrosocial *territories* expands this approach to explore the regional social and material networks that link urban and extra-urban environments (Hommes et al., 2019). Hydrosocial territories can be considered a spatial articulation of metabolic relations (complex interchanges) among humans and water that transcends distinctions between urban and rural.

Exploring urbanization beyond the boundaries of "the city," hydrosocial enquiry focuses on the material connections and social relations surrounding water that link urban and extra-urban regions. Exploring the case of Lima, Peru, Hommes and Boelens (2017) highlight how the construction of dams, canals, and hydropower infrastructure transformed the physical-ecological, sociocultural, and political-institutional relations in the city's hydro region. The introduction of modernist water infrastructure ideas in the early 20th century enabled engineers connected to hydropower and drinking water companies to dictate water infrastructure decision making, embedding a water management model that enclosed public waters (literally, behind locked gates) and enabled subsequent water extraction. With additional transformations in legal and administrative frameworks, new forms of water extraction from more distant watersheds were introduced. This culminated in an extensive water transfer network that supplies water to Lima from four different watersheds via a series of tunnels, canals, and reservoirs (Hommes & Boelens, 2017). Lima's urban water governance context is, therefore, dependent on a broader regional political ecology of hydrologic flows and control.

Such networks have prompted interest in three core dynamics: technologies and infrastructures; power and political influences; and contestation and links to extra-urban struggles. A focus on technologies and infrastructures highlights how water transfer mechanisms (such as extended pipelines, canal infrastructure, sluice technologies, etc.) can remake rural landscapes in line with powerful interests in urban areas. In Spain's Guadalhorce Valley hydrosocial territory (around Málaga), for example, rural–urban water transfers were implemented in line with the aspirations of urban elites seeking continuous water supplies. This resulted in the dispossession of water from rural populations, whose local water governance institutions – related to historical Arab waterworks – were dismantled in the process (Duarte-Abadia & Boelens, 2019).

Such unequal outcomes have prompted hydrosocial scholarship to focus on the political influences and water-related discourses that shape management decisions across hydrosocial territories. For example, the top-down design of payments for ecosystem services approaches to urban water management can ignore rural–urban linkages and subordinate upstream community needs to those of the city (Bleeker & Vos, 2019). Conversely, shifting water needs and emerging rural–urban water transfer projects can shift the balance of political action, leading to more politically engaged rural populations, such as those in rural areas surrounding Cochabamba, Bolivia, in the years following the famous "Water Wars" (Hoogendam, 2019). It has also been shown that hydrosocial territories extend beyond national borders. For example, Goldman and Narayan (2019) demonstrate how Bangalore's contemporary water scarcity crisis is constituted by networks of power, influence, and capital that stretch as far as private equity firms in New York (as local real estate and the city government are indebted to foreign corporate lenders and dependent on speculative investments).

Overall, these accounts reveal the limits of considering urban water governance as taking place within a preconceived and spatially defined unit ("the city"), highlighting instead the need to understand how technological, political, and social dynamics link urban and extra-urban areas in broader processes of transformation. Careful attention to rural–urban tensions is necessary to uncover some of the key structural issues that entrench the unequal division of water resources and related decision-making powers. As Hommes et al. (2019) emphasized, however, these same rural–urban linkages may also be fundamental to exploring alternative approaches to water control and management, particularly as water-related knowledge beyond urban areas might be mobilized in pursuit of water justice across urban–rural divides. Therefore, understanding hydrosocial urbanization means unpacking complex drivers of urbanizing water needs, highlighting the roles of multiple and often-competing actors (not just elite decision-makers), and exploring alternative policies and management arrangements via a range of governance techniques – all across multiple scales (Hommes et al., 2019).

22.3 Science, expertise, and the material politics of co-producing urban waters

A second key contemporary thread in political ecological investigations of urban water governance is evident in ongoing attention to infrastructure and a related connection to science and technology studies (STS). Influenced by Donna Haraway's (1991) foundational work on nature as both social and material, such accounts highlight how struggles over urban waters hinge on water's physical matter (Swyngedouw, 1999). Space and environmental requirements for pipe installation, for example, delimit where pipes can be placed, where leaks occur, and thus how water access is distributed (Uitermark & Tieleman, 2021). But pipe placement – just as with wetland restoration or stream culverting – is also a collective decision subject to fiscal accounting, risk management, property rights, and visions of local development. Pipes and wetlands may be material, but their generation, maintenance, and effects are social. When a pipe leaks, or water contamination occurs, the physical effects of these processes on people are further mediated through both material infrastructures (e.g., pipe location in network, temperature of water) and social relations (e.g., access to alternatives water sources, health care).

Truelove and O'Reilly (2020) encapsulate these relationships using the term "infrastructural intersectionality." This term refers to the way in which lived interactions with material infrastructure – everything from toilets to urban streams – intersect with other relationships between people and materials, often compounding existing patterns of scarcity and surplus in access to water and sanitation. Underpinning these accounts is a common concern for how water's materiality is intimately connected to the roles of key actors (particularly water "experts"), infrastructures, and forms of water-related science and knowledge production.

Urban political ecology considers the materiality of water as constitutive of its politics. The design of toilets can affect whether they are used by both women and men and under what circumstances and at what times of day (Caruso et al., 2017; Schmitt et al., 2018); the sounds of wind and water lapping at the side of a reservoir can motivate volunteers to engage in urban wetland restoration (Gearey et al., 2019); the quantity and quality of water in an urban river might be better explained by electricity prices over the course of a day than by regional precipitation (Soentgen, 2019). Water's materiality can therefore affect the politics of its distribution and access in surprising and counter-intuitive ways, producing varied and sometimes unplanned "infrastructure configurations" (Lawhon et al., 2018, p. 722).

Investigations into these configurations, and their effects on experiences of the materiality of water, often start by identifying the uneven outcomes of a material phenomenon, before explaining them through biophysical and social analysis. Bresnihan and Hesse (2020), for example, began their research after *Cryptosporidium* contamination occurred in Galway City, Ireland. They traced the bacteria to its source in rural agricultural land use, observing that *Cryptospyridium* levels are higher in rural waterways than in cities. Yet the response to the urban crisis was to treat the water in the city rather than prevent bacteria contamination at the source, thereby treating the symptoms rather than the cause of contamination and providing a solution only for urban residents.

Such uneven outcomes can lead to contestation around water infrastructures. Molden and Meehan (2018), for example, demonstrate that the revitalization of stone waterspout technologies in Nepal upholds alternative imaginaries of urban water, as they challenge modernist and technocratic supplies of water by reviving communal and symbolic forms of water access and shared distribution. The assertion of such alternative imaginaries through symbolic and material processes raises questions about the authority of water "experts" and the role of scientists in shaping water infrastructure development.

22.4 Science and urban water: Co-constructing urban water between science and society

Scientists, as translators of hydrosocial processes, engage in the governance of urban water by collecting and interpreting evidence about water, its use, and its benefactors. The processes of generating and interpreting scientific evidence involve choices that are shaped by social processes and norms (King & Tadaki, 2018); in turn, these choices shape how decisions are made about pipe networks, river culverts, roads, parks, and so on. Recognizing this iterative and dynamic relationship between the material environment and social processes, political ecologists have drawn critical attention to how science and scientists are engaged in political processes that underpin the production of urban environments. This allows us to examine how values are embedded within science and with what effects, for whom. This work engages critically with the politics *of* science, questioning scientific assumptions and biases, while seeking to improve scientific outcomes to transform water governance arrangements in line with environmental justice.

Scientists engage politically to frame urban environments. Frames "identify and give meaning to a situation by defining what it is, what facts about it are the most relevant, and what other situations it is related to" (Halffman, 2019, p. 37). For example, urban water can be framed as a flood hazard or, in scientific terms, as a problem of inefficient transmission of water out of the landscape (Lane et al., 2011). Such framing highlights specific features of water, such as its volume, timing, and spatial distribution – often in line with prevailing engineering and hydrological theories. Biophysical data can, in turn, be generated to represent these features and guide human decisions, thereby supporting the existing frame, which is presided over by experts with the knowledge and skills to generate and analyse the data. For over half a century, water resources scientists and engineers have exercised privileged status in defining the problems posed by urban water; and often, this has involved framing water in reductionist terms – as a volume in need of efficient conveyance through the landscape (Ashmore, 2015).

Frames also necessarily omit some aspects of urban hydrosocial processes, and more than one frame can be applied to a particular situation. For instance, if we depart from typical understandings of urban floods as a volume by emphasizing the spatiotemporal boundedness of the process of flooding (i.e., conceiving floods as too much water at a particular location within a specified period), it becomes possible to think about interventions that slow down water's travel across the landscape. Such interventions could range from bunds (small dams) on hills (Lane et al., 2011) to urban wetlands, porous pavement materials, and green infrastructures (Finewood, 2016; Finewood et al., 2019).

Frames also have ontological implications, determining what is and is not considered an urban water problem. For example, Young Rae Choi (2021) has shown how various scientific discourses compete over the "slippery ontology" of tidal mudflats in Korea. These shifting landscapes – called *getbol* – challenge modern and scientific knowledge, as agencies such as the Korea Hydrographic and Oceanographic Agency struggle to frame shifting ocean levels and terrestrial borders, often also failing to account for the tidal flat ontologies of local users (such as fishers). Questions therefore remain about what is considered land, sea, and a liminal in-between space – questions that have implications for urban coastal water management. Such studies highlight the ways in which political ecologies of urban hydrosocial spaces reveal the limits of scientific knowledge production. In the process, these studies ask questions about what alternative knowledge, framings, and ontologies might foster new opportunities for revising urban water governance for a more inclusive, deliberative, and just environmental politics (see Finewood et al., 2019; Holifield & Schuelke, 2015; Lane, 2017).

Political ecologies of urban water governance also explore how the domains of science and decision making are connected and debated. Wilfong and Pavao-Zuckerman (2020) for example, utilize a hydrosocial framework to understand urban water debates in Tucson and Pittsburgh in the United States. Understanding the history of urban water issues facing residents in the two cities reveals how repeated exposure to material conditions of water scarcity or temporary surplus has fostered diverse experiences and interpretations of water governance problems. Scientists, developers, urban planners, and others, therefore, need to consider how rainwater is considered "scarce" in Tucson, while stormwater is conceived as a "hazard" in Pittsburgh. The problem frames promoted by these actors support different claims and political demands regarding water and its management – often while occluding others, particularly those held among marginalised urban groups.

These insights show that political ecological analyses of urban water environments are not simply a critical exercise in dis-entangling power relationships; rather, they overlap with contributions from cognate subfields seeking to open scientific and decision-making processes concerning urban water governance. Critical physical geography, for example, is similarly dedicated to investigating human–environment intersections to "produce knowledge that transforms our communities, environments, and worlds" (Biermann et al., 2018, p. 559; Lave et al., 2018). It has also evolved from a concern with bridging political ecology and science and technology studies so as to highlight how political economic relations are central to scientific knowledge claims, education, authority, and policy concerning urban water issues (Lave, 2012). Political ecologists and allied scholars are therefore important actors in countering dominant forces and actors in producing environmental knowledge claims (such as those produced through the private sector and commercialized ventures), in the process creating opportunities for the democratization of knowledge for more than just hydrosocial governance (Forsyth, 2002; Goldman et al., 2010; Lave, 2015).

22.5 From biopolitical to relational urban waters

A third contemporary thread in political ecological investigations of urban water governance builds on post-structural insights concerning biopolitics to understand the impact of governance approaches on diverse experiences and engagements with water. Biopolitical readings of water have become an important focus for exploring the connected bodily aspects of urban water. Focusing on the ways in which governments regulate life in connection to water, a biopolitical perspective highlights the links between water's flows (through ecosystems and infrastructures) and the internalization of governing logics among individuals, creating new lived experiences in the process (Bakker, 2010; Gearey et al., 2019; Hellberg, 2017).

These approaches highlight how the materiality of water and related infrastructure converges with governance frameworks and governing approaches to shape how individuals experience water and, ultimately, how they live their water-related lives. In Cape Town, for example, water management devices converge with both a demand management approach to water supply and a regulatory framework for water access that is informed by the human right to water. The result is a contradictory lived experience of limited water access (due to a pay-before-you-consume model), despite claims to free basic water, resulting in new forms of individual self-management of water consumption.

Exploring these unfolding experiences, urban political ecologies often bring feminist and embodied approaches together to focus on the site and scale of the body as critical for understanding everyday urban water access and inequality. In doing so, they simultaneously highlight the intersectional contexts, relations, and (techno-)politics that shape embodied experiences and everyday practices (Hellberg, 2017; Kundu & Chatterjee, 2020; Truelove, 2019). Understanding the resulting socio-spatially varied effects requires a situated approach – one that foregrounds diverse water experiences instead of just reproducing dominant water governance approaches (Loftus, 2006). This body of work continues, as everyday materialities of water and infrastructure are problematized, not just to highlight contestation, inequality, and ambiguity in urban water distribution, but also to identify and initiate debate concerning alternative governance arrangements for improved water (re)distributions (Alba et al., 2021; Ramakrishnan et al., 2021; Truelove & Cornea).

Recent attention has also attempted to challenge the dualisms that underpin a Western biopolitics of water (that is, governance approaches driven by Western scientific understandings of water, which often separate water and society into two distinct and manageable sectors). Such accounts increasingly highlight a relational understanding of water, which begins from the premise of a connected socio-natural existence where water is the lifeblood of society and where human life and aquatic life are inseparable (Yates et al., 2017).

For example, focusing on Chemical Valley – an industrial zone located in southwestern Ontario, Canada – Wiebe (2016) demonstrates that this inseparability is often ignored by urban water policies. Such policies deny Indigenous understandings of water's agency and personality, thereby continuing the harm caused by ongoing contamination in the area. These experiences show that inequalities are not just about material consequences in terms of water access and exposure to contaminants, but also about the continued denial and suppression of Indigenous relational ontologies of water.

Settler-colonial water governance has traditionally adopted an extractive approach to water sources, viewing them primarily as an atomized, commodifiable resource – as a source of clean freshwater for settlements, industry, and farming and, paradoxically, a sewer and drain for waste. This approach to water has historically failed to respect Indigenous peoples' laws and water governance obligations, and this failure continues to reverberate as demonstrated by the infamous conflict at Standing Rock in North and South Dakota, USA. There, the Standing Rock Sioux community has sought to assert its sovereignty over its land and water on the Missouri River and to defend its capacity to carry out its obligation to protect it (see Estes, 2019; MacLean Lane, 2018). While sometimes perceived (and cast by settler states) as non-urban, these issues are intimately connected to urban water and resource governance issues, as demonstrated by protests and lawsuits by First Nations relating to the danger of water contamination from oil and gas pipeline developments in the Canadian city of Burnaby (Jonasson et al., 2019).

Reconceptualisations of human-water relations and related legal developments (often led by Indigenous peoples) have begun to disrupt this historic disconnection between settler-colonial and Indigenous water governance practices and their underlying legal orders (Curran, 2019; Martuwarra RiverOfLife et al., 2021; O'Donnell et al., 2020; Redvers et al., 2020). These recent developments – in policy, court rulings, and legislation – represent a fundamental shift in the law's treatment of water bodies, moving them from legal *objects* (relevant only in terms of their role in the legal relationship between human people) to legal *subjects* (relevant in and of themselves). The subjectivity of water is important here, not only because it provides human beings with the capacity to claim legal protection on behalf of water, but also because it provides a mechanism through which to claim legal recognition and protection of their relationships with water, particularly including their obligations. This represents a significant expansion in the capacity of settler-colonial law to respectfully engage with a wider range of approaches to nature (and our place within it), potentially transforming colonial legalities towards better understanding of multiple water ontologies, relational connections, and processes of water "worlding" (Furlong & Kooy, 2017; Yates et al., 2017).

Examples of ontological politics in transforming urban governance include recent cases of granting legal personhood to rivers. The well-known Te Awa Tupua (Whanganui Claims Settlement) Act 2017 (N.Z.) ("Whanganui River Act"), set a precedent in granting personhood to Te Awa Tupua (the Whanganui River) and creating the office of Te Pou Tupua as the river's "human face" in everyday governance. The act – and its related forms of water governance – does not entirely overcome incompatibilities between Māori and settler ontologies, particularly concerning relations among human and nonhuman nature (Charpleix, 2018). Yet it created conversations and opportunities for transforming water governance arrangements around an ontological politics of diversity and multiplicity, ultimately setting the stage for similar transformations elsewhere.

In 2017, the Australian State of Victoria passed the Yarra River Protection (Wilip-gin Birrarung murron) Act 2017 (VIC) (the "Yarra Birrarung Act"), as part of an ambitious project to reconceptualize and transform governance arrangements relating to the Yarra River (Birrarung), which runs through the city of Melbourne. In an Australian first, the Yarra Birrarung Act was co-named in both the English and Woi-wurrung languages (Woi-wurrung is the language of the Wurundjeri Traditional Owners). The Woi-wurrung name for the act, *Wilip-gin Birrarung murron*, means "keep the Birrarung [Yarra River] alive," emphasizing the reciprocal relationships and obligations of the *Wurundjeri* Traditional Owners towards the living river. This is further reflected in the preamble, which includes the Birrarung story in both English and Woi-wurrung, and starts with: "We, the Woi-wurrung, the First People, and the Birrarung, belong to this Country. This Country, and the Birrarung are part of us. The Birrarung is alive, has a heart, a spirit and is part of our Dreaming" (Yarra Birrarung Act). Here "Country" has a particular meaning, which has been described as the "grounded performativity of Aboriginal remembering in which the land itself is the repository of history, story and knowledge" (McAuley, 2009, p. 54). As Pelizzon and Kennedy (2012, p. 65) explain, this "Aboriginal concept of Country is rather distinct from the political sense of belonging (or being subject) to an abstract geopolitical entity identified by arbitrary, man-made boundaries, which forms the basis of the Western concept of nationhood."

Both the Whanganui Act and the Yarra Birrarung Act draw heavily on Indigenous law and ontology in reconceptualizing Te Awa Tupua and Birrarung in settler-colonial law. However, the processes that led to the drafting of the acts also highlighted the relational values and narratives that also existed among settler peoples living along the rivers (Clark et al., 2018, pp. 794, 804, 828). While this merging of Indigenous and settler values and law carries the risks of cultural appropriation, oversimplification, and environmental colonialism (O'Donnell et al., 2020; Rawson & Mansfield, 2018; Zenner, 2019, p. 43), it also opens the space for Indigenous peoples to repoliticize water governance by advocating and supporting multiple legal systems within the one state – a context referred to as legal pluralism (Curran, 2019; Ruru, 2018). Recent developments have demonstrated the scope of these possibilities, with the government of the state of Victoria announcing its plan to hand back two gigalitres of unallocated water in the Mitchell River as part of its commitment to recognizing the key role of traditional owners' connection to water in its management strategy (The Hon Daniel Andrews Premier ofVictoria, 2020). The mechanisms and possibilities generated by these transformations in urban hydrosocial relations create opportunities for countering the hegemony of ontological assumptions within modern water governance and scientific culture, simultaneously dispelling notions that Indigenous relationships with water cannot "be urban" (Jackson, 2018; McLean, 2017).

Interest in these developments among political ecologists reflects growing commitments to understanding urban waters as relational; to highlighting governance dilemmas as more than techno-managerial tasks; to embracing diversity in water-related concepts and experiences; and to recognizing Indigenous water rights claims in urban as well as extra-urban areas – all with the aim of pursuing more just water access and distribution to meet human and environmental needs (Jepson et al., 2017). Political ecologists continue to highlight the need to challenge the structures that prevent even access and distribution in urban contexts (Allen & Hofmann, 2017). Emphasis is also increasingly being placed on the opportunities brought about by such challenges when diverse experiences of water are brought to the fore in new governance arrangements. Highlighting decolonial urban waters, for example, means more than just advocating for the progressive legal transformations outlined above. It also means supporting the embedding of Indigenous water epistemologies and realities in everyday management – from Indigenous community-based monitoring to water-related knowledge sharing frameworks and reciprocal learning for community autonomy in urban water governance (Arsenault et al., 2018; Wilson & Inkster, 2018).

22.6 Conclusion: Towards plurality in urban water futures

This chapter has examined three contemporary threads of academic debate of particular concern in urban political ecology: a focus on metabolic relations evolving into attention to and hydrosocial territories beyond a rural–urban divide; critical studies of science, expertise, and infrastructure; and the biopolitics of water with shifting attention to relation urban waters. We aimed to illustrate the theoretical utility of these threads in grasping the nature of struggle and contestation in urban water governance. The political ecological concerns in these threads highlight particular arenas and processes in urban water governance that are worthy of theoretical and empirical attention and that are necessary for realizing an egalitarian agenda for urban water governance.

In political ecology, a widespread political economy influence has evolved from Marxist concerns with metabolic relations to encompass the multifaceted study of class-based tensions in urban water governance, underpinned by historically informed analyses of the processes that produce and sustain injustices in evolving urban water landscapes. The pursuit of hydrosocial justice in this vein will require going beyond arguing about the merit of specific infrastructures to support collective demands for the right to urban water to counter privatization and the technocratic vesting of water decision making in the hands of the few. Recent research on hydrosocial territorialization highlights the struggle for just governance of urban water needs to extend its spatial scale and scope. Widening the focus to include multiscalar and networked assemblages of actors, infrastructures, and ecologies that underpin rural–urban water connections and relations will help to unpack the complex drivers of urbanizing water needs, reveal the roles of multiple and often-competing actors (not just elite decision-makers), and explore alternative policies and management arrangements via a range of governance techniques (Hommes et al., 2019).

Influence from science and technology studies, including concepts of materiality and hybridity, places emphasis on water-related knowledge and expertise and aligns with broader critiques of scientific discourses in urban water management. By problematizing the issues of who gets to have a say over water-related decisions, and who gets to determine water governance arrangements, these approaches locate the who and how of political contestation in networks of science and policymaking. Revealing water science to be part of a process of framing urban water issues helps to identify diverse knowledge and water realities and illuminates how some knowledge and experiences are held above others within governance and management frameworks. Challenging the unequal processes in such arenas means countering dominant environmental knowledge claims and creating opportunities for the democratization of urban hydrosocial knowledge.

Finally, biopolitical readings of water highlight how management and governance mechanisms create conditions of water access and exposure for particular groups and individual bodies. Given that the domain of urban water governance extends beyond human actors to consider the agency of water, material infrastructure, and relational connections, we need to recognize the existence, location, and qualities of ontological contestation about what water *is*. Work in this vein is increasingly turning to relational understandings of water to highlight diverse ways of being with water and to decolonize water governance and legal frameworks.

Across these discussions, political ecologists carry forward a shared commitment to challenge representations of nature and water in urban environments, highlighting that urban nature is imagined, framed, and symbolically charged (Swyngedouw & Kaika, 2014). Other emerging concerns include issues such as planetary urbanism in the Anthropocene (Connolly, 2019; Swyngedouw & Kaika, 2014), the prospect of perpetual drought in some of the world's largest cities and urban areas (Millington, 2018; Millington & Scheba, 2020), and critically engaging with populist narratives of environmental change (Neimark et al., 2019). Topical evolutions such as these will continue to emerge according to empirical need and significance. We nonetheless suggest that they will often be approached with some of the analytical tools derived from the fields outlined here (which is not to imply that other approaches to urban water governance are categorically not political ecological). The task for political ecologists of urban water governance, we propose, is to expose, critically interrogate, and rework the complex power relations that underpin knowledge production, science, decision making, authority, ownership, and rights to water – all with a view to democratizing water governance processes for more just, equitable, and sustainable approaches to distribution and access.

References

Alba, R., Kooy, M., & Bruns, A. (2021). Conflicts, cooperation and experimentation: Analysing the politics of urban water through Accra's heterogeneous water supply infrastructure. *Environment and Planning E: Nature and Space*, *5*(1), 250–271. Forthcoming(online), 2514848620975342. https://doi.org/10.1177/2514848620975342

Allen, A., & Hofmann, P. (2017). Relational Trajectories of Urban Water Poverty in Lima and Dar es Salaam. In A. Lacey (Ed.), *Women, Urbanization and Sustainability: Practices of Survival, Adaptation and Resistance* (pp. 93–117). Palgrave Macmillan UK. https://doi.org/10.1057/978-1-349-95182-6_5

Arsenault, R., Diver, S., McGregor, D., Witham, A., & Bourassa, C. (2018). Shifting the Framework of Canadian Water Governance through Indigenous Research Methods: Acknowledging the Past with an Eye on the Future. *Water*, *10*(1), 49. https://doi.org/10.3390/w10010049

Ashmore, P. (2015). Towards a sociogeomorphology of rivers. *Geomorphology*, *251*, 149–156. https://doi.org/10.1016/j.geomorph.2015.02.020

Bakker, K. (2010). *Privatizing Water: Governance Failure and the World's Water Crisis*. Cornell University Press.

Biermann, C., Lane, S. N., & Lave, R. (2018). Critical reflections on a field in the making. In R. Lave, C. Biermann, & S. N. Lane (Eds.), *The Palgrave Handbook of Critical Physical Geography* (pp. 559–573). Palgrave Macmillan.

Bleeker, S., & Vos, J. (2019). Payment for ecosystem services in Lima's watersheds: Power and imaginaries in an urban-rural hydrosocial territory [Article]. *Water International*, *44*(2), 224–242. https://doi.org/10.1080/02508060.2019.1558809

Bresnihan, P., & Hesse, A. (2020). Political ecologies of infrastructural and intestinal decay. *Environment and Planning E: Nature and Space*, *4*(3), 778–798. 2021 (Available online), 2514848620902382. https://doi.org/10.1177/2514848620902382

Bryant, R. L. (Ed.). (2015). *The International Handbook of Political Ecology*. Edward Elgar Publishing. https://doi.org/10.4337/9780857936172.

Caruso, B. A., Clasen, T. F., Hadley, C., Yount, K. M., Haardörfer, R., Rout, M., Dasmohapatra, M., & Cooper, H. L. (2017). Understanding and defining sanitation insecurity: Women's gendered experiences of urination, defecation and menstruation in rural Odisha, India. *BMJ Glob Health*, *2*(4), e000414. https://doi.org/10.1136/bmjgh-2017-000414

Charpleix, L. (2018). The Whanganui River as Te Awa Tupua: Place-based law in a legally pluralistic society. *The Geographical Journal*, *184*(1), 19–30. https://doi.org/10.1111/geoj.12238

Choi, Y. R. (2021). Slippery ontologies of tidal flats. *Environment and Planning E: Nature and Space*, *5*(1), 340–361. 2022 Online First. https://doi.org/10.1177/2514848620979312

Clark, C., Emmanouil, N., Page, J., & Pelizzon, A. (2018). Can you hear the rivers sing: Legal personhood, ontology, and the nitty-gritty of governance. *Ecology Law Quarterly*, *45*(4), 787–844. https://heinonline.org/HOL/P?h=hein.journals/eclawq45&i=815. https://heinonline.org/HOL/PrintRequest?handle=hein.journals/eclawq45&collection=journals&div=38&id=815&print=section&sction=38 (844)

Connolly, C. (2019). Urban Political Ecology Beyond Methodological Cityism. *International Journal of Urban and Regional Research*, *43*(1), 63–75. https://doi.org/10.1111/1468-2427.12710

Curran, D. (2019). Indigenous processes of consent: Repoliticizing water governance through legal pluralism. *Water*, *11*(3), 571. https://doi.org/10.3390/w11030571

Duarte-Abadia, B., & Boelens, R. (2019). Colonizing rural waters: The politics of hydro-territorial transformation in the Guadalhorce Valley, Malaga, Spain [Article]. *Water International*, *44*(2), 148–168. https://doi.org/10.1080/02508060.2019.1578080

Estes, N. (2019). *Our History is the Future: Standing Rock versus the Dakota Access Pipeline, and the Long Tradition of Indigenous Resistance*. Verso.

Finewood, M. H. (2016). Green infrastructure, grey epistemologies, and the urban political ecology of Pittsburgh's water governance. *Antipode*, *48*(4), 1000–1021. https://doi.org/10.1111/anti.12238

Finewood, M. H., Matsler, A. M., & Zivkovich, J. (2019). Green Infrastructure and the Hidden Politics of Urban Stormwater Governance in a Postindustrial City. *Annals of the American Association of Geographers*, *109*(3), 909–925. https://doi.org/10.1080/24694452.2018.1507813

Forsyth, T. (2002). *Critical Political Ecology: The Politics of Environmental Science*. Routledge.

Furlong, K., & Kooy, M. (2017). Worlding water supply: Thinking beyond the network in Jakarta [Article]. *International Journal of Urban and Regional Research*, *41*(6), 888–903. https://doi.org/10.1111/1468-2427.12582

Gandy, M. (1997). The making of a regulatory crisis: Restructuring New York City's water supply. *Transactions of the Institute of British Geographers*, *22*(3), 338–358. https://doi.org/10.1111/j.0020-2754.1997.00338.x

Gandy, M. (1999). The Paris sewers and the rationalization of urban space. *Transactions of the Institute of British Geographers*, *24*(1), 23–44. https://doi.org/10.1111/j.0020-2754.1999.00023.x

Gandy, M. (2008). Landscapes of disaster: Water, modernity, and urban fragmentation in Mumbai. *Environment and Planning A*, *40*(1), 108–130. http://www.envplan.com/abstract.cgi?id=a3994

Gandy, M. (2014). *Fabric of Space: Water, Modernity, and the Urban Imagination*. MIT Press. <Go to ISI>://WOS:000358502400010

Gandy, M. (2021). Urban political ecology: A critical reconfiguration. *Progress in Human Geography*, *46*(1), 21–43 2022 Online First. https://doi.org/10.1177/03091325211040553

Gearey, M., Church, A., & Ravenscroft, N. (2019). From the hydrosocial to the hydrocitizen: Water, place and subjectivity within emergent urban wetlands. *Environment and Planning E: Nature and Space*, *2*(2), 409–428. https://doi.org/10.1177/2514848619834849

Goldman, M., & Narayan, D. (2019). Water crisis through the analytic of urban transformation: An analysis of Bangalore's hydrosocial regimes [Article]. *Water International*, *44*(2), 95–114. https://doi.org/10.1080/02508060.2019.1578078

Goldman, M., Nadasdy, P., & Turner, M. D. (2010). *Knowing Nature: Conversations at the Intersection of Political Ecology and Science Studies*. University of Chicago Press.

Halffman, W. (2019). Frames: Beyond facts versus values. In E. Turnhout, W. Halffman, & W. Tuinstra (Eds.), *Environmental Expertise: Connecting Science, Policy and Society* (pp. 36–57). Cambridge University Press. https://doi.org/10.1017/9781316162514.004

Haraway, D. (1991). *Simians, Cyborgs and Women: The Reinvention of Nature*. Routledge.

Hellberg, S. (2017). Water for survival, water for pleasure – A biopolitical perspective on the social sustainability of the basic water agenda *Water Alternatives*, *10*(1), 65–80.

Heynan, N., & Van Sant, L. (2015). Political ecologies of activism and direction action politics. In T. Perreault, G. Bridge, & J. McCarthy (Eds.), *The Routledge Handbook of Political Ecology* (pp. 169–178). Taylor and Francis. http://ebookcentral.proquest.com/lib/monash/detail.action?docID=3569469

Holifield, R., & Schuelke, N. (2015). The place and time of the political in Urban political ecology: Contested imaginations of a River's future. *Annals of the Association of American Geographers*, *105*(2), 294–303. https://doi.org/10.1080/00045608.2014.988102

Hommes, L., & Boelens, R. (2017). Urbanizing rural waters: Rural-urban water transfers and the reconfiguration of hydrosocial territories in Lima [Article]. *Political Geography*, *57*, 71–80. https://doi.org/10.1016/j.polgeo.2016.12.002

Hommes, L., Veldwisch, G. J., Harris, L. M., & Boelens, R. (2019). Evolving connections, discourses and identities in rural–urban water struggles. *Water International*, *44*(2), 243–253. https://doi.org/10.1080/02508060.2019.1583312

Hoogendam, P. (2019). Hydrosocial territories in the context of diverse and changing ruralities: The case of Cochabamba's drinking water provision over time. *Water International*, *44*(2), 129–147. https://doi.org/10.1080/02508060.2019.1551711

Jackson, S. (2018). Water and Indigenous rights: Mechanisms and pathways of recognition, representation, and redistribution. *WIREs Water*, *5*(6), e1314. https://doi.org/10.1002/wat2.1314

Jepson, W., Budds, J., Eichelberger, L., Harris, L., Norman, E., O'Reilly, K., Pearson, A., Shah, S., Shinn, J., Staddon, C., Stoler, J., Wutich, A., & Young, S. (2017). Advancing human capabilities for water security: A relational approach. *Water Security*, *1*, 46–52. https://doi.org/10.1016/j.wasec.2017.07.001

Jonasson, M. E., Spiegel, S. J., Thomas, S., Yassi, A., Wittman, H., Takaro, T., Afshari, R., Markwick, M., & Spiegel, J. M. (2019). Oil pipelines and food sovereignty: Threat to health equity for Indigenous communities. *Journal of Public Health Policy*, *40*(4), 504–517. https://doi.org/10.1057/s41271-019-00186-1

Kennedy, C., Pincetl, S., & Bunje, P. (2011). The study of urban metabolism and its applications to urban planning and design. *Environmental Pollution, 159*(8), 1965–1973. https://doi.org/10.1016/j.envpol.2010.10.022

King, L., & Tadaki, M. (2018). A framework for understanding the politics of science (core tenet #2). In R. Lave, C. Biermann, & S. N. Lane (Eds.), *The Palgrave Handbook of Critical Physical Geography* (pp. 67–88). Palgrave Macmillan.

Kundu, R., & Chatterjee, S. (2020). Pipe dreams? Practices of everyday governance of heterogeneous configurations of water supply in Baruipur, a small town in India [Article; Early Access]. *Environment and Planning C-Politics and Space, 39*(2), 318–335, 2021, 18, Article 2399654420958027. https://doi.org/10.1177/2399654420958027

Lane, S. N. (2017). Slow science, the geographical expedition, and critical physical geography. *The Canadian Geographer / Le Géographe canadien, 61*(1), 84–101. https://doi.org/10.1111/cag.12329

Lane, S. N., Odoni, N., Landström, C., Whatmore, S. J., Ward, N., & Bradley, S. (2011). Doing flood risk science differently: An experiment in radical scientific method. *Transactions of the Institute of British Geographers, 36*(1), 15–36. https://doi.org/10.1111/j.1475-5661.2010.00410.x

Lave, R. (2012). Bridging political ecology and STS: A field analysis of the Rosgen wars. *Annals of the Association of American Geographers, 102*(2), 366–382. https://doi.org/10.1080/00045608.2011.641884

Lave, R. (2015). The future of environmental expertise. *Annals of the Association of American Geographers, 105*(2), 244–252. https://doi.org/10.1080/00045608.2014.988099

Lave, R., Biermann, C., & Lane, S. N. (Eds.). (2018). *The Palgrave Handbook of Critical Physical Geography*. Palgrave Macmillan.

Lawhon, M., Nilsson, D., Silver, J., Ernstson, H., & Lwasa, S. (2018). Thinking through heterogeneous infrastructure configurations. *Urban Studies, 55*(4), 720–732. https://doi.org/10.1177/0042098017720149

Loftus, A. (2006). Reification and the dictatorship of the water meter. *Antipode, 38*(5), 1023–1045. https://doi.org/10.1111/j.1467-8330.2006.00491.x

Loftus, A. (2015). Political ecology as praxis. In T. Perreault, G. Bridge, & J. McCarthy (Eds.), *The Routledge Handbook of Political Ecology* (pp. 179–187). Taylor and Francis. http://ebookcentral.proquest.com/lib/monash/detail.action?docID=3569469

MacLean Lane, T. (2018). The frontline of refusal: Indigenous women warriors of standing rock. *International Journal of Qualitative Studies in Education, 31*(3), 197–214. https://doi.org/10.1080/09518398.2017.1401151

Martuwarra River Of Life, Pelizzon, A., Poelina, A., Akhtar-Khavari, A., Clark, C., Laborde, S., Macpherson, E., O'Bryan, K., O'Donnell, E., & Page, J. (2021). Yoongoorrookoo. *Griffith Law Review, 30*(3), 505–529. https://doi.org/10.1080/10383441.2021.1996882

McAuley, G. (2009). Unsettled country: Coming to terms with the past. *About Performance, 45*, 45–65.

McLean, J. (2017). Water cultures as assemblages: Indigenous, neoliberal, colonial water cultures in northern Australia. *Journal of Rural Studies, 52*, 81–89. https://doi.org/10.1016/j.jrurstud.2017.02.015

Millington, N. (2018). Producing water scarcity in Sao Paulo, Brazil: The 2014-2015 water crisis and the binding politics of infrastructure [Article]. *Political Geography, 65*, 26–34. https://doi.org/10.1016/j.polgeo.2018.04.007

Millington, N., & Scheba, S. (2021). Day zero and the infrastructures of climate change: Water governance, inequality, and infrastructural politics in Cape Town's water crisis [Article; Early Access]. *International Journal of Urban and Regional Research, 45*(1). 116–132 https://doi.org/10.1111/1468-2427.12899

Molden, O. C., & Meehan, K. (2018). Sociotechnical imaginaries of urban development: Social movements around "traditional" water infrastructure in the Kathmandu Valley [Article]. *Urban Geography, 39*(5), 763–782. https://doi.org/10.1080/02723638.2017.1393921

Neimark, B., Childs, J., Nightingale, A. J., Cavanagh, C. J., Sullivan, S., Benjaminsen, T. A., Batterbury, S., Koot, S., & Harcourt, W. (2019). Speaking power to "Post-Truth": Critical political ecology and the new authoritarianism. *Annals of the American Association of Geographers, 109*(2), 613–623. https://doi.org/10.1080/24694452.2018.1547567

O'Donnell, E., Poelina, A., Pelizzon, A., & Clark, C. (2020). Stop burying the lede: The essential role of indigenous law(s) in creating rights of nature. *Transnational Environmental Law, 9*(3), 403–427. https://doi.org/10.1017/S2047102520000242

Pelizzon, A., & Kennedy, J. (2012). Welcome to country: Legal meanings and cultural implications. *Australian Indigenous Law Review, 16*, 58–69.

Perreault, T., Bridge, G., & McCarthy, J. (Eds.). (2015). *The Routledge Handbook of Political Ecology*. Taylor and Francis. http://ebookcentral.proquest.com/lib/monash/detail.action?docID=3569469

Ramakrishnan, K., O'Reilly, K., & Budds, J. (2021). Between decay and repair: Embodied experiences of infrastructure's materiality. *Environment and Planning E: Nature and Space*, Forthcoming(Online), 2514848620980597. https://doi.org/10.1177/2514848620980597

Rawson, A., & Mansfield, B. (2018). Producing juridical knowledge: "Rights of Nature" or the naturalization of rights? *Environment and Planning E: Nature and Space*, *1*(1–2), 99–119. https://doi.org/10.1177/2514848618763807

Redvers, N., Poelina, A., Schultz, C., Kobei, D. M., Githaiga, C., Perdrisat, M., Prince, D., & Blondin, B. s. (2020). Indigenous natural and first law in planetary health. *Challenges*, *11*(2), 29. https://www.mdpi.com/2078-1547/11/2/29

Robbins, P. (2020). *Political Ecology: A Critical Introduction* (3rd ed.). Wiley-Blackwell.

Ruru, J. (2018). Listening to Papatūānuku: A call to reform water law. *Journal of the Royal Society of New Zealand*, *48*(2–3), 215–224. https://doi.org/10.1080/03036758.2018.1442358

Schmitt, M. L., Clatworthy, D., Ogello, T., & Sommer, M. (2018). Making the Case for a Female-Friendly Toilet. *Water*, *10*(9), 1193. https://www.mdpi.com/2073-4441/10/9/1193

Soentgen, J. (2019). The river Lech: A Cyborg. *Analecta Hermeneutica*, *10*. https://journals.library.mun.ca/ojs/index.php/analecta/article/view/2059

Swyngedouw, E. (1996). The city as a hybrid: On nature, society and cyborg urbanization. *Capitalism Nature Socialism*, 7(2), 65–80. https://doi.org/10.1080/10455759609358679

Swyngedouw, E. (1999). Modernity and hybridity: Nature, regeneracionismo, and the production of the Spanish waterscape, 1890–1930. *Annals of the Association of American Geographers*, *89*(3), 443 – 465.

Swyngedouw, E. (2006). Metabolic urbanization: The making of cyborg cities. In N. Heynen, M. Kaika, & E. Swyngedouw (Eds.), *In the Nature of Cities: Urban Political Ecology and the Politics of Urban Metabolism*. (pp. 21–40). Routledge

Swyngedouw, E. (2007). Technonatural revolutions: The scalar politics of Franco's hydro-social dream for Spain, 1939–1975. *Transactions of the Institute of British Geographers*, *32*(1), 9–28. https://doi.org/10.1111/j.1475-5661.2007.00233.x

Swyngedouw, E., & Kaika, M. (2014). Urban political ecology. Great promises, deadlock… and new beginnings? [urban political ecology; environmental politics; urban theory; socio-ecological conflict]. *Documents D Analisi Geografica*, *60*(3), 459–481. https://doi.org/10.5565/rev/dag.155

The Hon Daniel Andrews Premier ofVictoria. (2020, 11/11/2020). *Water For Traditional Owners A Victorian First*. https://www.premier.vic.gov.au/water-traditional-owners-victorian-first

Truelove, Y. (2019). Rethinking water insecurity, inequality and infrastructure through an embodied urban political ecology [Article]. *Wiley Interdisciplinary Reviews-Water*, *6*(3), 7, Article e1342. https://doi.org/10.1002/wat2.1342

Truelove, Y., & Cornea, N. Rethinking urban environmental and infrastructural governance in the everyday: Perspectives from and of the global South. *Environment and Planning C: Politics and Space*, *39*(2), 231–246. https://doi.org/10.1177/2399654420972117

Truelove, Y., & O'Reilly, K. (2021). Making India's cleanest city: Sanitation, intersectionality, and infrastructural violence. *Environment and Planning E: Nature and Space*, *4*(3), 718–735. https://doi.org/10.1177/2514848620941521

Uitermark, J., & Tieleman, J. (2021). From fragmentation to integration and back again: The politics of water infrastructure in Accra's peripheral neighbourhoods. *Transactions of the Institute of British Geographers*, *46*(2), 347–362. https://doi.org/10.1111/tran.12420

Wiebe, S. M. (2016). *Everyday Exposure: Indigenous Mobilization and Environmental Justice in Canada's Chemical Valley* [Book]. UBC Press. http://ezproxy.lib.monash.edu.au/login?url=https://search.ebscohost.com/login.aspx?direct=true&db=nlebk&AN=1353098&site=ehost-live&scope=site

Wilfong, M., & Pavao-Zuckerman, M. (2020). Rethinking stormwater: Analysis using the hydrosocial cycle. *Water*, *12*(5), 1273. https://doi.org/10.3390/w12051273

Wilson, N. J., & Inkster, J. (2018). Respecting water: Indigenous water governance, ontologies, and the politics of kinship on the ground. *Environment and Planning E: Nature and Space*, *1*(4), 516–538. https://doi.org/10.1177/2514848618789378

Yates, J. S., Harris, L., & Wilson, N. J. (2017). Multiple ontologies of water: Politics, conflict, and implications for governance. *Environment and Planning D: Society & Space*, *35*(5), 797–815. https://doi.org/10.1177/0263775817700395

Zenner, C. (2019). Turning to traditions: Three cultural-religious articulations of fresh waters' value(s) in contemporary governance frameworks. In F. Sultana & A. Loftus (Eds.), *Water Politics: Governance, Justice and the Right to Water* (pp. 42–53). Taylor and Francis. https://doi.org/10.4324/9780429453571

23

TERRITORIAL INTEGRATION AND INNOVATION FOR GOOD URBAN WATER GOVERNANCE

Susana Neto

23.1 Introduction: Recent and current challenges and evolving paradigm of water governance

To discuss the evolving paradigm of water governance, it is useful to understand the recent trends in global and contextual challenges that have an impact on water. Water challenges complexity has been increasing in the last decades due to mismanagement of available water resources, bad governance, and aggravated extreme events caused by climate change. In 2008, McEvoy presented some facts regarding the increase in frequency and severity of weather-related extreme events forecast, while stressing also the social and economic nature of this aggravated situation (McEvoy et al. 2008, 50). The threat of climate variability was perceived when between 2003 and 2007 several extreme phenomena impacted European regions and urban areas with heatwaves, outbreaks of forest fires, long drought periods in the Mediterranean region, and widespread and devastating flooding in northern Europe. In 2007 the International Panel for Climate Change (IPCC) Report called attention to the scientific consensus that these extreme events would become more common in the future with warmer climate (IPCC 2007). Although climate change gained recognition and gradual inclusion in the political agendas, the recently published IPCC Report (2021) clearly shows the lack of effective progress and the unlikelihood of attaining the necessary goals. It also calls attention to the fact that climate change is bringing multiple changes in different regions besides increasing temperature. Among those, it highlights the intensification of the water cycle with more intense rainfall and associated flooding and more intense drought in many regions. Another highlight relates to the impacts on cities, where some aspects of climate change may be amplified, including heat, flooding, and sea-level rise in coastal cities (IPCC 2021).

Societal and economic dimensions play a critical role in climate change adaptation strategies. One of the aims of water governance is to contribute to a more informed and conscious society regarding the responsibilities within a paradigm of governance of water towards the Sustainable Development Goals (SDGs). The pandemic that affected the whole world since 2020 aggravated the already complex situation created by climate change and by the demographic crisis. Urban areas continue to be the destination of many displaced populations, either domestic migrants or those seeking refuge from other countries. The drivers for change differ dramatically between rich urbanised areas of developed countries and the suburban or metropolitan deprived

DOI: 10.4324/9781003057574-29

neighbourhoods of poorer countries. In those contexts, the need to change is limited to survival and access to basic needs, such as water, food, and shelter. No unique way exists to address the urban context in the world. Living in the suburbs of Perth (Australia) resembles in no way living in the suburbs of Luanda (Angola). They are two "opposed worlds," although both are located in the Southern Hemisphere.[1]

Within the first known definition of water governance of the Global Water Partnership (GWP), the authors further stated that "governance looks at the balance of power and the balance of actions at different levels of authority," and that "usually improving governance means reform" (GWP 2002). To address this reform need, it is important to better understand what can "water governance" provide differently. The concept came to life with the realisation of the need to involve all social partners in the decision-making processes of water management with an integrated approach. Hofwegen defined integrated water resources management (IWRM) as "the management of surface and subsurface water in qualitative, quantitative, and ecological senses from a multidisciplinary perspective, and focussed on the needs and requirements of society at large regarding water" (Hofwegen, n.d.). He also advocates the creation of a platform for weighing all relevant interests on water uses in the river basin (Hofwegen 2001). A permanent dialogue among the stakeholders and the governmental agencies is advisable if the final aim is to sanction powers to protect the interest of society at large, since government has also the *care function* of managing water resources. First-order interests are considered the interests of society and therefore it is the government that represents them, and second-order interests are interests of individuals, groups, or parts of society, and these can best be represented by their stakeholders (Hofwegen 2001, 39). This idea is also reflected and included in the GWP definition of water governance that was proposed in 2002. Another relevant dimension asserted by GWP is the reinforcement of trust and dialogue, along with decentralisation (2003). Recognising the need for better communication between the decision-makers and stakeholders, a "dialogue" process was promoted by GWP in partnership with other United Nations (UN) agencies, focusing on the challenge of making water governance work by creating dynamic relations and accountability. Awareness building and bringing more local players into the process was a key outcome of many dialogues. The features advanced for achieving better distributed water governance included improved regulation, clearer definition of roles, better water allocation mechanisms to match society's changing needs, capacity building of individuals and institutions, improved use of existing financial resources, decentralisation, and shared river basin management. This dialogue has clearly demonstrated that water governance should not be seen apart from the broader governance framework, and, most importantly, it should consider political and social changes, which can create opportunities for improving water management (GWP 2003). Remarkably, the OECD reaffirmed 12 years later these two dimensions (trust and engagement) as one of the three axes around which the 15 Water Governance Principles for Water Governance were presented (OECD 2015).

To address the challenges of water governance, and the contribution to be made by them, the need also exists to reaffirm the critical role of water resource management and water services regarding basic human needs. The next section will address the social and political dimensions of water and the need to bring water management to a higher political level to guarantee access and security.

23.2 Water as a human right and a social and political issue

The UN officially recognised in 2010 a "human right to water," which entitles everyone without discrimination to sufficient, safe, acceptable, accessible, and affordable water. The UN

Resolution calls upon states and international organisations to provide financial resources, capacity building, and technology transfer through international assistance and cooperation to scale up efforts to provide drinking water and sanitation for all, particularly in developing countries (UN 2010). Regarding the societal dimension debate of water problems, water access and security go beyond ensuring sufficient clean water delivery and sanitation services and include the sociocultural aspects of water resources regarding sustainability, inclusiveness, and empowering people to develop their personal capabilities. Much more than what we drink or use, water is deeply related to values and social structures (Neto et al. 2019). It bears cultural and social specificities, clarifying roles and structure habits. In this sense, urban areas appear as an assembler of agents interacting within a network, while challenging the social and physical prejudices (Neto and Henriques 2016). Water services should be viewed as a social matter in the water governance spectrum, demanding a transdisciplinary approach where the urban water cycle is situated within the integrated and collaborative framework of water services and resources, including territorial and urban planning. The "right to water" and its translation into innovative policies and practices may be advanced as an adequate approach to the current complexity, bringing more diversity and a mix of varied knowledge inputs to water problems, therefore contributing to more sustainable solutions (ibid.).

Managing water resources to guarantee access to water within a secure context, on the other hand, requires a clear understanding of both external factors, such as climate change (e.g., aggravated extreme events like droughts and floods), demographic change (e.g., population ageing and migratory processes), and internal problems (e.g., inequality in access to water and social and economic water scarcity). These various factors can cause conflicts and serious blockages to adequate and equitable water management, and they call for having available robust and consistent multidisciplinary knowledge systems. Esteban Castro (2012) refers, for example, to the fragmentation created by "artificial epistemic divisions" in the production of scientific knowledge about water in general and water conflicts by classifying it as "hard" versus "soft" or "natural" versus "social" disciplinary approaches, creating more difficulties rather than facilitating understanding of the processes (Castro 2012, 140–141).

The combined challenges of climate change, population growth, ageing infrastructure, declining revenues, and a variety of other localised challenges can be addressed within an IWRM framework. Moreover, it is necessary to adopt a more holistic approach to water management, which means primarily to involve diverse views and perceptions of the problems (or challenges) in the diagnostic and in the search for solutions. This also means working locally and providing the necessary information to upscale those perceptions to broader contexts.

Approaching water issues within a broad framework of control, distribution, rights, and access, K. J. Joy (2014) claimed a need to "repoliticise" water debates towards explicit discussion of water in terms of justice (Joy et al. 2014). It is widely recognised that the human right to water is a basic requisite for human dignity and for the realisation of other rights. While the dimension of human rights introduced by the UN (2010) has recognised access to drinking water and sanitation as a human right, access to water must also be discussed in relation to the scarcity concept, as a human need, the satisfaction of which implies a key political objective (Neto and Camkin 2020). The Millennium Development Goals (MDGs) drove a global effort to tackle the indignity of poverty by establishing measurable, universally agreed objectives for dealing with extreme poverty and hunger, preventing deadly diseases, and expanding primary education for all children, among other development priorities.

Repoliticising water can avoid some "governance traps," such as relying on hierarchical governance arrangements of low complexity that produce a short-term vision (Cáñez-Cota and Pineda-Pablos 2019). Water policies are made and implemented in specific institutional contexts, but the need exists to focus on the larger institutional context beyond organisations and individuals (Merrey et al. 2012), and a central idea behind the concept of repoliticisation is the need to recognise that politics of water places power relations among the different actors (or stakeholders) of water governance (Mollinga 2008), and the final benefits of water and financial resources allocation need to be well justified, bringing also back to this discussion the need for increased trust in the political process of water governance. The example of Europe establishing in 2000 the Water Framework Directive (WFD) brought to the EU mainstream policy the core principle that "Water is not a commercial product like any other but, rather, a heritage which must be protected, defended and treated as such" (EC 2000). Previously, other agreements were reached among scientists, researchers, and activists for water rights, considering that the "access to this general interest equity or value must be recognized and guaranteed to all, as citizens," and water is an "essential good" and a "human right" (Petrella 1998).

23.3 Policy, planning, and integrating water territories

To link water challenges with the territories where they take place, it is necessary to observe the different institutional levels and see how they effectively collaborate in an integrated way or keep operating in silos.

The ecological processes occur and function at various scales from very small and localised to very large, with deferred effects in time, and which can present interscale variations due to the interference of various, often unpredictable, factors. Accordingly, interactions between the different territorial scales condition the results arising from those processes. When attention is focused exclusively on a specific scale, it can hinder the perception of these interactions, resulting in institutional responses that are not the most suitable because they do not cover the whole process (Alcamo et al. 2003). This territorial integration is not typically expressed in the current institutional systems, requiring various adaptations and articulations between organisations and between the powers of different levels. Therefore, a cross-scale approach is required, along with a multiple level intervention and coordination, namely by territorial (and urban) planning. The complexity of integrated water management emerges from the need to properly articulate a great diversity of subsystems at very different levels. Pritchard and Sanderson (2002) state that every element of every ecosystem, and every natural resource user, falls under multiple political jurisdictions that cover many scales. The search for the proper level for interventions affects the articulation between the biophysical system of "water" and the social-economic system of "land use," which includes a large diversity of social, economic, institutional, and cultural dimensions. Scale can also constitute an argument that empowers state institutions, while local and Indigenous knowledge is many times seen as less relevant and powerless. Sadoff et al. (2015) noted that political will and institutional capacity, as well as coordination across scales, are fundamental within this challenging context. Some problems must be addressed in a short-term perspective at the local level, whereas others (with strategic character) must be addressed at the global level within a long-term perspective.

Neto (2010) stressed that understanding and coping with the long-term effects of human occupation upon the ecological processes, as well as with its consequences and alterations to the natural systems, is a complex task for which various fields of knowledge are required. Evolving expectations of water resources use by human activities also require adaptive capacity of the organisations responsible for water management. This flexibility is necessary at both institutional and political levels, which is not always compatible with the existing structures (Neto 2010, 16–17). As Falkenmark et al. (1999, 6) wrote, "in earlier generations, water was seen as a technical issue and a question of organizing proper water supply for human societies and agricultural production." A greater recognition now exists that while water has a strong technical component, it is more about managing people's expectations and considering their interaction with the natural environment and the services that ecosystems provide. Effectively, water policy is where new perspectives for water management and planning should be framed.

Le Meur et al. (2005) define the distinction between water management and "water politics" based on the more comprehensive perspective and the inclusion of the social dimension of the last one, as the most relevant difference. They dispute that "speaking of water politics instead of water management is intended to shed light upon the contested nature of the water issue, involving social actors and groups endowed with unequal resources and skills and carrying diverse individual projects" and "such an approach is at odds with a managerial perspective supporting the idea that the right (technical and institutional) solution for any issue exists, and can be worked out with expertise and participation" (Le Meur et al. 2005, 51). The point of difference between a simplistic managerial perspective and a "water policy" is the way the common resources are shared and how this sharing is regulated by the state. According to Ostrom (1990), how "individuals or groups of individuals share water resources as a common property resource, people are connected in a socio-political, economic and ecological sense" (cited in Le Meur et al. 2005).

It is generally accepted in the literature that water management should involve the adoption of policies that include the conservation and recovery of water courses and surrounding ecosystems, and the protection of springs and underground waters, in a comprehensive and territorial perspective, following the need of all parties to recognise the ecological functions of the natural water cycle (Neto 2010, 58–59). On the other hand, allocation of the available resources, according to priorities defined by the state, and the control of these resources, or the appropriation of "the rights from the common property group to create state property with a lesser amount owned privately," is a prerogative of the state through its policies. This responsibility of how to deploy a resource to the national advantage is the key to water governance in the early 21st century regarding "how, through politics, the State can achieve this fairly and equitably, without reducing incentives for efficient use of the resource" (Le Meur et al. 2005, 51). However, fragmentation at administrative levels is a traditional problem of environment and water institutions, with different authorities seeking different organisational objectives and agendas to handle different issues, so that it becomes more difficult to promote and feed interdependencies.

As addressed by Falkenmark et al. (1999) though reality is multidimensional and complex, all our institutional organisations are oriented towards unidirectional and unidimensional interventions, leading to the fact that *individual institutions have specific and often opposing objectives.* The institutional framework is built to frame action in compartmentalised "worlds," often hindering any interaction and cooperation that would be crucial to address these problems with complex objectives. The *land/water dichotomy characterised by a mental image of land as opposed to water* referred by Falkenmark is still a common reality, with policy structures typically based on the separation between land and water to be managed separately and within non-dialoguing management systems. Institutions regulate the articulation between the different subsystems of water management at different levels, requiring effective integration mechanisms within the institutional framework. Brown and Farrely (2009), following Wakely's (1997) argument, have noted that institutions are often not "ready or willing" to adapt and therefore, to encourage institutional change, and that building capacity is determinant to assess the ability of an institution to perform effectively at its own (internal) tasks and in cooperation and coordination (external) with others in its field. These authors also argue that most existing impediments to coordination between institutions are predominantly institutionally embedded, systemic, relating to interorganisational capacity and external rules and incentives, and that they are *socioinstitutional* rather than technical (ibid., 845).

Discussion of what is or should be "good" water governance influence and effects in the urban sphere is still ongoing. Many decades after Falkenmark et al. (1989) published pioneer work on land-water integration, the call remains urgent for integrated land and water management at regional and urban scales. The need to develop analytical frameworks that effectively assess the impacts and the role of territorial planning in the urban water cycle has been established (Neto 2010). This need is not automatically reflected in the current institutional frameworks, and existing processes require several levels of adaptation as well as articulation between organisations (Falkenmark et al. 1989). There is a persistent need to adapt the profiles and the *modus operandis* of the organisations responsible for water management to overcome the siloed and sectoral nature of the existing institutional structures (Hofwegen and Jaspers 1999; Falkenmark et al. 1999; Alcamo et al. 2003; Le Meur et al. 2005; Sadoff et al. 2015; Neto 2010, 2016, 2018).

A set of selected analytical frameworks or major criteria for their implementation is presented in the following section. The aim is not to exhaust all the dimensions that innovative water governance should have (nor the many studies undertaken on this matter) but to highlight the central axis for action and change.

23.4 Analytical frameworks in action for water governance

The need is urgent for institutional frameworks that go beyond reforms in water policy and law and that effectively contribute to the implementation of water governance. Several analytical frameworks (AFW) have been developed that cover diverse components of IWRM, namely, regarding assessing governance of water utilities and services, conducting institutional analysis, assessing the performance of water institutions, assessing the relationship between water governance and poverty, assessing the governance of wastewater utilities, examining water security in the context of transboundary freshwater management, or making comparative analysis of governance in the provision of sanitation and water supply services. Nevertheless, no comprehensive analytical framework is in place to assess the coordination between water management and land use or territorial planning. On the one hand, the transversal nature of water calls for an interdisciplinary approach, with the need to clarify concepts that were gradually added to water management from different disciplines and used with different meanings. The "territorial" nature of water implies, on the other hand, consideration of different and evolving paradigms in planning theory, which gives additional complexity to this exercise (Neto 2010).

Falkenmark et al. (1999) identified the critical difficulty of bridging individual institutions living in compartmentalised "worlds" with specific or opposing objectives as the main obstacle to integrated approaches. Water management and related organisations are isolated from one another, and they apply governance mechanisms in silos, while an atmosphere of distrust prevails among stakeholders, resulting in a lack of coordination, inefficiency, and the loss of resources, including water. The framework proposed highlights the regulatory dimension and demands a dialogue among different agencies and institutions (Falkenmark et al. 1999).

Elinor Ostrom (1998) defined institutions as regularised rules-in-use that structure human behaviour, and she proposed a Social-Ecological Systems Framework (SESF), which seeks to provide a basis for "institutional crafting" in that context. Ostrom also stated that frameworks are not theory neutral, with structural variables that are present in all institutional arrangements and structure actors' patterns of interactions (at the constitutional, collective choice, and operational level of interactions) in so-called action situations and the resulting outcomes (Ostrom 1998, 2011).

Hofwegen and Jaspers developed guidelines to assess the required capacity building interventions to arrive at these conditions and to establish such platforms, as illustrated in Figure 23.1. A minimum set of institutional conditions should be met to allow IWRM platforms to operate successfully (Hofwegen and Jaspers 1999; Hofwegen 2001). This framework is based on a development process with and by the stakeholders to come from *an identified present water resources management situation* to the *desired integrated water resources management situation*. It is a compromise between the present and an "ideal IWRM situation," involving a negotiation process among policymakers, water resources and water utility managers, and stakeholders (Hofwegen 2001).

Sanford Berg (2016) proposed seven elements that affect infrastructure performance, framing these in a discussion of regulation needs in the water sector. These include institutions, interests, information, incentives, ideas, ideals, and individuals, and they are presented in Figure 23.2. Berg discussed regulatory governance, including the organisations, coordination, tools, disciplines, and practices that influence the quality of the regulatory frameworks. He also argued that water governance is broader since it considers not only all the tools of water policies and water resources

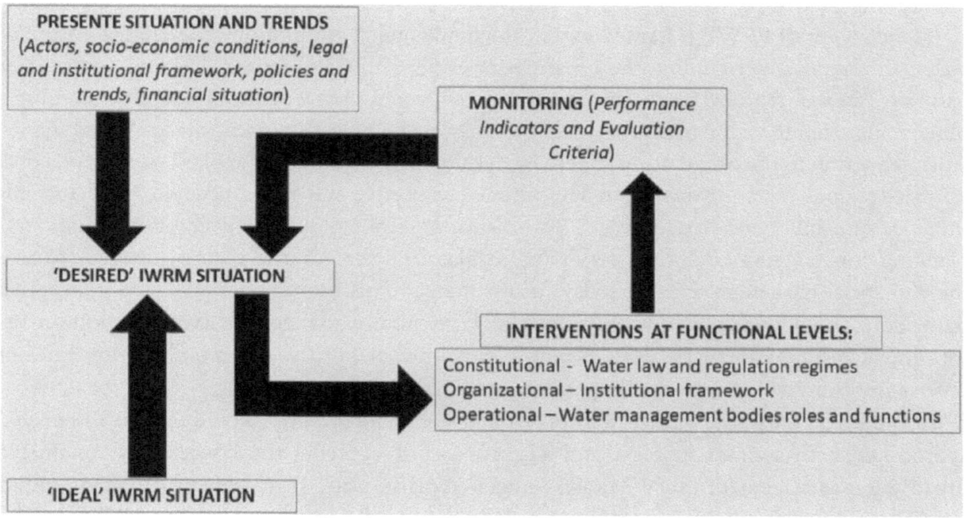

Figure 23.1 Analytical framework for the assessment of the institutional setting and capacity building requirements for integrated water resources management (adapted from Hofwegen 2001).

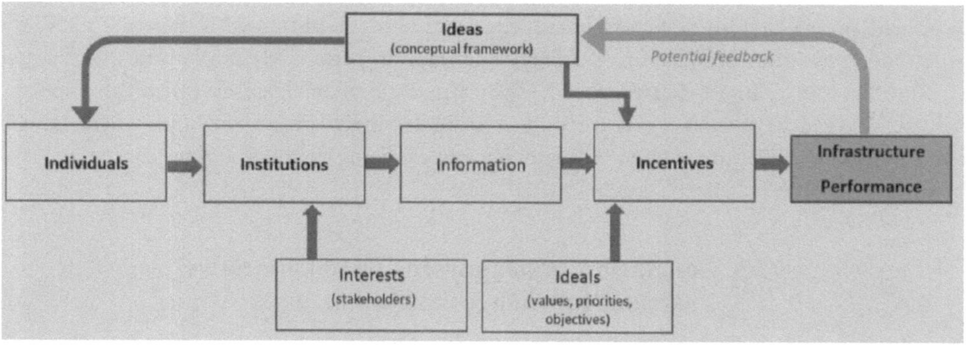

Figure 23.2 Seven elements affecting infrastructure performance (adapted form Berg 2016).

management, but also all the subsectors of water that require regulation, such as water services, agriculture, environment, health, and safety. Therefore, institutional arrangements characterising governance affect how the seven elements influence water sector performance, giving as a good example how capacity building within organisations affects the way ideas are generated and transmitted (ibid., 8–11).

Across the several AFW described, it is evident that water governance requires the reinforcement of the institutional capacity of integration at any level, or scale. Policy integration (both vertical: bottom-up, top-down, and horizontal: between planning instruments) of the multilevel strategies represents a major challenge and can be seen as an indicator of political will and effective success of the desired "integrated" water management. Building upon the previous discussion of water governance and territorial integration of water management by Neto (2010), incorporating the concepts of sustainable urban water management (SUWM) by Brown and Farrelly (2009), and those of integrated urban water management (IUWM) by Mitchell (2006), a conceptual framework based on "territorial integration," "sustainable urban development," and "integrated water management" was proposed by Neto (2016) to address the new and more complex challenges through an "integrated urban water management policy" (IUWP).

The concept of IUWP is framed within the fundamental assumption of territorial nature of water, aiming at understanding the urban water cycle within its territorial reference, the river basin processes at regional level. This assumption triggered the recognition of a new interdisciplinary approach in water-related sciences that may create a consistent knowledge-based framework for action in the urban water sphere by developing a new field of applied science, practical knowledge, and social involvement. The ultimate objective was to propose an AFW towards urban sustainability and "good" water governance, rethinking and discussing the "traditional" planning conceptual models (top-down and regulatory oriented), and their inadequacy to face the real social challenges existing today in urban areas. This more policy-oriented conceptual framework should consider not only the policy integration dimensions (vertical: bottom-up, top-down, and horizontal: between planning instruments), but also the multilevel strategies of diverse institutional agencies that impact on water (Neto 2016).

In the urban areas, the vertical and horizontal policy integration should lead to sustainable approaches, with a more strategic and adaptive vision, applying the concepts of sustainable urban water management (SUWM) and integrated urban water management (IUWM) within a policy-oriented and holistic understanding of the societal, natural, and economic dimensions. The IUWP framework proposed also requires the adoption of innovative territorial planning approaches, building on more operational and crosscutting agendas between different agencies and administrative levels. This approach can be intensified if effective collaborative actions are promoted among the diverse actors, and institutional levels, as illustrated in Figure 23.3. Water issues have been highlighted across the pandemic impacts, particularly in countries with water access difficulties, calling for increased adaptive capacity, in general, and in urban areas, specifically. IUWP can be seen as a way forward in an increasingly urbanised world, where human water needs are constantly stressed by unpredictable changing contexts and impacting external factors.

23.5 Conclusion: Water governance collaborative actions for change in urban areas

IIED (n.d.) argues that water is a politically contested resource and, as a result, water management institutions and policies are effects of political practices. The institutions are spaces of

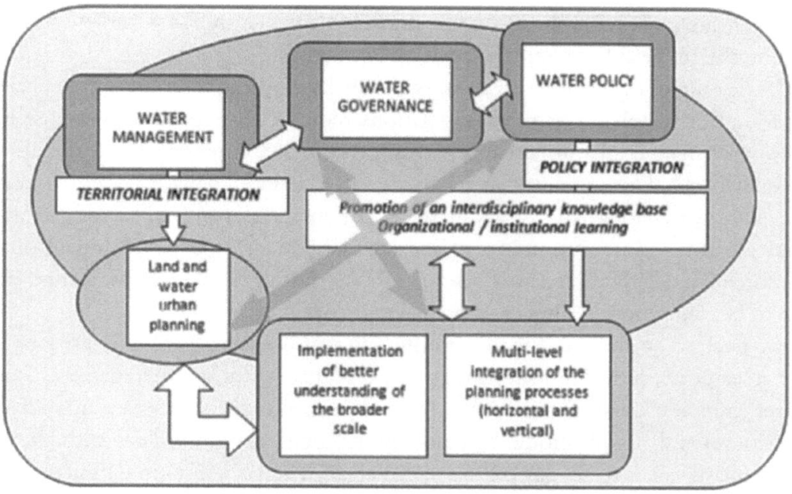

Figure 23.3 IUWP framework as a dynamic and evolving process (adapted from Neto 2016)

negotiation and power influence, with policies resulting from diverse social and political interactions. IIED refers to the important contrasts among developing countries in how they go about crafting new policies and implementation arrangements, from a top-down almost entirely bureaucratic approach, driven by government agencies as the major stakeholders, with processes driven by a combination of technical and economic concerns and interagency politics. There is no room in such approaches for less organised, "informal" interests, especially poor people, to participate in and gain access to water. Advocacy is present for water reforms leading to institutional change towards broad democratic processes, with boundaries of consent drawn from political choice (IIED n.d.).

The example of Europe with the WFD launched a new paradigm and constituted the main instrument for a new Water Policy in the European Union (EC 2000). By giving policies a more comprehensive orientation and striving to approach the available water as a good of high environmental and ecological value, it recognises that all users are responsible for its preservation. The WFD imposes on states sharing waters (surface or groundwater) to address jointly the critical issues that will be the object of respective river basin plans, triggering the need for collaboration between riparian states. This "common front" reinforces the principles of the two more important UNECE and UN Water Conventions (Helsinki in 1992, and New York in 1995), and effectively created a regional platform for international and bilateral water cooperation among the European member states as a whole and in subregional cases of shared surface and groundwater resources.

The IUWP framework proposed by Neto (2016) assumes the fundamental concept of territorial integration of water management, requiring the consideration of the dimensions illustrated in the previous section. It points towards the development of a consistent theoretical and a more solid knowledge-based framework for action. It calls for the implementation of innovative water plans based upon territorial integration and may be applied by more operational and crosscutting agendas between different organisations and administrative levels (vertical and horizontal policy integration). It may finally develop within a collaborative urban water governance framework that facilitates potential synergies between strategic and adaptive interventions for SUWM and IUWM. The collaborative perspective allows acting jointly at local or regional levels, learning together, and reapplying the lessons in an evolving paradigm for water and urban governance, enriching the previously identified "political" dimension.

As stated at the beginning of this chapter, in a reality of overlapping and interconnected challenges, space exists for effective collaboration among nations, regions, communities, or epistemic groups, and a way forward in an increasing urbanised world where the human needs are intrinsically linked with water needs. The sources of stress created by constant changing contexts and impacting external factors of various natures may be seen as opportunities and drivers of change. Within a collaborative framework that combines political, social, and environmental dimensions, it is possible to reinforce the adaptive capacity to reorganise (at institutional level), to change (at political level), and to become more resilient (at societal and community level).

As a final note, we need to be aware of the major difficulty that was here highlighted through the voices of many researchers, regarding factors that block, oppose, or delay the reorganisation of institutional water agencies. Understanding why this happens in each context and identifying the adequate measures and actions should be certainly the focus of water policy in the next years. A particular emphasis needs to be put on the urban areas, where the multiple dimensions are more difficult to address by individual institutions, and where the expected impacts of external factors will be more devastating without preventive networked action and without political and institutional change.

Note

1 A critique of the standard instrumental division Global North/South can be found in the approach proposed by Immanuel Wallerstein, with the "world-systems theory" (Wallerstein, 1974), as well as in the works of Boaventura Sousa Santos (1995) and Maria Paula Meneses (2008) with the "Epistemologies of the South."

References

Alcamo, Joseph, Bennett, Elena, and Program, Millennium. 2003. *Ecosystems and Human Well-being: A Framework for Assessment e Millennium Ecosystem Assessment (MA)*. Paris: UNESCO.

Berg, S.V. 2016. Seven elements affecting governance and performance in the water sector. *Utilities Policy* 43-A: 4–13.

Brown, R., and Farrelly, M. 2009. Delivering sustainable urban water management: A review of the hurdles we face. *Water Science and Technology: A Journal of the International Association on Water Pollution Research* 59: 839–46. DOI: 10.2166/wst.2009.028

Cáñez-Cota, A., and Pineda-Pablos, N. 2019. Breaking out of the governance trap in rural Mexico. *Water Alternatives* 12(1): 221–240.

Castro, E. 2012. In search of meaningful interdisciplinary: Understanding urban water conflicts in Mexico. In UNESCO-IHP (Ed.), *Urban Water Conflicts*. Paris and Leiden: Taylor and Francis: 313.

European Commission. 2000. Water Framework Directive (WFD) 2000/60/EC: Directive 2000/60/EC of the European Parliament and of the Council of 23 October 2000 establishing a framework for community action in the field of water policy.

Falkenmark, M., Lundqvist, J., and C. Widstrand. 1989. Macro-scale water scarcity requires micro-scale approaches. *Natural Resources Forum* 13: 258–267.

Falkenmark, M., Andersson, L., Castensson, R., and Sundblad, K. 1999. *Water: A Reflection of Land Use: Options for Counteracting Land and Water Mismanagement*. Stockholm: NFR for UNESCO (IHP).

Global Water Partnership. 2002. Effective water governance. *Peter Rogers and Alan W Hall for GWP. Technical Committee (TEC) No. 7*.

Global Water Partnership. 2003. Dialogue on effective water governance: Learning from the dialogues. *Status Report Prepared for Presentation at the 3rd World Water Forum*. Kyoto, Japan.

Hofwegen, P.J.M. 2001. Analytical framework for integrated water resources. In *Proceedings of the International Workshop on Integrated Water Management in Water-Stressed River Basins in Developing Countries*. Delft, the Netherlands: IHE, 137–158.

Hofwegen, P.J.M. van, and Frank G.W. Jaspers. 1999. *Analytical Framework for Integrated Water Resources Management Guidelines for Assessment of Institutional Frameworks. IHE Monograph 2*. Rotterdam/Brookfield: Balkema Publishers.

IIED. n.d. Water governance literature assessment. *Report by Charles Batchelor for IIED to develop a DFID Research Programme on Water Ecosystems and Poverty Reduction under Climate Change*.

IPCC. 2007. *Climate Change 2007: Synthesis Report. An Assessment of the Intergovernmental Panel on Climate Change*.

IPCC. 2021. Climate change 2021: The physical science basis. In *Contribution of Working Group I to the Sixth Assessment Report of the Intergovernmental Panel on Climate Change*. Cambridge: Cambridge University Press. In Press.

Joy, K.J., Kulkarni, S., Roth, D., and Zwarteveen, M. 2014. Repoliticising water governance: Exploring water re-allocations in terms of justice. *Local Environment: The International Journal of Justice and Sustainability*, 19(9), 954–973.

McEvoy, D., Cots, F., Lonsdale, K., Tabara, J. D., and Werners, S. 2008. The role of institutional capacity in enabling climate change adaptation: The case of the Guadiana River Basin. In *Transborder Environmental and Natural Resource Management*, CIAS Discussion Paper No.4. Chapter 3, 49–59, Wageningen University, Wageningen.

Menezes, M.P. 2008. Epistemologias do Sul. *Revista Critica de Ciências Sociais* 80: 5–10.

Merrey, D.J., and Cook, S. 2012. Fostering institutional creativity at multiple levels: Towards facilitated institutional bricolage. *Water Alternatives* 5(1): 1–19.

Le Meur, P.Y., Hauswirth, D., Leurent, T., and Lienhard, P. 2005. The local politics of land and water: Case studies from the Mekong Delta. In *Coll. Etudes et Travaux, serie en ligne 4*. Editions du Gret, Nogent sur Marne.

Mitchell, V.G. 2006. Applying integrated urban water management concepts: A review of Australian experience. *Environmental Management* 37(5): 589–605.

Mollinga, P. 2008. Water, politics and development: Framing a political sociology of water resources management. *Water Alternatives* 1(1): 7–23.

Neto, S. 2010. *Water, Territory and Planning. Contemporary Challenges: Towards a Territorial Integration of Water Management*. PhD Thesis.

Neto, S. 2016. Water governance in an Urban Age. *Utilities Policy*, 43(A), 32–41. ISSN:0957-1787.

Neto, S. 2018. Territorial integration of water management in the city. Chapter in Book *Water Challenges of an Urbanizing World.*, IntechOpen, London.

Neto, S., and Camkin, J. 2020. What rights and whose responsibilities in water? Revisiting the purpose and reassessing the value of water service tariffs. *Utilities Policy* 63, 101016.

Neto, S., and Henriques, C. 2016. Hybrid solutions to address complexity. Presented to the Conference Adapting to Climate Change: Water, Waste and other Local Infrastructure, 21–22 June, Lisbon, Portugal.

Neto, S., da Silva, M., Muller, L., and Weller, K. 2019. Social and technological innovation for water conservation: The project ECH2O-ÁGUA. In Proceedings of the 2nd International Congress on Engineering and Sustainability in the XXI Century, 1st ed. 2020, XXIV, 1200 p. 2 volumes.

OECD 2015. OECD principles on water governance centre for entrepreneurship, SMEs, regions and cities. Adopted by the OECD Regional Development Policy Committee on 11 May 2015.

Ostrom, E. 1998. A behavioral approach to the rational choice theory of collective action: Presidential address, American Political Science Association, 1997. *American Political Science Review* 92(1): 1–22.

Ostrom, E. 1990. *Governing the Commons: The Evolution for Collective Action*. Cambridge University Press, Cambridge.

Ostrom, E. 2011. Background on the institutional analysis and development framework. *The Policy Studies Journal* 39(1).

Petrella, R. 1998. *The Water Manifesto: Arguments for a World Water Contract*. Zed Books, London 2001 135.

Pritchard, L., and Sanderson, S.E., 2002. The dynamics of political discourse in seeking sustainability. In: Gunderson, L., Holling, C.S. (Eds.), *Panarchy*. Washington, DC: Island Press.

Sadoff, C.W., Hall, J.W., Grey, D., Aerts, J.C.J.H., Ait-Kadi, M., Brown, C., Cox, A., Dadson, S., Garrick, D., Kelman, J., McCornick, P., Ringler, C., Rosegrant, M., Whittington, D., and Wiberg, D., 2015. *Report of the GWP/OECD Task Force" on Water Security and Sustainable Growth*. Oxford: GWP/OECD.

Santos, B.S. 1995. *Toward a New Common Sense: Law, Science and Politics in the Paradigmatic Transition*. New York: Routledge.

United Nationals General Assembly Resolution 64/292. 2010. *The Human Right to Water and Sanitation.* Adopted by the General Assembly on 28 July 2010. A7RES/64/292, 2010.

Wakely, P. 1997. *Capacity Building for Better Cities. Journal of the Development Planning Unit.* University College London.

Wallerstein, I. 1974. *The Modern World-System.* New York/London: Academic Press.

24
URBAN WATER SECURITY

Joost Buurman

24.1 Introduction

Cape Town in South Africa faced a severe drought from 2015 to 2018. In early 2018, it came very close to "Day Zero," i.e., the day that there would be no more municipal water supply. While extremely dry weather was one reason, social, economic, and political reasons also contributed to the water shortage (Muller, 2018). Similarly, São Paulo in Brazil faced a severe drought between 2014 and 2015. One of the responses to dwindling water supplies was a significant reduction of water pressure in the supply system. This affected low-income families and those living in the urban periphery most (Millington, 2018). During the drought, in February and March 2015, local torrential storms flooded the streets of São Paulo but did little to replenish the city's main supply reservoirs, located 60 kilometres away (Davies, 2015a, b). A much larger flood happened in Thailand in 2011. Large parts of the capital Bangkok and its surroundings were flooded for months affecting global supply chains as automobile, electronics, and other factories could no longer operate (Haraguchi and Lall, 2015).

The disasters cited are just a few examples of cases where too little water or too much water affects well-being, livelihoods, and socio-economic activities of urban residents. Similarly, too dirty water can be a problem. The Flint water crisis in the United States, where drinking water was polluted with lead and affected residents' health (Grossman and Slusky, 2019), is a high-profile example, yet all over the world poor sanitation, wastewater discharge, and other sources of pollution affect the quality of freshwater resources.

While disasters may hit the front page, all cities face larger and smaller water problems and issues. As the world is urbanising and climate change increases the probability of extreme weather events, the challenges for urban water management are mounting; not only in fast-growing cities in developing countries, where many people do not have access to clean water and sanitation, but also in developed cities where economic growth leads to higher water demands, ageing infrastructure requires massive investments in maintenance, rehabilitation and renewal of water systems, and demands for more sustainable water systems require changing conventional approaches.

In the past decades, a number of approaches, concepts, and frameworks have appeared to address urban water problems, such as urban water security, integrated urban water management, water-sensitive cities, water-sensitive urban design, and sustainable urban drainage systems.

DOI: 10.4324/9781003057574-30

Some of these are derived from approaches applied at the basin or country scale, while others are specific for cities. Some are broad approaches covering the entire urban water management cycle, while others are techniques focusing on specific areas, such as stormwater management. Water security takes a central position among these approaches, concepts, and frameworks because it is all-encompassing. It is generally used in a very broad sense covering all aspects and dimensions of urban water management (Hoekstra, Buurman and van Ginkel, 2018). The term "security" is also popular and urban water security fits with all kinds of security challenges of our times. With water security becoming a widely adopted concept (Staddon and Scott, 2018), it is important to have a good understanding of what it means and how it can be used to better manage water in cities.

This chapter discusses urban water security. First, we look at defining urban water security, what is special about cities when it comes to water security, and different ways to look at urban water security. We will then look at the different dimensions of urban water security from a systems perspective. Lastly, we look at how urban water security can be measured and used in urban water management.

24.2 What is urban water security?

Water security is a water management concept – a framework for the whole process of planning, organising, coordinating, leading, decision making, controlling, etc. of water resources – that can be used at different levels, from global water security to household water security. Before we look at urban water security, we should first look at water security in general. Although the term "water security" had been used earlier, it really took off from the year 2000 onwards when it was used in the Ministerial Declaration of The Hague on Water Security in the 21st Century of the 2nd World Water Forum and in two influential publications: "A Water Secure World" by the World Water Council (WWC, 2000) and "Towards Water Security: A Framework for Action" by the Global Water Partnership (GWP, 2000b). In the following two decades, the number of academic publications and mentions of water security in reports and policy documents grew rapidly. The concept of water security evolved from its original alignment with national and military security threats to an all-encompassing concept that seemingly incorporates the previous dominant concepts of integrated water resources management (IWRM) and sustainable water management.

However, important differences in focus between the concepts are found. IWRM is an established water management paradigm that has been promoted and used since the 1980s (Biswas, 2008; Cook and Bakker, 2012). It can be defined as "a process which promotes the coordinated development and management of water, land and related resources, in order to maximise the resultant economic and social welfare in an equitable manner without compromising the sustainability of vital eco-systems" (GWP, 2000a). Similar to water security, IWRM emphasises that water problems can be addressed only through a more holistic, interdisciplinary approach. Water systems need to be analysed as a whole, as water resources are connected in the water cycle, problems with water quality and quantity issues are often interrelated, and water systems interact with other social, economic, and environmental systems. However, as we will see below, water security differs from IWRM in the emphasis on goals rather than the process. The other concept, sustainable water management, gained traction as a concept in the 1990s following the emergence of the sustainable development paradigm (Brundtland Commission, 1987). It is also a broad, multi-objective water management concept that emphasises the ability to meet the water needs of the present generation without compromising the water needs of future generations (Loucks, 2000). The evolution of terminology is still ongoing with new concepts emerging, such as adaptive water management, alluding to adaptation to climate change and emphasising the need for flexible solutions in the face of uncertainty (Pahl-Wostl, 2007).

Water security is an elusive concept with many different interpretations. A widely cited definition of water security is:

> The capacity of a population to safeguard sustainable access to adequate quantities of acceptable quality water for sustaining livelihoods, human well-being, and socio-economic development, for ensuring protection against water-borne pollution and water-related disasters, and for preserving ecosystems in a climate of peace and political stability.
>
> *(UN-Water, 2013)*

An earlier definition from the Global Water Partnership is:

> Water security, at any level from the household to the global, means that every person has access to enough safe water at affordable cost to lead a clean, healthy and productive life, while ensuring that the natural environment is protected and enhanced.
>
> *(GWP, 2000b)*

These definitions are very broad and ensure that nothing important is left out. However, they also give little clarity on inevitable trade-offs that have to be made (Hoekstra, Buurman and van Ginkel, 2018). Water is a renewable, but limited resource: water polluted by industry can no longer be used for agriculture; over-extracted groundwater can no longer be used by others or by future generations, water used in agriculture is no longer available for the environment. From the perspective of these definitions, water security is a holistic and integrative concept, somewhat similar to IWRM, but not a well-defined measurable concept. Yet, there are many framings of water security explicitly focused on measuring and evaluating water security, which is discussed in the section on measuring urban water security below.

Another aspect of the concept of water security is that it is goal oriented: the goal of achieving water security of a country, city, or other entity. The Global Water Partnership (2000b) considers working towards water security as an overarching common goal to tackle the global water crisis. The goal orientation is a strong aspect of water security as it focuses attention on addressing water problems. In this respect, water security closely aligns with the Sustainable Development Goals (SDGs). In SDG 6 – to ensure access to water and sanitation for all – targets that should be achieved by 2030 and indicators to measure these are defined. SDG 6 is central to achieving other SDGs, such as SDG 1 (no poverty), SDG 2 (zero hunger), SDG 3 (good health and well-being), SDG 14 (life below water), and SDG 15 (life on land) (Mugagga and Nabaasa, 2016). This highlights the broad reach and integrative aspect of water security.

Although the SDGs define measurable goals for 2030, goals in water security are not static. Goals change over time with changes in preferences, socio-economic conditions, climate change, etc. Goals also differ among places and people. This also leads to different perspectives on what water security is (Hoekstra, Buurman, and van Ginkel, 2018), see Figure 24.1. People working in different disciplines will frame water security differently. Engineers may frame water security as protection against water-related hazards, such as floods, droughts, and contamination, while public health officials may frame water security in terms of safe and reliable water supply and sanitation, and politicians may frame water security in terms of accessibility and availability of water resources and equitable distribution among residents. Perspectives on what water security is could also be framed around the three specific problems of too little, too much, and too dirty

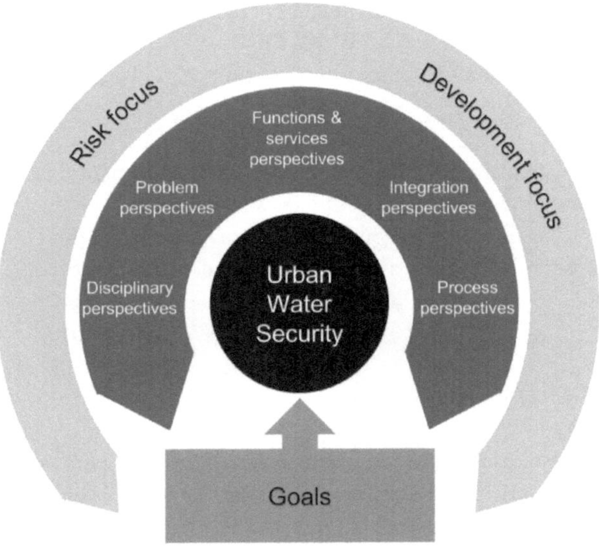

Figure 24.1 Focuses and perspectives in urban water security (based on Hoekstra, Buurman, and van Ginkel, 2018).

water, or around the functions that water systems fulfil and ecosystem services they provide, such as drinking water supply, recreational values, and flood protection. There are also different framings on the type of integration that water security requires: an integrated water perspective, which considers all water issues comprehensively, or a water-integrated perspective, which considers water as part of environmental and other policies. Finally, different framings on the process towards water security could be distinguished, with a policy-analytical framing focusing on instruments and solutions and a governance perspective framing focusing on the institutions and processes required.

Now that we have an overview of the concept of water security in general, we turn to urban water security, i.e., water security applied to the territory of a city, municipality, or urban agglomeration. What is special about urban water security and how can it help in urban water management? Cities are spatial concentrations of people. These people need water for their well-being, livelihoods, and socio-economic activities. Hence, cities concentrate water demands and as most cities do not have water resources locally available, water from the region is used to supply cities. Globally the average distance water is transported to cities ranges from 27 km for low-income countries to 57 km for high-income countries (McDonald et al., 2014). Cities depend even more on external water resources for the food consumed and products used by their citizens that are produced elsewhere. This is called the external water footprint of urban consumption (Hoekstra et al., 2011). (The water footprint is discussed in chapter 32 of Hoekstra et al., 2011.) The dependency on the region and other areas is a defining feature of urban water security. It means that cities have to manage imported water risks; over many of these risks, they do not have direct control.

Concentration of people and activities also means concentration of emissions and discharges, making urban water resources vulnerable to pollution. Discharge of polluted water can lead to problems in downstream regions. Moreover, flooding risks in cities can be high because natural hydrology has been severely altered with many hard surfaces and because many cities are located in areas with high flood hazards, such as river deltas and coastal zones. The concentration of people and economic activities also means there is a high potential exposure to flood damage, requiring high protection standards. There can be a dependency between cities and the regions they are located in for flood risks as well: flood infrastructure to protect cities could affect rural areas, for instance, if they are used as flood buffer areas.

The governance setting to address urban water security is different from those at other levels. Typically, different municipal departments are responsible for different tasks in the urban water cycle, such as drainage, water supply, and pollution prevention, and tasks that are indirectly related to water, such as spatial planning. Urban water management is governed through local, regional, and national institutions, policies, and regulations. This creates a complex environment with many policy processes and stakeholders.

Similar to the general concept of water security, understanding of what urban water security is differs among people, places, and times. In most cases where urban water security is discussed or analysed, it is not clearly defined; it seems that urban water security is used as an umbrella term for all kinds of urban water issues. Allen, Kenway, and Head (2018) found only three urban water security definitions after an extensive literature search. They proposed a definition that focuses on the needs of community stakeholders and use different themes to guide what the objectives of urban water security might be, such as balancing competing demands, providing sufficient water volumes, and the community's risk appetite.

Many water security interpretations have a strong risk focus. Risk-based approaches are common in water management (Garrick and Hall, 2014). In this view, ensuring water security is a matter of managing risks, which is basically a matter of managing variability (van Beek and Lincklaen Arriens, 2014), such as variability in rainfall, variability in water levels, or variability in water quality. Of particular interest are events that can cause urban water systems to fail, for instance, a drought that would cause water supply to be interrupted or heavy rainfall that would cause flooding. While many of these risks can be quantified through models and data analysis, it is not possible to directly quantify water security goals based on risks. What people define as "secure" depends on their interpretation and acceptance of risks. For instance, some communities consider a 1 in 100 years flood probability as secure, while other communities consider a 1 in 10,000 years flood probability as secure. As such, the community's risk appetite features in the definition of urban water security of Allen, Kenway, and Head (2018).

Risk is a combination of hazard, exposure, and vulnerability (see Figure 24.2). In every city, exposure is relatively high due to the concentration of people and economic activities. The hazard and vulnerability can vary significantly between cities (and also within cities). Hazards are threatening events, such as floods, droughts, and poor water quality. While they are fundamentally natural events, hazards are increasingly affected by human activities, such as reduction in flood plain storage and reducing the permeability of urban surfaces, which increases flood hazards. Vulnerability is determined by preparedness and ability to cope. It can be reduced through investments in infrastructure, human capital, legislation, preparation of plans, etc. Figure 24.2 shows that while urban water security risks can be similar in different cities, significant differences can exist between hazard and vulnerability. Figure 24.2 shows only the situation for water supply; for other water issues, such as flooding, the situation in the same cities could be quite different. However, as we will see in the next section, the capacity and resources of a city to address urban water security issues greatly affect its water security situation.

There is a complementary development focus besides the risk focus of urban water security. The part of the UN-Water definition "for sustaining livelihoods, human well-being, and socio-economic development" alludes to this development focus. In water insecure places, risks are high, which inhibits development. Globally millions of people, often women and girls, spend much of their time collecting water, which is not productive and takes time away from

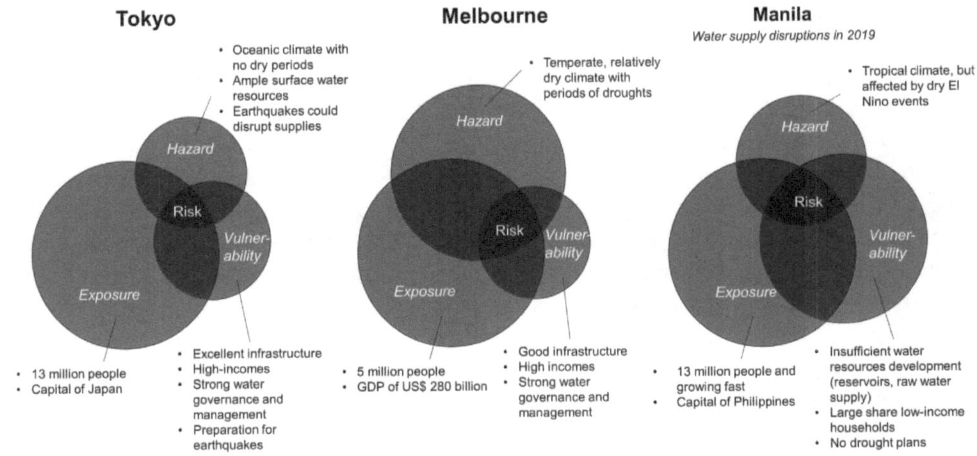

Figure 24.2 Indication of water supply risk in three cities as a combination of hazard, exposure, and vulnerability.

education. Pollution and poor sanitation cause diseases and time lost from income-generating activities. Flood damage affects growth and development in multiple ways: money is spent on replacement and repairs rather than productive investments; floods result in unproductive time and people are not willing to invest in flood-prone areas. A development focus on urban water security means that water security increases economic welfare, improves social equity, and is sustainable in the long term (Hoekstra, Buurman, and van Ginkel, 2018). Water and welfare are clearly linked. In water insecure cities, water problems negatively affect economic growth, while in water-secure cities, water resources contribute to growth (Hall and Borgomeo, 2013). Improving water security should address equity concerns: water insecurity often affects the poor the most; they are the least serviced by water supply and sanitation services, are located in the most flood-prone areas, and are most affected by water pollution. In the same city, the richer part of the population may be much more water-secure than the poor part of the population. Finally, similar to the focus of sustainable water management, water-secure cities are those that take into account future needs into today's optimal allocation of water resources.

24.3 A systems approach to urban water security

Understanding urban water security is complex because of the many dimensions and interpretations. Urban water systems link to other socio-economic, technical, and environmental systems and change over time. A system dynamics approach can help in understanding how urban water security could be improved over time by analysing stocks, flows, and feedback loops in complex urban water systems. An apt approach is to use the Pressure, State, Impact, Response (PSIR) schematisation, which is a variant of the schematisations often used for environmental sustainability assessments (OECD, 1993; EEA, 1999; Meyar-Naimi and Vaez-Zadeh, 2012). In this schematisation, first, the pressures on the urban water system are identified. The pressures drive the changes that occur over time in the urban water system: they work on, and change, the state of the system. This has impacts on the functions that the urban water system fulfils. To address these impacts, policymakers and stakeholders will respond by intervening in the pressures, states, or impacts (see Figure 24.3).

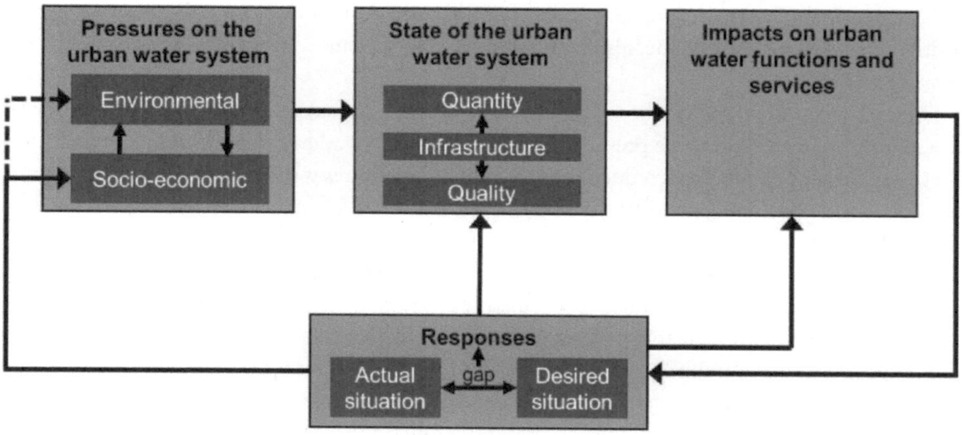

Figure 24.3 Pressure-state-impact-response feedback.

The pressures on the urban water system can be grouped into environmental and socio-economic pressures. Environmental pressures are caused by the hydrological and geographical conditions and changes. Variability in precipitation and water levels are the main pressures and climate change causes long-term changes in these. Droughts can exert severe pressure on urban water systems, as shown by the examples of Cape Town and São Paulo. Similarly, floods caused by intense rainfall or storm surges can exert severe pressure, as shown by the example of Bangkok. Socio-economic pressures are manifold and include population size and population growth, urbanisation leading to more built-up areas, income growth and inequality, and water use characteristics of the population and industry. These socio-economic pressures increase the exposure to hazards and affect the vulnerability. Environmental and socio-economic pressures can also interact. For instance, land use changes caused by a growing population can affect the hydrology in a water catchment and climate change is also caused by human activities (think of the urban heat island effect at the city level).

The current properties of the water system define the state. The quantity of water available to a city is defined by stocks and flows of water within the city boundaries and the region from which a city sources its water. Conventional freshwater resources are surface water, such as rivers and lakes, groundwater, and rainfall collected from rooftops. Increasingly, unconventional water resources are used, such as salt and brackish water as input for desalination, air moisture for fog harvesting, and wastewater for water recycling (UN-Water, 2020). The quality of the water resources is affected by natural conditions (e.g., the presence of arsenic in groundwater and sediments in rivers) and, increasingly, by pollution from human activities. Water infrastructure, such as reservoirs, water supply plants, desalination plants, pipelines, sewer systems, and water treatment plants, should be taken into account when defining the state, as it affects the stocks, flows, and quality of water in urban areas. The presence of drainage and flood protection infrastructure determines the state of the flood risks. The state of the water infrastructure in terms of reliability, efficiency, and resilience to shocks is also of importance.

The impacts describe how well the urban water system is fulfilling its functions and providing services. This takes a functional perspective (see Figure 24.1) on urban water security. Examples of functions are a reliable supply of good quality water to citizens, minimising pollution of surface and groundwater resources, provision of recreational services, and absence of floods. These functions can be quantified and measured as goals, which are further discussed in the next section. Changes in pressure and state impact the functions, both positive and negative. Positive changes are usually the result of a response as pressures are generally associated with negative impacts.

Society's response to the gap between the actual and the desired situation can take many forms. The response can aim to reduce pressures, improve the state, or address the impacts. From the perspective of urban policymakers, reducing environmental pressures is the most difficult. While some socio-economic pressures could be addressed, these are often changes over which policymakers have little control. In most cases, the easiest response is infrastructure development, yet almost all responses have technical, institutional, and organisational dimensions. Societies do not necessarily need to respond through governmental organisations: citizens can also respond, though this does not always happen in a sustainable manner. For instance, in Jakarta, a city with an inadequate piped water supply system, households and companies pump groundwater from their own wells, causing the city to subside below sea level. The response is the most complex to analyse in the PSIR schematisation as it deals with human behaviour and the functioning of governance and management systems. Nevertheless, an adequate response is key to improving urban water security.

24.4 Measuring urban water security

As we discussed earlier, urban water security is usually not clearly defined, and it is often used in a qualitative way. However, quantification can help to obtain insights into the complex matter of urban water security (van Beek and Lincklaen Arriens, 2014; Jensen and Wu, 2018). Particularly, the quantification of urban water security goals through indicators and indices could help in stimulating policy action through carrying out assessments, prioritising measures and investments, and monitoring and evaluating effectiveness of policy interventions. Quantifying urban water security could focus decision-makers' attention and facilitate discussions among different stakeholders. It could also be used to reflect on the differences between cities and to develop strategies to address areas of water insecurity. It should be added that the use of indicators can also be criticised (Jensen and Wu, 2018). Specifically, reducing complex problems and policies to a single or a few numbers could lead to oversimplification and loss of information in decision making. Another criticism is that indicators could also be misused for political reasons, such as securing support for policies that have already been decided on. Nevertheless, indicators that are relevant accurately capture the different elements and dimensions of urban water security, and they are accepted by policymakers and stakeholders and can be meaningful and support evidence-based policymaking.

A large number of indices and indicators exist for all kinds of dimensions of water security in general. For instance, for drought alone more than 100 indicators have been proposed (Zargar et al., 2011) and a systematic literature review found 50 water vulnerability tools that employed a total of 710 indicators (Plummer, de Loë and Armitage, 2012). For measuring different facets of urban water security, a large number of indices and indicators are also available. For instance, benchmarks have been developed for measuring the operational and financial performance of water utilities (Berg, 2010), which include indicators that overlap with urban water security, such as network coverage and reliability. However, only a few studies exist that attempt to develop comprehensive measurements that cover the myriad of dimensions and perspectives present in urban water security. One of the measurements that is relatively broad is the City Blueprint, which assesses the sustainability of the urban water cycle (Koop and van Leeuwen, 2015). It consists of 12 trends and pressures indicators divided into three dimensions and 25 performance indicators divided into seven dimensions. The trends and pressure indicators are factors over which water authorities in a city do not have any control, such as the urbanisation rate, water scarcity, and poverty rate. The City Blueprint's focus is on the urban IWRM performance, i.e., those factors that water authorities can use to transition towards "water-sensitive" or "water-wise" cities, and, as such, it is not a comprehensive water security index. Another example is the Sustainable Cities Water Index, which examines the water sustainability of 50 cities (Arcadis, 2016). It consists of 19 indicators divided into the dimensions: resilience, efficiency, and quality. However, the 19 indicators do not cover the full urban water security spectrum; for instance, governance and equity aspects are missing. A third example is the Urban Water Security Dashboard, which is probably the most comprehensive measurement of urban water security to date. The Urban Water Security Dashboard uses the PSIR framework described in the previous section (van Ginkel et al., 2018). A total of 54 indicators are categorised into pressure, state, impact, and response indicators and a further grouping in each of those four categories exist; for instance, pressure indicators are divided into environmental pressures related to water scarcity, environmental pressures related to flooding, and socio-economic pressures. The indicators can be aggregated at four different levels, from no aggregation (tier 1) to aggregation by categories (or dimensions of water security, tier 2) to the PSIR stages (tier 3) to a single index for a city (tier 4). Figure 24.4 shows an example of the Urban Water Security Dashboard for Almaty, Kazakhstan.

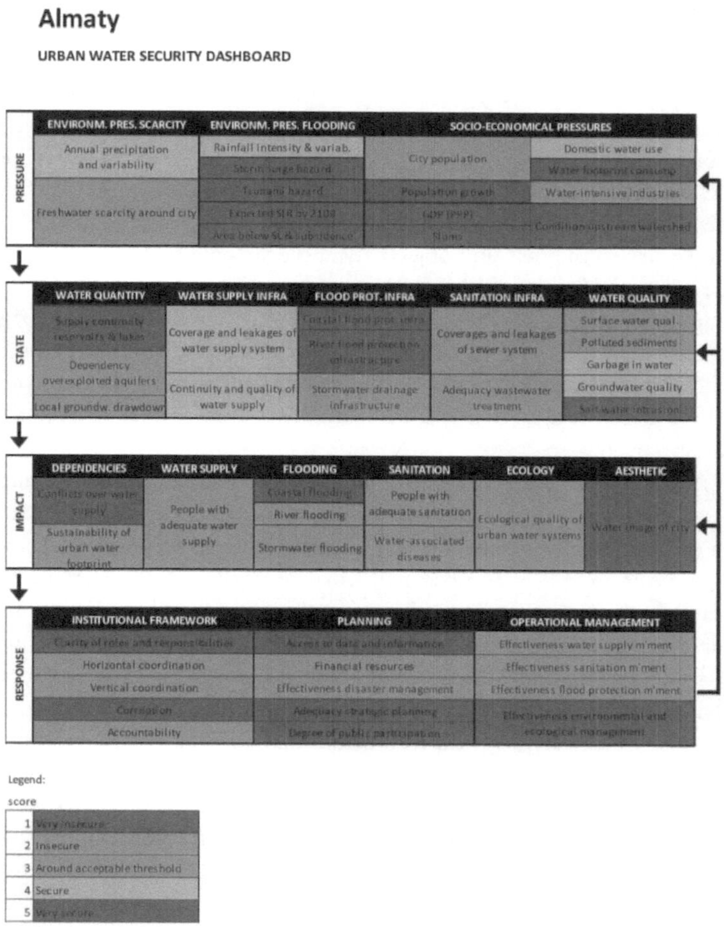

Figure 24.4 Example of the Urban Water Security Dashboard for Almaty, Kazakhstan.

In addition to these three examples of relatively comprehensive measurements of urban water security, there are other examples (Jensen and Wu, 2018; Aboelnga et al., 2019; Krueger, Rao, and Borchardt, 2019; Khan et al., 2020; Maurya et al., 2020). The studies have in common that to capture the different dimension of water security, a composite index is created consisting of indicators that are grouped into different dimensions of water security. The dimensions often reflect the water security perspective and vision or goal of the index. More comprehensive water security perspectives typically have more indicators. Indicators need to be easy to understand for policymakers and stakeholders, based on up-to-date data, relevant for the specific issues involved, credible, transparent, and accurate (Van Beek and Lincklaen Arriens, 2014).

Finding data to score the individual indicators is a challenge that all measurements of urban water security have. Most studies' measurements depend on publicly available data, which may not be available for all indicators. Sometimes country-level data are used if city-level data are not available, which could lead to incorrect scoring, as characteristics of variables that make up country averages are often different in cities. In practice, the selection and definition of indicators are partly driven by the availability of data to score the indicators. The City Blueprint uses a combination of data collection by means of a questionnaire and by using publicly available data. Data collection by questionnaire is possible only if the city's water authorities would like to collaborate and provide the data. When collecting data for different cities, different sources of data may be inconsistent as things are measured and defined in different ways. For instance, a simple number like water use per capita per day depends on the way the population is defined (e.g., including or excluding temporary residents and tourists) and the way water use is measured (e.g., water use in common areas of residential buildings could be classified under "services" or "residential" water use). The Urban Water Security Dashboard developed a detailed guide with formalisation of indicators, which describes how indicator data need to be collected and how they need to be scored, to ensure consistency and objectivity and overcome some of the issues caused by inconsistent data. The guide includes Internet search procedures to score information that is hard to compare among cities or of a more qualitative nature (see van Ginkel et al., 2018 for a link to download the guide). The annex to this chapter provides a brief summary of the formalisation of each indicator. Expert opinion can also be used if data are not available. Both in the City Blueprint and the Urban Water Security Dashboard this is done for the governance and response dimension, respectively.

The quantitative data for the individual indicators are usually scaled to compare them and to be able to aggregate scores to a composite index. The Urban Water Security Dashboard scores all indicators on a scale from 1 (very insecure) to 5 (very secure). The City Blueprint converts everything to a scale from 0 (a very poor performance needing further attention by managers and politicians) to 10 (an excellent performance that requires no further attention). Aggregation of individual indicators to a composite index further requires careful consideration of procedures for averaging and normalisation. Taking the Urban Water Security Dashboard (Figure 24.4) as an example, it can be seen that simply adding the individual indicator scores in a dimension or PSIR category will lead to unbalanced scores as the number of indicators in each category differs. Hence, the arithmetic mean of the scores is used here to aggregate categories and PSIR stages and to determine the final score for the cities. It is, however, recommended to look at the disaggregated scores to understand and analyse water security issues and avoid making simplistic conclusions. Moreover, some water functions or dimensions may be prioritised over others, which would justify applying different weights. In practice, most measurements of urban water security apply equal weights as assigning weights is subjective and should be left to policymakers and stakeholders. Nevertheless, multicriteria decision-making methods (see, for instance, Ishizaka and Nemery, 2013) could be used for scoring, weighting, and aggregating indicators in applications of the measurement frameworks and obtain consensus among stakeholders with different perspectives on what is important in water security.

The measurement studies that result in rankings of cities (Koop and van Leeuwen, 2015; Arcadis, 2016; van Ginkel et al., 2018; Krueger, Rao, and Borchardt, 2019) all show that cities in high-income countries are more water-secure than cities in low-income countries. Cities in low-income countries typically face higher socio-economic pressures and lack the means for an adequate response to pressures. An interesting observation is that environmental pressures, such as water resources availability and variability and flood risk, are of much lesser importance for the urban water security situation. This shows that looking at only a single or few dimensions can lead to wrong conclusions. For instance, water stress according to WRI's Aqueduct tool (WRI, 2019), which is at the regional scale, is extremely high in the regions where Brussels (Belgium) and Rome (Italy) are located and low in the regions where Jakarta (Indonesia) and Lagos (Nigeria) are located. However, it is obvious that the water security in Brussels and Rome is much better than in Jakarta and Lagos, where large parts of the citizens' health, well-being, and productivity are affected by poor access to water, water pollution, and flooding. Hence, the broad perspective that urban water security provides is required to understand a city's water situation and improve urban water resources management.

24.5 Conclusions

Urban water security is a broad concept that aims to cover all dimensions of urban water management. Different people have different perspectives on urban water security, and while perspectives can be broad and holistic, people usually narrow down the interpretation of urban water security. This can be useful to focus attention on specific problems, but as many water problems are interconnected and interconnected with other urban problems, a broad perspective is needed to improve a city's water security.

Urban water security is, in essence, not very different from IWRM, sustainable water resources management, or adaptive water management as it encompasses these concepts. These concepts also emphasise a holistic, integrative approach to urban water management. Yet, urban water security is the broadest urban water management concept. To provide some structure to the broad and complex analysis of urban water security, a systems analysis can be used; a PSIR schematisation can provide insights in environmental and socio-economic pressures, the state of the water system, impacts of changes, and the societal responses. In addition, the risk focus and the development focus provide substance to the urban water management concept. The risk focus is congruent with existing risk approaches in water management, and the development focus emphasises that changes over time should lead to improved water security for all citizens.

Similar to Biswas's (2008) question for IWRM, we can ask: urban water security, is it working? How can urban water security help in urban water management? Urban water security lacks a strong theoretical framework or clear implementable strategies or elements. It is not prescriptive and does not provide recipes to improve urban water security. Nevertheless, it has its merits as a concept for urban water management. The concept is universal and flexible enough to apply in different contexts, cultures, and situations. It helps to better understand urban water systems, and implementation of the concept through indicators and indices is useful in focusing policy efforts and supporting evidence-based policymaking. Urban water security is a concept that is still in development and future research could strengthen practical application to help cities become more water-secure.

Acknowledgement

This chapter draws heavily on the research that was conducted for the paper "Urban Water Security: A Review" by Arjen Hoekstra, Joost Buurman, and Kees van Ginkel (2018). Unfortunately, the research collaboration on urban water security came to an abrupt end with the sudden passing of Arjen Hoekstra on 18 November 2019. This chapter is dedicated to his memory.

References

Aboelnga, H.T. *et al.* (2019) 'Urban water security: Definition and assessment framework', *Resources*, 8(4), p. 178. doi:10.3390/resources8040178.

Allan, J.V., Kenway, S.J. and Head, B.W. (2018) 'Urban water security: What does it mean?', *Urban Water Journal*, 15(9), pp. 899–910. doi:10.1080/1573062X.2019.1574843.

Arcadis (2016) *Sustainable Cities Water Index*. Amsterdam: Arcadis. Available at: https://www.arcadis.com/en/global/our-perspectives/which-cities-are-best-placed-to-harness-water-for-future-success-/ (Accessed: 1 May 2016).

Berg, S. (2010) *Water Utility Benchmarking*. Iwa Publishing, London.

Biswas, A.K. (2008) 'Integrated water resources management: Is it working?', *International Journal of Water Resources Development*, 24(1), pp. 5–22.

Brundtland Commission (1987) *Our Common Future*. Oxford: Oxford University Press.

Cook, C. and Bakker, K. (2012) 'Water security: Debating an emerging paradigm', *Global Environmental Change*, 22(1), pp. 94–102. doi:10.1016/j.gloenvcha.2011.10.011.

Davies, R. (2015a) *1 Hour of Rain Floods Sao Paulo, Brazil: FloodList*. Available at: http://floodlist.com/america/1-hour-rain-floods-sao-paulo-brazil (Accessed: 1 December 2020).

Davies, R. (2015b) *Heavy Rain Floods Sao Paulo Again But Drought Persists: FloodList*. Available at: http://floodlist.com/america/heavy-rain-floods-sao-paulo-drought-persists (Accessed: 1 December 2020).

EEA (1999) *Environmental Indicators: Typology and Overview*. Copenhagen: European Environmental Agency.

Garrick, D. and Hall, J.W. (2014) 'Water security and society: Risks, metrics, and pathways', *Annual Review of Environment and Resources*, 39(1), pp. 611–639.

van Ginkel, K.C.H. *et al.* (2018) 'Urban water security dashboard: Systems approach to characterizing the water security of cities', *Journal of Water Resources Planning and Management*, 144(12), p. 04018075. doi:10.1061/(ASCE)WR.1943-5452.0000997.

Grossman, D.S. and Slusky, D.J.G. (2019) 'The impact of the flint water crisis on fertility', *Demography*, 56(6), pp. 2005–2031. doi:10.1007/s13524-019-00831-0.

GWP (2000a) *Integrated water resources management. 4*. Stockholm: Global Water Partnership. Available at: http://dlc.dlib.indiana.edu/dlc/bitstream/handle/10535/4986/TACNO4.PDF?sequence=1 (Accessed: 6 December 2020).

GWP (2000b) *Towards Water Security: A Framework for Action*. Stockholm: Global Water Partnership.

Hall, J. and Borgomeo, E. (2013) 'Risk-based principles for defining and managing water security', *Philosophical Transactions. Series A, Mathematical, Physical, and Engineering Sciences*, 371(2002): 20120407. doi:10.1098/rsta.2012.0407.

Haraguchi, M. and Lall, U. (2015) 'Flood risks and impacts: A case study of Thailand's floods in 2011 and research questions for supply chain decision making', *International Journal of Disaster Risk Reduction*, 14, pp. 256–272. doi:10.1016/j.ijdrr.2014.09.005.

Hoekstra, A.Y. *et al.* (2011) *The Water Footprint Assessment Manual: Setting the Global Standard*. Routledge.

Hoekstra, A.Y., Buurman, J. and van Ginkel, K.C.H. (2018) 'Urban water security: A review', *Environmental Research Letters*, 13(5), p. 053002. doi:10.1088/1748-9326/aaba52.

Ishizaka, A. and Nemery, P. (2013) *Multi-criteria Decision Analysis: Methods and Software*. John Wiley & Sons.

Jensen, O. and Wu, H. (2018) 'Urban water security indicators: Development and pilot', *Environmental Science & Policy*, 83, pp. 33–45. doi:10.1016/j.envsci.2018.02.003.

Khan, S. *et al.* (2020) 'A comprehensive index for measuring water security in an urbanizing world: The case of Pakistan's capital', *Water*, 12(1), p. 166. doi:10.3390/w12010166.

Koop, S.H.A. and van Leeuwen, C.J. (2015) 'Assessment of the Sustainability of Water Resources Management: A Critical Review of the City Blueprint Approach', *Water Resources Management*, 29(15), pp. 5649–5670. doi:10.1007/s11269-015-1139-z.

Krueger, E., Rao, P.S.C. and Borchardt, D. (2019) 'Quantifying urban water supply security under global change', *Global Environmental Change*, 56, pp. 66–74. doi:10.1016/j.gloenvcha.2019.03.009.

Loucks, D.P. (2000) 'Sustainable Water Resources Management', *Water International*, 25(1), pp. 3–10. doi:10.1080/02508060008686793.

Maurya, S.P. *et al.* (2020) 'Identification of indicators for sustainable urban water development planning', *Ecological Indicators*, 108, p. 105691. doi:10.1016/j.ecolind.2019.105691.

McDonald, R.I. *et al.* (2014) 'Water on an urban planet: Urbanization and the reach of urban water infrastructure', *Global Environmental Change*, 27, pp. 96–105.

Meyar-Naimi, H. and Vaez-Zadeh, S. (2012) 'Sustainable development based energy policy making frameworks, a critical review', *Energy Policy*, 43, pp. 351–361. doi:10.1016/j.enpol.2012.01.012.

Millington, N. (2018) 'Producing water scarcity in São Paulo, Brazil: The 2014-2015 water crisis and the binding politics of infrastructure', *Political Geography*, 65, pp. 26–34. doi:10.1016/j.polgeo.2018.04.007.

Mugagga, F. and Nabaasa, B.B. (2016) 'The centrality of water resources to the realization of Sustainable Development Goals (SDG). A review of potentials and constraints on the African continent', *International Soil and Water Conservation Research*, 4(3), pp. 215–223. doi:10.1016/j.iswcr.2016.05.004.

Muller, M. (2018) 'Cape Town's drought: Don't blame climate change', *Nature*, 559, pp. 174–176.

OECD (1993) *OECD Core Set of Indicators for Environmental Performance Reviews*. Paris: Organisation for Economic Development and Cooperation.

Pahl-Wostl, C. (2007) 'Transitions towards adaptive management of water facing climate and global change', *Water Resources Management*, 21, pp. 49–62.

Plummer, R., de Loë, R. and Armitage, D. (2012) 'A systematic review of water vulnerability assessment tools', *Water Resources Management*, 26(15), pp. 4327–4346. doi:10.1007/s11269-012-0147-5.

Staddon, C. and Scott, C.A. (2018) 'Putting water security to work: Addressing global challenges', *Water International*, 43(8), pp. 1017–1025. doi:10.1080/02508060.2018.1550353.

UN-Water (2013) *Water Security and the Global Water Agenda*. Ontario, Canada: United Nations University. Available at: https://www.unwater.org/publications/water-security-global-water-agenda/ (Accessed: 7 December 2020).

UN-Water (2020) *Unconventional Water Resources*. Geneva, Switzerland: UN-Water. Available at: file:///C:/Users/sppjjgb/Downloads/UN-Water-Analytical-Brief-Unconventional-Water-Resources.pdf.

Van Beek, E. and Lincklaen Arriens, W. (2014) *Water Security: Putting the Concept into Practice*. Stockholm: Global Water Partnership Technical Committee. Available at: https://www.gwp.org/globalassets/global/toolbox/publications/background-papers/gwp_tec20_web.pdf (accessed 11 May 2022).

WRI (2019) *17 Countries, Home to One-Quarter of the World's Population, Face Extremely High Water Stress*. Available at: https://www.wri.org/blog/2019/08/17-countries-home-one-quarter-world-population-face-extremely-high-water-stress (Accessed: 14 December 2020).

WWC (2000) *A Water Secure World : Vision for Water, Life, and the Environment*. Marseille: World Water Council, p. ix, 70 p.: 8 boxes, 2 tab.

Zargar, A. *et al.* (2011) 'A review of drought indices', *Environmental Reviews*, 19, p. 333+.

Annex: Indicator formalisations for the Urban Water Security Dashboard

The table below gives a brief summary on how each indicator in the Urban Water Security Dashboard was operationalised. The values derived for each indicator were converted to a 5-scale ranking as described in the supplementary materials ofVan Ginkel et al. (2018).

Name	*Measurement*	*Primary source*
Pressures		
Annual precipitation and variability	Millimetres of annual rainfall and presence of a dry period	Local data and Peel et al., 2007
Freshwater scarcity around city	Annual averaged monthly blue water scarcity at 30 x 30 arc minute resolution	Mekonnen & Hoekstra, 2016
Rainfall intensity and variability	Millimetres of annual rainfall and presence of a dry period	Local data and Peel et al., 2007
Storm surge hazard	The extreme sea level of an event with a return period of 100 years	Muis et al., 2016 or Internet search
Tsunami hazard	Potential wave height of an event with a return period of 500 years	Peduzzi et al., 2009 or Internet search
Expected SLR by 2100	SLR by 2100 predicted by the IPCC	Church et al., 2013
Area below MSL +1 m and subsidence	Exposed urban area below the threshold of 1 m above current sea level and subsidence rate in mm/year	United States Geological Survey and Google Earth Engine, 1996 and Internet search
Total city population	Population of the city proper	Internet search
Population growth	Average annual population growth over the past 10 years	Internet search
GDP (PPP)	GDP in USD/capita following the PPP method	Internet search; if there is no city data available country data from World Bank, 2017 is used
Slums	Percentage of people living in slums	National figures from United Nations Statistical Divison, 2015, qualitatively adjusted based on an Internet search if no city-specific quantitative data can be found
Domestic water use	Litres per capita per day	IB-NET, 2017 or Internet search
Water footprint of consumption	Total (green + blue + grey) water footprint of national consumption per capita	Mekonnen and Hoekstra, 2011
Water-intensive industries	Qualitative assessment	Internet search procedure
Condition upstream watershed	Qualitative assessment	Internet search procedure

(*Continued*)

Name	*Measurement*	*Primary source*
State		
Supply continuity reservoirs and lakes	Maximum storage capacity to daily urban water demand in days	Global Reservoir and Dam Database, Lehner et al., 2011 or Internet search
Dependency overexploited aquifers	Ratio between natural recharge and withdrawal and the depletion time of the aquifer	Internet search or Gleeson et al., 2012
Local groundwater drawdown	Qualitative assessment	Internet search procedure
Coverage and leakages water supply system	The total percentage of people with access to piped water, the percentage with direct service connections at the households, and the percentage of non-revenue water	Internet search or IB-NET, 2017
Continuity and quality of water supply	Supply continuity in hours per day and qualitative evaluation of water quality of supplied tap water	Internet search or IB-NET, 2017
Coastal flood protection infrastructure	Estimated (maximum) coastal flood protection level in terms of return period (years)	Hallegatte et al., 2013 or Internet search
River flood protection infrastructure	Estimated (maximum) river flood protection level in terms of return period (years)	FLOPROS (Scussolini et al., 2016)
Stormwater drainage infrastructure	Recurrence time of the design storm event in ordinary residential areas	Internet search
Coverages and leakages of sewer system	Percentage of households in the city that is connected to a sewage system which is connected to a treatment plant and the current state of the sewer system (qualitative)	Internet search
Adequacy wastewater treatment	Qualitative assessment	Internet search procedure
Surface water quality	Qualitative assessment	Internet search procedure
Polluted sediments	Qualitative assessment	Internet search procedure
Garbage in surface water	Qualitative assessment	Internet search procedure (images)
Groundwater quality	Qualitative assessment	Internet search procedure
Salt water intrusion in groundwater	Qualitative assessment	Internet search procedure
Impacts		
Conflicts over water supply	Qualitative assessment	Internet search procedure
Sustainability of urban water footprint	Unsustainable fraction of the blue water footprint	Mekonnen and Hoekstra, 2016, Hoekstra and Mekonnen, 2012

(Continued)

Name	*Measurement*	*Primary source*
People with adequate water supply	Share of urban population with access to improved water sources, share of urban population with access to piped water on the premises, supply continuity	WHO & UNICEF, 2015
Coastal flooding River flooding Stormwater flooding	Casualties and extent of damage	Dartmouth Flood Observatory Database (Brakenridge, 2016)
People with adequate sanitation	National share of the urban population with improved sanitation facilities on the premises and share of urban population with shared (unimproved) sanitation facilities	WHO & UNICEF, 2015
Water-associated diseases	Qualitative assessment	Internet search procedure
Ecological quality of urban water	Qualitative assessment	Internet search procedure
Water image of city	Evaluate first 50 relevant water images from Google Images search; qualitative scoring	Google Images search
Response		
All response indicators are measured using a questionnaire sent to experts in each city.		

Data sources:

- Brakenridge, G.R., 2016. Global Active Archive of Large Flood Events [WWW Document]. URL http://floodobservatory.colorado.edu/Archives/index.html (accessed 12.20.16).
- Church, J.A., Clark, P.U., Cazenave, A., Gregory, J.M., Jevrejeva, S., Levermann, A., Merrifield, M.A., Milne, G.A., Nerem, R.S., Nunn, P.D., Payne, A.J., Pfeffer, W.T., Stammer, D., Unnikrishnan, A.S., 2013. 2013: Sea level change, in: Stocker, T.F., Qin, D., Plattner, G.-K., Tignor, M., Allen, S.K., Boschung, J., Nauels, A., Xia, Y., V., B., Midgley, P.M. (Eds.), Climate Change 2013: The Physical Science Basis. Contribution of Working Group I to the Fifth Assessment Report of the Intergovernmental Panel on Climate Change. Cambridge University Press, Cambridge and New York, pp. 1137–1216. doi:10.1017/CBO9781107415315.026
- Gleeson, T., Wada, Y., Bierkens, M.F.P., van Beek, L.P.H., 2012. Water balance of global aquifers revealed by groundwater footprint. Nature 488, 197–200. doi:10.1038/nature11295
- Hoekstra, A.Y., Mekonnen, M.M., 2012. The water footprint of humanity. Proc. Natl. Acad. Sci. 109, 3232–3237. doi:10.1073/pnas.1109936109
- IB-NET, 2017, International Benchmarking Network for Water and Sanitation Utilities. IB-NET database [WWW Document]. URL http://database.ib-net.org (accessed 6.19.17).
- Lehner, B., Liermann, C.R., Revenga, C., Vörömsmarty, C., Fekete, B., Crouzet, P., Döll, P., Endejan, M., Frenken, K., Magome, J., Nilsson, C., Robertson, J.C., Rödel, R., Sindorf, N., Wisser, D., 2011. High-resolution mapping of the world's reservoirs and dams for sustainable river-flow management. Front. Ecol. Environ. 9, 494–502. doi:10.1890/100125

- Mekonnen, M.M., Hoekstra, A.Y., 2011. National water footprint accounts: The green, blue and grey water footprint of production and consumption. Value of Water Research Report Series No. 50. Delft, The Netherlands.
- Mekonnen, M.M., Hoekstra, A.Y., 2016. Four billion people facing severe water scarcity. Sci. Adv. 2, e1500323. doi:10.1126/sciadv.1500323
- Muis, S., Verlaan, M., Winsemius, H.C., Aerts, J.C.J.H., Ward, P.J., 2016. A global reanalysis of storm surges and extreme sea levels. Nat. Commun. 7, 11969. doi:10.1038/ncomms11969
- Peduzzi, P., Deichmann, U., Maskrey, A., Nadim, F.A., Dao, H., Chatenoux, B., Herold, C., Debono, A., Giuliani, G., Kluser, S., 2009. Global disaster risk: patterns, trends and drivers, in: Global Assessment Report on Disaster Risk Reduction. United Nations, Geneva, pp. 17–57.
- Peel, M.C., Finlayson, B.L., McMahon, T.A., 2007. Updated world map of the Köppen-Geiger climate classification. Hydrol. Earth Syst. Sci. 11, 1633–1644. doi:10.5194/hess-11-1633-2007
- Scussolini, P., Aerts, J.C.J.H., Jongman, B., Bouwer, L.M., Winsemius, H.C., De Moel, H., Ward, P.J., 2016. FLOPROS: an evolving global database of flood protection standards. Nat. Hazards Earth Syst. Sci. 16, 1049–1061. doi:10.5194/nhess-16-1049-2016
- United Nations Statistical Divison, 2015. Slum population as percentage of urban population [WWW Document]. Millenn. Dev. Goals Indic. URL https://mdgs.un.org/unsd/mdg/Data.aspx (accessed 6.14.17).
- United States Geological Survey, Google Earth Engine, 1996. Global 30 Arc-Second Elevation [WWW Document]. URL explorer.earthengine.google.com (accessed 8.22.17).
- WHO, UNICEF, 2015. Joint Monitoring Program for water supply and sanitation [WWW Document]. URL https://www.wssinfo.org/data-estimates/tables/ (accessed 6.29.17).
- World Bank, 2017. GDP per capita, PPP (current international dollar) [WWW Document]. URL http://data.worldbank.org/indicator/NY.GDP.PCAP.PP.CD (accessed 6.14.17).

25

URBAN WATER QUALITY AND CHEMICAL POLLUTION

New emerging contaminants, nanomaterials, and microplastics

Serge Stoll and Stéphan Ramseier Gentile

25.1 Introduction

Water supply and quality are the major components of urban water security. In particular, water pollution and contamination by chemicals pose serious human health risks and threats to the sustainability and biodiversity of aquatic ecosystems. With the increase in population, increase in water demand for human consumption, and human activities in confined areas, urban areas are generating important volumes of wastewaters and, thus, exacerbating chemical pollution issues. Urban areas enhance the emission, transport, and circulation of significant quantities of chemical pollutants, including emerging contaminants, with the potential to contaminate urban waters and natural aquatic systems in many ways. As a result, surface waters as well as groundwaters used for water supply can be polluted from a diversity of urban water effluents, such as the leaking and diffusion of chemical substances related to toxic waste chemical storage, intentional dumping of hazardous substances, industrial and household activities, events of accidental natures, etc. Urban runoffs from streets also have a great potential to transport various compounds, such as oils, hydrocarbons from gasoline, heavy metals, or ionic species from road salt. On the other hand, indirect contamination is also possible via air, which, in some circumstances, can ultimately lead to acid rain pollution and can help in the transport and transformation of chemicals and particles, which at the end and after deposition, will directly alter water quality of surface waters. Furthermore, untreated or poorly treated urban wastewaters can have a low content in dissolved oxygen, high concentration in microbiological pollutants, high content in chemicals such as nitrates and phosphorous, or in organic matter, and they will have a strong impact on the functioning and sustainability of natural aquatic systems in which they are usually discharged. Finally, emerging pollutants released in urban waters, such as pharmaceuticals, microplastics, and nanomaterials, which are not all fully regulated at the moment, are today presenting new and important challenges that have to be addressed in view of their potential negative impacts on water quality and threats to the environment and human health. Thus, in order to understand and define challenges related to urban water quality and chemical pollution, we first begin by discussing the water treatment processes used to produce drinking water and treat wastewater. Then we discuss the major sources and pathways of chemical pollutants in the urban context.

 DOI: 10.4324/9781003057574-31

After that, emerging pollutants are discussed with a particular emphasis on nanomaterials and microplastics, which are today raising important debates.

25.2 Drinking and wastewater treatment processes

Urban water management involves the design, planning, and use of complex infrastructures to satisfy the demand not only for drinking water and sanitation, but also for the control of water distribution via the use of pumping, storage, distribution, and maintenance and sustainability of various urban infrastructures, for example, for recreational activities (Tolba Alboelnga, 2019). Drinking water and wastewater treatment infrastructures are of particular importance in urban water security, quality, and understanding their functioning, limitations, and future challenges needs to be addressed (Wei, 2018). In particular, potential human health risks of emerging pollutants through drinking water require special attention. However, water treatment is a complex process. In the following, we describe these two types of infrastructures by focusing on chemical aspects.

Before water can be used for human consumption, biological and chemical contaminants need to be removed from the raw water in **water treatment plants**. Raw water usually contains microorganisms, pathogens, and a variety of non-desirable or excessive organic and inorganic chemicals and compounds containing arsenic, chromium, copper, and lead as well as various micropollutants and emerging contaminants. To remove (or transform) these contaminants, a typical or conventional municipal water treatment plant involves a series of physical and chemical processes as shown in **Figure 25.1**. The different steps described here correspond to the processes used in the main drinking water treatment plant in Geneva (Switzerland), providing drinking water for 500,000 consumers. The number and nature of these water treatment processes have to be adjusted and calibrated to raw water physico-chemical properties and desired drinking water quality and quantities. Financial considerations must also be considered since they contribute to determining what is feasible or not. Here the water treatment process is a conventional one and involves a multistage procedure, including raw water pre-treatment, acidification, flocculation, sand filtration, ozonation, activated carbon filtration, pH neutralization, and final disinfection.

Pumping and initial filtering of raw water (abstracted from Lake Geneva) are carried out to remove debris and suspended material, such as plastic bags, leaves, branches, ropes, rags, and fish. Filtering is performed via a simple screening physical filtration process (strainer). Then, coagulation using chemicals (namely coagulants), such as iron or alum-based chemicals, here poly-aluminum chloride, at a concentration of about 0.36mg/L (Al/L), is added to promote aggregation of the suspended colloidal particulate matter. Before coagulation, pH adjustment

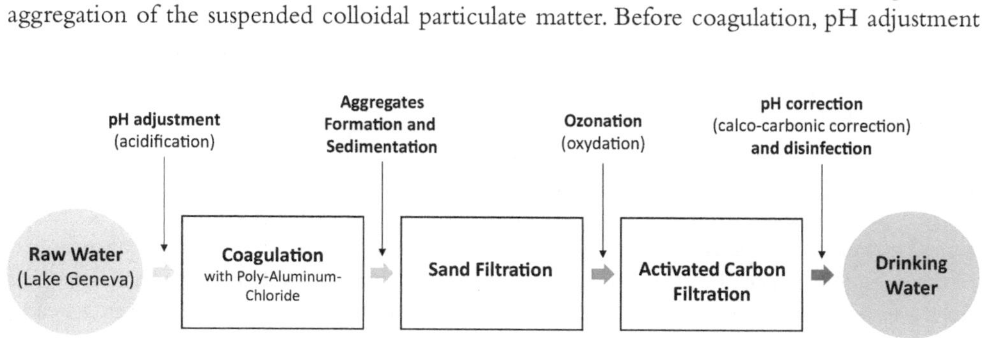

Figure 25.1 Geneva (Switzerland) drinking water treatment plant description. A conventional water treatment is made to provide tap water to 500,000 consumers.

is made to optimize coagulant efficiency. Hetero-aggregates made of a complex mixture of coagulated particles, including microorganisms and suspended materials, are then formed rapidly and removed by sedimentation and subsequent filtration. Then, water is filtered through porous media made of sand and/or other granular materials (quartz sand and pumice stone here) to essentially remove the remaining aggregates. Then, the filtered water is treated with a chemical disinfectant to kill any remaining pathogens (viruses, bacteria, protozoa, parasites, fungi). For that purpose, an effective disinfectant, such as chlorine, could be considered but its use can cause the formation of potentially dangerous substances, such as trihalomethanes, in particular, in the presence of organic matter. An alternative to chlorine, here concerns the use of ozone. Ozone (O_3) is a gas that is particularly efficient in killing viruses, and it promotes oxidation processes, hence favoring the elimination, precipitation, or transformation of many dissolved chemical substances, such as metals, pharmaceuticals, and proteins. Ozone converts dissolved molecules into small fragments; it also promotes coagulation. Water passes afterward through a layer of granular activated carbon. Activated carbon is well known to improve water color, taste, and odor, which are easily detected by consumers. Color can be due to organics, such as tannic acids from the degradation of biomass, or inorganics, such as high concentrations of ferric iron. Activated carbon also removes organics very well, such as chlorine or radon, but only marginally inorganics in particular metals and nitrates. Every excess of ozone, which has a sweet smell, is rapidly converted to oxygen (O_2) by activated carbon. However, unlike chlorine, ozone does remain in the water distributed in the network, so it offers no long-lasting killing effect against bacteria that might be in storage tanks and water pipes of the water distribution system. This is the reason why chlorine (or chlorine dioxide), after pH adjustment for calco-carbonic correction, is carefully added at the end for preventing possible re-contaminations. Water could also be disinfected using ultraviolet radiation to kill microorganisms, but this has the same limitation as ozone.

It should be noted that after purification, water can travel under pressure for several hours and days through dozens of kilometers and kilometers of distribution pipes before reaching domestic taps. Important travel times can be a major source of drinking water degradation, due to disinfection by-products, bacterial regrowth, and the release of corrosion products, such as Fe, Pb, Cu, Zn as well as micro and nanoplastics and nanoparticles released from pipe material corrosion and pipe coating abrasion (sand abrasion) or after maintenance operation periods.

Pollution in terms of untreated wastewater is one of the major risks to water quality, public health, and urban security if wastewater is not treated via **wastewater treatment plants** and then discharged properly in natural aquatic systems. Wastewater contains high amounts of inorganic and organic dissolved and suspended solid materials (concentrations in untreated sewage range from 100 to 350 mg/L) resulting in high turbidity. Moreover, pathogens are also present at very high concentrations. Furthermore, treated wastewater also contains large quantities of nutrients, such as ammonia and phosphorus, ranging, respectively, from 2 mg/L to 50 mg/L and from 1 mg/L to 20 mg/L.

In wastewater treatment, plants biodegradation is expected to play important role and one measure of the biodegradable component concentration in the wastewater is the biochemical oxygen demand, namely the BOD5. It represents the amount of dissolved oxygen (O_2) consumed by microorganisms after a given time, here five days, via their metabolization processes of the organic materials. Untreated raw wastewater typically has a BOD5 concentration ranging from a few hundred mg/L, which has to be reduced to a few mg/L before discharging treated wastewater effluents in the environment.

As shown in Figure 25.2, wastewater treatment is also a multistage process with the objective to reduce (and not totally remove) solids, organic matter, nutrients, disease-causing microorganisms,

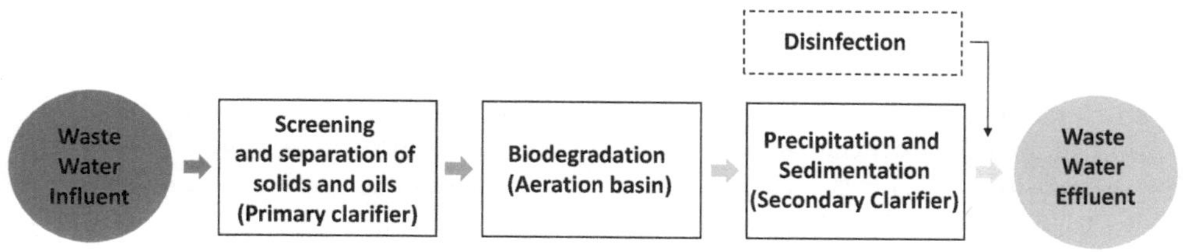

Figure 25.2 Layout of a waste water treatment plant.

and other pollutants before releasing it in natural aquatic systems, mainly rivers, lakes, and coastal zones. Treated wastewaters are also used in irrigation in certain countries as fertilizers or reused as an additional source of water. It should be noted here that, despite the fact that raw wastewater presents a high risk of biological and chemical pollution, it can also provide opportunities if considered as an additional source of water and if properly managed to mitigate water scarcity.

The first treatment process (primary treatment) consists of a preliminary screening treatment, which removes solid materials, such as large particles, sand, gravel, rags, and a variety of objects flushed, for example, in toilets. Then, the wastewater passes through clarifier tanks, which are used to separate remaining suspended particulate matter solids and oils. Particles are then allowed to settle and the oil to float. The clarified wastewater is then flowing to the next, secondary stage of wastewater treatment. It consists typically of a biodegradation treatment process to degrade and consume the dissolved organic matter. For that purpose, specific and cultivated microorganisms (bacteria) are added to the wastewater for the biodegradation of nutriments (ammonia, organic phosphorus) and organic matter. Microorganisms develop rapidly in number and size and form large biological flocs, which settled out as activated sludge. However, additional treatment is sometimes necessary to improve the removal of some nutriments (like phosphorus) by adding chemicals, such as iron-based chemicals, or for denitrification. Then, before discharge in aquatic systems, in some cases a treatment is necessary to control pathogens and microorganisms. To achieve this, chlorine is added or ultraviolet radiation is used. Activated carbon adsorption and ozone and membrane filtration can also be considered for the removal of specific microorganisms and emerging pollutants, such as pharmaceuticals, pesticides, and cosmetics.

25.3 Major sources and chemical pollutants in the urban context

Urban waters, including natural freshwater and groundwater used for water potabilization, wastewater, drains, and runoff can be contaminated by a large amount of chemical (and biological) pollutants from a variety of specific or more diffuse sources. All harmful substances that induce the deterioration of water quality, the biological community, and sediment quality in natural aquatic systems can be considered water pollutants. These sources usually involve industrial, residential, and commercial effluents, agricultural wastes, and water runoff as well as more mobile and diffuse sources, such as those related to road traffic (Müller, 2020).

Industrial pollution sources result from waters used in industrial processes. Owing to the industrial usages and nature of these processes, waters may contain important amounts of organic matter (food processing), inorganic substances and heavy metals (mining, bar turning), toxic chemicals and solvents (chemical industries), and pathogens (biological products). Sources of industrial pollution are mainly concentrated in dying and leather, paper, food, and electroplating industries, and other industries with high pollution. Furthermore, cooling water from thermal power plants can also be a source of pollution having impacts on aquatic systems.

Residential and commercial pollution sources correspond to all types of wastewater generally related to the use of tap water for daily life needs, such as cooking, washing, bathing, and sanitation. As mentioned before, wastewater usually exhibits high turbidity and odor, and it contains an important concentration of organic matter and nutriments, oil, and also detergent residues, such as surfactants. Wastewater can also contain a series of emerging contaminants, such as microplastic particles and fibers resulting from clothes washing, as well as pharmaceuticals and nanomaterials. After treatment in wastewater treatment plants, a decrease in organic matter concentration, suspended particulate matter, specific chemicals such as phosphorus and nitrogen, and of the BOD5 is expected before discharge into water bodies.

Agricultural drains can also be a source of contamination and are generally related to the improper use of fertilizers or pesticides, such as organochlorine pesticides and organic mercury pesticides that unfortunately induce the pollution of both surface and groundwater bodies often used as a supply for the production of drinking water.

Finally, urban runoff and, in particular, urban stormwater runoff represent a significant source of urban water pollution, and the large-scale, global impacts due to climate variability and change could increase these risks. Runoff usually consists of a heterogeneous mixture of anthropogenic suspended particles, debris, and chemical pollutants, such as metals, pesticides, microplastics, and nanomaterials that are washed form the urban landscape and particularly roads during rainfall. Road traffic is regarded as an important source of pollution. For example, due to vehicle road abrasion and tires and road paint abrasion, a large amount of microplastics and hydrophobic substances are produced, contaminating runoff with hazardous components, including heavy metals coming from automobile brake pads and catalytic converters.

Many different classes of pollutants contaminate urban waters. Major and conventional pollutants are mostly include organic and inorganic pollutants, heavy metals and plant nutriments. Organic pollutants are mainly present in municipal effluents and industrial wastewater and include polysaccharides, proteins, sugars, fats, and oils. Domestic effluents mainly consist of human solid and liquid excreta resulting in high biological oxygen demand in addition to pathogens and microorganisms, and detergents that contain surfactants and in some cases phosphates. In natural aquatic systems, the process of biological oxidation and decomposition of these pollutants requires a large amount of dissolved oxygen. Once the oxygen is almost totally consumed and its concentration is insufficient, the oxidation will stop, hence causing anaerobic fermentation of organic matter, contamination of the environment, and poisoning of aquatic organisms, and, as a result, a decrease in the water quality. Plant nutriments mainly refer to nitrogen and phosphorus compounds, which can be present under various chemical and oxidation states. They induce eutrophication, hence resulting in a decrease of the water quality of potential water supply sources. Heavy metals, which generally refer to those elements having an atomic number greater than 20 and atomic density above 5 $g.cm^{-3}$ and must exhibit the properties of metal (mercury, cadmium, lead), exhibit significant biological ecotoxicity and also refer to metals with a certain toxicity, such as zinc, copper, and iron. The main problem is that they are not degraded by microorganisms, and they accumulate and persist in the environment producing toxic effects. Moreover, because of their bioaccumulation and amplification in the environment they can impact the food chain and can also be transformed into more toxic compounds (metabolites) via chemical and biological processes.

25.4 Emerging pollutants

Emerging pollutants, also referred as new contaminants, in urban waters are posing a potential serious risk to humans, water resources, and aquatic ecosystems. They represent a new challenge that must be urgently addressed before their irreversible accumulation and concentration increase (Pena-Guzman, 2019). These chemicals, which for most of them were not considered originally as pollutants when introduced in our environment, are a source of concern due to their continuous discharge, often at very low concentrations, and subsequent bioaccumulation into water systems. They are often related to new technological development, medicine progress, and changes in consumer consumption. They are part of the development of our modern societies along with the intensive use of plastics. Emerging pollutants consist of mainly synthetic chemicals produced for and used in our daily lives, such as pharmaceuticals, personal care products, industrial and household chemicals, food additives, and innovative materials. For example, nanoparticles used as UV filters in cosmetics or incorporated in paints and rubbers and even in our food have been introduced into our daily life in recent years with the development of nanotechnologies, and they constitute major emerging contaminants. Emerging pollutants are mainly found in municipal wastewater, industrial effluents, and urban runoff. Unfortunately, many of these pollutants are not easily eliminated by conventional wastewater treatment processes, and, consequently, they persist in the environment, are toxic, and bioaccumulate. It should be remembered that wastewater treatment plants were originally intended mainly to treat organic matter, nitrogen, and phosphorus. The situation is particularly acute in developing countries where large quantities of untreated wastewater and industrial effluents are directly discharged into surface water and coastal zone waters or used for agricultural purposes as fertilizers or reused as an additional source of drinking water.

Current scientific knowledge and understanding of the sources, fate, transport, and risks of emerging pollutants, individually or in combination with other contaminants, and their accumulation in the environment is relatively limited and, in many cases, their effects on ecosystems and human health are still to be evaluated. In particular, pharmaceuticals, which are used both in humans and animals for therapeutic and diagnostic purposes, are present in rivers, lakes, and coastal zones even at very low concentrations and wastewater at significant concentrations. These pollutants belong to the therapeutic groups of anti-inflammatory and analgesics, cardiovascular and central nervous system agents and antimicrobial substances, and they are a source of concern due to their increasing and continuous discharge and accumulation into the environment. Some of them are capable, at very low concentrations, to cause endocrine disruption in humans and aquatic systems and the development of bacterial pathogen resistance (Al Salah, 2020). It should be noted that pharmaceuticals have also been detected in drinking water, highlighting the fact that water purification processes are not always fully effective in removing these synthetic chemical compounds. Despite the fact that advanced technologies, such as membrane filtration, ultrafiltration, nanofiltration and reverse osmosis, partially, or even completely eliminate some endocrine-disrupting chemicals and pharmaceutically active compounds, their application is still limited both in developing and in developed countries due to high costs, and the efficiency of these advanced technologies in the removing of microplastics or manufactured nanoparticles is still under debate.

25.5 Microplastics

Over years and especially since the 1950s, plastics have been widely used in daily life and for many industrial and commercial purposes due to their versatile properties, such as their light weight, low cost, resistance to water and chemicals, durability, mechanical properties, and the possibility they offer to produce objects having complex forms, colors, and properties. From this perspective, plastics are fantastic and unique. Unfortunately, many of these plastic objects have become waste and most of them have accumulated in the environment as plastic debris. Due to their property of durability, the very first plastic objects are probably still existing somewhere in our environment. As a result, today, we currently face unprecedented increasing levels of plastic fragments in our food, water, and air, and plastics as a contaminate has become a global concern (Negrete Velasco et al., 2020; Faure, 2015; Turner, 2019).

With time, in all environmental compartments, plastics are broken down into small-sized plastics under the influence of physical, chemical, and biological factors, such as photolysis, thermal oxidation, chemical degradation, hydrolysis, and biodegradation. Fragments or plastics of sizes less than 5 mm are defined as microplastics, whereas plastics having sizes less than 1000 nm are commonly defined as nanoplastics. Microplastics can be produced intentionally at their sizes, called primary microplastics, or produced by breakdown and fragmentation of larger plastic particles, called secondary microplastics. Once released into the environment, due to their sizes, microplastics are usually difficult to remove. Microplastics include a diversity of polymer types (polyethylene, polypropylene, polystyrene) and shapes (fragments, pellets, and fibers) that can originate from different sources and materials. Considered as one of the main final sinks for plastic waste, water supply bodies have therefore received over time significant amounts of microplastics, potentially endangering water quality and human health.

To obtain a full picture of the relationship between microplastic pollution sources and transport in urban area and human activities, correlations between urban characteristics and anthropogenic activities and microplastic pollution can be established. Urban characteristics can include population density, residential and commercial activities, urbanization rate, industrial activities, water supply properties, and wastewater treatment efficiency (Zhou, 2020). Concerning the pathways of plastic contamination in urban areas and, in particular, in urban waters, it is shown that plastic debris and microplastics enter the environment mainly via treated wastewater effluents, from the landscape by precipitation, and via (storm)water runoff. Effluents from wastewater treatment plants and urban runoff are therefore important components of the global microplastic urban water cycle despite the fact that, in some cases, other sources, such as atmospheric deposition, illegal dumping, and agricultural runoff, can be significant. Buoyancy, surface properties, and, in particular, durability make microplastics a new contaminant type because of the resulting potential long distances they can be transported in the environment and the potential dispersal of chemicals as well as biological contaminants that are associated with microplastics (McCormick, 2016). More studies are needed to quantify the importance of stormwater and wastewater on the transport pathways of microplastics to and via urban waters. It was found (Werbowski, 2021) that stormwater may be a more significant pathway for microplastics than treated wastewater. In this study, fibers and black rubbery fragments, tire fragments, and road wear particles were the most frequently occurring morphologies, indicating that inputs due to vehicle road abrasion and tires as well as road paint abrasion are particularly important as major sources of microplastics in urban runoff, which can also be contaminated with heavy metals coming from vehicle catalytic converters or brakes.

Regarding wastewater produced by residential and commercial activities, it has been found that microplastic fibers due to washing machines and the washing process of synthetic textiles and pellets as abrasives in personal care products are abundant both in untreated and in treated wastewater and are now considered as a point source for surface water contamination. This also indicates that microplastics may not be captured by wastewater treatment plants due to their small sizes and specific properties.

Studies have also revealed the occurrence of microplastics in raw water that supplies drinking water treatment plants (Kirstein, 2021; Wang, 2020; Negrete Velasco et al., 2022). Despite the fact that negligible amounts were detected in raw water from groundwater, high quantities of microplastics were reported in the case of surface water used for potabilization. As a result, microplastics have also emerged as an issue in water from tap and public fountains as well as in bottled water. Investigations into microplastic contamination in drinking water indicate that contamination can vary greatly (Pivokonsky, 2020) and will depend on raw water contamination, processes used for water purification, design of the infrastructure, use of plastic materials, etc. It should be noted that variability in the obtained results is not only related to effective microplastic contamination, but also to the lower size boundaries of analyzed microplastics, protocols, and characterization techniques drawing comparisons between the different studies. However, when considering treated water originating from surface water bodies with high microplastic concentrations, the removal efficiency based on conventional processes, including coagulation, sand, and activated carbon filtration, was found higher than 80% for microplastics with sizes higher than 63 μm (Negrete Velasco et al., 2022). Therefore, drinking water treatment plants are expected to provide a relatively effective barrier preventing microplastics from entering water for human consumption.

Regarding the impact of microplastics, microplastics are likely to be eaten by aquatic organisms by mistake, and the significant bioaccumulation and bioamplification of microplastics in aquatics organisms are expected to result in potential biological toxicity and cause environmental issues. In addition, microplastics exhibit large specific surface area and strong hydrophobicity and consequently can adsorb other contaminants, such as heavy metals and persistent organic pollutants, that are expected to amplify microplastics biological toxicity. In particular, the reduced size of nanoplastics makes them susceptible of being ingested by microorganisms that are at the base of the food chain (Saavedra Vargas, 2019; Pochelon, 2021). Concerning drinking waters, microplastics increase health concerns due not only to possible mechanical effects associated with particle ingestion, but also to the presence of monomers or additives, or the capability of microplastics to adsorb and desorb environmental toxic chemicals, which may exacerbate their toxicity. Microplastics and associated chemicals could lead to oxidative stress, cellular damage, inflammatory and immune responses, and neurotoxic and metabolic changes. Despite the urgent need to assess this issue, it should be noted that the effects of microplastics on human health are still largely unknown and no prescriptions are in place today regulating the content of microplastics in drinking water.

25.6 Nanomaterials

Due to their specific intrinsic properties (size, surface area, surface reactivity, charge, shape) and clear benefits, nanomaterials and, in particular, engineered nanoparticles have been incorporated in many daily live products. Engineered nanoparticles are generally defined as objects with at least one dimension in the size range of 1–100nm. From titanium dioxide used in cosmetics and personal care products to silver nanoparticles that are incorporated in textiles or multiwalled carbon nanotubes that are used in automotive parts, the use of nanosized materials (a nanometer is one million of a millimeter) has also raised concerns due to their environmental and health risks. They are present in urban water at variable concentrations, ranging from ng/L to μg/L, with the possibility of altering or modifying the metabolism of living organisms. They have been detected in drinking water, indicating that conventional water treatment processes are not always effective and designed for the removal of these contaminants (Enfrin, 2019).

Despite the fact that their incorporation in a wide variety of consumer products is questionable, they create new opportunities in the field of air purification and in the field of water treatment and remediation via the use of nanosized adsorbents, photocatalysts, and nanomembranes

(Mehrnoosh, 2020). In particular, nanomaterial adsorbents exhibit an extremely large reactive surface area at low concentrations and remove pollutants more efficiently when comparison is made with activated carbon. Moreover, in some cases, the adsorption process can be complemented by degradation processes, such as reductive degradation and photocatalysis that allow the decomposition of organic pollutants into nontoxic metabolites. For example (Xie, 2016; Cao, 2005), chitosan-stabilized nano Zero-valent iron was used to degrade perchlorate (ClO_4^-), which can impede the endocrine function by blocking iodide from entering the thyroid gland, in water into a less detrimental compound, chlorine (Cl^-). Perchlorate has emerged as a widespread contaminant in groundwater and surface water because of its unique properties (water solubility, non-complexing, non-volatile, and chemical stability), it is highly challenging to remove perchlorate from water by traditional water treatment approaches.

The major nanomaterial sources, mainly as nanoparticles, in the urban context are home products such as cleaning products, food additives, clothing, and personal care products (toothpaste, creams, sunscreens) that are released in wastewater. Industrial processes manufacturing nanoparticles that are used for catalysis, fuel additives, industrial polishing processes, and cleaning processes can also be a significant source of emission in the urban context, and hospitals via medical nanoformulations can also represent a significant emission source via wastewater effluents. Traffic via exhaust emissions and the formation of ultrafine particles, tire abrasion, brake wear, car wax and paints, building materials such as paints and surface coatings that are used outdoors also represent an important source in the urban context (Rönkkö, 2019). Studies conducted on ambient air influenced by important road traffic report very high nanoparticle concentrations with the potential to contaminate urban water bodies. Thus, it can be concluded that large numbers of people are exposed to large numbers of atmospheric traffic-originated nanoparticles (Yang, 2016). Also, the presence of agricultural areas using nanofertilizers and nanopesticides constitutes another emission source and a potential source of contamination of rivers, lakes, and groundwater. Concerning nanoplastics, which are a special class of nanoparticles, the breakdown of larger plastic items results in the release of synthetic fibers during the washing of clothes, mechanical abrasion, and disintegration of plastics as well as industrial activities, such as thermal cutting, 3D printing, and biomedical applications due to the use of polymeric nanocapsules (Pinto da Costa, 2016).

Nanomaterials can enter the environment via several emission scenarios (Domercq, 2018). In urban areas, they can be released either unintentionally or intentionally during the production of raw materials, released during use, and released after the disposal of products containing nanomaterials. One of the most relevant release pathways to urban surface waters is via wastewater, runoff, and direct release. Recent studies have provided evidence that leached Ag (silver) nanoparticles from exterior building facades and TiO_2 contained in sunscreens are important sources of Ag and TiO_2 nanoparticles into runoff waters, producing threats to aquatic systems that may serve as water supplies. Nanoparticles are likely emitted to urban water via traffic emission runoff and leaching, and specific sources via surface water runoff, wastewater treatment, and hospitals and industry effluents. Release patterns and concentrations depend on the nanoparticle types and their applications. For example, the main emission pathway of TiO_2 has been found to occur via wastewater and from accumulation in sewage sludges during wastewater treatment, which are in many countries ultimately released onto soils (Bundschuh, 2018). Therefore, a more systematic approach is urgently needed to understand the role of soil properties and thus the risk of groundwater contamination by nanoparticles. It should be noted that studies suggest that stormwater ponds act as an important conveyer for a wide range of nanoparticles in the urban environment (Baaloushaa, 2016). Concerning the nanoparticle concentrations found in natural systems, including Ag nanoparticles and oxides of Ce, Ti, and Zn, they were found typically ranging from 1 μg/L to 10 μg/L and can even reach up to 100 μg/L.

Concerning their fate and impact, nanomaterials in the environment can undergo ageng processes such as chemical transformation, aggregation, and disaggregation, and the interplay between these processes and nanoparticle transport will determine their fate and ecotoxicological potential. For example, both in natural water and in drinking water treatment processes aggregation plays a key role and depends on nanoparticles surface properties, coating, concentration of natural organic matter, pH, oxidative potential, concentration of divalent ions, etc. Such features will determine their ability to form aggregates and their ability to be eliminated from water systems by sedimentation processes (Ramirez Arenas, 2019, 2020). It should also be noted that the solubility of the nanoparticle is also expected to significantly influence its behavior and extent of environmental concentrations.

Interactions of nanoparticles with living organisms, including humans, are expected to induce toxicity via different mechanisms (Chen, 2022). They can interact via cells via adhesion, penetration through cell membranes, and endocytosis, inducing damages and dysfunction of biochemical functions or components, such as proteins oxidation, genotoxicity, energy production blockage, generation of reactive oxygen species (ROS), and toxic substances, or promoting oxidative stress.

Concerning nanoplastics, due to their size definition (<1000 nm), chemical composition (polymers and presence of additives), physical properties (buoyancy), shape, and surface properties, they clearly constitute a distinct category compared to engineered nanoparticles (<100 nm), and they are expected to interact differently with biological interfaces (Reynaud, 2021). Compared to microplastics, due to their higher surface-to-volume ratio, behavior in the environment will also be quite different (Boyle, 2020). Additives will be more easily released from nanoplastics than microplastics and will behave as more efficient sorbents regarding hydrophobic contaminants, such as polycyclic aromatic hydrocarbons, which are the most-cited persistent pollutants that adsorb or are adsorbed to plastics in the environment. Finally, nanoplastics, according to their sizes, and regardless of their aggregation ability, will diffuse and penetrate more easily into natural interfaces and barriers. However, since nanoplastics are probably the least known and characterized type of nanomaterials, a precautionary approach and a reasonable period of time will be necessary to evaluate and understand their ecological effects and impact on urban water. In that context, the need is urgent to develop accurate and harmonized methods for detecting and characterizing nanoplastics in the urban context, particularly in wastewater and drinking water, and for their efficient removal in drinking and wastewater treatment (Ramirez Arenas, 2021).

25.7 Conclusions

Although research is growing into emerging pollutants in the urban water context, effective tools and policies to monitor, regulate, and control these pollutants in water resources are lacking and have to be considered. For that purpose, the need is urgent to adopt precautionary measures and to increase our scientific knowledge and data as well as to adopt appropriate technological and regulation approaches to monitor emerging pollutants in drinking water and wastewater and to assess their potential human health and environmental risks so as to prevent and limit their disposal in water resources and the environment. Solutions involve finding sustainable ways for urban areas to reduce both their dependence on potential chemical pollutants and the amount of pollutants they produce and to properly dispose of, recycle, or reuse pollutants before they contaminate water as well as air and soils. In particular, measures to reduce the input of emerging contaminants should involve all steps of the product lifecycle from manufacturing to consumption to waste management. The need also exists for more studies to quantify the relative importance of urban water as a transport pathway to aquatic ecosystems.

The need is urgent as well to reverse declining water quality and to improve wastewater management and safe reuse. Considering the complexity of urban areas, potabilization, and water quality, the use of conventional water treatment processes becomes increasingly challenging with the removal of emerging contaminants and expected increasing concentrations in the future. Consequently, advanced treatment technologies, including membrane filtration (ultrafiltration, nanofiltration, reverse osmosis) and advanced oxidation processes (AOPs) are expected to provide alternatives for better protection of public health and the environment but with a higher cost as a consequence. New technical solutions can be applied in wastewater treatment plants, mainly as advanced treatment methods. The treatment methods that could be used to enhance the removal of pharmaceuticals are ozonization and other advanced oxidation processes; adsorptive methods using, for example, activated carbon, membrane, and nanofiltration; and reverse osmosis. Oxidation, adsorption, and filtration methods could also be used for the pre-treatment of hospital wastewater and industrial effluents prior to discharging. Nonetheless, even if these technologies exist, their use is relatively limited due to high costs.

Among these new emerging contaminants, the challenges for a comprehensive risk assessment are numerous, but more than ever, research on engineered nanoparticles, microplastics, and nanoplastics is critical to accurately characterize the potential hazard of small and ultra-small particles to human health by considering their ecotoxicity and health impact to particle characteristics, such as size, shape, composition, presence of additive and co-contaminants, and surface properties as well as characterization routes of exposures of humans via aquatic food chains and translocation in human organs and tissues.

References

Al Salah, Dhafer M.M., Ngweme, Georgette N., Laffite, Amandine, Otamonga, Jean-Paul, Mulaji, Crispin, Pote-Wembonyama, John. 2020. Hospital wastewaters: A reservoir and source of clinically relevant bacteria and antibiotic resistant genes dissemination in urban river under tropical conditions. *Ecotoxicology and Environmental Safety*, 200, 110767.

Baalousha, Mohammed, Yang, Yi, Vance, Marina E., Colman, Benjamin P., McNeal, Samantha, Xu, Jie, Blaszczak, Joanna, Steele, Meredith, Bernhardt, Emily, Hochella Jr, Michael F. 2016. Outdoor urban nanomaterials: The emergence of a new, integrated, and critical field of study. *Science of the Total Environment*, 740–753, 557–558.

Boyle, Kellie, Örmeci, Banu. 2020. Microplastics and nanoplastics in the freswater and terrestrial environment: A review. *Water*, 12, 2633–2662.

Bundschuh, Mirco, Filser, Juliane, Lüderwald, Simon, McKee, Moira S., Metreveli, George, Schaumann, Gabriele E., Schulz, Ralf, Wagner, Stephan. 2018. Nanoparticles in the environment: Where do we come from, where do we go to? *Environmental Sciences Europe*, 30, 6.

Cao, Jiasheng, Elliott, Daniel, Zhang, Wie-xian. 2005. Perchlorate reduction by nanoscale iron particles. *Journal of Nanoparticle Research*, 7, 499–506.

Chen, Yuying, Liu, Wei, Leng, Xiaojing, Stoll, Serge. 2022. Toxicity of selenium nanoparticles on Poterioochromonas malhamensis algae in Waris-H culture medium and Lake Geneva water: Effect of nanoparticle coating, dissolution, and aggregation. *Science of the Total Environment*, 808, 152010.

Domercq, Prado, Praetorius, Antonia, Boxall, Alistair B.A. 2018. Emission and fate modelling framework for engineered nanoparticles in urban aquatic systems at high spatial and temporal resolution. *Environmental Science Nano*, 5, 533–543.

Enfrin, Marie, Dumée, Ludovic F., Lee, Judy. 2019. Nano/microplastics in water and wastewater treatment processes: Origin, impact and potential solutions. *Water Research*, 161, 621–638.

Faure, Florian, Demars, Colin, Wieser, Olivier, Kinz, Manuel, Felippe de Alencastro, Luiz. 2015. Plastic pollution in Swiss surface waters: Nature and concentrations, interactions with pollutants. *Environmental Chemistry*, 12, 582–591.

Kirstein, Inga V., Hensel, Fides, Gomerio, Alessio, Iordachescu, Lucian, Vianello, Alvise, Wittgren, hans B., Vollersten, Jes. 2021. Drinking plastics?: Quantification and qualification of microplastics in drinking water distribution systems by µFTIR and Py-GCMS. *Water Research*, 188, 116519.

McCormick, Amanda R., Hoellein, Timothy J., London, Maxwell G., Hittie, Joshua, Scott, John W., Kelly, John J. 2016. Microplastic in surface waters of urban rivers: Concentration, sources, and associated bacterial assemblages. *Ecosphere*, 7, e01556.

Müller, Alexandra, Österlund, Helene, Marsalek, Jiri, Viklander, Maria. 2020. The pollution conveyed by urban runoff: A review of sources. *Science of the Total Environment*, 709, 136125.

Negrete Velasco, Angel De Jesus, Rard, Lionel, Blois, Wilfried, Lebrun, David, Lebrun, Frank, Pothe, Frank, Stoll, Serge. 2020. Microplastic and fibre contamination in a remote mountain lake in Switzerland, *Water*, 12, 2410–2426.

Negrete Velasco, Angel de Jesus, Ramseier, Stéphan, Zimmermann, Stéphane, Stoll, Serge. 2022. Contamination and removal efficiency of microplastics and synthetic fibres in a conventional drinking water treatment plant. *Frontiers*, 4, 835451.

Pena-Guzman, Carlos, Ulloa-Sachez, Stephanie, Mora, Karen, Helena-Bustos, Rosa, Lopez-Barrera, Ellie, Alvarez, Johan, Rodriguez-Pinzon, Manuel. 2019. Emerging pollutants in the urban water cycle in Latin America: A review of the current literature. *Journal of Environmental Management*, 237, 408–423.

Pinto da Costa, João, Santos, Patrícia S.M., Duarte, Armando C., Rocha-Santos, Teresa. 2016. (Nano)plastics in the environment: Sources, fates and effects. *Science of the Total Environment*, 556–557, 15–26.

Pivokonský, Martin, Pivokonská, Lenka, Novotna, Katerina, Čermáková, Lenka, Klimtova, Martina. 2020. Occurrence and fate of microplastics at two different drinking water treatment plants within a river catchment. *Science of the Total Environment*, 741, 140236.

Pochelon, Alexis, Stoll, Serge, Slaveykova, Vera. 2021. Polystyrene nanoplastic behavior and toxicity on Crustacean Daphnia magna: Media composition, size, and surface charge effects. *Environments*, 8, 101–113.

Ramirez Arenas, Lina Marcela, Ramseier Gentile, Stéphan, Zimmermann, Stéphane, Stoll, Serge. 2019. Behavior of TiO2 and CeO_2 nanoparticles and polystyrene nanoplastics in bottled mineral, drinking and Lake Geneva waters. Impact of water hardness and natural organic matter on nanoparticle surface properties and aggregation. *Water*, 11, 721–734.

Ramirez Arenas, Lina Marcela, Ramseier Gentile, Stéphan, Zimmermann, Stéphane, Stoll, Serge. 2020. Coagulation of TiO2, CeO_2 nanoparticles, and polystyrene nanoplastics in bottled mineral and surface waters. Effect of water properties, coagulant type, and dosage. *Water Environment Research*, 92, 1184–1194.

Ramirez Arenas, Lina Marcela, Ramseier Gentile, Stephan, Zimmermann, Stéphane, Stoll, Serge. 2021. Nanoplastics adsorption and removal efficiency by granular activated carbon used in drinking water treatment process. *Science of the Total Environment*, 791, 148175.

Reynaud, Stephanie, Aynard, Antoine, Grassi, Bruno, Gigault, Julien. 2021. Nanoplastics: From model materials to colloidal fate. *Current Opinion in Colloid & Interface Science*, 57, 101528, 101528.

Rönkkö, Topi, Timonen, Hilkka. 2019. Overview of sources and characteristics of nanoparticles in urban traffic-influenced areas. *Journal of Alzheimer's Disease*, 72, 15–28.

Saavedra Vargas, Juan, Stoll, Serge, Slaveykova, Vera. 2019. Influence of nanoplastic surface charge on eco-corona formation, aggregation and toxicity to freshwater zooplankton. *Environmental Pollution*, 252, 715–722.

Tolba Aboelnga, Hassan, Ribbe, Lars, Frechen, Franz-Bernd, Saghir, Jamal. 2019. Urban water security: Definition and assessment framework. *Resources*, 8, 178.

Turner, Simon, Horton, Alice A., Rose, Neil L., Hall, Charlotte. 2019. A temporal sediment record of microplastics in an urban lake, London, UK. *Journal of Paleolimnology*, 61, 449–462.

Wang, Zhifeng, Lin, Tao, Chen, Wei. 2020. Occurrence and removal of microplastics in an advanced drinking water treatment plant (ADWTP). *Science of the Total Environment*, 700, 134520.

Wei, Huaibin, Wang, Yimin, Wang, Mingna. 2018. Characteristic and pattern of urban water cycle: Theory. *Desalination and Water Treatment*, 110, 349–354.

Werbowski, Larissa M., Gilbreath, Alicia N., Munno, Keenan, Zhu, Xia, Grbic, Jelena, Wu, Tina, Sutton, Rebecca, Sedlak, Margaret D., Deshpande, Ashok D., Rochman, Chelsea M. 2021. Urban stormwater runoff: A major pathway for anthropogenic particles, black rubbery fragments, and other types of microplastics to urban receiving waters. *ACS EST Water*, 1(6), 1420–1428.

Xie, Yanhua, Yi, Yan, Qin, Yinhong, Wang, Lanting, Liu, Guoming, Wu, Yulan, Diao, Zhiqiang, Zhou, Tingheng, Xu, Mo. 2016. Perchlorate degradation in aqueous solution using chitosan-stabilized zera-valent iron particles. *Separation and Purification Technology*, 171, 164–173.

Yang, Yi, Vance, Marina, Tou, Feiyun, Tiwari, Andrea, Liua, Min, Hochella Jr., Michael F. 2016. Nanoparticles in road dust from impervious urban surfaces: Distribution, identification, and environmental implications. *Environmental Science: Nano*, 3, 534–544.
Zhou, Guanyu, Wang, Qingguo, Zhang, Jing, Li, Qiansong, Wang, Yunqi, Wang, Meijing, Huang, Xue. 2020. Distribution and characteristics of microplastics in urban waters of seven cities in the Tuojiang River basin, China. *Environmental Research*, 189, 109893.

INDEX

Note: Page numbers in *italics* indicate figures, **bold** indicate tables in the text, and references following "n" refer endnotes.